License Plates　　California has more passenger vehicles than any ~~other~~　　　that the standard format cannot account for all of the license plates. A second format was introduced. But then California introduced yet a third format. Is this an instance of irrational behavior in the Golden State, or is there a mathematical justification? The Case Study will reveal the answer.

Jane Blalock and the LPGA　　When Jane Blalock, star professional golfer, was banned from participating in tournaments by the Ladies' Professional Golf Association, she sued the organization under the federal antitrust laws. How much money would she have earned? Her court battle was based on probability theory and the Case Study gives the surprising result of her suit.

The Federalist Papers　　The 13 colonies in America joined forces from 1776 to 1783 to fight for their independence from Great Britain, but then a family squabble broke out as the independent colonies wrestled with their own self-governance. The most instrumental documents supporting the ratification of the Constitution were The Federalist papers, but for almost 200 years the identity of the authors of some of The Federalist papers remained unknown. Conclusive evidence was first published in 1964 using statistical methods.

Allowance for Doubtful Accounts　　In business an unpaid bill is called a bad debt, and some bad debts are never collected. Large retail firms must account for a percentage of money due that they never will collect. The Case Study shows how accountants use Markov chains to give firms a structured means of predicting the amount of bad debts to account for.

Game Theory and the Normandy Invasion　　Taken from a classic paper written by Colonel O. G. Haywood, Jr. in 1954, the Case Study shows that game theory can be used to analyze military decisions by centering on a key battle in World War II.

The Norway Cement Company Merger　　A very stormy merger of Norway's three cement corporations took place in the 1960s. Even though three groups were involved in the negotiation, the conflict was resolved by using a two-person-game-theoretic approach.

Finite Mathematics with Applications

David E. Zitarelli
Raymond F. Coughlin

Temple University

SAUNDERS COLLEGE PUBLISHING

Philadelphia   New York   Chicago
San Francisco   Montreal   Toronto
London   Sydney   Tokyo

Text Typeface: Times Roman
Compositor: General Graphic Services, Inc.
Acquisitions Editor: Robert B. Stern
Project Editor: Sally Kusch
Copy Editor: Elaine Honig
Art Director: Carol Bleistine
Art Assistant: Doris Bruey
Text Designer: Caliber Design Planning, Inc.
Cover Designer: Lawrence R. Didona
Text Artwork: Vantage Art, Inc.
Layout Artist: York Production Services
Production Manager: Merry Post

Cover Credit: A computer-generated image of the inside of a "Wente-Torus" of constant mean curvature. It was produced by James T. Hoffman of the Geometry, Analysis, Numerics and Graphics group at the University of Massachusetts, Amherst © 1987, 1988.

Printed in the United States of America

FINITE MATHEMATICS WITH APPLICATIONS

0–03–011292–3

Library of Congress Catalog Card Number: 88-043 523

01 015 9876543

PREFACE

Students in engineering and the physical sciences rarely ask the question "Why do we have to study this subject?" But other students are asking it in increasing numbers as more and more programs require more and more mathematics. The goal of this book is to answer the question for students who are majoring in business, a social science, or a biological science. Although the table of contents is fairly standard, the means of achieving the goal differ from other texts in two notable ways.

First, every section contains several referenced applications of mathematics in the form of examples or exercises. Each one of these applications is accompanied by a footnote so that the student can pursue it in greater depth or the instructor can assign it as part of a special project. Because such applications deal with real-life situations, calculators are frequently necessary to carry out the computations. Here is a representative list of the referenced exercises, along with the section where they can be found.

Application	Source	Section
stereo speakers	Teledyne Acoustic Research	1–1
archaeological deposits	*American Antiquity*	2–1
crop rotations	*American Journal of Agricultural Economics*	3–3
dairy farming	*Journal of the Operational Research Society*	4–1
athletes' contracts	text *Financial Management*	5–2
pitch classes	*Journal of Music Theory*	6–1
homicide and deterrence	*Southern Economics Journal*	7–5
economic forecasts	*Management Science*	8–3
urban status	*American Journal of Sociology*	9–1
auditing procedures	*The Accounting Review*	10–2

In addition, every chapter has a Case Study that treats a particular application in greater detail than do the referenced examples. The topics have been chosen not only to illustrate instances in which mathematics has played a crucial role in solving an important problem, but to highlight areas of particular human interest.

Even if the instructor does not have time to cover the referenced examples or Case Studies, it is hoped that their relevance and wealth (in both diversity and number) will make a lasting impression.

Organization

The text can be divided into three parts: linear mathematics (Chapters 1 to 4), probability and statistics (Chapters 6, 7, and 8), and applications (Chapters 5, 9 and 10).

Linear Mathematics

Chapter 1 introduces the Cartesian coordinate system and linear equations. Systems of two equations in two variables are presented separately from systems of three equations in three variables for those instructors who want to proceed directly to the geometric solution of linear programming problems in Chapter 3.

Chapter 2 introduces matrices and their connection to systems of linear equations. Section 1–4 on Gaussian elimination is a prerequisite for Section 2–3 on the Gauss–Jordan method. The matrix inverse method is applied to input-output analysis.

Chapters 3 and 4 deal with linear programming. Chapter 3 illustrates the geometric method of solving linear programming problems with two variables. In Chapter 4 the simplex method is used for solving linear programming problems with more than two variables.

Probability and Statistics

Chapter 6 lays the foundation for this part. It reviews the basic aspects of set theory, including Venn diagrams and tree diagrams, and presents counting techniques, permutations and combinations, and the Binomial Theorem. In Chapter 7 the basic aspects of probability are motivated by a study of automobile accidents on a major highway in Boston. After the set-theoretic foundations are described, laws governing the addition, subtraction, multiplication, and division of probabilities are derived. Chapter 8 presents the rudiments of statistics, including graphical methods, measures of central tendency, measures of spread from the central tendency, the normal curve, and binomial experiments.

Applications

Chapter 5 is independent of the rest of the book. It deals primarily with the mathematics of finance, but it also treats inflation, cost of living and population growth.

Chapters 9 and 10 apply matrices and probability to Markov chains and game theory. The last section in the book ties this material to linear programming, and the epilogue discusses the historical meeting that produced the confluence of these seemingly unrelated fields.

Chapter Dependence and Text Organization

The accompanying display gives a schematic for chapter dependence. The linear mathematics and the probability and statistics can be covered in either order.

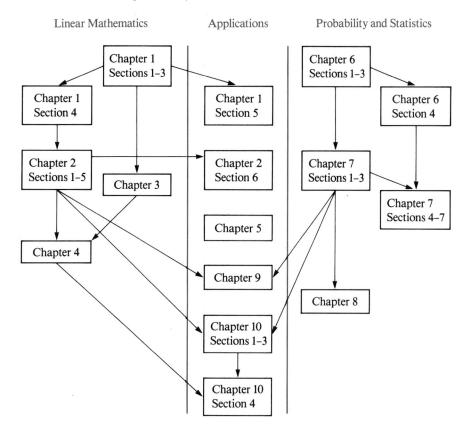

Linear Mathematics   Applications   Probability and Statistics

We have tried to keep the length of each section to material that can be covered in a typical 50-minute class. Sometimes, however, the topic has dictated more extensive coverage. Each numbered section has been partitioned into subsections.

There are two kinds of examples. One explains a new skill that is being encountered for the first time and is labeled simply EXAMPLE. The other illustrates a skill which has been explained previously. It uses the following format:

EXAMPLE | **Problem**
SOLUTION

The student should be able to make a good attempt at solving the problem independently before reading its solution.

Each chapter ends with a review of the terms, notation, and formulas that were introduced in the chapter. It also contains review problems that test a student's ability to solve problems in a cumulative manner. These problems can also be used profitably as a review for tests.

Exercises

The exercises are divided into three sets by color bullets. There is a slight distinction between the problems in the first two sets. The first set of problems provides practice on the skills introduced in the examples; they appear in the same order as the examples. The second set requires either new thinking or additional skills. Their order need not match the order of the presentation in the text.

The remaining problems serve several different purposes. Most are drawn from areas outside mathematics and are accompanied by complete references to the literature. Other problems present theoretical material for those instructors who want to provide a deeper insight into the subject, while some are aimed at challenging the better students.

The answers to all odd-numbered problems are provided in the back of the book.

Supplements

A number of supplements are available for use with this text. They are as follows:

The **Instructor's Manual** contains answers to all odd- and even-numbered problems along with lecture suggestions for every chapter.

A **Student Solutions Manual and Study Guide** contains answers and solutions to every other odd-numbered problem as well as helpful suggestions, examples, and additional problems for the student to work.

The **computer supplement** available for this text covers many of the important topics in each chapter with graphical and computational support. The software requires an IBM or IBM-compatible computer with at least 512K. It will operate with either a color or monochrome graphics adapter.

Also available for this book is a **computerized test bank** which permits the user to choose, edit, add, or delete items. The test bank consists of approximately 1,250 test items, and is available for the IBM-PC and IBM-compatible computers with 256K, dual disk drives, graphics card, graphics monitor, and graphics printer. The printer can be either parallel or serial if the serial printer has a parallel driver.

In addition to the computerized testbank a **complete booklet of prepared tests** has been written with five forms per chapter of 25 questions each. Both open response as well as multiple choice questions are included.

Acknowledgments

It is a pleasure to acknowledge the help we received from many people. Our biggest support came from our wives (Anita and Judy) and our children (Paul and Nicole; Virginia, Christina, and Sara), who have tolerated our total immersion in this project for several years.

Our reviewers played an important part in transforming a manuscript used at Temple University into a book that is appropriate for many types of institutions of higher learning. Several reviewers deserve special thanks for their unstinting, constructive criticism of the entire manuscript:

Stephen Andrilli, LaSalle University
Ron Davis, DeKalb College
Garrett Etgen, University of Houston
Alec Ingraham, New Hampshire College
Terry Jenkins, University of Wyoming
Giles Maloof, Boise State University
Bill Pannell, Sierra College
Duane Porter, University of Wyoming
Daniel Symancyk, Anne Arundel Junior College
Melvin Tuscher, West Valley College
Lee Witt, Western Michigan University

A very important part of the reviewing process is the checking of errors in problems and examples. For their aid in this process, we would like to thank the following people:

Louis Hoelzle, Bucks County Community College
Diana Snyder, Philadelphia Community College
Daniel Symancyk, Anne Arundel Junior College

We extend special gratitude to our friend Vincent Damiano, who is a Senior Project Engineer for Chevron USA, Inc. He read the entire manuscript and suggested several realistic linear programming problems.

The staff at Saunders College Publishing has been a pleasure to work with. We especially thank our editor, Robert Stern, for the interest, encouragement, guidance, and persistence he has bestowed upon this project. We also thank the project editor, Sally Kusch, and the art director, Carol Bleistine, not only for their professional expertise but for their patience and humor as well. Our appreciation also goes to Elaine Honig for her excellent copyediting and for compiling the Index of Applications.

David E. Zitarelli
Raymond F. Coughlin

Philadelphia, PA
November 1988

Contents

INDEX OF APPLICATIONS

General Topics

1 Linear Equations

CHAPTER OVERVIEW   A linear equation in two variables is an equation whose graph is a straight line. The association of a curve (like a line) with an equation (like a linear equation) was a revolutionary idea that occurred during the early part of the 17th century. The first two sections discuss the correspondence between curves in two dimensions and equations in two variables. Section 1–1 introduces the Cartesian coordinate system and sketches the graphs of various equations. Section 1–2 discusses three standard forms of linear equations.

The next two sections deal with systems of linear equations. Section 1–3 presents several methods for solving systems of two linear equations in two variables. It emphasizes the method of elimination and discusses its advantages. Section 1–4 introduces Gaussian elimination to solve systems of three equations in three variables.

Section 1–5, which can be covered after Section 1–3, discusses least squares. It investigates voting trends in recent presidential elections and lays the foundation for the Case Study.

CASE STUDY PREVIEW

Records for the mile run have been kept since 1864. Walter George set the world record four times between 1880 and 1884. His strategy revolutionized the way that milers run the race. Since then the world has experienced many drastic changes, yet the record for the mile run has been lowered with remarkable regularity.

The Case Study examines this surprising phenomenon. It derives an equation that projects the year in which the world record will be lowered to 3 minutes. The equation also predicts what the world record will be in any given year. You can test its validity for the year you are reading this book.

1–1 THE CARTESIAN COORDINATE SYSTEM

About 1630 two Frenchmen, René Descartes and Pierre Fermat, independently revolutionized the way of expressing scientific laws. The idea they hit upon has come to be called "analytic geometry" or "algebraic geometry," because it is a blend of algebra and geometry. Descartes is usually given credit for being the founder of this field, and the term "Cartesian coordinate system" reflects human-kind's debt to him, but Fermat deserves at least as much credit for his pioneering work.* Sections 1–1 and 1–2 are devoted to their discovery.

The Graph of an Equation

Consider the formula $y = 3x + 1$. It is an algebraic expression that describes the relationship between the variables x and y, but it can also be viewed as a geometric object, namely, a line in the plane. To accomplish this both Descartes and Fermat introduced a *coordinate system,* which we will describe briefly.

We start with the **real line.** It is a straight line with one point marked as the **origin,** corresponding to the number 0 (see Figure 1a). A point to the right of 0 is marked and it corresponds to the number 1. The distance from 0 to 1 is called a **unit distance.** Points that are located at one-unit intervals to the right of 1 are marked 2, 3, and so on. Negative numbers are located to the left of 0. In this way, each real number a corresponds to the point that is $|a|$ units to the right of 0 if a

*For the historical details, see Chapter XVII, Section 11, of Carl B. Boyer, *A History of Mathematics,* Princeton, N.J., Princeton University Press, 1985.

FIGURE 1

(a)

(b)

(a)   (b)

FIGURE 2

is positive or to the left if a is negative. Additional points, corresponding to some rational and irrational numbers, are marked off in Figure 1b.

Next construct a vertical line perpendicular to the horizontal real line at the origin, and mark off a unit length on it. The positive direction is upward. Usually the horizontal line is called the **x-axis** and the vertical line the **y-axis.** Every point in the plane is represented by an ordered pair of numbers. The ordered pair (1, 2) corresponds to the point that is one unit to the right of the y-axis and two units above the x-axis. (See Figure 2a.) In the ordered pair (1, 2), 1 is called the **x-coordinate** and 2 the **y-coordinate.**

In Figure 2b additional points are plotted in the coordinate system. The numbers in the ordered pairs are also called the first and second **coordinates.** This coordinate system is called the **Cartesian coordinate system** after Descartes. The choice of the letters x and y is arbitrary; we will change them when a particular problem dictates. The Cartesian coordinate system divides the plane into four distinct **quadrants** denoted by the Roman numerals shown in Figure 2a.

Consider the equation $y = 3x + 1$. If $x = 1$, then $y = 4$. These values are treated as the point (1, 4) in the Cartesian coordinate system. The point (1, 4) is said to "satisfy" the equation $y = 3x + 1$. The set of all points that satisfy an equation is called the **graph** of the equation. The graph is obtained from a table of values of x and the corresponding values of y. Example 1 illustrates the procedure for drawing the graph of an equation.

EXAMPLE 1

Consider the equation $y = 3x + 1$. To find its graph, make a table of some values of x and the corresponding values of y. If $x = -2$, then $y = 3x + 1 = 3(-2) + 1 = -5$. This means that the point $(-2, -5)$ lies on the graph. The table lists four additional points corresponding to $x = -1, 0, 1$, and 2.

x	-2	-1	0	1	2
y	-5	-2	1	4	7
point	$(-2, -5)$	$(-1, -2)$	$(0, 1)$	$(1, 4)$	$(2, 7)$

4   1 Linear Equations

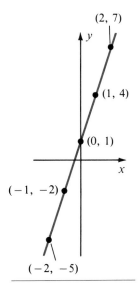

(2, 7)

(1, 4)

(0, 1)

(−1, −2)

(−2, −5)

FIGURE 3

Next, plot the points $(-2, -5)$, $(-1, -2)$, $(0, 1)$, $(1, 4)$, and $(2, 7)$ on a Cartesian coordinate system. By letting x be *any* real number in the equation $y = 3x + 1$, there will be an infinite number of points $(x, 3x + 1)$ on the graph. They constitute the graph of $y = 3x + 1$, which is the line drawn in Figure 3.

The values for x in the table were chosen arbitrarily. The same line would result if other choices were made.

Example 1 illustrates how algebra and geometry were molded into one mathematical discipline called analytic geometry. The essence is that equations in two variables correspond to curves in two dimensions. The procedure for sketching the graph of an equation is called the **plotting method.** It consists of three steps:

1. Make a table of values of x and y.
2. Treat the values as ordered pairs on a coordinate system and plot them.
3. Connect the points with a smooth curve.

The graph of the equation is the curve that results from step 3. In Section 1–2 we will see how to proceed in the opposite direction, from the graph to the equation.

This simple idea allows algebraic equations to be pictured geometrically, reflecting the old adage that "a picture is worth a thousand words." The next example illustrates it.

EXAMPLE 2

Problem

Sketch the graph of the equation $4x + 2y = 1$ by the plotting method.

SOLUTION Construct a table for the equation $4x + 2y = 1$ by choosing a value for x and substituting it into the equation to get the corresponding value of y. If $x = 1$, then $4(1) + 2y = 1$, so $y = -\frac{3}{2}$. Therefore $(1, -\frac{3}{2})$ is a point on the graph of $4x + 2y = 1$. The table provides four additional points.

x	-2	-1	0	1	2
y	$\frac{9}{2}$	$\frac{5}{2}$	$\frac{1}{2}$	$-\frac{3}{2}$	$-\frac{7}{2}$
point	$(-2, \frac{9}{2})$	$(-1, \frac{5}{2})$	$(0, \frac{1}{2})$	$(1, -\frac{3}{2})$	$(2, -\frac{7}{2})$

Plot these points and draw the curve through them. The graph is shown in Figure 4.

An equation in two variables whose graph is a line is called a **linear equation.** Examples 1 and 2 show that $y = 3x + 1$ and $4x + 2y = 1$ are linear equations.

Applications

For many businesses the total cost C (in dollars) of producing a commodity and the number x of units produced can be written as a linear equation. For instance,

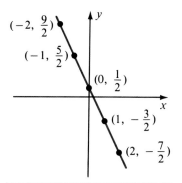

FIGURE 4

let

$$C = 500 + 4x$$

If no units are produced, then $x = 0$, so $C = \$500$. This amount is called the fixed cost. If $x = 25$ units are produced, then the total cost is $C = \$600$.

Often the total revenue R (in dollars) derived from selling the commodity can be written as a linear equation. For instance, let

$$R = 8x$$

If 25 units are produced, then $R = \$200$, in which case the revenue is $400 less than the cost. The business will have to produce more units in order to turn a profit. In Section 1–3 we will explain how to compute the number of units needed to derive a profit from the sales, but for now we lay the foundation for the solution.

EXAMPLE 3

Problem

Sketch the graphs of the cost equation and the revenue equation on the same coordinate axis.

SOLUTION Use the plotting method to sketch the graph of the cost $C = 500 + 4x$.

x	0	50	100	150
C	500	700	900	1100
point	(0, 500)	(50, 700)	(100, 900)	(150, 1100)

Sketch the graph of $R = 8x$ for the same values of x.

x	0	50	100	150
R	0	400	800	1200
point	(0, 0)	(50, 400)	(100, 800)	(150, 1200)

The graphs of the two equations are shown in Figure 5.

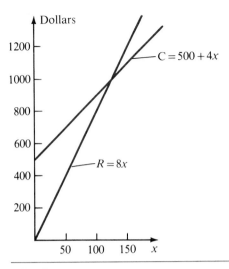

FIGURE 5

Most large firms have a planning group which constructs sales forecasts. These forecasts are critical for production scheduling, plant design, and financial planning. Graphs are often used for viewing the relationship between sales growth and external financial requirements. Example 4 examines the Arinson Products Company, whose owner, John Arinson, sought funds to maintain exclusive control of the company.†

EXAMPLE 4

Let x represent the sales growth rate of the Arinson Products Company and let y represent the external funds required (in thousands of dollars), where

$$y = 494x - 16$$

Problem

(a) Draw the graph of the equation. (b) If the sales growth rate is 5%, what external funding is required? (c) If $107,500 is available for external funding, what is the sales growth rate?

SOLUTION (a) If $x = 0.1$ (corresponding to a 10% sales growth rate), then $y = 494(0.1) - 16 = 33.4$. The table lists several other values of x and the corresponding values of y.

x	−0.2	−0.1	0	0.1	0.2
y	−114.8	−65.4	−16	33.4	82.8

The graph is sketched in Figure 6.

---

†B. Campsey and E. Brigham, *Introduction to Financial Management*, Hinsdale, Ill., The Dryden Press, 1985, pp. 190–204.

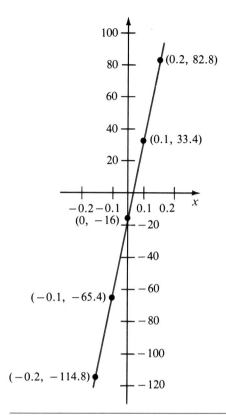

FIGURE 6

(b) For a sales growth rate of 5% set $x = 0.05$. Then $y = 494(0.05) - 16 = 8.7$. Since the units for y represent thousands of dollars, the amount of external funds required is $8700.

(c) If $107,500 is available for external funding, set $y = 107.5$. The sales growth rate is obtained by substituting $y = 107.5$ into the equation and solving for x.

$$494x - 16 = y$$
$$494x - 16 = 107.5$$
$$x = \frac{107.5 + 16}{494} = 0.25$$

The sales growth rate is $x = 25\%$.

What's in a Name?

All of the graphs in this section have been lines. Of course, these are not the only graphs of algebraic equations according to the method of Descartes and Fermat;

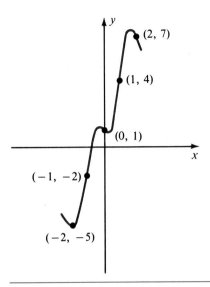

FIGURE 7

there are also circles, parabolas, and many others. It is useful to be able to picture a graph when its equation is given.

So far, graphs have been drawn by the plotting method, but this method has its shortcomings. Given the table of five points for the equation $y = 3x + 1$ in Example 1, it is conceivable that its graph could be the curve in Figure 7. In Section 1–2 we will see why it must be a straight line.

EXERCISE SET 1–1

• In Problems 1 to 4 fill in the table for the given values. Supply your own values of x in Problems 3 and 4.

1. $y = 2x - 3$

x	-2	-1	0	1	2	3
y						
point						

2. $y = x - 2$

x	-2	-1	0	1	2	3
y						
point						

3. $y = 4 - x$

x	
y	
point	

4. $3x - y = 2$

x	
y	
point	

In Problems 5 to 10 use the plotting method to sketch the graph of the equation.

5. $y = 2x - 3$ 6. $y = x - 2$

7. $y = 4 - x$

8. $3x - y = 2$

9. $2x + y = 6$

10. $4x + y = 7$

In Problems 11 to 14 the graphs of the given equations are straight lines. Plot two points and then sketch the graph.

11. $x - y = 3$

12. $2x + y = 3$

13. $3x - y = -1$

14. $3x + y = 1$

In Problems 15 to 18 sketch the graphs of the total cost C and the total revenue R on the same coordinate axis.

15. $C = 25 + 3x$
    $R = 8x$

16. $C = 40 + x$
    $R = 5x$

17. $C = 100 + 3x$
    $R = 13x$

18. $C = 300 + x$
    $R = 31x$

Problems 19 and 20 refer to the equation $y = 494x - 16$ given in Example 4.

19. (a) If the sales growth rate is 15%, what external funding is required? (b) If $33,400 is available for external funding, what is the sales growth rate?

20. (a) If the sales growth rate is -5%, what external funding is required? (b) If $82,800 is available for external funding, what is the sales growth rate?

•• In Problems 21 to 24 use the plotting method to sketch the graph of the equation.

21. $x + 4y = 7$

22. $x + 3y = 3$

23. $3x + 4y = -12$

24. $3x - 4y = 6$

In Problems 25 to 34 the equations are nonlinear. Sketch the graph using the plotting method.

25. $y = x^2 - 3$

26. $y = x^2 + 1$

27. $y = -x^2$

28. $y = -2x^2$

29. $y = 1 - 2x^2$

30. $y = 1 - x^2$

31. $y = x^3$

32. $y = x^3 - 8$

33. $y = \sqrt{x}$

34. $y = \sqrt{x} - 1$

In Problems 35 and 36 sketch the graphs of the two equations on the same coordinate axis.

35. $\begin{aligned} x + y &= 120 \\ 100x + 40y &= 6000 \end{aligned}$

36. $45x + 9y = 45$
    $10x + 6y = 20$

Problems 37 to 40 refer to the equation $5F - 9C = 160$, where F is degrees Fahrenheit and C is degrees Celsius.

37. Sketch the graph of the equation.

38. Find the degrees Celsius when $F = 68$ degrees.

39. Find the degrees Celsius when $F = 98.6$ degrees.

40. Find the degrees Fahrenheit when $C = 20$ degrees.

Problems 41 and 42 refer to a company whose total cost is $C = 0.5x + 20,000$, where x is the amount of sale and the fixed cost is $20,000.

41. If the company sells $60,000 worth of goods, what is the total cost?

42. If the company sells $30,000 worth of goods, what is the total cost?

43. A person's height can be estimated by the length of the tibia bone, which runs from the knee to the ankle. The formula for the height y [in centimeters (cm)] in terms of the length x (in cm) of the tibia is

$$3y - 8x = 211$$

(a) Draw the graph of the equation.
(b) How tall is a person whose tibia is 37 cm long?
(c) How long is the tibia if the person is 185 cm tall?

44. A person's height can be estimated by the length of the radius bone, which runs from the wrist to the elbow. The formula for the height y (in cm) in terms of the length x (in cm) of the radius is

$$2y - 7x = 166$$

(a) Draw the graph of the equation.
(b) How tall is a person whose radius bone is 24 cm long?

(c) How long is the radius bone if the person is 174 cm tall?

45. What is the meaning of the value of x at which the cost line and the revenue line meet in Figure 5?

●●● 46. The amplifier in a stereo system must be strong enough to produce loud peak volumes level (106 decibels) in the speakers. The strength needed depends on the character of the room. For instance, a "live" room is one in which a hand clap will cause a fluttering echo. For Acoustic Research loudspeakers,* the linear relationship between the power y (in watts) and the volume x of a "live" room (in cubic feet) is given by

$$y = \frac{x}{180} + \frac{80}{9}$$

(a) What power is needed for a room measuring 2000 cubic feet?

(b) What power is needed for a room measuring 6500 cubic feet?

(c) If a system produces 40 watts, what size room will allow it to produce loud peak levels?

47. The percentage share of the national radio audience for FM radio stations (contrasted with AM radio stations) in any year since 1972 can be approximated by the formula

$$S = 3.514286Y - 226.2953$$

where S is a percent and Y is the last two digits in the year.† (For instance, for the year 1989, $Y = 89$.)

(a) What was the approximate percentage share in 1985?

(b) In what year was the percentage share about 50%?

*Source: Teledyne Acoustic Research of Canton, Massachusetts. This company manufactures the Acoustic Research loudspeakers.

†This equation was derived from data supplied by Statistical Research, Incorporated.

1–2 LINES

There are two basic aspects to the Descartes–Fermat invention of analytic geometry. One is a fixed frame of reference, called the Cartesian coordinate system. The other is the relationship between equations and curves. Section 1–1 showed how to begin with a linear equation in two variables and draw its graph. This section illustrates the opposite direction: given the graph of a line in some form, construct its equation.

General Form

Recall from Section 1–1 that a linear equation in two variables is an equation whose graph is a line. We examine three forms of linear equations.

It can be proved that an equation in two variables x and y is linear if and only if it can be written in the form $Ax + By = C$, where A, B, C are numbers such that A and B are not both zero. This form of a linear equation is called the *general form*. The equation $4x + 3y = 12$ is linear, but $x^2 - y = 0$ and $\sqrt{x} - y = 0$ are nonlinear.

_____ **Definition** _____

The **general form** of a linear equation is

$$Ax + By = C$$

where A, B, and C are real numbers, with A and B not both equal to 0.
_____

Example 1 illustrates a method of graphing linear equations in general form that is more effective than the plotting method shown in Example 2 of Section 1–1. It makes use of the intercepts of a line. The **x-intercept** of a line is the point where the line crosses the x-axis, and the **y-intercept** is the point where it crosses the y-axis.

EXAMPLE 1

Sketch the graph of the linear equation

$$3x + 2y = 4$$

SOLUTION Find the y-intercept by setting $x = 0$. Then $y = 2$, so the y-intercept is $(0, 2)$. Find the x-intercept by setting $y = 0$. Then $x = \frac{4}{3}$, so the x-intercept is $(\frac{4}{3}, 0)$. Draw the line through the two intercepts. This is shown in Figure 1.

Slope-Intercept Form

Every equation of the form $y = mx + b$, where m and b are numbers, is linear since it can be written in the general form $mx - y = -b$. Therefore the graph of the equation $y = 3x + 1$ is a line with $m = 3$ and $b = 1$; its graph cannot be the curve shown in Figure 7 of Section 1–1. The expression "the line $y = mx + b$" refers to the graph of the linear equation $y = mx + b$.

The numbers m and b in the linear equation $y = mx + b$ play an important role. The property measured by m is intuitively described as "steepness." It is called the _slope_ of the line. To measure the slope, start at some point (x_1, y_1) on

FIGURE 1 Graph of $3x + 2y = 4$.
_____

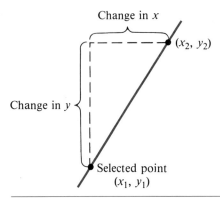

FIGURE 2   Slope.

the line, proceed in the vertical direction some distance, then proceed in the horizontal direction until the line is intersected at a point (x_2, y_2). (See Figure 2.) The vertical distance is called the "change in y" (or the "rise") and the horizontal distance is called the "change in x" (or the "run"). The slope is then defined to be the ratio of the change in y to the change in x (or the ratio of the rise to the run).

_____ **Definition** _____

Let (x_1, y_1) and (x_2, y_2) be distinct points on a line with $x_1 \neq x_2$. The **slope** of the line is the number

$$m = \frac{\text{change in } y}{\text{change in } x} = \frac{y_2 - y_1}{x_2 - x_1}$$

_____

The points $(1, -1)$ and $(4, 5)$ lie on the line $y = 2x - 3$. (See Figure 3.) The slope of the line is

$$m = \frac{\text{change in } y}{\text{change in } x} = \frac{5 - (-1)}{4 - 1} = \frac{6}{3} = 2$$

Figure 4 shows the graphs of several lines. As the values of x increase from left to right, those with positive slope tend upward (or increase) while those with negative slope tend downward (or decrease).

Example 2 investigates the slope of two pairs of points on the same line.

EXAMPLE 2 | **Problem**

Find the slope of the line $y = 3x + 1$ through these pairs of points.
(a) $(0, 1)$ and $(3, 10)$
(b) $(-1, -2)$ and $(4, 13)$

SOLUTION (a) The slope of the line through the points $(0, 1)$ and $(3, 10)$ is

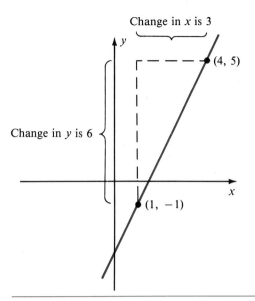

FIGURE 3   Slope of line $y = 2x - 3$.

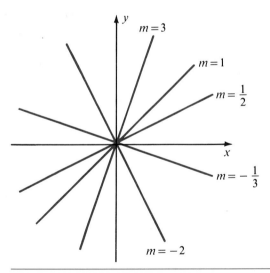

FIGURE 4   Lines with various slopes.

$$m = \frac{10 - 1}{3 - 0} = \frac{9}{3} = 3$$

(b) The slope of the line through the points $(-1, -2)$ and $(4, 13)$ is

$$m = \frac{13 - (-2)}{4 - (-1)} = \frac{13 + 2}{4 + 1} = \frac{15}{5} = 3$$

In general the slope of a line is independent of the two points chosen on the line. Example 2 verifies this fact. Another way of viewing the slope can be seen from a table of points.

x	-2	-1	0	1	2
y	-5	-2	1	4	7

Each time the value of x increases by 1, the value of y increases by the slope 3.

The number b in the linear equation $y = mx + b$ can be obtained by setting $x = 0$. Then, $y = m \cdot 0 + b = b$, so the line passes through the point $(0, b)$. The number b is referred to as the y-intercept of the line $y = mx + b$ even though formally the y-intercept is the point $(0, b)$.

The linear equation $y = mx + b$ is called the *slope-intercept form* of a line because m is the slope and b is the y-intercept.

_____ **Definition** _____

The **slope-intercept form** of a line is

$$y = mx + b$$

where m is the slope and b is the y-intercept.

EXAMPLE 3 | **Problem**

Sketch the graph of the equation $y = -2x + 3$.

SOLUTION The y-intercept is $b = 3$, so the line passes through $(0, 3)$. The slope is $m = -2$. Since m is negative the line slopes downward (from left to right). The graph is sketched in Figure 5.

FIGURE 5 Graph of $y = -2x + 3$.

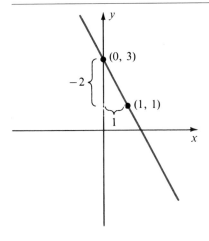

Parallel Lines

Two lines are **parallel** if they do not meet in either direction when extended indefinitely. Since parallel lines have the same steepness, it follows that two lines are parallel if and only if they have the same slope.

EXAMPLE 4 | **Problem**

Write the equation of the line that is parallel to the line $y = -2x + 2$ and passes through the origin.

SOLUTION The graph of the linear equation $y = -2x + 2$ is a line with slope $m = -2$. The new line will have the same slope $m = -2$. Since the new line passes through $(0, 0)$, its y-intercept is $b = 0$. Substituting the values $m = -2$ and $b = 0$ into the slope-intercept form $y = mx + b$ yields $y = -2x$. This is the equation of the line parallel to $y = -2x + 2$ and passing through the origin. The lines are sketched in Figure 6.

There are two special kinds of lines. **Horizontal lines** are parallel to the x-axis and **vertical lines** are parallel to the y-axis. The slope of a horizontal line can be determined by choosing any two points $(x_1, 0)$ and $(x_2, 0)$ on the x-axis. By definition the slope of the line through these points is

$$m = \frac{0 - 0}{x_2 - x_1} = 0$$

Thus the slope of the x-axis is 0. Since parallel lines have the same slope, it follows that *the slope of any horizontal line is zero.*

FIGURE 6 Parallel lines.

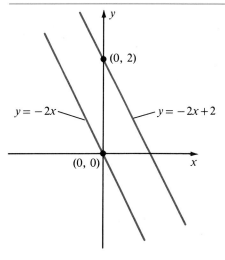

Next, consider vertical lines. If $(0, y_1)$ and $(0, y_2)$ are two points on the y-axis, the slope is

$$m = \frac{y_2 - y_1}{0 - 0}$$

Since division by 0 is not permitted, the slope of the y-axis is not defined. It follows that *the slope of any vertical line is not defined.*

EXAMPLE 5

Problem

Write the equations of the horizontal and vertical lines that pass through the point $(2, 3)$.

SOLUTION The slope of the horizontal line is $m = 0$. Since the horizontal line passing through $(2, 3)$ is parallel to the x-axis, its y-intercept is $b = 3$. Substituting $m = 0$ and $b = 3$ into the equation $y = mx + b$ yields $y = 3$.

The equations of vertical lines cannot be obtained from the linear equation $y = mx + b$ because m is not defined. The vertical line through $(2, 3)$ is located two units to the right of the y-axis. Some additional points on it are $(2, 2)$, $(2, 1)$, $(2, 0)$, and $(2, -1)$. The value of x is always 2. Hence the equation of the line is $x = 2$.

Both lines are sketched in Figure 7.

Example 5 suggests the following standard forms for all horizontal and vertical lines:

horizontal lines: $y = c$,   a number
vertical lines: $x = k$,   a number

Table 1 summarizes the results on the slopes of lines.

FIGURE 7   Horizontal and vertical lines.

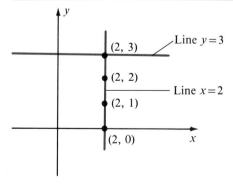

Table 1

Line:	Rising	Falling	Horizontal	Vertical
Slope:	Positive	Negative	Zero	Not defined

Graph

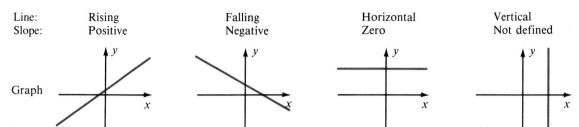

Point-Slope Form

The equation of a line that is neither vertical nor horizontal can be expressed in several forms. We have already indicated that the slope-intercept form gives the equation of a line when its slope and y-intercept are known. But what if the slope m and another point (x_1, y_1) on the line are known? Then an arbitrary point (x, y) on the line can be used with (x_1, y_1) to compute the slope $m = (y - y_1)/(x - x_1)$. So $y - y_1 = m(x - x_1)$. This form is aptly named the *point-slope form*.

_____ **Definition** _____

The **point-slope form** of a line is

$$y - y_1 = m(x - x_1)$$

where m is the slope and (x_1, y_1) is the given point on the line.

_____

Example 6 shows how to convert from the point-slope form to the slope-intercept form.

EXAMPLE 6 | **Problem**

Find the point-slope form and the slope-intercept form of the line that passes through the point (3, 2) and has slope 4.

SOLUTION We have $x_1 = 3$, $y_1 = 2$, and $m = 4$. The equation of the line in point-slope form is

$$y - 2 = 4(x - 3)$$

Use the properties of algebra indicated on the right to write the equation of the line in slope-intercept form.

$y - 2 = 4(x - 3)$	(point-slope form)
$y = 4(x - 3) + 2$	(add 2 to each side)
$y = 4x - 12 + 2$	(distributive law)
$y = 4x - 10$	(add -12 and 2)

Therefore $m = 4$ and $b = -10$. The graph is sketched in Figure 8.

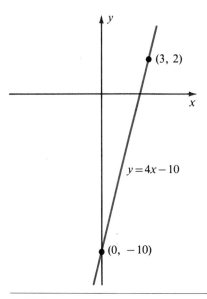

FIGURE 8   Line through (3, 2) with $m = 4$.

Example 7 shows how to determine the equation of a line when two points on the line are known.

EXAMPLE 7

Problem

Find an equation of the line that passes through the points $(1, 2)$ and $(-1, 5)$.

SOLUTION The slope of the line is

$$m = \frac{5 - 2}{-1 - 1} = \frac{3}{-2} = -\frac{3}{2}$$

Using the point-slope form with $(1, 2)$ (we could have chosen $(-1, 5)$ instead) the equation is

$$y - 2 = -\frac{3}{2}(x - 1)$$

The equation in slope-intercept form is found by solving for y.

$$y = -\frac{3}{2}x + \frac{7}{2}$$

Here is a summary of the main items in this section.

1. The slope is $m = \dfrac{y_2 - y_1}{x_2 - x_1}$ (unless $x_1 = x_2$).

2. The slope measures steepness. As x increases from left to right, y increases if the slope is positive and y decreases if the slope is negative.

3. Horizontal lines have form $y = c$; their slope is 0. Vertical lines have form $x = k$; their slope is undefined.

4. Three principal forms for the equation of a line follow.

Name of Form	Equation	Variables
general	$Ax + By = C$	A, B, C real numbers A, B not both 0
slope-intercept	$y = mx + b$	$m = $ slope $b = y$-intercept
point-slope	$y - y_1 = m(x - x_1)$	$m = $ slope $(x_1, y_1) = $ point

EXERCISE SET 1–2

• In Problems 1 to 4 sketch the graph of the line.

1. $5x - 2y = 10$ 2. $5x + 3y = 15$

3. $3x + 4y = 12$ 4. $3x - y = 6$

In Problems 5 to 10 find the slope of the line passing through the two points.

5. $(1, 2), (2, 4)$ 6. $(3, 1), (4, 3)$

7. $(-1, 2), (3, -4)$ 8. $(-3, 8), (3, -8)$

9. $(1.6, 4.1), (2.5, 5.9)$ 10. $(\frac{3}{2}, 5), (-1, -\frac{3}{2})$

In Problems 11 to 14 sketch the graph of the line.

11. $y = 4x - 5$ 12. $y = 2x + 1$

13. $y = -2x + 3$ 14. $y = -3x - 2$

In Problems 15 and 16 determine if the lines are parallel.

15. Line 1 passes through $(0, 1)$ and $(1, 2)$; line 2 passes through $(-1, 6)$ and $(6, -1)$.

16. Line 1 passes through $(1, 3)$ and $(5, -1)$; line 2 passes through $(1, 1)$ and $(2, 0)$.

In Problems 17 and 18 determine the slope of the line.

17. (a) $x = 2$ (b) $y = -7$

18. (a) $x = 15$ (b) $y = 2$

In Problems 19 and 20 find the slope-intercept form of the line passing through the given points.

19. $(2, 1), (3, -6)$ 20. $(2, -1), (-3, 6)$

•• In Problems 21 to 34 find an equation of the line from the information given. Sketch the graph of the line.

21. $m = 2, b = 7$ 22. $m = -2, b = 7$

23. The horizontal line that passes through $(-5, -3)$.

24. The vertical line that passes through $(-5, -3)$.

25. The slope is undefined and the x-intercept is -2.

26. The slope is 0 and $b = -4$.

27. $m = -2$; the line passes through $(6, -1)$.

28. $m = 1$; the line passes through $(6, -1)$.

29. $b = -1$ and the line is parallel to $y = 2x + 1$.

30. The line passes through $(1, 1)$ and is parallel to the line $x - 3y = 4$.

31. The line passes through $(-2, 0)$ and $(0, 2)$.

32. The line passes through $(2, 0)$ and $(0, -2)$.

33. The line passes through $(-2, 0)$ and $(-2, -2)$.

34. The line passes through $(0, 4)$ and $(2, 4)$.

In Problems 35 and 36 determine if the slope of the line is positive, negative, zero, or undefined.

35. (a)   (b)

(c)

36. (a)   (b)

(c)

37. (a) Put the equation $3x + 2y = 12$ in slope-intercept form. (b) What is the slope? (c) What is the y-intercept?

38. (a) Put the equation $5x + 3y = 15$ in slope-intercept form. (b) What is the slope? (c) What is the y-intercept?

In Problems 39 and 40 determine the equation of the line passing through the two intercepts.

39. (a)   (b)

40. (a)   (b)

          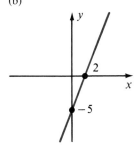

In Problems 41 and 42 write the equation (in point-slope form) of the line that has the given table.

41.

x	-1	0	1
y	5	1	-3

42.

x	-2	0	2
y	5	1	-3

43. For tax purposes, some firms assume that the current value y of an asset is related to the asset's age x by a linear equation. Suppose a firm buys a work station with ten microcomputers and assumes that after x years the work station's worth is given by

$$y = 60{,}000 - 15{,}000x$$

(a) Draw the graph of the equation.
(b) What is the work station's worth after three years?
(c) What economic interpretation can be given to the y-intercept?

••• 44. Prove that the slope of the line $y = mx + b$ is m. (*Hint:* Let (x_1, y_1) and (x_2, y_2) be points on the line with $x_1 \neq x_2$. Show that

$$\frac{(y_2 - y_1)}{(x_2 - x_1)} = m$$

45. Is it true that every equation in the variables x and y in which the highest power of x is 1 and the highest power of y is 1 must be linear?

46. Figure 9 shows the power necessary to achieve loud peak volume levels (106 decibels) for a pair of Acoustic Research loudspeakers.* The three different types of rooms ("dead," "average," and "live") require amplifiers of different sizes to produce loud peak volume levels. Let y be the power (in watts) and x be the volume of the room (in cubic feet). For each type of room, express y in terms of x in slope-intercept form.

47. The petroleum industry uses "decline curve analysis" to estimate reserves of oil and gas wells, even though more sophisticated techniques are available. The reason for the continued emphasis on this method is that the required data are almost always available. Recently two petroleum engineers developed a simple graphing technique for evaluating production data from a linear equation.† Let y represent the loss ratio (a technical term measured here in months) and x represent the time (also in months). The line which the engineers discovered for a particular

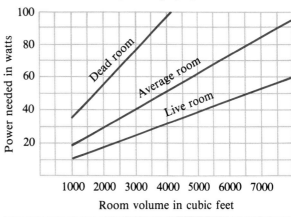

Power per channel required for a pair of acoustic, research loudspeakers playing at 106 dB S.P.L.

FIGURE 9 Problem 46.

set of data has slope 0.5 and y-intercept 7.5. (a) What is the equation of the line? (b) Sketch the line on the interval of 0 months to 72 months. (c) What is the loss ratio after 6 years?

48. This problem deals with four "rules of thumb" relating the height and weight of adults to those of children.‡ Write each relationship as an equation. (This requires the introduction of appropriate variables.) (a) The height of a male adult is double the height at age 2. (b) The height of a female adult is double the height at age $1\frac{1}{2}$. (c) The weight of a male adult is five times the weight at age 2. (d) The weight of a female adult is five times the weight at age $1\frac{1}{2}$.

*Source: Teledyne Acoustic Research of Canton, Massachusetts.
†D. A. Rowland and C. Lin, "New Linear Method Gives Constants of Hyperbolic Decline," *Oil and Gas Journal*, January 14, 1985, pp. 86–90.
‡Wilton M. Krogman, *Child Growth*, University of Michigan Press, 1972, p. 42.

1–3 INTERSECTION OF LINES

Many applications of mathematics involve more than one linear equation at the same time. In this section we describe several methods of solving systems of two linear equations in two variables. The next section extends these techniques to systems having more than two equations or more than two variables.

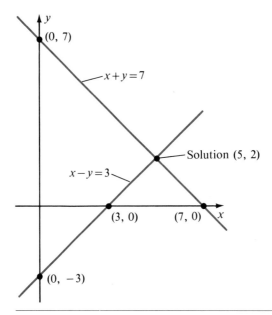

FIGURE 1   Intersection of lines.

The Solution of a System

Consider the two linear equations

$$x + y = 7$$
$$x - y = 3$$

Taken together they form a **system of two linear equations in two variables.** A **solution** is an ordered pair (x, y) of numbers that satisfy both equations at the same time, so $(5, 2)$ is a solution because

$$5 + 2 = 7$$
$$5 - 2 = 3$$

However, the pair $(6, 1)$ is not a solution because it does not satisfy the second equation, even though it does satisfy the first equation.

The solution $(5, 2)$ was obtained by inspection; it is fairly obvious. Another way of solving the system is to draw the graphs of both equations on the same coordinate system. Since both graphs are lines, the solution of the system is the point of intersection of the two lines because the point $(5, 2)$ satisfies both equations. (See Figure 1.) Example 1 illustrates the graphing method.

EXAMPLE 1

Problem

Solve this system of linear equations by graphing.

$$x + 3y = 5$$
$$x - y = 1$$

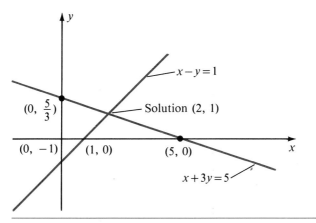

FIGURE 2 Graphing method.

SOLUTION The graph of each equation is a line. The two lines are drawn on the same coordinate system in Figure 2. It appears from the figure that the lines intersect at the point (2, 1). To verify this, substitute $x = 2$ and $y = 1$ into the equations

$$2 + 3(1) = 5$$
$$2 - 1 = 1$$

Since the two equations are satisfied, (2, 1) is the solution of the system.

The graphical method is valuable for approximating a solution to a system of two equations, but often it is not precise enough to produce exact solutions.

Elimination

The method of elimination is a very efficient method of soiving systems of linear equations. It consists of eliminating one variable and then solving for the other. We illustrate it on this system of linear equations.

$$2x + 3y = 4$$
$$3x - 2y = 6$$

Choose a variable to eliminate. Sometimes it is easier to eliminate one variable than the other, but here it does not matter so we choose y. Multiply the first equation by 2 and the second by 3.

$$4x + 6y = 8$$
$$9x - 6y = 18$$

The coefficients of y are negatives of each other. To eliminate y, add the two equations.

$$13x = 26$$

A solution to this equation is also a solution to the original system of equations. Solve for x.

$$x = 2$$

Substitute $x = 2$ into either one of the original equations to get y. From the first equation

$$2x + 3y = 4$$
$$2(2) + 3y = 4$$
$$y = 0$$

Since $x = 2$ and $y = 0$, the solution to the system is $(2, 0)$.

The variable y was eliminated by producing the numbers 6 and -6 as coefficients of y and then adding the two equations. This two-step procedure is the essence of the method of elimination.

Two Steps in Eliminating a Variable

1. Multiply the equations by appropriate numbers to produce coefficients of a variable such that one is the negative of the other.
2. Add the two resulting equations.

EXAMPLE 2 | **Problem**

Solve this system of linear equations by elimination.

$$3x - 4y = 1$$
$$4x - 2y = 3$$

SOLUTION Eliminating y can be accomplished with only one multiplication. Multiply the second equation by -2 and add the resulting equations.

$$
\begin{array}{rr}
3x - 4y = & 1 \\
-8x + 4y = & -6 \\
\hline
-5x \qquad = & -5 \\
x = & 1
\end{array}
$$

Substitute $x = 1$ into the first equation of the given system to get $y = \frac{1}{2}$. The solution is $(1, \frac{1}{2})$.

The next example applies elimination to a system in which some variable has the same coefficient in both equations.

EXAMPLE 3 | **Problem**

Solve this system of linear equations.

$$3x + 2y = 2$$
$$3x - y = 5$$

SOLUTION Subtract the second equation from the first.

$$3x + 2y = 2$$
$$\underline{-3x + y = -5}$$
$$3y = -3$$

Solve for y.

$$y = -1$$

Substitute $y = -1$ into the first equation to get $x = \frac{4}{3}$. The solution is $(\frac{4}{3}, -1)$.

Substitution

The method of substitution involves solving one equation for one variable and then substituting it into the other equation. This method also leads to one equation in one variable.

EXAMPLE 4

Problem

Solve this system of linear equations.

$$2x + 4y = 1$$
$$x - 5y = -2$$

SOLUTION Solve the second equation for x.

$$x = 5y - 2$$

Substitute this expression into the first equation.

$$2x + 4y = 1$$
$$2(5y - 2) + 4y = 1$$
$$14y - 4 = 1$$
$$y = \frac{5}{14}$$

To find x, substitute $y = \frac{5}{14}$ to get $x = 5y - 2 = -\frac{3}{14}$. The solution is $(-\frac{3}{14}, \frac{5}{14})$.

Systems Without Unique Solutions

Every system that we have encountered so far contains exactly one solution. This is not always the case, however, as can be seen by examining the geometric meaning of a system of linear equations in two unknowns. The graph of each equation is a line, so the system consists of two lines, l_1 and l_2. Figure 3 shows that such a system can have one solution, no solutions, or infinitely many solutions.

A system with no solutions is called **inconsistent.** Figure 3b shows that the graph of such a system consists of two parallel lines. Example 5 shows how to detect such a system algebraically.

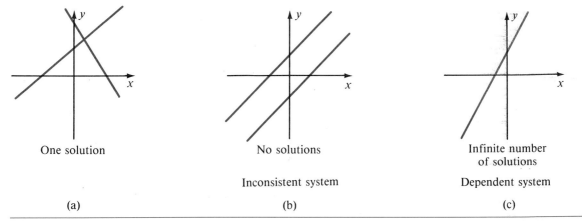

One solution

No solutions

Inconsistent system

Infinite number
of solutions

Dependent system

(a)   (b)   (c)

FIGURE 3 Systems of two equations.

EXAMPLE 5 | **Problem**

Solve this system of linear equations.

$$x - y = 1$$
$$2x - 2y = 3$$

SOLUTION To solve this system by elimination, multiply the first equation by -2 and add it to the second equation.

$$-2x + 2y = -2$$
$$\underline{2x - 2y = 3}$$
$$0 = 1$$

If an ordered pair (x, y) were to satisfy the system, then $0 = 1$, so a contradiction would result. Therefore the original system cannot have any solutions.

This can be seen geometrically by writing each equation in slope-intercept form: $y = x - 1$ and $y = x - (\frac{3}{2})$. The lines have equal slopes but different y-intercepts, so the equations represent distinct parallel lines.

A system with infinitely many solutions is called **dependent.** Dependent systems occur when the two equations represent the same line. (See Figure 3c.) Example 6 shows how to uncover a dependent system algebraically.

EXAMPLE 6 | **Problem**

Solve this system of linear equations.

$$x - y = 1$$
$$2x - 2y = 2$$

SOLUTION To solve this system by elimination, multiply the first equation by -2 and add it to the second equation.

$$-2x + 2y = -2$$
$$\underline{2x - 2y = 2}$$
$$0 = 0$$

This is true for *all* ordered pairs (x, y) that satisfy the first equation. The reason is that the second equation has the same set of solutions as the first.

Breakeven Analysis

In financial management the **breakeven point** is the level of operations where neither profits nor losses are incurred. Above this point, each unit sold contributes to profits; below it, each unit sold contributes to losses. Therefore the breakeven point is the intersection of the sales revenue and the total cost. The revenue is derived from the sales of the units. The total cost usually involves a fixed cost (like salaries and rental expenses) and a variable cost (like sales commissions and advertising). Example 7 illustrates these terms.

EXAMPLE 7

A candy company specializes in making mints. Each box of mints sells for $2 and has a variable cost of $1.20. The fixed cost is $40,000.

Problem

Determine where the company's breakeven point occurs.

SOLUTION The sales revenue is derived from the $2 price for each box of mints. Let x be the number of boxes sold. If R is the sales revenue, then $R = 2x$. The variable cost is $1.20 per box, so it is $1.2x$. The total cost C, which includes the $40,000 fixed cost, is given by $C = 1.2x + 40,000$.

The breakeven point occurs at a value of x where $R = C$. Therefore, at this point, the equation for total cost can be written $R = 1.2x + 40,000$. The breakeven point is found by solving the system of linear equations

$$R = 2x$$
$$R = 1.2x + 40,000$$

The solution is $x = 50,000$ and $R = 100,000$. Therefore the breakeven point occurs when 50,000 boxes of mints are sold.

EXERCISE SET 1–3

• In Problems 1 and 2 solve the system by inspection. (Think, do not write.)

1. $x + y = 5$
   $x + 2y = 6$

2. $x + y = 2$
   $x - y = 2$

In Problems 3 and 4 solve the system by graphing.

3. $2x + y = 4$
   $x - y = 2$

4. $x + 3y = 1$
   $2x - y = 9$

In Problems 5 to 10 solve the system by elimination.

5. $2x + y = 3$
   $3x + 4y = 7$

6. $2x + 5y = 2$
   $3x - 2y = 3$

7. $5x + 2y = 1$
   $2x - 3y = 8$

8. $8x - 3y = 1$
   $3x + 4y = 9$

9. $2x - 3y = 1$
   $3x - 2y = 3$

10. $-4x + 2y = -3$
    $3x - 2y = 3$

In Problems 11 to 14 solve the system by substitution.

11. $x + y = -2$
    $x - y = 4$

12. $2x - 3y = 2$
    $3x - y = 3$

13. $2x - 4y = 1$
    $x - y = 2$

14. $x + 3y = 4$
    $2x - 6y = 1$

In Problems 15 to 18 determine whether the system is inconsistent or dependent.

15. $3x - y = 1$
    $6x - 2y = 5$

16. $8x + y = 2$
    $8x + y = 3$

17. $2x - 4y = 2$
    $3x - 6y = 3$

18. $12x - 6y = 3$
    $8x - 4y = 2$

In Problems 19 and 20 determine the breakeven point of a candy company that specializes in making mints.

19. Each box of mints sells for \$3 and has a variable cost of \$1.50. The fixed cost is \$30,000.

20. Each box of mints sells for \$5 and has a variable cost of \$2.00. The fixed cost is \$30,000.

•• In Problems 21 to 36 solve the system by any method. If there is not a unique solution, state whether there are no solutions or infinitely many solutions.

21. $2x - y = 3$
    $4x + y = 1$

22. $-x + 2y = 3$
    $2x + 6y = 1$

23. $4x + y = 18$
    $4x - 3y = 10$

24. $2x + y = 6$
    $x - 2y = 3$

25. $x + y = 120$
    $100x + 40y = 6000$

26. $45x + 9y = 45$
    $10x + 6y = 20$

27. $2x + 3y = 4$
    $3x - 2y = 6$

28. $x + 2y = 2$
    $x - y = 5$

29. $25x - 50y = 1500$
    $30x - 15y = 2700$

30. $1.2x - 3.5y = 23$
    $8.5x + 7y = 1.5$

31. $x + 1.5y = 0$
    $y = 8.5x$

32. $15x - 110y = 4000$
    $10x + 55y = 1000$

33. $x + y = 5$
    $2x + 2y = 8$

34. $x - y = 3$
    $3x - 3y = 9$

35. $2x + 5y = 40$
    $2x + y = 24$

36. $2x + 4y = 40$
    $50x + 200y = 1600$

37. A travel agency is planning charter flights to Hawaii for exactly 1400 passengers aboard two types of jets. Each 747 jet holds 400 passengers and each 707 holds 150 passengers. The rental fee is \$15,000 for each 747 and \$9000 for each 707. If exactly \$66,000 must be used for renting the jets, how many of each type should be ordered?

38. A dietician is preparing a meal of chicken and corn that must provide exactly 63 grams (g) of protein and 21 milligrams (mg) of iron. Each serving of chicken contains 45 g of protein and 9 mg of iron; each serving of corn contains 9 g of protein and 6 mg of iron. How many servings of each ingredient should be prepared?

39. Suppose that \$5000 is invested in two securities, school bonds and public utility bonds. School bonds pay 6% interest and public utility bonds pay 8% interest. If an interest of \$370 is required, how much money should be invested in each security?

40. Suppose that regular gasoline costs \$1 per gallon and premium gasoline costs \$1.20 per gallon. How many gallons of each gasoline should be blended to produce 400 gallons of mixture that is worth \$1.05 per gallon?

••• Problems 41 and 42 are adapted from an article on the ability of the Big Three American automakers to reduce their breakeven point substantially to become more competitive with foreign competition.* The variable x represents the number of cars sold (in millions of cars) and y represents the sales revenue and the total cost (in millions of dollars).

*Adapted from Lynn Adkins, "Detroit Gets Lean and Mean," *Dun's Business Month*, January 1983, pp. 56–58.

41. (a) Determine the breakeven point if the sales revenue is $y = 1.5x$ and the total cost is $y = x + 6.1$.
    (b) Determine the breakeven point if the sales revenue is $y = 2x$ and the total cost is $y = 0.5x + 13.35$.
    (c) The solution in (a) was the breakeven point for the Big Three in 1979. The solution in (b) was the breakeven point in 1983. By how many cars did the Big Three drop the breakeven point in four years?

42. (a) Determine the breakeven point if the sales revenue is $y = 0.5x$ and the total cost is $y = 0.25x + 0.6$.
    (b) Determine the breakeven point if the sales revenue is $y = 0.8x$ and the total cost is $y = 0.6x + 0.28$.
    (c) The solution in (a) was the breakeven point for Chrysler in 1981. The solution in (b) was the breakeven point in 1983. By how many cars did Chrysler drop the breakeven point in two years?

43. Figure 4 shows the relationship between short-term interest rates and the volume of funds to be raised in credit markets. The graph is due to Peter Crawford, who is in charge of financial analysis for Citicorp Information Services of New York, NY.† (a) At what rate of interest did the supply of credit equal the demand for credit in 1985? (b) At what rate of interest did the supply of credit equal the demand for credit in 1983?

44. In 1939 L. F. Richardson proposed a mathematical model of an arms race that is still valid today.‡ The model is based upon equations from calculus called differential equations, but we have adapted it to a system of linear equations.

Supply and demand for credit, early 1985 versus early 1983

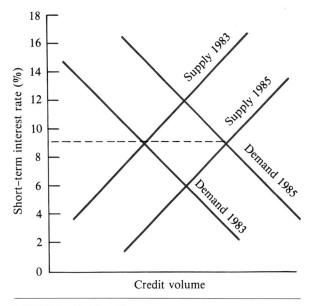

FIGURE 4   Problem 43.

Let X and Y be two countries with arms expenditures x and y, respectively. Denote each change in expenditure by Δx and Δy. Then

$$\Delta x = \underbrace{ay}_{\text{spurring term}} - \underbrace{bx}_{\text{breaking term}} + \underbrace{c}_{\text{grievance term}} \quad (a > 0, b > 0)$$

$$\Delta y = dx - ey + f \quad (d > 0, e > 0)$$

A state of equilibrium occurs when there is no change in expenditures, that is, when $\Delta x = \Delta y = 0$. Find the equilibrium point for $a = 0.01$, $b = 0.02$, $c = 2$, $d = 0.0125$, $e = 0.1$, and $f = 6.25$.

†Peter Crawford, "Interest Rate Outlook: Focus on Credit Supply," *Business Economics,* July 1985, pp. 34–39.

‡L. F. Richardson, "Generalized Foreign Politics," *British Journal of Psychology,* Monograph Supplement XXIII, Cambridge University Press, 1939. A BASIC program for solving this system is given in the article by Leonard M. Wapner, "Modeling with Difference Equations: Two Examples," *The Mathematics Teacher,* February 1984, pp. 136–140.

1–4 SYSTEMS OF LINEAR EQUATIONS

Around 1800 the German mathematician Karl Friedrich Gauss (1777–1855) introduced an effective method for solving systems of linear equations by eliminating the variables one at a time. This method was known in China 500 years earlier, but it remained unknown in the West, where it is now called Gaussian elimination. This section applies Gaussian elimination to systems of three linear equations in three variables.

Gaussian Elimination

A **linear equation in three variables** is an equation of the form $ax + by + cz = k$, where a, b, c, and k are constants such that a, b, and c are not all 0. A similar definition applies to linear equations in four or more variables. The equation $2x - 3y + 0.5z = 4$ is a linear equation in three variables. Its graph is a plane in three-dimensional space.

Consider this system of three linear equations in three variables.

$$
\begin{aligned}
x + y + z &= 10 \\
x - y &= 2 \\
y - z &= 1
\end{aligned}
$$

A solution is the ordered triple of numbers (5, 3, 2) because all three equations are satisfied when the values $x = 5$, $y = 3$, and $z = 2$ are substituted into the three equations. However (6, 4, 0) is not a solution because it does not satisfy the third equation, even though it does satisfy the first two equations.

It is possible to use the methods of inspection, graphing, and substitution to solve systems with three or more variables, but in general these methods are not very practical. However, Gaussian elimination can be performed easily on a system of three equations in three variables, can be extended to systems of other sizes, and can be applied to whatever numbers are involved in the system (fractions or decimals). Although only a slight variation of elimination, it is very efficient. We will describe it in detail on the following system, whose equations are denoted by E1, E2, and E3.

$$x + y + 2z = 9 \tag{E1}$$

$$2x + 4y - 3z = 1 \tag{E2}$$

$$3x + 6y - 5z = 0 \tag{E3}$$

The first step is to eliminate the variable x from E2 by multiplying E1 by -2, adding it to E2, and replacing E2 by this new equation. The number -2 is called the **multiplier** of E2. The notation

$$(-2)(\text{E1}) + (\text{E2})$$

represents this procedure. The multiplier for the third equation is -3. The original system is replaced by a simpler system with the new equations in place of the original second and third equations.

$$x + y + 2z = 9$$
$$2y - 7z = -17 \qquad (-2)(\text{E1}) + (\text{E2})$$
$$3y - 11z = -27 \qquad (-3)(\text{E1}) + (\text{E3})$$

Notice that the variable x has been eliminated from the second and third equations.

The set of solutions of a system of equations is called the **solution set.** Two systems of linear equations are called **equivalent** if their solution sets are equal. The new system above is equivalent to the original system because multiplying an equality by a number produces another equality, and adding two equalities also produces an equality. The next step eliminates the variable y from the third equation.

$$x + y + 2z = 9$$
$$2y - 7z = -17$$
$$-\tfrac{1}{2}z = -\tfrac{3}{2} \qquad (-\tfrac{3}{2})(\text{E2}) + (\text{E3})$$

This system of equations is called the **reduced system** of the original system.

The purpose of Gaussian elimination is to convert the original system of equations to an equivalent, reduced system, then to solve the reduced system. The first part, called **elimination,** reduces the original system to an equivalent, reduced system. The second part, called **back-substitution,** solves the reduced system of equations by working from the bottom equation up. The solution to the third equation is $z = 3$. Substitute it into the second equation and solve for y.

$$2y - 7(3) = -17$$
$$y = 2$$

Finally, the values $y = 2$ and $z = 3$ are substituted into the first equation.

$$x + 2 + 2(3) = 9$$
$$x = 1$$

The reduced system has one solution: $x = 1, y = 2, z = 3$. Since the original system has the same solution set, the triple (1, 2, 3) is the only solution of the original system too. It should be checked by substituting it into the original system.

Check:

$$1 + 2 + 2(3) = 9$$
$$2(1) + 4(2) - 3(3) = 1$$
$$3(1) + 6(2) - 5(3) = 0$$

The accompanying display describes the entire method for solving systems of three linear equations in three variables x, y, and z.

Gaussian Elimination

Elimination
1. Eliminate x by replacing E2 by E2 plus its multiplier times E1, and by replacing E3 by E3 plus its multiplier times E1.
2. Eliminate y from the resulting bottom equation by replacing E3 with the sum of E3 and its multiplier times E2.

> **Back-Substitution**
> 3. Solve the third equation for z.
> 4. Substitute the value of z into the second equation and solve for y.
> 5. Substitute the values of y and z into the first equation and solve for x.
>
> **Check**
> 6. Substitute the triple (x, y, z) into the original system.

Example 1 illustrates the method of Gaussian elimination by solving the system that began this section.

EXAMPLE 1 | **Problem**

Solve this system of linear equations.

$$\begin{aligned} x + y + z &= 10 \\ x - y \phantom{{}+z} &= 2 \\ y - z &= 1 \end{aligned}$$

SOLUTION In step 1 the multiplier for equation 2 is -1. The variable x is already eliminated from equation 3, so no operation is necessary. The system becomes

$$\begin{aligned} x + y + z &= 10 \\ -2y - z &= -8 \qquad (-1)(\text{E1}) + (\text{E2}) \\ y - z &= 1 \end{aligned}$$

In step 2 the multiplier for equation 3 is $\frac{1}{2}$. Thus

$$\begin{aligned} x + y + \phantom{(\tfrac{3}{2})}z &= 10 \\ -2y - \phantom{(\tfrac{3}{2})}z &= -8 \\ -(\tfrac{3}{2})z &= -3 \qquad (\tfrac{1}{2})(\text{E2}) + (\text{E3}) \end{aligned}$$

Now apply back-substitution. Step 3 yields $z = 2$. By step 4

$$\begin{aligned} -2y - z &= -8 \\ -2y - 2 &= -8 \\ y &= 3 \end{aligned}$$

Apply step 5 with $y = 3$ and $z = 2$.

$$\begin{aligned} x + y + z &= 10 \\ x + 3 + 2 &= 10 \\ x &= 5 \end{aligned}$$

The solution is $(5, 3, 2)$, which confirms the solution noted earlier.

Simplifications

The efficiency of Gaussian elimination is due to the fact that it uses only one operation repeatedly: multiply an equation by a number, add it to a second equation, and replace the second equation by the new equation. Gaussian elimination is the method that is used on almost all computers to solve systems of linear equations because it is fast and accurate. However, when we humans solve such a system it is advantageous to simplify some of the computations. There are two additional operations on equations that are easy to carry out by hand and do not change the solution set: interchange equations and multiply an equation by a nonzero number.

Interchanging two equations is sometimes helpful in avoiding fractions (which often cause numerical errors), but at other times it is absolutely necessary. For instance, consider this system.

$$y - z = 1 \tag{E1}$$

$$x - y = 2 \tag{E2}$$

$$x + y + z = 10 \tag{E3}$$

It is impossible to find a multiplier for x in E1 because the coefficient of x is 0. The operation E1 $\leftrightarrow$ E3 leads to the system solved in Example 1. The operation E1 $\leftrightarrow$ E2 could have been used instead.

Here is a list of the three operations that preserve the solution set of a system of linear equations.

Elementary Operations on a System of Linear Equations

1. Interchange two equations.
2. Multiply each number in an equation by a nonzero number.
3. Replace an equation by the sum of that equation and a multiple of another equation.

EXAMPLE 2

Problem

Solve the system of equations

$$-2x \qquad + 3z = -5$$
$$2x + 3y - z = 9$$
$$x - y + 2z = -3$$

SOLUTION It is helpful to interchange the top and bottom equations so the coefficient of x is 1.

$$x - y + 2z = -3 \qquad \text{E1} \leftrightarrow \text{E3}$$
$$2x + 3y - z = 9$$
$$-2x \qquad + 3z = -5$$

The multiplier for equation 2 is -2 and the multiplier for equation 3 is 2.

$$x - y + 2z = -3$$
$$5y - 5z = 15 \qquad (-2)(\text{E1}) + (\text{E2})$$
$$-2y + 7z = -11 \qquad (2)(\text{E1}) + (\text{E3})$$

To avoid fractions, reduce the second equation.

$$
\begin{aligned}
x - y + 2z &= -3 \\
y - z &= 3 \qquad (\tfrac{1}{5})(\text{E2}) \\
-2y + 7z &= -11
\end{aligned}
$$

The multiplier for equation 3 is 2.

$$
\begin{aligned}
x - y + 2z &= -3 \\
y - z &= 3 \\
5z &= -5 \qquad (2)(\text{E2}) + (\text{E3})
\end{aligned}
$$

By back-substitution, the solution is $(1, 2, -1)$. It checks when substituted into the original system.

Example 3 illustrates these operations in a practical setting.

EXAMPLE 3   A software firm makes disks in these sizes: $3\tfrac{1}{2}''$, $5\tfrac{1}{4}''$, and $8''$. They are tested in three phases: A, B, and C. The $3\tfrac{1}{2}''$ disks require exactly 1 second in each of the three phases. The $5\tfrac{1}{4}''$ disks require 1 second in phase A, 2 seconds in phase B, and 3 seconds in phase C. The $8''$ disks require 4 seconds in phase A, 8 seconds in phase B, and 8 seconds in phase C. Phase A is available for exactly 27 seconds, phase B for 49 seconds, and phase C for 63 seconds.

Problem

How many disks of each size can be tested?

SOLUTION Let

$$
\begin{aligned}
x &= \text{the number of } 3\tfrac{1}{2}'' \text{ disks} \\
y &= \text{the number of } 5\tfrac{1}{4}'' \text{ disks} \\
z &= \text{the number of } 8'' \text{ disks}
\end{aligned}
$$

In phase A the $3\tfrac{1}{2}''$ disks require 1 second, the $5\tfrac{1}{4}''$ disks require 1 second, and the $8''$ disks require 4 seconds. There are exactly 27 seconds available, so

$$
x + y + 4z = 27
$$

Similarly, the conditions on phases B and C lead to the equations

$$
\begin{aligned}
x + 2y + 8z &= 49 \\
x + 3y + 8z &= 63
\end{aligned}
$$

This system reduces to the equivalent system

$$
\begin{aligned}
x + y + 4z &= 27 \\
y + 4z &= 22 \\
-4z &= -8
\end{aligned}
$$

By back-substitution, the solution is $x = 5$, $y = 14$, and $z = 2$. Therefore five $3\tfrac{1}{2}''$ disks, fourteen $5\tfrac{1}{4}''$ disks, and two $8''$ disks can be tested.

Number of Solutions

All of the 3×3 systems in this section have had precisely one solution. However, some 3×3 systems have infinitely many solutions and some have no solution, just like the 2×2 case. Are there any other possibilities? For instance, is it possible for a system to have exactly two solutions? Or between five and ten solutions? To answer these questions, we turn to the geometrical interpretation of a linear system.

Equations of the form $ax + by + cz = d$ represent planes in space. Three such equations correspond to three planes, and their simultaneous solution corresponds to the intersection of planes. Three distinct planes can intersect in one point (view the floor, front wall, and a side wall of the room you are sitting in); no points (view the floor, the ceiling, and the front wall); or infinitely many points (the front wall, a side wall, and an imaginary wall that runs through the line where the front and side walls meet). There are no other possibilities. Thus every system of three linear equations in three variables has either no solution, a unique solution, or infinitely many solutions.

EXERCISE SET 1–4

• In Problems 1 to 4 a system of equations and three ordered triples are given. Determine whether any of the triples are solutions.

1. $\begin{aligned} x + y + z &= 3 \\ x - y - 2z &= -2 \\ 2x + y - z &= 2 \end{aligned}$
    $(1, 0, 2), (3, 1, -1), (1, 1, 1)$

2. $\begin{aligned} 2x + y - z &= 2 \\ 3x - 2y + z &= 2 \\ x - 3y + 2z &= 0 \end{aligned}$
    $(1, 1, 1), (1, 0, 1), (0, 4, -6)$

3. $\begin{aligned} 2x - y &= 3 \\ x - y + z &= 3 \\ 3x \quad - 8z &= 3 \end{aligned}$
    $(3, 1, 2), (0, -3, 0), (2, -1, 0)$

4. $\begin{aligned} 3x - y &= 0 \\ y - 2z &= -2 \\ x - y &= 0 \end{aligned}$
    $(1, 3, 1), (0, 0, 1), (0, 0, 0)$

The systems in Problems 5 to 8 have unique solutions. Find them by inspection. (Think, don't write.)

5. $\begin{aligned} x + y + z &= 3 \\ x \quad - z &= 0 \\ y - z &= 0 \end{aligned}$

6. $\begin{aligned} x + y + z &= 0 \\ x \quad - z &= 0 \\ y - z &= 0 \end{aligned}$

7. $\begin{aligned} x + y + z &= 0 \\ x + y \quad &= 0 \\ x \quad + z &= 0 \end{aligned}$

8. $\begin{aligned} x + y + z &= 6 \\ x + y \quad &= 3 \\ x \quad + z &= 4 \end{aligned}$

In Problems 9 to 16 use Gaussian elimination to find the unique solution. Make sure to check your answers.

9. $\begin{aligned} x + y - z &= 2 \\ 2x + 3y - z &= 6 \\ 3x + 2y - z &= 7 \end{aligned}$

10. $\begin{aligned} x - 2y - z &= -3 \\ 2x - 3y \quad &= -2 \\ -x + y + z &= 1 \end{aligned}$

11. $\begin{aligned} 2x - y + 3z &= 5 \\ x - y - z &= 2 \\ -3x \quad - 2z &= -9 \end{aligned}$

12. $\begin{aligned} 2y - z &= 3 \\ 3x + 5y - z &= 6 \\ x + 2y - 3z &= 0 \end{aligned}$

13.
$$\begin{aligned} 2x + 4y - 2z &= -2 \\ 3x + 5y - z &= 3 \\ -2x + 3y - z &= -7 \end{aligned}$$

14.
$$\begin{aligned} 3x - 6y + 9z &= -24 \\ 4x + 2y &= -2 \\ -x + y - 3z &= 5 \end{aligned}$$

15.
$$\begin{aligned} -x + y + 2z &= 5 \\ x + y + z &= 3 \\ 2x + 3y + z &= 5 \end{aligned}$$

16.
$$\begin{aligned} 2x + 3y + z &= 5 \\ -x + y + 2z &= 0 \\ x + y + z &= 1 \end{aligned}$$

17. A software firm tests small, medium, and large disks. The small disks require 1 second in phase A, 2 seconds in phase B, and 3 seconds in phase C. The medium disks require 1 second in phase A, 3 seconds in phase B, and 4 seconds in phase C. The large disks require 2 seconds in phase A, 6 seconds in phase B, and 9 seconds in phase C. Phase A is available for exactly 14 seconds, phase B for 37 seconds, and phase C for 55 seconds. How many disks of each size can be tested?

18. A software firm tests small, medium, and large disks. The small disks require 1 second in phase A, 4 seconds in phase B, and 3 seconds in phase C. The medium disks require 2 seconds in phase A, 9 seconds in phase B, and 6 seconds in phase C. The large disks require 1 second in phase A, 3 seconds in phase B, and 2 seconds in phase C. Phase A is available for exactly 21 seconds, phase B for 89 seconds, and phase C for 60 seconds. How many disks of each size can be tested?

•• In Problems 19 to 28 solve each system of equations by finding the unique solution when it exists. Otherwise, state that there is not a unique solution.

19.
$$\begin{aligned} x + y - z &= 4 \\ 2x + 3y - z &= 9 \\ 3x + 2y - z &= 8 \end{aligned}$$

20.
$$\begin{aligned} x - 2y - z &= -5 \\ 2x - 3y &= -4 \\ -x + y + z &= 3 \end{aligned}$$

21.
$$\begin{aligned} 2x - y + 3z &= 5 \\ x - y - z &= 5 \\ -3x - 2z &= -10 \end{aligned}$$

22.
$$\begin{aligned} 2y - z &= 1 \\ 3x + 5y - z &= -2 \\ x + 2y - 3z &= 4 \end{aligned}$$

23.
$$\begin{aligned} 2x + 4y - 2z &= -2 \\ 3x + 5y - z &= 3 \\ -2x + 3y - z &= -7 \end{aligned}$$

24.
$$\begin{aligned} 2x + 4y - 2z &= 0 \\ 3x + 5y - z &= 0 \\ -2x + 3y - z &= 0 \end{aligned}$$

25.
$$\begin{aligned} x - 2y + z &= 1 \\ 3x - y + 2z &= 2 \\ 5x - 5y + 4z &= 4 \end{aligned}$$

26.
$$\begin{aligned} 2x - y + z &= 1 \\ 3x + 3y - z &= 11 \\ x + 4y - 2z &= 10 \end{aligned}$$

27.
$$\begin{aligned} 2x + 4y + 6z &= 2 \\ x - 2y - 5z &= 4 \\ x + 2y + 3z &= 0 \end{aligned}$$

28.
$$\begin{aligned} 3x - 6y - 9z &= 12 \\ x + 2y - 5z &= 4 \\ x - 2y - 3z &= 0 \end{aligned}$$

29. A sound studio uses special equipment for mixing and dubbing the recording of solo artists and groups. The equipment is available exactly 16 hours a day for mixing and 13 hours a day for dubbing. Each solo artist requires 2 hours for mixing and 5 hours for dubbing, while each group requires 7 hours for mixing and 4 hours for dubbing. How many solo artists and groups can be recorded each day?

30. A meat manufacturer mixes beef and pork into two types of sausage links, regular and all-beef. Each pound (lb) of regular links contains 0.6 lb of beef and 0.4 lb of pork, while each pound of all-beef links contains 0.8 lb of beef and 0.2 lb of pork. There are exactly 24 lb of beef and 11 lb of pork in stock. How many pounds of each type of sausage links can be made?

31. A hospital nutritionist prepares a meal whose primary ingredients are fish, corn, and rice. Each ounce of fish contains 100 calories (cal), 2 grams (g) of fat, and 3 milligrams (mg) of iron; each ounce of corn contains 200 cal, 5 g of fat, and 7 mg of iron; and each ounce of rice contains 200 cal, 6 g of fat, and 9 mg of iron. The meal must provide exactly 1000 cal, 24 g of fat, and 35 mg of iron. How many ounces of each ingredient should be used?

32. A shoe company makes special shoes for runners, hikers, and bikers on three machines. Each running shoe requires 1 hour on each machine; each hiking shoe requires 2 hours on machine A, 1 hour on machine B, and 2 hours on machine C; each biking shoe requires 2 hours on machine A, 1 hour on machine B, and 1 hour on machine C. Machine A is available exactly 26 hours a week, machine B 21 hours a week, and machine C 23 hours a week. How many shoes of each type can be made each week?

●●● 33. Apply Gaussian elimination to solve a general system of two equations in two variables.

$$ax + by = e$$
$$cx + dy = f$$

where a, b, c, d, e, and f are real numbers with $ad - bc \neq 0$.

34. The Chinese were the first society to solve systems of linear equations. Their method, which is strikingly similar to Gaussian elimination, was based on computing board techniques that used red rods for positive numbers and black rods for negative numbers. This explains the origin of the financial phrases ''to be in the red'' and ''to be in the black.'' Solve the following problem from the book *Chiu chang suan shu* (*Nine Chap-*

ters on the Mathematical Art), which was written about 200 B.C.*

> Of three classes of cereal plants, 3 bundles of the first, 2 of the second, and 1 of the third will produce 39 *tou* of corn after threshing; 2 bundles of the first, 3 of the second, and 1 of the third will produce 34 *tou*; while 1 of the first, 2 of the second, and 3 of the third will produce 26 *tou*. Find the measure of corn contained in 1 bundle of each class. (1 *tou* = 10.3 liters)

A system of linear equations is called **homogeneous** if all of the numbers on the right-hand side are equal to zero. In Problems 35 and 36 solve the stated problems with the same coefficients on the left-hand side but with all zeros on the right-hand side.

35. Problems 19, 25, and 27.

36. Problems 20, 26, and 28.

In Problems 37 to 40 determine whether the statement is true or false, using Problems 35 and 36 as a guide.

37. Every homogeneous system of linear equations has exactly one solution.

38. Every homogeneous system of linear equations has at least one solution.

39. The triple (0, 0, 0) is always a solution to a homogeneous system of linear equations.

40. The triple (0, 0, 0) is the unique solution to every homogeneous system of linear equations.

41. In a pari-mutuel wagering system such as horse racing, *dutching* means to wager on more than one horse to win in the same race. This problem deals with dutching three horses.† (Two prob-

---

*The Chinese solution to this problem is described on page 14 of the article by Frank Swetz titled ''The Evolution of Mathematics in Ancient China,'' which appeared in the January 1979 issue of the journal *Mathematics Magazine*.

†See, for example, William L. Scott, *Investing at the Race Track*, Simon & Schuster, 1981. The article ''Some Mathematical Applications of Pari-mutuel Wagering'' by Frank P. Soler (which appeared in *The Mathematics Teacher*, May 1987, pp. 394–399) presents a method in which the better can use a hand-held calculator to carry out the computations on dutching three horses in 15 seconds.

lems in Sections 2–3 and 2–5 will consider other aspects of dutching five horses and two horses, respectively.) If a bettor seeks to win $w on three horses whose odds of winning are a:1 (read "a to 1"), b:1, and c:1, then the amounts which should be bet on each horse are x, y, and z, respectively, where

$$
\begin{aligned}
ax - y - z &= w \\
-x + by - z &= w \\
-x - y + cz &= w
\end{aligned}
$$

If the odds on three particular horses are 3:1, 3:1, and 6:1, how much money should be bet on each horse in order to win $100?

1–5 LEAST SQUARES

Many applications of mathematics are concerned with the relationship between one quantity x and another quantity y. For instance, x can be a person's height and y can be that person's shoe size. If you record the height and shoe size of ten people and then plot the ten ordered pairs of points (x, y) on a coordinate system, you will most likely find that the points do not lie on a straight line. However, they will appear to be distributed along a line, in the sense that a line can be drawn to pass very near to all of them. This line is called the line of best fit, and the method of least squares produces its equation.

This section shows how to construct the equation for the line of best fit. It is applied to recent voting trends in presidential elections in the United States.

The Line of Best Fit

TABLE 1

x	y
2	1
3	6
4	9
5	13
6	15

Table 1 lists five points in the plane. Regard them as ordered pairs (x, y) and graph them on a coordinate system. (See Figure 1.) The points do not lie on the same line, but seem to be spaced in a linear fashion. The method of least squares produces the line of best fit for these points, where by "best fit" we mean that the sum of squares of the distances from the points to the line are minimal. Since distances are measured in terms of squares and least is synonymous with minimal, the method is called "least squares."

The method of least squares involves two equations in two unknowns, m (slope) and b (y-intercept). The Greek letter sigma, Σ, denotes the "sum."

$$Pb + (\Sigma\, x)m = \Sigma\, y \tag{LS1}$$

$$(\Sigma\, x)b + (\Sigma\, x^2)m = \Sigma\, xy \tag{LS2}$$

where

$$
\begin{aligned}
P &= \text{the number of points} \\
\Sigma\, x &= \text{the sum of the } x \text{ values} \\
\Sigma\, y &= \text{the sum of the } y \text{ values} \\
\Sigma\, x^2 &= \text{the sum of the squares } x^2 \\
\Sigma\, xy &= \text{the sum of the products } xy
\end{aligned}
$$

We refer to the equations as LS1 and LS2. Their derivation involves calculus so we will illustrate them without deriving them.

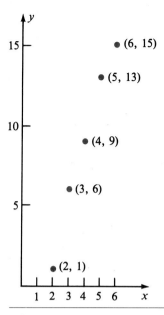

FIGURE 1 Five points.

    For the five points in Table 1, set $P = 5$. To obtain the four sums, expand Table 1 by two more columns. Place a column for x^2 to the left of the x column and a column for xy to the right of the y column. This is shown in Table 2, where the row of totals yields $\Sigma\, x = 20$, $\Sigma\, y = 44$, $\Sigma\, x^2 = 90$, and $\Sigma\, xy = 211$. The least squares equations are

$$5b + 20m = \quad 44$$
$$20b + 90m = 211$$

This system can be solved by elimination. Multiply the top equation by -4, then add the two equations.

$$-20b - 80m = -176$$
$$\underline{20b + 90m = \quad 211}$$
$$10m = \quad 35$$

TABLE 2

	x^2	x	y	xy
	4	2	1	2
	9	3	6	18
	16	4	9	36
	25	5	13	65
	36	6	15	90
Totals	90	20	44	211

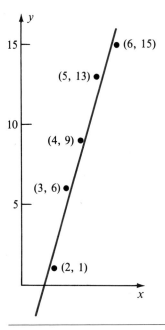

FIGURE 2   Line of best fit.

The slope of the line of best fit is $m = 3.5$. Substitute this value into the equation

$$5b + 20m = 44$$

Then

$$5b + 20(3.5) = \quad 44$$
$$b = -5.2$$

The y-intercept is $b = -5.2$. The line of best fit is uniquely determined by its slope and y-intercept.

$$y = 3.5x - 5.2$$

The line and the five points are graphed in Figure 2.

The Method of Least Squares

For a given set of data points (x, y) the **line of best fit** is

$$y = mx + b$$

where m and b are obtained by solving the system of **least squares equations**

$$Pb + (\Sigma\ x)m = \Sigma\ y \qquad\qquad \text{(LS1)}$$

$$(\Sigma\ x)b + (\Sigma\ x^2)m = \Sigma\ xy \qquad\qquad \text{(LS2)}$$

where

$$P = \text{number of points}$$
$$\Sigma\, x = \text{sum of the } x \text{ values}$$
$$\Sigma\, y = \text{sum of the } y \text{ values}$$
$$\Sigma\, x^2 = \text{sum of the squares } x^2$$
$$\Sigma\, xy = \text{sum of the products } xy$$

EXAMPLE 1 | **Problem**

Find the line of best fit for these points.

$$(1, 1), \qquad (2, 2), \qquad (4, 3), \qquad (5, 6)$$

SOLUTION Write a vertical table of x values and y values. Place a column of the squares x^2 to the left of the x values and a column of the products xy to the right of the y values.

	x^2	x	y	xy
	1	1	1	1
	4	2	2	4
	16	4	3	12
	25	5	6	30
Totals	46	12	12	47

There are four points, so $P = 4$. The row of Totals shows that

$$\Sigma\, x^2 = 46, \qquad \Sigma\, x = 12, \qquad \Sigma\, y = 12, \qquad \Sigma\, xy = 47$$

The least squares equations are

$$4b + 12m = 12 \tag{LS1}$$

$$12b + 46m = 47 \tag{LS2}$$

Multiply LS1 by -3, then add it to LS2. The resulting equation is $10m = 11$, so $m = 1.1$. Substituting $m = 1.1$ into LS1 yields $b = -0.3$, so the line of best fit is

$$y = 1.1x - 0.3$$

It is graphed in Figure 3.

FIGURE 3 Line of best fit $\mathbf{y} = 1.1\mathbf{x} - 0.3$.

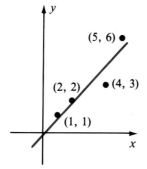

Voting Trends

Example 2 applies the method of least squares to presidential elections. We will then discuss one of the drawbacks to basing predictions of future outcomes on historical results.

EXAMPLE 2

Table 3 lists the percentage of eligible voters who have voted in the presidential elections from 1960 to 1980.

TABLE 3

Year	Percentage
1960	62.8
1964	61.9
1968	60.9
1972	55.5
1976	54.4
1980	52.3

Find the equation of the line of best fit.

SOLUTION Set x = the year of the presidential election and y = the percentage of voters who voted. Write the years without the prefix 19, so that $x = 60$ stands for 1960. Expand Table 3 to include the two additional columns.

	x^2	x	y	xy
	3600	60	62.8	3768.0
	4096	64	61.9	3961.6
	4624	68	60.9	4141.2
	5184	72	55.5	3996.0
	5776	76	54.4	4134.4
	6400	80	52.3	4184.0
Totals	29,680	420	347.8	24,185.2

The system of least squares equations is

$$6b + \quad 420m = \quad 347.8 \qquad \text{(LS1)}$$

$$420b + 29680m = 24185.2 \qquad \text{(LS2)}$$

Multiply LS1 by -70, then add it to LS2.

$$
\begin{aligned}
-420b - 29400m &= -24{,}346.0 \\
420b + 29680m &= \quad 24{,}185.2 \\
\hline
280m &= -160.8
\end{aligned}
$$

The slope is $m = -1608/2800 = -201/350$. Substituting m into LS1 yields the y-intercept b.

$$6b + 420\left(\frac{-201}{350}\right) = 347.8$$

$$6b = 347.8 + \left(\frac{84{,}420}{350}\right)$$

$$6b = \frac{206{,}150}{350} = 589$$

$$b = \frac{589}{6}$$

Thus the line of best fit is $y = -(201/350)x + (589/6)$.

The primary importance of the line of best fit is for projecting future values. To obtain the projection for 1984, substitute $x = 84$ and solve for y.

$$y = -(201/350)(84) + (589/6) \approx 49.93$$

The projection was that slightly less than 50% of the voters would vote in the 1984 election. However, 53.3% voted. Does this reveal a limitation on basing projections on historical evidence, or is the 1984 figure an exception to the rule of continuing decline? Future elections will answer the question.

EXERCISE SET 1–5

• In Problems 1 to 8 find the equation of the line of best fit for the given points.

1. $(1, 3)$, $(2, 1)$, $(3, 4)$, $(4, 3)$

2. $(1, 3)$, $(2, 7)$, $(4, 9)$, $(5, 12)$

3. $(1, 3)$, $(2, 4)$, $(3, 1)$, $(4, 2)$

4. $(1, 12)$, $(2, 9)$, $(4, 7)$, $(5, 3)$

5. $(-2, -2)$, $(-1, 2)$, $(1, 4)$, $(2, 7)$

6. $(-1, 0)$, $(0, -2)$, $(1, 1)$, $(2, 0)$

7. $(2, 15)$, $(3, 13)$, $(4, 9)$, $(5, 6)$, $(6, 1)$

8. $(-1, -5)$, $(0, 0)$, $(1, 3)$, $(2, 7)$, $(3, 9)$

•• In Problems 9 to 14 find the equation of the line of best fit for the given points.

9. $(0.3, 0.4)$, $(0.4, 0.3)$, $(0.1, 0.2)$, $(0.3, 0.1)$

10. $(0.1, 0.5)$, $(0.9, 0.4)$, $(0.7, 0.2)$, $(0.3, 0.1)$

11. $(-3, 0)$, $(-2, 2.5)$, $(-1, 4.5)$, $(0, 6)$, $(1, 7.5)$, $(2, 10)$, $(3, 13)$

12. $(-3, -4)$, $(-2, -4)$, $(-1, -3)$, $(0, -3)$, $(1, -2.5)$, $(2, -2)$, $(3, -1)$

13. $(-3, 1)$, $(-2, 0)$, $(-1, 0)$, $(0, -0.5)$, $(1, -1)$, $(2, -1.5)$, $(3, -2)$

14. $(-3, 9)$, $(-2, 4.5)$, $(-1, 0)$, $(0, -3.5)$, $(1, -8)$, $(2, -12)$, $(3, -16)$

Problems 15 to 18 refer to figures compiled by the Department of Health and Human Services. The average 5-year-old boy is 43 inches (in.) tall, the average 10-year-old boy is 54 in. tall, and the average 15-year-old boy is 67 in. tall.

15. (a) Plot these values on a coordinate system with age along the x-axis and height along the y-axis.
    (b) Find the equation for the line of best fit that links age to height.

16. Use the equation for the line of best fit in Problem 15 to approximate the height of an average 8-year-old boy?

17. Use the equation for the line of best fit in Problem 15 to approximate the height of an average 17-year-old boy?

18. Use the equation for the line of best fit in Problem 15 to approximate the height of an average 12-year-old boy?

••• 19. Samuel Messick, an official of the Educational Testing Service, admitted that "coaching" for the SATs could be effective.* He revealed that a typical student will increase the SAT-Math score by 10 points after 8 hours of coaching, 20 points after 19 hours, 30 points after 45 hours, and 40 points after 107 hours.
    (a) Plot these points with hours on the x-axis and points on the y-axis.
    (b) Do the points appear to be spaced linearly?
    (c) What are the two equations for the method of least squares?
    (d) What is the equation for the line of best fit?

*Edward B. Fiske, "Kicking and Boosting the S.A.T. Scores," *New York Times*, December 23, 1980, p. C4.

(e) Use part (d) to project how many points the score will be raised with 200 hours of coaching.

(f) Use part (d) to project how many hours of coaching will be required to raise the score 100 points.

20. The origin of three meteorites, which are called Shergottites, has confounded the world's best geologists.† Table 4 relates the concentrations of certain elements found in the Shergottites to those found on Mars.

   (a) What are the two equations for the method of least squares?

   (b) What is the equation of the line of best fit?

   (c) Does this evidence suggest that the Shergottites have origins similar to the origins of Mars?

TABLE 4

Mineral		Shergotty Meteorite (%)	Martian Soil (adjusted %)
Silicon dioxide	SiO_2	50.4	53.9
Ferrous oxide	FeO	19.3	19.7
Calcium oxide	CaO	9.6	6.7
Magnesium oxide	MgO	9.3	10.0
Aluminum oxide	Al_2O_3	7.0	6.8
Titanium dioxide	TiO_2	.9	1.0
Potassium oxide	K_2O	.2	.1

21. Table 5 has been brought to you in part by Delta Air Lines. Draw a coordinate system with the altitude (ft) as the x-axis and the temperature (degrees F) as the y-axis. Plot the points to see if they are linear. Then compute the line of best fit.

22. Use Problem 21 to project
   (a) the temperature at 40,000 ft.
   (b) the temperature at 4,000 ft.
   (c) the altitude at 32 degrees.
   (d) the altitude at −30 degrees.

TABLE 5   How Cold Is Up?

There you are, bareheaded and with no gloves and in the lightest of summer clothes, eating a cold salad at 69° below (−56°C) zero!

It's routine on Delta jets, summer or any other season, for you to travel in the upper levels of air far from surface heat. It's a case of the higher, the lower. Perhaps you'd like to know exactly how the temperature changes with altitude, how it can be 104°F (40°C) in the sun of the parking lot and well below zero just four miles from the lot, straight up!

The seasons influence lower level temperatures, of course, but on a "standard" day (59°F) (15°C) at sea level, the mercury would drop this way as you went up:

Feet	Meter	F	C
1,000	300	56°	13°
5,000	1,500	41°	5°
10,000	3,000	23°	− 5°
15,000	4,500	5°	− 15°
20,000	6,000	− 15°	− 26°
30,000	9,000	− 47°	− 44°
36,087	10,826	− 69°	− 56°
	(Stratosphere)		

Source: Courtesy of Delta Air Lines.

23. In financial management, an area called "trend analysis" deals with the construction of forecasting models to predict a firm's cash flow (expenditures and receipts). Figure 4 is taken from a study of a particular firm's cash receipts, where it is called Exhibit 3.‡ The answers to parts (a) and (b) can be obtained directly from the figure. (a) What kind of mathematical method is being used for the forecast model? (b) What is the equation of the line of best fit? (c) How many receipts does this model forecast after $t = 20$ time periods? (d) How many time periods does this model forecast for 33 receipts?

24. Figure 5 (see p. 46) depicts the relationship between fat intake and death from breast cancer for the inhabitants of many countries around the world.§ Does the relationship appear to be linear?

†Harry Y. McSween, Jr. and Edward M. Stolper, "Basaltic meteorites," *Scientific American,* June 1980, pp. 54–63.

‡James L. Pappas and George P. Huber, "Probabilistic Short-term Financial Planning," *Financial Management,* Autumn 1973, pp. 36–44.

§Leonard A. Cohen, "Diet and cancer," *Scientific American,* November 1987, pp. 42–48.

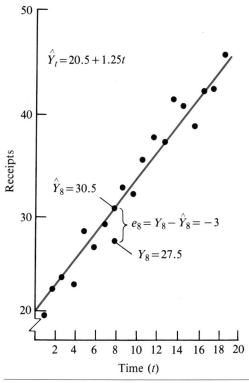

Problems 25 and 26 refer to Table 6, which lists the percentage of doctorates awarded in mathematics to U.S. citizens over a 15-year period.‖

25. (a) Find the equation of the line of best fit for the percentage of women who received doctorates.
    (b) Use the line to project the percentage in the year 2000–2001.

26. (a) Does the total number of doctorates awarded in mathematics to U.S. citizens seem to be linear over this period?
    (b) Find the equation of the line of best fit for the total number of U.S. citizens who received doctorates.
    (c) Use the line to project the total number in the year 2000–2001.

FIGURE 4 Graph of the least squares forecasting model.

TABLE 6 U.S. Citizen Doctorates, Male and Female

	Doctorates who are U.S. Citizens	Male	Female	% Female
1972–1973	774	696	78	10
1973–1974	677	618	59	9
1974–1975	741	658	83	11
1975–1976	722	636	86	12
1976–1977	689	602	87	13
1977–1978	634	545	89	14
1978–1979	596	503	93	16
1979–1980	578	491	87	15
1980–1981	567	465	102	18
1981–1982	519	431	88	17
1982–1983	455	366	89	20
1983–1984	433	346	87	20
1984–1985	396	315	81	20
1985–1986	386	304	82	21
1986–1987	362	289	73	20

‖Edward A. Connors, "Report on the 1987 Survey of New Doctorates," *Notices of the American Mathematical Society*, November 1987, pp. 1081–1086.

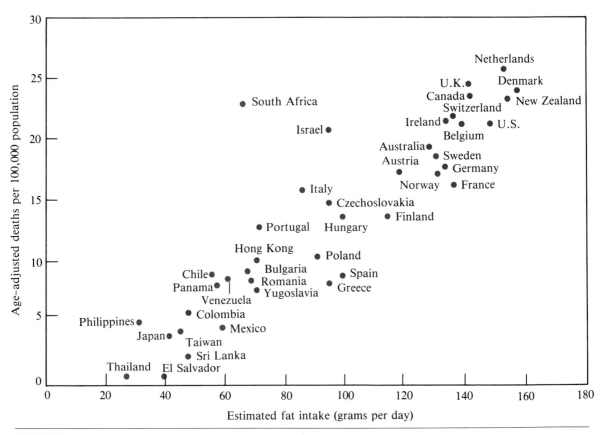

FIGURE 5

Of all track-and-field events, the mile run is perhaps the most international. It is the only nonmetric event that is still contested. The first world record for the mile run was recorded in 1864 with a time of 4:56. Within a year that standard was lowered to 4:36.5.

Walter George of Great Britain broke the record four times between 1880 and 1884. His 1884 record of 4:18.4 lasted ten years. However, George's greatest gift to running is his race strategy, which consists of a fast first quarter-mile (when the runner is fresh), a slower second quarter, a still slower third quarter, and then as fast a final quarter as possible.

Most milers adopted George's strategy, and by 1945 the world record was lowered to 4:01.4. Then the "barrier" set in. It was felt that the human body was incapable of running a sub-4-minute mile. In spite of the increased popularity after World War II of track-and-field sports throughout the world, no miler was able to better Gunder Haegg's time of 4:01.4.

Then on May 6, 1954 an English medical student named Roger Bannister set out to smash the barrier in a dual meet at Oxford University. Following George's strategy, he ran the first three quarters in 57.5, 60.7, and 62.3 seconds, respectively. His time going into the final quarter was 3:00.5. Bannister was known as a "kick runner," one who preferred to use a great finishing kick to win a race in the stretch. But could he use his great kick to defeat the clock instead of another runner? Yes. Bannister ran the last quarter in 58.9 seconds. His world record of 3:59.4 proved that the 4-minute-mile barrier was psychological, not physiological. Further proof was obtained less than two months later when John Landy ran the mile in 3:58.

The world record has been lowered with surprising regularity since then, and as of this writing (June 1988), it is 3:46.31. Take a look at Table 1, which lists the world record for the mile run since Roger Bannister's historic race. Of course, the times get lower and lower. Is it possible to say more? Is there a pattern to the decrease in times that is obscured by the apparent chaos of facts?

One way to answer these questions is to take a geometric approach. Figure 1 shows the graph of the world record for the mile run when the x-axis is the year and the y-axis is the time of the record in minutes and seconds. For instance, when $x = 1975$, the value $y = 3:50$ reflects Filbert Bayi's world record of 3 minutes 50 seconds.

Figure 1 reveals a remarkable pattern: the points seem to lie along a line. This means that the world record for the mile run is approximately linear, and if this is true, then the future of the event can be predicted with some degree of confidence. It means that we can project what the mile record will be in 1989. And in 1990. And even in the year 2000. In fact, we can project when the mile will be run in 3 minutes flat. Beyond that, we can project when it will be run in 0 seconds flat.

TABLE 1   World Record for the Mile Run

Name (country)	Year	Time
Roger Bannister (Great Britain)	1954	3:59.4
John Landy (Australia)	1954	3:58
Derek Ibbotson (Great Britain)	1957	3:57.2
Herb Elliott (Australia)	1958	3:54.5
Peter Snell (New Zealand)	1962	3:54.4
Peter Snell (New Zealand)	1964	3:54.1
Michel Jazy (France)	1965	3:53.6
Jim Ryun (United States)	1966	3:51.3
Jim Ryun (United States)	1967	3:51.1
Filbert Bayi (Tanzania)	1975	3:50
John Walker (New Zealand)	1975	3:49.4
Sebastian Coe (Great Britain)	1979	3:49.1
Steve Ovett (Great Britain)	1980	3:48.8
Sebastian Coe (Great Britain)	1981	3:48.53
Steve Ovett (Great Britain)	1981	3:48.4
Sebastian Coe (Great Britain)	1981	3:47.33
Steve Cram (Great Britain)	1985	3:46.31

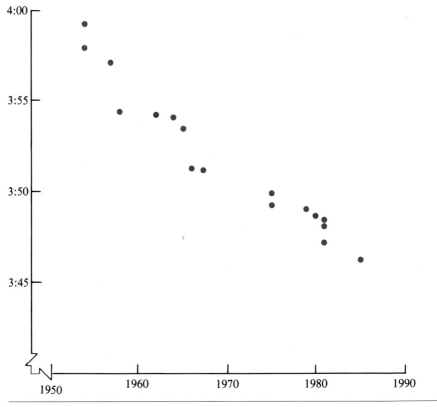

FIGURE 1 World records in the mile run.

Zero seconds flat! No way.

Let us investigate these intriguing possibilities.

EXAMPLE 1 | **Problem**

Determine the line of best fit for the mile run based on the world records for the years 1954–1985.

SOLUTION Revise Table 1 by writing the years without the prefix 19, so that 54 stands for 1954, and by converting the time to seconds, so that 3:59.4 is written as 239.4 seconds. See the top portion of Table 2.

The record was set 17 times during this period, so $P = 17$. The sums of the columns are

$$\Sigma\, x = 1184 \qquad \Sigma\, y = 3941.47$$
$$\Sigma\, x^2 = 84{,}254 \qquad \Sigma\, xy = 273{,}869$$

The system of equations is

$$17b + 1{,}184m = 3{,}941.47$$
$$1184b + 84{,}254m = 273{,}869.00$$

TABLE 2

x^2	x	y	xy
2916	54	239.40	12927.6
2916	54	238.00	12852.0
3249	57	237.20	13520.4
3364	58	234.50	13601.0
3844	62	234.40	14532.8
4096	64	234.10	14982.4
4225	65	233.60	15184.0
4356	66	231.30	15265.8
4489	67	231.10	15483.7
5625	75	230.00	17250.0
5625	75	229.40	17205.0
6241	79	229.10	18098.9
6400	80	228.80	18304.0
6561	81	228.53	18510.9
6561	81	228.40	18500.4
6561	81	227.33	18413.7
7225	85	226.31	19236.4
84254	1184	3941.47	273869.0

Year	Record	Year	Record
1980	3:48.14	1991	3:44.19
1981	3:47.78	1992	3:43.83
1982	3:47.42	1993	3:43.48
1983	3:47.06	1994	3:43.12
1984	3:46.70	1995	3:42.76
1985	3:46.34	1996	3:42.40
1986	3:45.99	1997	3:42.04
1987	3:45.63	1998	3:41.68
1988	3:45.27	1999	3:41.32
1989	3:44.91	2000	3:40.96
1990	3:44.55	2001	3:40.61

The solution to this system can be obtained by elimination.

$$b = 256.835 \qquad m = -0.358709$$

The line of best fit is

$$y = -0.358709x + 256.835$$

Consider the meaning of the slope $m = -0.358709$ in Example 1. The negative sign means that the record is being lowered, that the time of the event is decreasing. The number 0.358709 is about $\frac{1}{3}$, so the record will drop about $\frac{1}{3}$ second per year, or about 1 second every three years. Since the record in 1985 was 3:46.31, it should be about 3:45.31 in 1988 and about 3:44.31 in 1991. It will not be long before we see how accurate these projections are.

Example 2 improves these projections.

EXAMPLE 2 | **Problem**

Use the line of best fit to project what the world record will be in the years 1989 and 2000.

SOLUTION To find the exact times that the line of best fit projects for the future, substitute the desired year into the equation $y = -0.358709x + 256.835$. Remember that the values for x omit the prefix 19. For the year 1989 set $x = 89$. Then $y = -0.358709(89) + 256.835 = 224.91$ so the world record in 1989 will be 3:44.91.

To find the projection for the world record in the year 2000, set $x = 100$. Then $y = 220.96$ so the projected time is 3:40.96.

Table 2 gives the projected times for the years 1980–2001 based on the line of best fit in Example 1. Note that the projected figure for 1985 was just 0.03 seconds from the world mark.

Sometimes it is necessary to determine when a performance will reach a certain level; for instance, when the mile will first be run in 3 minutes flat. To do this, substitute the appropriate time for the value of y and then solve for x. Since the time is given in terms of seconds, $y = 180$. Then

$$y = -0.358709x + 256.835$$
$$180 = -0.358709x + 256.835$$
$$x = 214.2$$

Since x denotes the number of years after 1900, the year is 2114. Thus the mile will first be run in 3 minutes flat in the year 2114. Probably none of us will witness this event.

Let us extend this result to what now seems an inhuman feat. For $y = 120$ the line of best fit yields $x = 381$, so the mile will be run in 2 minutes flat in 2281. It will be run in 1 minute flat in 2448. Carrying these projections to an extreme reveals that the mile will be run in 0 seconds flat sometime during the year 2616. This is patently absurd. Let us discuss the ramifications.

The least squares method can be carried out on any set of data, but it is most effective when the points appear to lie along a straight line. If the points are not linear, then the method of least squares will not provide reasonable projections, so curves other than lines have to be used. And that is what will eventually occur with the mile run. Since the world record for that event was set in 1865, progress has been remarkably linear. We have isolated the years since 1954 to illustrate this fact. At some time in the future the world record will not continue to follow a linear path, and at that time we will have to apply another curve to model the situation.

For now, though, it provides us with a fascinating glimpse into the near future. And it will allow you to see just how accurate our projections will be.

Case Study Exercises

1. Use Table 1 to make a table of the world records for the mile in the period 1967–1980. What are the two equations for the method of least squares?

2. Use Table 1 to make a table of the world records for the mile in the period 1957–1965. What are the two equations for the method of least squares?

3. Use Problem 1 to determine the line of best fit for the mile run based on the years 1967–1980.

4. Use Problem 2 to determine the line of best fit for the mile run based on the years 1957–1965.

5. Use Problem 3 to project the world record for the mile run in the year 2000 based on the period 1967–1980.

6. Use Problem 4 to project the world record for the mile run in the year 2000 based on the period 1957–1965.

7. Use Problem 3 to project the first year in which the mile will be run in 3 minutes flat, based on the years 1967–1980.

8. Use Problem 4 to project the first year in which the mile will be run in 3 minutes flat, based on the years 1957–1965.

9. The line of best fit for the mile run based on the period 1975–1985 is roughly

$$y = -0.32x + 254$$

Use this line to project
(a) what the world record will be in the year 2000.
(b) what year the world record will be 3:40.

10. Repeat Problem 9 for the period 1954–1964. The line of best fit is roughly

$$y = -0.5x + 264$$

11. Problem 9 describes the line of best fit for the years 1975–1985.
(a) What is the slope of the line?
(b) How many seconds will the record improve each year?
(c) How many seconds will the record improve in a century?

12. Repeat Problem 11, using the line of best fit described in Problem 10.

CHAPTER REVIEW

Key Terms

1–1 The Cartesian Coordinate System
Real Line
Origin
Unit Distance
x-Axis; y-Axis
x-Coordinate; y-Coordinate
Coordinates

Cartesian Coordinate System
Quadrants
Graph
The Plotting Method
Linear Equation

1–2 Lines

General Form	Parallel Lines
x-Intercept	Horizontal Lines
y-Intercept	Vertical Lines
Slope	Point-Slope Form
Slope-Intercept Form	

1–3 Intersection of Lines

System of Equations	Dependent System
Solution of a System	Breakeven Point
Inconsistent System	

1–4 Systems of Linear Equations

Linear Equation in	Reduced System
Three Variables	Elimination
Multiplier	Back-Substitution
Solution Set	Gaussian Elimination
Equivalent Systems	Homogeneous System

1–5 Least Squares

Line of Best Fit	Least Squares Equations

REVIEW PROBLEMS

1. For the equation $y = 3 - 2x$ find the values of y for $x = -2, -1, 0, 1,$ and 2. Plot them and sketch the graph of the equation.

2. Find the slope of the line passing through the points $(-2, 3)$ and $(1, -3)$.

In Problems 3 to 5 sketch the graph of the stated line.

3. $y = -x - 5$

4. $y - 2 = 3(x + 1)$

5. $2x - 4y = 8$

6. Find the slope-intercept form of the line that is parallel to the line $x + y = 0$ and whose x-intercept is equal to -2.

7. Write an equation of the line whose y-intercept is -3 and whose slope is 3.

8. Write the equation of the horizontal line that passes through the point $(-2, -5)$.

9. Write the equation for one day's car rental cost (C) in terms of miles driven (x) if the charges are \$25 per day and 22 cents for each mile the car is driven.

In Problems 10 and 11 determine whether the system has a solution. If it does, write the solution. If it does not, tell whether it is inconsistent or dependent.

10. $\begin{aligned} 15x - 5y &= 20 \\ -12x + 4y &= -16 \end{aligned}$

11. $\begin{aligned} x + 2y &= 1 \\ x - y &= -3 \end{aligned}$

12. Determine the point of intersection of the lines $y = 2x - 1$ and $y = -x + 5$.

In Problems 13 and 14 determine how many solutions the system has. If it has one, state it; if there are more than one, state two of them.

13. $\begin{aligned} x - y &= 2 \\ 2x - 2y &= 4 \end{aligned}$

14. $\begin{aligned} 2x + 4y &= -6 \\ -3x - 5y &= 12 \end{aligned}$

15. Use Gaussian elimination to find the unique solution to this system of equations.

$$\begin{aligned} x - 2y - 3z &= 5 \\ 2x + y - z &= -5 \\ -x - 3y + 2z &= 2 \end{aligned}$$

16. Computer chips are manufactured for micro-, mini-, and mainframe computers. Suppose that each chip for a microcomputer requires 1 A-transistor, 1 B-transistor, and 2 C-transistors; each chip for a minicomputer requires 1 A-transistor, 3 B-transistors, and 7 C-transistors; and each chip for a mainframe computer requires 2 A-transistors, 7 B-transistors, and 13 C-transistors. How many computer chips of each kind can be manufactured if there are 29 A-transistors, 59 B-transistors, and 126 C-transistors available?

17. Find the equation of the line of best fit for the points $(-1, 5)$, $(0, 3)$, $(1, 6)$, and $(2, 5)$.

18. (a) Plot the data from the table on a coordinate system in which the altitude (in thousands of meters) is the x-axis and the temperature (in degrees Celsius) is the y-axis.

altitude	1	2	3	4	6
temperature	50	35	23	10	-15

   (b) Write the least squares equations.
   (c) Find the equation of the line of best fit and graph it on the same coordinate system.
   (d) What will the temperature be at 5000 meters?
   (e) At what altitude will the temperature be 0 degrees?

2 Matrices

CHAPTER OVERVIEW This chapter defines the fundamental concept of a matrix and applies it to various settings. Sections 2–1 and 2–2 introduce the three basic operations on matrices: addition, scalar multiplication, and matrix multiplication.

The next three sections return to the theme of solving systems of linear equations. Section 2–3 explains the Gauss–Jordan method, which is used in Section 2–4 to compute the inverse of a matrix. Section 2–5 solves systems of equations by the matrix inverse method, contrasting it with the methods presented earlier.

Section 2–6 applies matrices to a branch of economics called input-output analysis. It sets the background for the Case Study.

CASE STUDY PREVIEW

Wassily W. Leontief entered college at age 15 and graduated with a doctorate at age 22. While a young faculty member at Harvard University he single-handedly introduced matrix methods into economics. He was later awarded the Nobel prize in economics for his pioneering work.

Leontief's theory correctly predicted that World War II would not be followed by massive unemployment and a depression. His theory also helped avert a crisis in that war by forewarning the government that a massive infusion of silver would be

55

required to build 50,000 badly needed airplanes. The Case Study examines the background of this eminent octogenarian and describes one of his articles on an input-output model of the U.S. economy.

2−1 ADDITION AND SCALAR MULTIPLICATION

In this section we introduce matrices, define the two basic operations stated in the section title, and strengthen the link between matrices and systems of equations.

Notation

The Pressman-Gutman Company is a fabric-finishing firm that deals primarily with two fabrics, cotton and synthetics. The company buys fabric from a mill, colors it, and gives it a smooth finish. Then the company sells the fabric to a manufacturing company whose end product is a finished garment, like a pair of jeans.

The Pressman-Gutman Company uses finishing plants throughout the country, with three main plants in Fall River (MA), Passaic (NJ), and Los Angeles (CA). The company stores its records in a computer as a matrix or an array. A **matrix** is a rectangular array of numbers with a specific number of rows and columns. The numbers in a matrix are called **entries.** Consider the following matrix:

$$\begin{array}{cc} & \text{cotton} \quad \text{synthetics} \\ \begin{array}{l} \text{Fall River} \\ \text{Passaic} \\ \text{Los Angeles} \end{array} & \begin{bmatrix} 2.0 & 4.5 \\ 2.5 & 4.0 \\ 3.5 & 1.5 \end{bmatrix} \end{array}$$

Each entry represents the number of linear yards of material that was shipped to that plant in the month of January 1988. The units are in millions of yards. Look at the first column of numbers. They show that 2 million yards of cotton were shipped to the Fall River plant, 2.5 million yards of cotton to the Passaic plant, and 3.5 million yards of cotton to the Los Angeles (LA) plant. The second column shows that 4.5 million yards of synthetics were shipped to Fall River, 4 million yards to Passaic, and 1.5 million to LA. By eliminating the labels from the rows and columns, we arrive at the matrix P.

$$P = \begin{bmatrix} 2.0 & 4.5 \\ 2.5 & 4.0 \\ 3.5 & 1.5 \end{bmatrix}$$

We generally use uppercase letters to denote matrices. Consider this matrix.

$$M = \begin{bmatrix} 1 & -1 & 2 & -3 \\ 2 & 3 & -1 & 9 \\ -2 & 0 & 3 & -5 \end{bmatrix}$$

This notation means that M is the name of the matrix whose entries are the numbers inside the brackets. We also say that M is a 3×4 matrix (read ''3 by 4''), because M consists of 3 rows and 4 columns. In general an $m \times n$ matrix has m rows and n columns. This is referred to as the **size** of the matrix. If $m = n$ the matrix is a **square matrix.**

   The entries of a matrix are denoted by the corresponding lowercase letter with two subscripts. The first subscript refers to the row that the entry lies in and the second subscript refers to the column. For the matrix M, m_{23} refers to the entry in row 2 and column 3, so $m_{23} = -1$. Sometimes m_{23} is called the $(2, 3)$ entry.

$$\text{row 2} \longrightarrow \begin{bmatrix} 1 & -1 & 2 & -3 \\ 2 & 3 & -1 & 9 \\ -2 & 0 & 3 & -5 \end{bmatrix} \quad \text{entry } m_{23} = -1$$

column 3

The general entry in a matrix M is written m_{ij}. This denotes the number that is located in row i and column j.

   A matrix with one row is called a **row matrix** and a matrix with one column is called a **column matrix.** The matrix R is a 1×4 row matrix, while C is a 3×1 column matrix.

$$R = \begin{bmatrix} -3 & 4 & 6 & 8 \end{bmatrix} \qquad C = \begin{bmatrix} 2 \\ 9 \\ -1 \end{bmatrix}$$

Addition

Two matrices are **equal** if they have the same size and their corresponding entries are equal. For instance, let

$$R = \begin{bmatrix} 1 & 2 \\ 0 & -1 \end{bmatrix} \qquad S = \begin{bmatrix} 1 & 2 \\ (-3)1 + 3 & (-3)2 + 5 \end{bmatrix}$$

$$T = \begin{bmatrix} 1 & 2 & 0 \\ 0 & -1 & 0 \end{bmatrix}$$

Then $R = S$ because all four entries are equal. However, $R \neq T$ since they have different sizes, even though the four entries of R are equal to the four left entries in T.

   Consider the two matrices E and F.

$$E = \begin{bmatrix} -5 & 2 & 3 \\ 6 & z & 8 \end{bmatrix} \qquad F = \begin{bmatrix} x & 2 & 3 \\ 6 & -1 & y \end{bmatrix}$$

If $E = F$, then $x = -5$, $y = 8$, and $z = -1$.

   The **sum** of two matrices M and N of the same size is the matrix $M + N$ of the same size and whose entries are the sums of the corresponding entries in M and N. The operation is called **addition.** Matrices of different sizes cannot be added.

For instance, consider the matrices A, B, and C:

$$A = \begin{bmatrix} 2 & -1 \\ 3 & -6 \end{bmatrix} \quad B = \begin{bmatrix} 4 & -1 \\ -3 & 8 \end{bmatrix} \quad C = \begin{bmatrix} 4 & -1 & 0 \\ -3 & 8 & 0 \end{bmatrix}$$

Then

$$A + B = \begin{bmatrix} 2+4 & -1+(-1) \\ 3+(-3) & -6+8 \end{bmatrix} = \begin{bmatrix} 6 & -2 \\ 0 & 2 \end{bmatrix}$$

However, the sum $A + C$ cannot be formed because A and C have different sizes.

The Pressman-Gutman example shows how addition arises in applications. Recall that P represents the amounts of fabrics that were shipped to the three finishing plants in January 1988. Let Q represent the corresponding shipments in February 1988.

$$P = \begin{bmatrix} 2.0 & 4.5 \\ 2.5 & 4.0 \\ 3.5 & 1.5 \end{bmatrix} \quad Q = \begin{bmatrix} 3.0 & 5.0 \\ 2.5 & 4.5 \\ 4.5 & 1.0 \end{bmatrix}$$

The total amounts of material shipped in January and February is the sum $P + Q$.

$$P + Q = \begin{bmatrix} 2.0+3.0 & 4.5+5.0 \\ 2.5+2.5 & 4.0+4.5 \\ 3.5+4.5 & 1.5+1.0 \end{bmatrix} = \begin{bmatrix} 5.0 & 9.5 \\ 5.0 & 8.5 \\ 8.0 & 2.5 \end{bmatrix}$$

The $(1, 1)$ entries show that since 2 million yards of cotton were shipped to Fall River in January and 3 million yards were shipped in February, 5 million yards were shipped in both months.

Example 1 combines equality and addition.

EXAMPLE 1

Problem

Find values for u, v, w, x, y, and z such that

$$\begin{bmatrix} x & 1 \\ -2 & 3 \\ 4 & y \end{bmatrix} + \begin{bmatrix} 4 & 1 \\ 2 & -3 \\ -3 & 6 \end{bmatrix} = \begin{bmatrix} 8 & u \\ v & 0 \\ 1 & 1 \end{bmatrix}$$

SOLUTION First add the two matrices on the left-hand side of the equation. Then set the sum equal to the right-hand side:

$$\begin{bmatrix} x+4 & 2 \\ 0 & 0 \\ 1 & y+6 \end{bmatrix} = \begin{bmatrix} 8 & u \\ v & 0 \\ 1 & 1 \end{bmatrix}$$

By the definition of equality

$$x + 4 = 8 \qquad 2 = u$$
$$0 = v \qquad y + 6 = 1$$

Then $u = 2$ and $v = 0$. The remaining values are obtained by solving the respective equations algebraically.

$$x + 4 = 8 \qquad \text{implies} \qquad x = 4$$
$$y + 6 = 1 \qquad \text{implies} \qquad y = -5$$

Scalar Multiplication

The matrix P represents the linear yards of cotton and synthetics shipped to three finishing plants by the Pressman-Gutman Company in January 1988. To determine the number of linear feet of material shipped, multiply each number in P by 3. For example, the 2 million yards of cotton that were shipped to the Fall River plant convert to 6 million feet. The matrix obtained by multiplying each number of P by 3 is called the **scalar product** and is denoted by $3P$.

$$3P = \begin{bmatrix} 3(2.0) & 3(4.5) \\ 3(2.5) & 3(4.0) \\ 3(3.5) & 3(1.5) \end{bmatrix} = \begin{bmatrix} 6.0 & 13.5 \\ 7.5 & 12.0 \\ 10.5 & 4.5 \end{bmatrix}$$

Since numbers are sometimes called scalars, the term "scalar multiplication" refers to multiplication of a matrix by a number.

The next example shows how addition and scalar multiplication together lead to an equation whose constant terms are matrices.

EXAMPLE 2

Problem

Find values of x and y such that

$$x\begin{bmatrix} 2 \\ 5 \end{bmatrix} + y\begin{bmatrix} 1 \\ 3 \end{bmatrix} = \begin{bmatrix} 1 \\ -3 \end{bmatrix}$$

SOLUTION Here x and y are unknown scalars. The left-hand side is

$$x\begin{bmatrix} 2 \\ 5 \end{bmatrix} + y\begin{bmatrix} 1 \\ 3 \end{bmatrix} = \begin{bmatrix} 2x \\ 5x \end{bmatrix} + \begin{bmatrix} y \\ 3y \end{bmatrix} = \begin{bmatrix} 2x + y \\ 5x + 3y \end{bmatrix}$$

Set the left-hand side equal to the right-hand side.

$$\begin{bmatrix} 2x + y \\ 5x + 3y \end{bmatrix} = \begin{bmatrix} 1 \\ -3 \end{bmatrix}$$

The entries are equal, so

$$2x + y = 1$$
$$5x + 3y = -3$$

The solution to this system is $x = 6$ and $y = -11$.

It is advisable to check the answer.

$$6\begin{bmatrix} 2 \\ 5 \end{bmatrix} + (-11)\begin{bmatrix} 1 \\ 3 \end{bmatrix} = \begin{bmatrix} 12 \\ 30 \end{bmatrix} + \begin{bmatrix} -11 \\ -33 \end{bmatrix} = \begin{bmatrix} 1 \\ -3 \end{bmatrix}$$

Example 2 solved the matrix equation $Ax + By = C$, where

$$A = \begin{bmatrix} 2 \\ 5 \end{bmatrix} \quad B = \begin{bmatrix} 1 \\ 3 \end{bmatrix} \quad \text{and} \quad C = \begin{bmatrix} 1 \\ -3 \end{bmatrix}$$

This is the matrix analog of the general form of a linear equation introduced in Section 1–2.

Properties

We examine some properties of addition and scalar multiplication on the set of all 2×3 matrices. These properties hold for all $m \times n$ matrices.

The set is **closed** under addition because if A and B are 2×3 matrices, then so is $A + B$. This follows from the definition of addition.

Matrix addition is **associative** because

$$(A + B) + C = A + (B + C)$$

Here the parentheses indicate the particular order in which addition is carried out. The left-hand side means that the matrix $A + B$ is formed first and then added to C. On the right-hand side the matrix $B + C$ is computed first and then added to A. Example 3 verifies the associative law but it does not constitute a general proof.

EXAMPLE 3

Problem

Verify that $(A + B) + C = A + (B + C)$ for

$$A = \begin{bmatrix} -1 & 2 & 4 \\ 3 & 8 & 9 \end{bmatrix} \quad B = \begin{bmatrix} -6 & 4 & 2 \\ 3 & 0 & -2 \end{bmatrix} \quad \text{and} \quad C = \begin{bmatrix} 0 & 0 & 1 \\ 1 & 0 & 0 \end{bmatrix}$$

SOLUTION Compute both sides of the equality separately and then show that they are equal. The left-hand side is $(A + B) + C$. First form $A + B$.

$$A + B = \begin{bmatrix} -7 & 6 & 6 \\ 6 & 8 & 7 \end{bmatrix}$$

Then add $A + B$ to C.

$$(A + B) + C = \begin{bmatrix} -7 & 6 & 6 \\ 6 & 8 & 7 \end{bmatrix} + \begin{bmatrix} 0 & 0 & 1 \\ 1 & 0 & 0 \end{bmatrix} = \begin{bmatrix} -7 & 6 & 7 \\ 7 & 8 & 7 \end{bmatrix}$$

Next evaluate the right-hand side $A + (B + C)$. First form $B + C$.

$$B + C = \begin{bmatrix} -6 & 4 & 3 \\ 4 & 0 & -2 \end{bmatrix}$$

Then add A to $B + C$.

$$A + (B + C) = \begin{bmatrix} -1 & 2 & 4 \\ 3 & 8 & 9 \end{bmatrix} + \begin{bmatrix} -6 & 4 & 3 \\ 4 & 0 & -2 \end{bmatrix} = \begin{bmatrix} -7 & 6 & 7 \\ 7 & 8 & 7 \end{bmatrix}$$

All six entries in $A + (B + C)$ are equal to the corresponding six entries in $(A + B) + C$. Thus

$$(A + B) + C = A + (B + C)$$

The associative law dispenses with the need for parentheses. Thus the expression $A + B + C$ is meaningful because it can be evaluated as either $(A + B) + C$ or $A + (B + C)$. Addition is also **commutative:** $A + B = B + A$. Both the associative and commutative properties hold because they hold for real numbers.

There is a special matrix called the **zero matrix.**

$$Z = \begin{bmatrix} 0 & 0 & 0 \\ 0 & 0 & 0 \end{bmatrix}$$

It behaves like the number zero, in that

$$M + Z = M \qquad \text{and} \qquad Z + M = M$$

for every 2×3 matrix M.

Given a matrix A, the matrix $B = -A$ satisfies

$$A + B = Z \qquad \text{and} \qquad B + A = Z$$

B is called the **additive inverse** of A. Subtraction can be defined by $X - Y = X + (-Y)$.

There are two properties linking scalar multiplication to addition

$$s(A + B) = (sA) + (sB)$$
$$(r + s)A = rA + sA$$

where r and s are scalars and A and B are matrices.

EXERCISE SET 2–1

• Problems 1 to 6 refer to these matrices.

$$A = \begin{bmatrix} 1 & -6 & 7 \\ -2 & 4 & 3 \end{bmatrix} \qquad B = \begin{bmatrix} 2 & -6 \\ 5 & 0 \\ -1 & 6 \end{bmatrix}$$

$$C = \begin{bmatrix} 1 \\ 9 \\ 7 \end{bmatrix} \qquad D = \begin{bmatrix} 1 & 2 & 3 \\ 4 & 5 & 6 \\ 7 & 8 & 9 \end{bmatrix}$$

$$E = \begin{bmatrix} 1 & 2 \\ 3 & 4 \end{bmatrix} \qquad F = [1 \quad 0 \quad 1 \quad 2]$$

1. What are the sizes of A, C, and E?

2. What are the sizes of B, D, and F?

3. Find (a) a_{22} (b) b_{31} (c) d_{13} (d) e_{11}.

4. Find (a) d_{33} (b) e_{12} (c) b_{22} (d) a_{21}.

5. What notation describes these entries in the matrix A?
   (a) -6 (b) 3   (c) 4

6. What notation describes these entries in the matrix B?
   (a) 5   (b) 6   (c) 0

In Problems 7 and 8 find values for x, y, and z such that $A = B$.

7.
$$A = \begin{bmatrix} x & 2 \\ 3 & y \\ 4 & 7 \end{bmatrix} \qquad B = \begin{bmatrix} 5 & 2 \\ 3 & 6 \\ 4 & z \end{bmatrix}$$

8.
$$A = \begin{bmatrix} 0 & 9 \\ x & 3 \\ 5 & y \end{bmatrix} \qquad B = \begin{bmatrix} 0 & z \\ 7 & 3 \\ 5 & 8 \end{bmatrix}$$

Problems 9 to 16 refer to the matrices R, S, and T.

$$R = \begin{bmatrix} 1 & -1 \\ 2 & -2 \\ 3 & -3 \end{bmatrix} \qquad S = \begin{bmatrix} -3 & 6 \\ 10 & 8 \\ 4 & 0 \end{bmatrix}$$

$$T = \begin{bmatrix} 11 & -4 \\ -3 & 10 \\ -5 & -1 \end{bmatrix}$$

9. Compute (a) $R + S$ (b) $(R + S) + T$ (c) $R + S + T$.

10. Compute (a) $S + R$ (b) $T + (S + R)$ (c) $T + S + R$.

11. Compute (a) $3R$ (b) $(-2)S$ (c) $-T$.

12. Compute (a) $2R$ (b) $(-3)T$ (c) $-S$.

13. Compute (a) $2R + 3S$ (b) $2R - 3S$.

14. Compute (a) $3S + T$ (b) $3S - T$.

15. Verify the equality $R + S = S + R$.

16. Verify the equality $3(R + S) = 3R + 3S$.

In Problems 17 to 20 the matrices A, B, D, and E are defined in Problems 1 to 6 and Z is a zero matrix.

17. What is the size of Z if (a) $A + Z = A$?
    (b) $D + Z = D$?

18. What is the size of Z if (a) $B + Z = B$?
    (b) $E + Z = E$?

19. Find a matrix X such that $A + X = Z$.

20. Find a matrix X such that $B + X = Z$.

21. Suppose that in March the Pressman-Gutman Company shipped 3 million yards of cotton and 1.5 million yards of synthetics to the Fall River plant, 0.5 million yards of cotton and 3 million yards of synthetics to the Passaic plant, and 5 million yards of cotton and 2 million yards of synthetics to the LA plant. Write the shipment matrix M for this month.

22. Suppose that in April the Pressman-Gutman Company shipped 5 million yards of cotton and 2 million yards of synthetics to the Fall River plant, 3.5 million yards of cotton and 2 million yards of synthetics to the Passaic plant, and 1.5 million yards of cotton and 2.5 million yards of synthetics to the LA plant. Write the shipment matrix A for this month.

23. Write the shipment matrix for the period of March through April. (Refer to Problems 21 and 22.)

24. If the shipments in May double those in April, what is the shipment matrix for May?

•• In Problems 25 and 26 find values for x and y such that $E = F$.

25. $E = \begin{bmatrix} -2 & 2x \\ 35 & y-2 \end{bmatrix}$ $F = \begin{bmatrix} -2 & 12 \\ 35 & 2 \end{bmatrix}$

26. $E = \begin{bmatrix} 21 & 5 \\ -3 & 9 \end{bmatrix}$ $F = \begin{bmatrix} 21 & x-2 \\ -3 & -3y \end{bmatrix}$

Problems 27 to 36 refer to the matrices R, S, and T defined in Problems 9 to 16. The matrix U is

$$U = \begin{bmatrix} u+2 & x \\ v & y+1 \\ w-3 & z+5 \end{bmatrix}$$

Find values of x, y, and z so the equality holds.

27. $R = U$ 28. $T = U$

Find values of x, y, and z so the equality holds.

29. $R - S = T + U$ 30. $R + S = T - U$

31. $(-2)R = U$ 32. $(-3)S = U$

Compute the stated operations.

33. (a) $R - (S + T)$ (b) $(R - S) - T$

34. (a) $R - (S - T)$ (b) $(R - S) + T$

35. What do you conclude from Problem 33?

36. What do you conclude from Problem 34?

In Problems 37 to 40 determine whether it is possible to find values of x and y that satisfy the equality. If so, find x and y.

37. $x \begin{bmatrix} 1 \\ 1 \end{bmatrix} + y \begin{bmatrix} 2 \\ -1 \end{bmatrix} = \begin{bmatrix} 5 \\ 2 \end{bmatrix}$

38. $x \begin{bmatrix} 1 \\ 1 \end{bmatrix} + y \begin{bmatrix} 1 \\ -1 \end{bmatrix} = \begin{bmatrix} 10 \\ 2 \end{bmatrix}$

39. $x \begin{bmatrix} 1 \\ 2 \end{bmatrix} + y \begin{bmatrix} 2 \\ 4 \end{bmatrix} = \begin{bmatrix} 0 \\ 1 \end{bmatrix}$

40. $x \begin{bmatrix} 2 \\ 4 \end{bmatrix} + y \begin{bmatrix} 3 \\ 6 \end{bmatrix} = \begin{bmatrix} 1 \\ 0 \end{bmatrix}$

41. The Clone Corporation makes three computer models (PC, XT, and AT) that are manufactured in Korea and assembled in the United States. The total costs (in dollars) are described by the matrices K and U, which stand for Korea and the United States, respectively.
    (a) Write the matrix of the total costs.
    (b) Write the matrix of the difference in costs between Korea and the United States.
    (c) Write the matrix of the situation where the costs in Korea are cut in half while the costs in the United States are doubled.

$$
K = \begin{matrix} \text{PC} \\ \text{XT} \\ \text{AT} \end{matrix} \begin{bmatrix} \overset{\text{manufacture}}{150} & \overset{\text{ship}}{12} \\ 325 & 15 \\ 700 & 18 \end{bmatrix}
$$

$$
U = \begin{matrix} \text{PC} \\ \text{XT} \\ \text{AT} \end{matrix} \begin{bmatrix} \overset{\text{assemble}}{15} & \overset{\text{ship}}{6} \\ 20 & 8 \\ 30 & 10 \end{bmatrix}
$$

42. A company owns two branch stores, Barney's and Carol's. The amounts of the retail sales and discount sales (in $1000) in three divisions (furniture, appliances, and clothing) for each branch store are given by the matrices B and C, respectively.
   (a) Write the matrix of the total sales for the stores.
   (b) If management wants Barney's to double its sales and Carol's to triple its sales, what matrix will represent the total sales?

$$
B = \begin{matrix} \\ \\ \end{matrix} \begin{bmatrix} \overset{\text{furniture}}{38} & \overset{\text{appliances}}{47} & \overset{\text{clothing}}{19} \\ 62 & 153 & 23 \end{bmatrix} \begin{matrix} \text{retail} \\ \text{discount} \end{matrix}
$$

$$
C = \begin{bmatrix} 17 & 18 & 67 \\ 25 & 22 & 75 \end{bmatrix} \begin{matrix} \text{retail} \\ \text{discount} \end{matrix}
$$

●●● 43. The matrix gives the shortest navigable distances in nautical miles between the ports of San Francisco, Vancouver, and Panama City.* Write the corresponding matrix in statute miles, where 1 nautical mile = 1.15 statute miles.

$$
\begin{matrix} \text{SF} \\ \text{V} \\ \text{PC} \end{matrix} \begin{bmatrix} \overset{\text{SF}}{0} & \overset{\text{V}}{812} & \overset{\text{PC}}{3246} \\ 812 & 0 & 4021 \\ 3246 & 4021 & 0 \end{bmatrix}
$$

44. An important part of an archaeologist's work is to assign dates to deposits and finds. The Brainerd-Robinson study investigated eight stratified deposits taken from three trenches.† The pieces of pottery found in the deposits were sorted into eight types. The 8×8 matrix B is defined by $b_{ij} = 1$ if deposit i contains an artifact of type j, and $b_{ij} = 0$ otherwise.

$$
B = \begin{bmatrix} 0 & 0 & 1 & 1 & 1 & 1 & 1 & 0 \\ 1 & 1 & 1 & 1 & 1 & 1 & 1 & 1 \\ 1 & 0 & 1 & 0 & 0 & 1 & 0 & 1 \\ 1 & 0 & 1 & 1 & 1 & 1 & 1 & 0 \\ 1 & 0 & 1 & 1 & 1 & 1 & 1 & 1 \\ 1 & 0 & 1 & 0 & 0 & 0 & 1 & 1 \\ 1 & 1 & 0 & 0 & 0 & 0 & 1 & 0 \\ 1 & 1 & 1 & 0 & 0 & 1 & 0 & 1 \end{bmatrix}
$$

   (a) Explain the meaning of the entry b_{24}.
   (b) Explain the meaning of the entry b_{73}.

45. Correspondence analysis is a technique for graphing tables for marketing purposes. It has many applications in marketing research, and involves the scaling of the rows and columns of a data matrix. This problem illustrates its use with an example representing soft drink consumption.‡ The matrix in Table 1 shows the result of a survey of the beverage preferences of 34 MBA students at Columbia. It is a "binary matrix" because each entry is either 1 (the beverage was consumed at least once over a two week period) or 0.
   (a) What is the size of the matrix?
   (b) Find the "correspondence matrix" P, where $P = (1/99)X$ and X is the binary matrix in Table 1.
   (c) Write the 8×8 matrix D_c, where d_{jj} is equal to the sum of the entries in the jth

*Source: "Distances Between Ports," Defense Mapping Agency, Hydrographic/Topographic Center.

†This work was done jointly, but published in two separate articles: W. S. Robinson, "A Method for Chronologically Ordering Archaeological Deposits," *American Antiquity* 16:293–301, 1951, and G. W. Brainerd, "The Place of Chronological Ordering in Archaeological Analysis," *idem*, 301–313. A very readable account of the work can be found in the paper by Alan Shuchat, "Matrix and Network Models in Archaeology," *Mathematics Magazine*, January 1984, pp. 3–14.

‡Donna L. Hoffman and George R. Franke, "Correspondence Analysis: Graphical Representation of Categorical Data in Marketing Research," *Journal of Marketing Research*, August 1986, pp. 213–227.

TABLE 1 The 34 × 8 Binary Indicator Matrix of Beverage Purchase and Consumption

Individual	Coke	Diet Coke	Diet Pepsi	Diet 7Up	Pepsi	Sprite	Tab	7Up
1	1	0	0	0	1	1	0	1
2	1	0	0	0	1	0	0	0
3	1	0	0	0	1	0	0	0
4	0	1	0	1	0	0	1	0
5	1	0	0	0	1	0	0	0
6	1	0	0	0	1	1	0	0
7	0	1	1	1	0	0	1	0
8	1	1	0	0	1	1	0	1
9	1	1	0	0	0	1	1	1
10	1	0	0	0	1	0	0	1
11	1	0	0	0	1	1	0	0
12	0	1	0	0	0	0	1	0
13	0	0	1	1	0	1	0	1
14	1	0	0	0	0	1	0	0
15	0	1	1	0	0	0	1	0
16	0	0	0	0	1	1	0	0
17	0	1	0	0	0	1	0	0
18	1	1	0	0	1	0	0	0
19	1	0	0	0	0	0	0	1
20	1	1	1	0	1	0	0	0
21	1	0	0	0	1	0	0	0
22	1	0	0	0	1	0	0	0
23	0	1	0	1	0	0	1	0
24	1	1	0	0	1	0	0	0
25	0	1	1	1	0	0	0	0
26	0	1	0	1	0	0	1	0
27	0	1	0	0	0	0	1	0
28	1	0	0	0	0	1	0	1
29	1	0	0	0	0	1	0	0
30	0	1	1	0	0	0	1	0
31	1	0	0	0	1	0	0	1
32	0	1	1	0	0	0	1	0
33	1	0	0	0	1	0	0	1
34	0	1	1	1	0	0	1	0

column of P and all other entries are equal to 0.

(d) Write the 34 × 34 matrix D_r, where d_{jj} is equal to the sum of the entries in the jth row of P and all other entries are equal to 0.

46. Let A and B be $m \times n$ matrices. Prove that $A + B = B + A$.

47. Let M be an $m \times n$ matrix and let Z be the $m \times n$ zero matrix. Prove that there always exists an $m \times n$ matrix X such $M + X = Z$.

2–2 MATRIX MULTIPLICATION

Matrix multiplication is quite different from addition and scalar multiplication. It might seem artificial and contrived at first, yet it is one of the most useful operations in all of mathematics. In this section we define matrix multiplication and illustrate its properties. The link between systems of equations and matrices shows why the definition of multiplication is the one that is commonly used.

Row-by-Column Multiplication

Recall that the Pressman-Gutman Company is a fabric-finishing firm that buys cotton and synthetics from a mill and finishes them into smooth, colored products. Three of its plants are located in Fall River, Passaic, and Los Angeles. The matrix P represents the number of linear yards of each material that was shipped to each plant in January 1988.

$$
\begin{array}{c}
\quad\quad\quad\text{cotton}\quad\text{synthetics} \\
\begin{array}{l}
\text{Fall River} \\
\text{Passaic} \\
\text{Los Angeles}
\end{array}
\begin{bmatrix}
2.0 & 4.5 \\
2.5 & 4.0 \\
3.5 & 1.5
\end{bmatrix}
\end{array}
$$

Suppose that it costs \$2 to ship each linear yard of unfinished cotton and \$1 to ship each linear yard of unfinished synthetics. The cost involved in shipping these fabrics to each of the finishing plants is computed by multiplying each number in the first column of P by \$2 and each number in the second column by \$1 and then adding these products. For instance, the cost of shipping the fabrics to Fall River is

$$(2.0)(\$2) + (4.5)(\$1) = \$8.5$$

Since the numbers in P are in terms of millions of yards, the cost of shipping the fabrics to Fall River is \$8.5 million, or \$8,500,000.

There is a convenient way of organizing the arithmetic in this problem by putting the two costs in a column matrix.

$$C = \begin{bmatrix} 2 \\ 1 \end{bmatrix}$$

Regard the first row of P as a row matrix. We define "row-by-column multiplication" to model the Pressman-Gutman problem.

$$[2.0 \quad 4.5] \begin{bmatrix} 2 \\ 1 \end{bmatrix} = [(2.0)(2) + (4.5)(1)] = [8.5]$$

This can be extended to the multiplication of a matrix times a column matrix by including the remaining rows of P.

$$
PC = \begin{bmatrix} 2.0 & 4.5 \\ 2.5 & 4.0 \\ 3.5 & 1.5 \end{bmatrix} \begin{bmatrix} 2 \\ 1 \end{bmatrix} = \begin{bmatrix} (2.0)(2) + (4.5)(1) \\ (2.5)(2) + (4.0)(1) \\ (3.5)(2) + (1.5)(1) \end{bmatrix} = \begin{bmatrix} 8.5 \\ 9.0 \\ 8.5 \end{bmatrix}
$$

Notice that the top entry in PC is equal to 8.5, which corresponds to the \$8.5 million cost of shipping to the Fall River plant.

It is not necessary for matrix multiplication to have a physical interpretation like the Pressman-Gutman example, nor for one matrix to be a column matrix. The display gives the formal definition, which is illustrated in Figure 1.

$$
\overset{A}{
\begin{bmatrix}
 & & & \\
\text{row } i \longrightarrow a_{i1}\, a_{i2} \cdots a_{in} & \\
 & & &
\end{bmatrix}}
\overset{B}{
\begin{bmatrix}
b_{1j} \\
b_{2j} \\
\vdots \\
b_{nj}
\end{bmatrix}}
=
\overset{C}{
\begin{bmatrix}
\\
c_{ij} \\
\\
\end{bmatrix}} \longleftarrow i
$$

column j j

where $c_{ij} = a_{i1}b_{1j} + a_{i2}b_{2j} + \cdots + a_{in}b_{nj}$.

FIGURE 1

_____ **Definition** _____

Let A be an $m \times n$ matrix and B an $n \times p$ matrix. The **product** of A and B is the $m \times p$ matrix C defined by

$$
c_{ij} = a_{i1}b_{1j} + a_{i2}b_{2j} + \cdots + a_{in}b_{nj}
$$

_____

Example 1 carries out the details of this definition. The operation is called **matrix multiplication.** Sometimes it is called row-by-column multiplication.

EXAMPLE 1 | **Problem**

Compute the product AB, where

$$
A = \begin{bmatrix} 7 & 0 \\ 8 & 1 \\ 9 & 2 \end{bmatrix}
\quad \text{and} \quad
B = \begin{bmatrix} 1 & 2 & 3 \\ 4 & 5 & 6 \end{bmatrix}
$$

SOLUTION Multiply each row of A by each column of B. The first row of A times the first column of B is

$$
\begin{bmatrix} 7 & 0 \end{bmatrix} \begin{bmatrix} 1 \\ 4 \end{bmatrix} = [(7)(1) + (0)(4)] = [7]
$$

The first row of A times the second column of B is

$$
\begin{bmatrix} 7 & 0 \end{bmatrix} \begin{bmatrix} 2 \\ 5 \end{bmatrix} = [(7)(2) + (0)(5)] = [14]
$$

The first row of A times the third column is equal to 21. This completes the first row of the product AB. The details of the entire multiplication are shown here.

$$AB = \begin{bmatrix} 7 & 0 \\ 8 & 1 \\ 9 & 2 \end{bmatrix} \begin{bmatrix} 1 & 2 & 3 \\ 4 & 5 & 6 \end{bmatrix}$$

$$= \begin{bmatrix} (7\cdot1) + (0\cdot4) & (7\cdot2) + (0\cdot5) & (7\cdot3) + (0\cdot6) \\ (8\cdot1) + (1\cdot4) & (8\cdot2) + (1\cdot5) & (8\cdot3) + (1\cdot6) \\ (9\cdot1) + (2\cdot4) & (9\cdot2) + (2\cdot5) & (9\cdot3) + (2\cdot6) \end{bmatrix}$$

$$= \begin{bmatrix} 7 & 14 & 21 \\ 12 & 21 & 30 \\ 17 & 28 & 39 \end{bmatrix}$$

It is not always possible to complete row-by-column multiplication. For instance, let

$$A = \begin{bmatrix} 1 & 3 & 4 \\ 4 & 5 & 6 \end{bmatrix} \quad \text{and} \quad B = \begin{bmatrix} 7 & 8 & 9 \\ 0 & 1 & 2 \end{bmatrix}$$

The product AB cannot be formed because row 1 of A has three entries but column 1 of B has only two. We say that the product AB is **undefined.** In general if A is an $m \times n$ matrix and B is a $q \times p$ matrix, then AB is an $m \times p$ matrix if $n = q$ and AB is undefined otherwise. In brief,

$$\begin{matrix} (m \times n) & (n \times p) & = (m \times p) \\ (m \times n) & (q \times p) & \text{is undefined if } n \neq q \end{matrix}$$

$$\underbrace{\qquad}_{\text{size of } A} \quad \underbrace{\qquad}_{\text{size of } B} \quad \underbrace{\qquad}_{\substack{\text{size of} \\ \text{product } AB}}$$

Example 2 shows why caution must be exercised when extending properties of the multiplication of numbers to matrix multiplication.

EXAMPLE 2

Problem

Compute BC and CB, where

$$B = \begin{bmatrix} -1 & 0 \\ 6 & 8 \end{bmatrix} \quad \text{and} \quad C = \begin{bmatrix} 0 & 4 & 1 \\ 6 & -1 & 0 \end{bmatrix}$$

SOLUTION The product BC is equal to

$$BC = \begin{bmatrix} -1 & 0 \\ 6 & 8 \end{bmatrix} \begin{bmatrix} 0 & 4 & 1 \\ 6 & -1 & 0 \end{bmatrix}$$

$$= \begin{bmatrix} -1\cdot0 + 0\cdot6 & -1\cdot4 + 0\cdot-1 & -1\cdot1 + 0\cdot0 \\ 6\cdot0 + 8\cdot6 & 6\cdot4 + 8\cdot-1 & 6\cdot1 + 8\cdot0 \end{bmatrix}$$

$$= \begin{bmatrix} 0 & -4 & -1 \\ 48 & 16 & 6 \end{bmatrix}$$

However, CB is undefined because C has three columns and B has two rows.

Systems of Equations

Suppose the Pressman-Gutman Company determines that to break even the Fall River plant must produce $15,000,000 worth of material, the Passaic plant must produce $15,500,000, and the Los Angeles plant $13,500,000. The company wants to find the price per yard of cotton and of synthetic material that it must charge in order to break even. To solve the problem, let x be the price per yard of cotton, y be the price per yard of synthetic material, and

$$X = \begin{bmatrix} x \\ y \end{bmatrix} \quad \text{be the price matrix.}$$

Then PX represents the income from the material finished at each plant. In order to break even, x and y must satisfy

$$\begin{bmatrix} 2.0 & 4.5 \\ 2.5 & 4.0 \\ 3.5 & 1.5 \end{bmatrix} \begin{bmatrix} x \\ y \end{bmatrix} = \begin{bmatrix} 15.0 \\ 15.5 \\ 13.5 \end{bmatrix}$$

This matrix equation is solved by multiplying the matrices on the left-hand side and setting the product equal to the right-hand side. Thus

$$\begin{bmatrix} 2.0x + 4.5y \\ 2.5x + 4.0y \\ 3.5x + 1.5y \end{bmatrix} = \begin{bmatrix} 15.0 \\ 15.5 \\ 13.5 \end{bmatrix}$$

The unknown quantities x and y must satisfy the system of equations

$$2.0x + 4.5y = 15.0$$
$$2.5x + 4.0y = 15.5$$
$$3.5x + 1.5y = 13.5$$

By Gaussian elimination the solution is $x = 3$ and $y = 2$. It can be written as the column matrix $\begin{bmatrix} 3 \\ 2 \end{bmatrix}$.

This illustrates why matrix multiplication is defined the way it is. The **coefficient matrix** of the system is the Pressman-Gutman matrix P. The **variable matrix** is the column matrix X. The **constant matrix** is the column matrix of numbers on the right-hand side:

$$C = \begin{bmatrix} 15.0 \\ 15.5 \\ 13.5 \end{bmatrix}$$

The product PX becomes

$$PX = \begin{bmatrix} 2.0 & 4.5 \\ 2.5 & 4.0 \\ 3.5 & 1.5 \end{bmatrix} \begin{bmatrix} x \\ y \end{bmatrix} = \begin{bmatrix} 2.0x + 4.5y \\ 2.5x + 4.0y \\ 3.5x + 1.5y \end{bmatrix}$$

Its entries are the left-hand side of the given system of equations. Thus the entire system can be written as the matrix equation

$$PX = C$$

In this equation P and C are given matrices (that is, their entries are numbers) and X is the unknown (or variable) matrix that is sought. In this sense the matrix equation $PX = C$ resembles the linear equation $ax = b$ for numbers.

> Every system of linear equations can be written as a matrix equation of the form $AX = C$. Conversely, every such matrix equation describes a system of linear equations.

EXAMPLE 3 | **Problem**

Write this system of equations as a matrix equation.

$$\begin{array}{rcr} x - 2y & = & 8 \\ 4y - 3z & = & -7 \end{array}$$

SOLUTION The coefficient matrix is

$$A = \begin{bmatrix} 1 & -2 & 0 \\ 0 & 4 & -3 \end{bmatrix}$$

Be sure to include the zeros. Since the system has three variables, the variable matrix is

$$X = \begin{bmatrix} x \\ y \\ z \end{bmatrix}$$

The constant matrix has two entries.

$$C = \begin{bmatrix} 8 \\ -7 \end{bmatrix}$$

The matrix equation is then $AX = C$.

Sections 2–3 and 2–5 will further develop the link between systems of equations and matrices.

Algebra of Square Matrices

The set of 2×3 matrices is not closed under matrix multiplication because of the rule

$$(2 \times 3) \quad (2 \times 3)$$

not defined because $3 \neq 2$

For square matrices the situation is different. If A and B are $n \times n$ matrices, then so is AB because

$$(n \times n)(n \times n) = (n \times n)$$

equal

The set of $n \times n$ matrices is closed under multiplication.

The multiplication of real numbers and the multiplication of matrices have some properties in common. For instance, the associative law holds.

$$(AB)C = A(BC)$$

whenever the sizes of the matrices permit them to be multiplied. But there are several differences that show that care must be exercised when dealing with matrices. For instance, the **cancellation law** does not hold, meaning that $AB = AC$ does not necessarily imply that $B = C$. Example 4 shows that the set of $n \times n$ matrices is not commutative under multiplication.

EXAMPLE 4 | Let

$$A = \begin{bmatrix} 1 & 2 & 3 \\ 4 & 5 & 6 \\ 7 & 8 & 7 \end{bmatrix} \quad \text{and} \quad B = \begin{bmatrix} 1 & 4 & 7 \\ 2 & 5 & 8 \\ 3 & 6 & 9 \end{bmatrix}$$

Problem

Determine whether $AB = BA$.

SOLUTION The definition of the product AB shows that the $(1, 1)$ entry is equal to 14. Similarly, the $(1, 1)$ entry of BA is equal to 66. In order for AB to be equal to BA their entries must all be equal, respectively. Since the $(1, 1)$ entries differ, they are not equal. Therefore $AB \neq BA$.

The **main diagonal** of an $n \times n$ matrix A is the diagonal of numbers running from the $(1, 1)$ entry to (n, n). The main diagonal of this matrix is shaded.

$$A = \begin{bmatrix} -5 & 4 & 0 \\ 6 & -3 & 7 \\ 9 & -1 & 4 \end{bmatrix}$$

The matrix with 1's down the main diagonal and zeros elsewhere is called the **identity matrix** and is denoted by I_n. The 3×3 identity matrix is

$$I_3 = \begin{bmatrix} 1 & 0 & 0 \\ 0 & 1 & 0 \\ 0 & 0 & 1 \end{bmatrix}$$

$AI_n = A$ and $I_nA = A$ for every $n \times n$ matrix A.

The operations of addition and multiplication of matrices are tied together by the same property that ties them together for numbers, the **distributive laws**

$$A(B + C) = AB + AC$$

and

$$(D + E)F = DF + EF$$

EXERCISE SET 2–2

• In Problems 1 to 6 find the product of the two matrices whenever it is defined.

$$A = \begin{bmatrix} 1 & -1 & 0 \\ 0 & 10 & 0 \\ 0 & 1 & 1 \end{bmatrix} \qquad B = \begin{bmatrix} -2 & 1 & 0 \\ 0 & 6 & 3 \\ 0 & 0 & -2 \end{bmatrix}$$

$$C = \begin{bmatrix} 1 & 0 & -1 & 6 \\ 2 & 5 & 2 & 3 \\ 4 & 1 & 3 & 1 \end{bmatrix} \qquad D = \begin{bmatrix} 0 \\ -1 \\ -2 \end{bmatrix}$$

$$E = \begin{bmatrix} 6 & 3 & 1 & 4 \\ 5 & 7 & 0 & -4 \end{bmatrix}$$

$$F = \begin{bmatrix} -1 & 10 \\ 6 & -7 \end{bmatrix} \qquad G = \begin{bmatrix} 16 & 3 \\ -2 & 0 \end{bmatrix}$$

1. AB

2. FG

3. AC

4. FE

5. AD

6. DA

In Problems 7 to 10 compute the stated entry in the product $J = KL$, where

$$K = \begin{bmatrix} -1 & 6 & 7 \\ 2 & 4 & 1 \\ 3 & 0 & 6 \\ 0 & 0 & 0 \\ -6 & 3 & 0 \end{bmatrix}$$

$$L = \begin{bmatrix} 1 & 6 & 0 & -1 \\ 6 & -2 & 0 & -3 \\ 0 & 6 & 0 & 6 \end{bmatrix}$$

7. j_{12}

8. j_{42}

9. j_{22}

10. j_{33}

In Problems 11 and 12 determine whether the product AB is defined. If it is, state its size.

11. (a) A is 4×6, B is 6×3
   (b) A is 4×3, B is 4×3

12. (a) A is 2×4, B is 2×4
   (b) A is 1×4, B is 4×9

In Problems 13 and 14 a system of equations is given. Write the coefficient matrix A, the variable matrix X, and the constant matrix C for each system.

13. $-x + 4y - 3z = 11$
    $2x - 2y + 7z = -4$

14. $4x - 6y = 8$
    $-x + 6y = 0$
    $2x - 3y = 4$

In Problems 15 to 18 write the matrix equation $AX = Z$ as a system of equations, where Z is the zero matrix and

$$X = \begin{bmatrix} x \\ y \\ z \end{bmatrix}$$

15.
$$A = \begin{bmatrix} 1 & 0 & 6 \\ -2 & 3 & 4 \\ 1 & 2 & 0 \end{bmatrix}$$

16.
$$A = \begin{bmatrix} 2 & 6 & 7 \\ 0 & 3 & -1 \\ 0 & 0 & 4 \end{bmatrix}$$

17.
$$A = \begin{bmatrix} 2 & 6 & 7 \\ 3 & 5 & -2 \end{bmatrix}$$

18.
$$A = \begin{bmatrix} 1 & 1 & -1 \\ 2 & 5 & 4 \end{bmatrix}$$

In Problems 19 to 22, determine whether $AB = BA$, where

$$B = \begin{bmatrix} 10 & 2 \\ -1 & 4 \end{bmatrix}$$

19. $A = \begin{bmatrix} 1 & 2 \\ 3 & 4 \end{bmatrix}$ 20. $A = \begin{bmatrix} -1 & 0 \\ 0 & 3 \end{bmatrix}$

21. $A = \begin{bmatrix} 3 & 0 \\ 0 & -2 \end{bmatrix}$ 22. $A = \begin{bmatrix} 0 & 3 \\ -2 & 0 \end{bmatrix}$

Problems 23 and 24 refer to this Pressman-Gutman matrix.

	cotton	synthetics
Fall River	2.0	4.5
Passaic	2.5	4.0
Los Angeles	3.5	1.5

23. If it costs \$1 to ship each linear yard of cotton and \$1.50 to ship each linear yard of synthetics,

what matrix represents the cost of shipping these fabrics to each of the finishing plants?

24. If it costs $4 to ship each linear yard of cotton and $3.50 to ship each linear yard of synthetics, what matrix represents the cost of shipping these fabrics to each of the finishing plants?

•• Problems 25 to 28 refer to these matrices.

$$A = \begin{bmatrix} -1 & 0 \\ 2 & 3 \end{bmatrix} \qquad B = \begin{bmatrix} 4 & -1 \\ 0 & 2 \end{bmatrix}$$

$$C = \begin{bmatrix} 6 & -9 \\ -4 & 5 \end{bmatrix}$$

25. (a) Compute $(AB)C$.
    (b) Compute $A(BC)$.
    (c) What do you conclude from parts (a) and (b)?
    (d) Compute ABC.

26. (a) Compute $AA - BB$.
    (b) Compute $(A + B)(A - B)$.
    (c) What do you conclude from parts (a) and (b)?

27. (a) Compute $A(B + C)$.
    (b) Compute $AB + AC$.
    (c) What do you conclude from parts (a) and (b)?

28. (a) Compute $(A + B)C$.
    (b) Compute $AC + BC$.
    (c) What do you conclude from parts (a) and (b)?

In Problems 29 and 30 let E_{ij} be the 3×3 matrix with the number 1 in the (i, j) entry and 0's elsewhere. Let

$$A = \begin{bmatrix} 1 & 2 & 3 \\ 4 & 5 & 6 \\ 7 & 8 & 9 \end{bmatrix}$$

29. (a) Compute $E_{12}A$.
    (b) Compute $E_{23}A$.
    (c) What do you conclude from parts (a) and (b)?

30. (a) Compute AE_{12}.
    (b) Compute AE_{23}.
    (c) What do you conclude from parts (a) and (b)?

31. Let A be a 3×3 matrix whose bottom row consists entirely of 0's and let B be any 3×4 matrix. What can you conclude about the product AB?

32. Let B be a 3×3 matrix whose middle column consists entirely of 0's and let A be any 4×3 matrix. What can you conclude about the product AB?

33. A researcher for a dairy company is experimenting with blends of low-fat milk and skim milk. The number of grams per cup of protein, carbohydrate, and fat contained in each kind of milk is given by the matrix

$$A = \begin{matrix} & \text{skim} & \text{low-fat} \\ & \begin{bmatrix} 9 & 9 \\ 13 & 13 \\ 1 & 5 \end{bmatrix} & \begin{matrix} \text{protein} \\ \text{carbohydrate} \\ \text{fat} \end{matrix} \end{matrix}$$

The number of cups of milk fed to people in three age categories is given by the matrix

$$B = \begin{matrix} \text{children} & \text{adolescents} & \text{adults} \\ \begin{bmatrix} 3 & 1 & 0 \\ 2 & 1 & 1 \end{bmatrix} & & \begin{matrix} \text{skim} \\ \text{low-fat} \end{matrix} \end{matrix}$$

What do the entries in the matrix AB tell the researcher?

34. A company makes records, tapes, and compact disks in Los Angeles and Nashville. Each item must be manufactured and quality tested. The time (in hours) for each operation is given by the matrix A.

$$A = \begin{matrix} & \text{manufacture} & \text{test} \\ & \begin{bmatrix} 1 & \frac{1}{3} \\ 3 & \frac{1}{2} \\ 8 & 1 \end{bmatrix} & \begin{matrix} \text{record} \\ \text{tape} \\ \text{compact disk} \end{matrix} \end{matrix}$$

The labor costs (in dollars per hour) in each city for each operation is given by the matrix B.

$$B = \begin{matrix} \text{Los Angeles} & \text{Nashville} \\ \begin{bmatrix} 7 & 6 \\ 10 & 5 \end{bmatrix} & \begin{matrix} \text{manufacture} \\ \text{test} \end{matrix} \end{matrix}$$

What do the entries in the matrix AB mean?

35. An automobile dealership has three franchises (F1, F2, and F3) in Florida and the same three in Georgia. The October sales volume, in cars sold, is shown in the matrix C.

$$C = \begin{array}{c} \\ F1 \\ F2 \\ F3 \end{array} \overset{\begin{array}{cc} FL & GA \end{array}}{\begin{bmatrix} 126 & 92 \\ 76 & 87 \\ 45 & 78 \end{bmatrix}}$$

Each car sold for $17,000 in Florida and $16,200 in Georgia. Use matrix multiplication to determine the total sales for each franchise in October.

36. An automobile dealership had two franchises in each state: Texas, Oklahoma, and Nebraska. The November sales volume, in cars sold, is shown in the matrix C.

$$C = \begin{array}{c} \\ F1 \\ F2 \end{array} \overset{\begin{array}{ccc} TX & OK & NB \end{array}}{\begin{bmatrix} 423 & 68 & 52 \\ 186 & 45 & 47 \end{bmatrix}}$$

Each car sold for $14,500 in Texas, $15,000 in Oklahoma, and $16,200 in Nebraska. Use matrix multiplication to determine the total sales for each franchise in November.

37. A bank pays interest rates of 10, 7, and 5% on certificates of deposit (CDs) of $5000, $2000, and $1000, respectively. Use row-by-column matrix multiplication to determine the total interest earned on the CDs.

38. A bank pays interest rates of 12, 10, 7, and 5% on certificates of deposit (CDs) of $25,000, $5000, $2000, and $1000, respectively. Use row-by-column matrix multiplication to determine the total interest earned on the CDs.

39. Two juices (A, B) are blended from four ingredients (W, X, Y, Z). The numbers of gallons of each ingredient in a 20-gallon container of each juice are given by the matrix J.

$$J = \begin{array}{c} \\ A \\ B \end{array} \overset{\begin{array}{cccc} W & X & Y & Z \end{array}}{\begin{bmatrix} 8 & 6 & 4 & 2 \\ 9 & 7 & 3 & 1 \end{bmatrix}}$$

Use matrix multiplication to find the total cost of making 20 gallons of each juice if W, X, Y, and Z cost $0.25, $0.50, $1, and $2 per gallon, respectively.

40. Use Problem 39 to compute the cost of making 60 gallons of juice A and 80 gallons of juice B under the same conditions given there.

••• 41. Let A be an arbitrary 2×2 matrix and let S be a scalar matrix, where s, a, b, c, and d are any numbers.

$$A = \begin{bmatrix} a & b \\ c & d \end{bmatrix} \qquad S = \begin{bmatrix} s & 0 \\ 0 & s \end{bmatrix}$$

(a) Verify that $SA = AS$.
(b) Prove that if B satisfies $AB = BA$ for all 2×2 matrices A, then $B = S$ for some number s.

42. Find 2×2 matrices A, B, and C, with C not the zero matrix, such that $AC = BC$ but $A \neq B$.

43. Find two 2×2 matrices A and B, different from the zero matrix Z, such that $AB = Z$.

44. Find a 2×2 matrix A, different from the zero matrix Z, such that $AA = Z$.

Decision theory makes use of matrices to organize and examine data that can be helpful to individuals when making personal decisions. Problems 45 and 46 apply decision theory to two unrelated situations. In both problems a matrix X describes the relative preferences of each option with respect to each criterion. The absolute preferences are obtained from the rows of the product XY.

45. A study was conducted for various universities in the Province of Ontario after the government ended its practice of administering universal exams (called "departmentals") for admission into the universities, because a period of grade inflation ensued at some of the high schools.* There were three options (G = to continue to accept high school grades, C = to calibrate high school grades, E = to construct new admission exams) and four criteria (F = fairness, P =

*Ed Barbeau, "Perron's Result and a Decision on Admissions Tests," *Mathematics Magazine*, February 1986, pp. 12–22.

predictability, L = low cost, and A = acceptability).

$$
\begin{array}{cccc}
& F & P & L & A \\
X = & \begin{bmatrix} .105 & .297 & .644 & .674 \\ .258 & .540 & .271 & .101 \\ .637 & .163 & .085 & .225 \end{bmatrix} & \begin{matrix} G \\ C \\ E \end{matrix}
\end{array}
$$

$$
Y = \begin{bmatrix} .543 \\ .085 \\ .213 \\ .159 \end{bmatrix}
$$

(a) What is the sum of the entries in each column of X? Why?

(b) Which option was most preferable?

(c) Which option was least preferable?

46. A plant pathologist observes a leaf wilt in her neighbor's tobacco plant; the wilt can be caused by either virus A or virus B.[†] Virus A cannot be treated, so that if the plant has this virus it will die, causing a loss of $15. There is a $5 treatment that will save the plant if it has virus B. If the plant has virus B and is not treated it will die. There are two options (T = treat the plant and N = do not treat the plant) and two criteria (H = healthy plants and L = low cost).

$$
\begin{array}{cc}
& H \quad L \\
X = & \begin{bmatrix} \frac{4}{5} & \frac{1}{3} \\ \frac{1}{5} & \frac{2}{3} \end{bmatrix} \begin{matrix} T \\ N \end{matrix}
\end{array}
\qquad
Y = \begin{bmatrix} \frac{1}{4} \\ \frac{3}{4} \end{bmatrix}
$$

Which option is more preferable?

In the 1970s Wendy's International, Inc. grew from a single store to a conglomerate of 1800 stores in the United States, Europe, and Japan. Today Wendy's is the third largest fast-food hamburger chain in the United States. The top managers at Wendy's faced three important strategy decisions: (1) whether to emphasize expansion of its domestic market or its foreign market, (2) whether to continue to appeal to the 82% of its customers who were over 25 years of age, and (3) whether to utilize stock or debt to finance new construction.

Problems 47–49 introduce a technique for dealing with these strategies called QSPM (Quantitative Strategic Planning Matrix).[‡] The "sum total attractiveness scores" are the products of a rating matrix r and a matrix of the values for various factors. The best strategy is the option with the highest sum total attractiveness score.

47. Where should Wendy's concentrate its markets, domestic D or foreign F? The rating matrix is

$$
r = \begin{bmatrix} 3 & 1 & 2 & 2 & 1 & 2 & 4 & 1 & 1 \end{bmatrix}
$$

and the value matrices are

$$
D = \begin{bmatrix} 4 \\ 4 \\ 2 \\ 3 \\ 4 \\ 2 \\ 4 \\ 4 \\ 3 \end{bmatrix}
\qquad
F = \begin{bmatrix} 3 \\ 2 \\ 1 \\ 2 \\ 1 \\ 3 \\ 2 \\ 1 \\ 2 \end{bmatrix}
$$

48. Which age group should Wendy's appeal to, U = 25 or under, or O = over 25? The rating matrix is

$$
r = \begin{bmatrix} 4 & 1 & 2 & 2 & 4 & 1 & 2 & 4 & 1 & 1 \end{bmatrix}
$$

and the value matrices are

$$
U = \begin{bmatrix} 3 \\ 4 \\ 1 \\ 3 \\ 3 \\ 2 \\ 4 \\ 2 \\ 2 \\ 2 \end{bmatrix}
\qquad
O = \begin{bmatrix} 4 \\ 2 \\ 4 \\ 2 \\ 2 \\ 4 \\ 2 \\ 4 \\ 3 \\ 3 \end{bmatrix}
$$

[†] JoAnne Growney, "Personal Decisions/Medical Decisions: Consider the Analytic Hierarchy Process," *UMAP Journal*, Winter 1985, pp. 7–35.

[‡] Fred R. David, "The Strategic Planning Matrix—A Quantitative Approach," *Long Range Planning* (*The International Journal of Strategic Management and Corporate Planning*), October 1986, pp. 102–107.

49. How should Wendy's finance its construction, by debt D or common stock S? The rating matrix is

$$r = [2 \quad 1 \quad 1 \quad 1]$$

and the value matrices are

$$D = \begin{bmatrix} 2 \\ 1 \\ 2 \\ 3 \end{bmatrix} \qquad S = \begin{bmatrix} 4 \\ 4 \\ 4 \\ 2 \end{bmatrix}$$

2–3 THE GAUSS–JORDAN METHOD

The word *echelon* in French means "a rung of a ladder." In English, echelon refers to a formation whose units are arranged in steplike fashion, like this formation of airplanes.

The mathematical use of the word describes a matrix in a form resembling an echelon formation. In this section we describe how the echelon form of a matrix is used to solve systems of linear equations. The procedure is called the Gauss–Jordan method because it is a refinement of Gaussian elimination due to Jordan.* We present this method for three reasons: (1) the Gauss–Jordan method is somewhat easier than Gaussian elimination when solving a system by hand, (2) it is indispensable for computing the inverse of a matrix in Sections 2–4 and 2–5, and (3) it forms an integral part of the simplex method (Chapter 4).

---

*Traditionally this refinement has been ascribed to the French mathematician Camille Jordan (1838–1922). But recent evidence suggests that the refinement is due to the German geodesist Wilhelm Jordan (1842–1899) instead. Renate McLaughlin presented the discovery (due to Steven Althoen and Renate McLaughlin) in a paper titled "Will the Real Jordan Please Stand Up?" at the International Congress of Mathematicians at Berkeley in 1986. Their results were published in the February 1987 issue of the *American Mathematical Monthly*, pp. 130–142.

Elementary Operations

In solving a system of linear equations, the coefficients are the essential ingredient and matrices are used as a device for keeping the coefficients aligned. By definition the **coefficient matrix** is the matrix of coefficients of the variables, and the **augmented matrix** consists of the coefficient matrix and the column of constants. The notation (R_) is used in place of (E_) because the operations are performed on rows instead of equations.

The permissible operations on the augmented matrix of a system of linear equations are called **elementary row operations.** They correspond to the permissible operations on the system of equations.

Elementary Row Operations

1. Interchange two rows.
2. Multiply each number in a row by a nonzero number.
3. Replace a row by the sum of that row and a multiple of another row.

We describe how to use the Gauss–Jordan method to solve a system of equations that was solved by Gaussian elimination in Section 1–4.

$$x + y + z = 10 \qquad\qquad\qquad (E1)$$

$$x - y = 2 \qquad\qquad\qquad (E2)$$

$$y - z = 1 \qquad\qquad\qquad (E3)$$

First form the augmented matrix. Write each variable atop the column of coefficients that it refers to.

$$
\begin{array}{ccc}
x & y & z
\end{array}
$$
$$
\left[\begin{array}{ccc|c}
1 & 1 & 1 & 10 \\
1 & -1 & 0 & 2 \\
0 & 1 & -1 & 1
\end{array}\right]
\qquad
\begin{array}{l}
(R1) \\
(R2) \\
(R3)
\end{array}
$$

The Gauss–Jordan method moves from the first column to the second column to the third column. For each column the aim is to produce a 1 as the first nonzero number of a row and to produce 0's above and below that 1. For the preceding system of equations we begin with the first column in the same way that we proceeded with Gaussian elimination.

$$
\left[\begin{array}{ccc|c}
1 & 1 & 1 & 10 \\
0 & -2 & -1 & -8 \\
0 & 1 & -1 & 1
\end{array}\right]
\qquad (-1)(R1) + (R2)
$$

Move to the second column to produce a 1 in the second row and 0's elsewhere. The 1 can be obtained by interchanging R2 and R3. Do *not* interchange R2 and R1 because the first column would no longer be in proper form.

$$\begin{bmatrix} 1 & 1 & 1 & | & 10 \\ 0 & 1 & -1 & | & 1 \\ 0 & -2 & -1 & | & -8 \end{bmatrix} \quad (R2) \leftrightarrow (R3)$$

$$\begin{bmatrix} 1 & 0 & 2 & | & 9 \\ 0 & 1 & -1 & | & 1 \\ 0 & 0 & -3 & | & -6 \end{bmatrix} \quad \begin{matrix} (-1)(R2) + (R1) \\ \\ 2(R2) + (R3) \end{matrix}$$

The final step is to transform the third column so that its third entry is 1 and the other entries are 0.

$$\begin{array}{ccc} x & y & z \end{array}$$
$$\begin{bmatrix} 1 & 0 & 2 & | & 9 \\ 0 & 1 & -1 & | & 1 \\ 0 & 0 & 1 & | & 2 \end{bmatrix} \quad (-\tfrac{1}{3})(R3)$$

$$\begin{bmatrix} 1 & 0 & 0 & | & 5 \\ 0 & 1 & 0 & | & 3 \\ 0 & 0 & 1 & | & 2 \end{bmatrix} \quad \begin{matrix} (-2)(R3) + (R1) \\ (R3) + (R2) \end{matrix}$$

Now the augmented matrix is in a form that yields the solution by inspection: $x = 5, y = 3, z = 2$

EXAMPLE 1

Problem

Solve the following system of equations by the Gauss–Jordan method.

$$\begin{aligned} x + 2y + 3z &= 1 \\ 2x + 5y + 7z &= 2 \\ 3x + 5y + 7z &= 4 \end{aligned}$$

SOLUTION Form the augmented matrix of the system

$$\begin{array}{ccc} x & y & z \end{array}$$
$$\begin{bmatrix} 1 & 2 & 3 & | & 1 \\ 2 & 5 & 7 & | & 2 \\ 3 & 5 & 7 & | & 4 \end{bmatrix}$$

Convert the first column to the desired form.

$$\begin{bmatrix} 1 & 2 & 3 & | & 1 \\ 0 & 1 & 1 & | & 0 \\ 0 & -1 & -2 & | & 1 \end{bmatrix} \quad \begin{matrix} (-2)(R1) + (R2) \\ (-3)(R1) + (R3) \end{matrix}$$

The desired second column is obtained as follows:

$$\begin{bmatrix} 1 & 0 & 1 & | & 1 \\ 0 & 1 & 1 & | & 0 \\ 0 & 0 & -1 & | & 1 \end{bmatrix} \quad \begin{matrix} (-2)(R2) + (R1) \\ \\ (R2) + (R3) \end{matrix}$$

Next produce a 1 in the third row, third column.

$$\begin{bmatrix} 1 & 0 & 1 & | & 1 \\ 0 & 1 & 1 & | & 0 \\ 0 & 0 & 1 & | & -1 \end{bmatrix} \quad (-1)(R3)$$

Convert the third column to the desired form.

$$\begin{array}{ccc} x & y & z \end{array}$$
$$\left[\begin{array}{ccc|c} 1 & 0 & 0 & 2 \\ 0 & 1 & 0 & 1 \\ 0 & 0 & 1 & -1 \end{array}\right] \qquad \begin{array}{l} (-1)(R3) + (R1) \\ (-1)(R3) + (R2) \end{array}$$

The solution can be read directly.

$$x = 2 \qquad y = 1 \qquad z = -1$$

The next example shows what can happen when a column cannot be converted to the desired form.

EXAMPLE 2 | **Problem**

Solve this system of linear equations by the Gauss–Jordan method.

$$\begin{array}{rcl} x - 3y & = & 4 \\ -3x + 8y - z & = & -10 \\ 2x - 7y - z & = & 10 \end{array}$$

SOLUTION Form the augmented matrix.

$$\left[\begin{array}{ccc|c} 1 & -3 & 0 & 4 \\ -3 & 8 & -1 & -10 \\ 2 & -7 & -1 & 10 \end{array}\right]$$

Convert the first column to the desired form.

$$\left[\begin{array}{ccc|c} 1 & -3 & 0 & 4 \\ 0 & -1 & -1 & 2 \\ 0 & -1 & -1 & 2 \end{array}\right] \qquad \begin{array}{l} 3(R1) + (R2) \\ (-2)(R1) + (R3) \end{array}$$

Put the second column in proper form.

$$\left[\begin{array}{ccc|c} 1 & 0 & 3 & -2 \\ 0 & 1 & 1 & -2 \\ 0 & 0 & 0 & 0 \end{array}\right] \qquad \begin{array}{l} -3(R2) + (R1) \\ (-1)(R2) \\ (-1)(R2) + (R3) \end{array}$$

Notice that the third column cannot be converted to the desired form. The corresponding system of equations is

$$\begin{array}{rcl} x + 3z & = & -2 \\ y + z & = & -2 \end{array}$$

Since this system of equations has infinitely many solutions, the original system does too.

The **general solution** to the system of equations in Example 2 is written in the form

$$x = -2 - 3z$$
$$y = -2 - z$$
$$z = \text{any number}$$

A **particular solution** is obtained from the general solution by choosing a particular number for z and then substituting to find values of x and y. For $z = 0$ we get $x = -2$ and $y = -2$. This solution is written $(-2, -2, 0)$. Another particular solution is $(-5, -3, 1)$. The remainder of this section shows how to complete the solutions to systems of equations like the one in Example 2.

Echelon Form

A matrix is said to be in **echelon form** if it satisfies these conditions.

1. If a row has a nonzero number, then the leftmost nonzero number in the row is 1. It is called the **leading 1.**
2. Rows with all zeros lie beneath the rows having a nonzero number.
3. When two rows have a leading 1, the leading 1 in the upper row is to the left of the leading 1 in the lower row.
4. Each column that contains a leading 1 has zeros everywhere else.

Echelon form resembles the airplane formation that began this section, with the leading 1's being the airplanes. The reduced matrices in Examples 1 and 2 are in echelon form.

$$\left[\begin{array}{ccc|c} 1 & 0 & 0 & 2 \\ 0 & 1 & 0 & 1 \\ 0 & 0 & 1 & -1 \end{array}\right] \qquad \left[\begin{array}{ccc|c} 1 & 0 & 3 & -2 \\ 0 & 1 & 1 & -2 \\ 0 & 0 & 0 & 0 \end{array}\right]$$

A variable is called a **free variable** if it can be any number and a **basic variable** if it depends on a free variable. Basic variables correspond to columns with leading 1's in echelon form. In Example 2 x and y are basic variables and z is a free variable.

EXAMPLE 3 | Let the augmented matrix of a system of linear equations be

$$M = \begin{array}{cccc} x & y & u & v \\ \left[\begin{array}{cccc|c} 1 & 0 & 1 & 2 & 1 \\ 0 & 1 & 0 & 3 & 0 \end{array}\right] \end{array}$$

Problem

Determine which variables are free and which are basic. Derive the general solution to the system of equations and find two particular solutions.

SOLUTION The matrix M is already in echelon form, so x and y are basic variables while u and v are free. The system of equations is

$$x + \quad u + 2v = 1$$
$$y + \quad 3v = 0$$

Solving them for the basic variables gives $x = 1 - u - 2v$ and $y = -3v$. The general solution is written

$$x = 1 - u - 2v$$
$$y = -3v$$
$$u = \text{any number}$$
$$v = \text{any number}$$

Particular solutions are obtained by choosing specific numbers for u and v. Two particular solutions are

$$x = 1 \qquad y = 0 \qquad u = 0 \qquad v = 0$$

and

$$x = 0 \qquad y = 0 \qquad u = 1 \qquad v = 0$$

Example 4 solves a system of equations by reducing it to echelon form.

EXAMPLE 4 | **Problem**

Solve the system of linear equations

$$x + 4y - 5z = 2$$
$$3x + 12y - 11z = 2$$
$$x + 4y - 3z = 0$$

SOLUTION Form the augmented matrix

$$
\begin{array}{ccc}
x & y & z \\
\end{array}
$$
$$
\left[\begin{array}{ccc|c}
1 & 4 & -5 & 2 \\
3 & 12 & -11 & 2 \\
1 & 4 & -3 & 0
\end{array}\right]
$$

Then

$$
\left[\begin{array}{ccc|c}
1 & 4 & -5 & 2 \\
0 & 0 & 4 & -4 \\
0 & 0 & 2 & -2
\end{array}\right]
\begin{array}{l}
\\
(-3)(\text{R1}) + (\text{R2}) \\
(-1)(\text{R1}) + (\text{R3})
\end{array}
$$

$$
\left[\begin{array}{ccc|c}
1 & 4 & -5 & 2 \\
0 & 0 & 1 & -1 \\
0 & 0 & 1 & -1
\end{array}\right]
\begin{array}{l}
\\
(\tfrac{1}{4})(\text{R2}) \\
(\tfrac{1}{2})(\text{R3})
\end{array}
$$

$$
\left[\begin{array}{ccc|c}
1 & 4 & 0 & -3 \\
0 & 0 & 1 & -1 \\
0 & 0 & 0 & 0
\end{array}\right]
\begin{array}{l}
5(\text{R2}) + (\text{R1}) \\
\\
(\text{R3}) - (\text{R2})
\end{array}
$$

This matrix is in echelon form. The variable y is free; x and z are basic. The top row gives $x + 4y = -3$ and the second row gives $z = -1$, so the general solution is

$$x = -3 - 4y$$
$$y = \text{any number}$$
$$z = -1$$

For $y = 0$ a particular solution is $(-3, 0, -1)$, and for $y = 1$ a particular solution is $(-7, 1, -1)$.

The discussion at the conclusion of Section 1–4 used the intersection of planes to show that a system of three equations in three variables will have exactly one solution, infinitely many solutions, or no solutions. Example 5 illustrates one such possibility.

EXAMPLE 5

Problem

Solve the system of linear equations

$$x - 2y + z = 4$$
$$3x + 4y - 2z = 2$$
$$2x + 6y - 3z = -4$$

SOLUTION

$$\begin{bmatrix} 1 & -2 & 1 & | & 4 \\ 3 & 4 & -2 & | & 2 \\ 2 & 6 & -3 & | & -4 \end{bmatrix}$$

$$\begin{bmatrix} 1 & -2 & 1 & | & 4 \\ 0 & 10 & -5 & | & -10 \\ 0 & 10 & -5 & | & -12 \end{bmatrix}$$

$$\begin{bmatrix} 1 & -2 & 1 & | & 4 \\ 0 & 1 & -\frac{1}{2} & | & -1 \\ 0 & 0 & 0 & | & 2 \end{bmatrix}$$

We stop the Gauss–Jordan method at this point because the equation corresponding to the bottom row is $0 = 2$, which is impossible. Therefore the system has no solution. Geometrically, this means that the three planes do not intersect because at least two of them are parallel.

EXAMPLE 6

A hospital nutritionist prepares a menu for patients on a low-fat, high-protein diet. The primary ingredients are chicken and rice. The meal must contain exactly 6 grams (g) of fat, 16.5 milligrams (mg) of iron, and 550 calories (cal). Each 3-ounce serving of chicken contains 4 g of fat, 9 mg of iron, and 500 cal. Each 1-cup serving of rice contains 2 g of fat, 6 mg of iron, and 150 cal.

Problem

How many servings of chicken and rice should be included in the meal?

SOLUTION Let x be the number of servings of chicken and y be the number of servings of rice. Then

$$\begin{aligned}
4x + 2y &= 6 && \text{(fat)} \\
9x + 6y &= 16.5 && \text{(iron)} \\
500x + 150y &= 550 && \text{(calories)}
\end{aligned}$$

The augmented matrix is

$$\left[\begin{array}{cc|c} 4 & 2 & 6 \\ 9 & 6 & 16.5 \\ 500 & 150 & 550 \end{array}\right]$$

The steps in the reduction to echelon form are

$$\left[\begin{array}{cc|c} 1 & 0.5 & 1.5 \\ 9 & 6 & 16.5 \\ 10 & 3 & 11 \end{array}\right] \quad \begin{array}{l} (\tfrac{1}{4})(R1) \\ \\ (\tfrac{1}{50})(R3) \end{array}$$

$$\left[\begin{array}{cc|c} 1 & 0.5 & 1.5 \\ 0 & 1.5 & 3 \\ 0 & -2 & -4 \end{array}\right] \quad \begin{array}{l} \\ (-9)(R1) + (R2) \\ (-10)(R1) + (R3) \end{array}$$

$$\left[\begin{array}{cc|c} 1 & 0.5 & 1.5 \\ 0 & 1 & 2 \\ 0 & 1 & 2 \end{array}\right] \quad \begin{array}{l} \\ (\tfrac{2}{3})(R2) \\ (-\tfrac{1}{2})(R3) \end{array}$$

$$\left[\begin{array}{cc|c} 1 & 0 & 0.5 \\ 0 & 1 & 2 \\ 0 & 0 & 0 \end{array}\right] \quad \begin{array}{l} (-0.5)(R2) + (R1) \\ \\ (R2) - (R3) \end{array}$$

Then $x = 0.5$ and $y = 2$. The meal should consist of a half serving of chicken and two servings of rice.

EXERCISE SET 2–3

• In Problems 1 to 6 use the Gauss–Jordan method to find the unique solution. Make sure to check your answers.

1. $\begin{aligned} x + y - z &= 2 \\ 2x + 3y - z &= 6 \\ 3x + 2y - z &= 7 \end{aligned}$

2. $\begin{aligned} x - 2y - z &= -3 \\ 2x - 3y &= -2 \\ -x + y + z &= 1 \end{aligned}$

3. $\begin{aligned} 2x - y + 3z &= 5 \\ x - y - z &= 2 \\ -3x - 2z &= -9 \end{aligned}$

4. $\begin{aligned} 2x + 4y - 2z &= -2 \\ 3x + 5y - z &= 3 \\ -2x + 3y - z &= -7 \end{aligned}$

5. $\begin{aligned} 3x - 6y + 9z &= -24 \\ 4x + 2y &= -2 \\ -x + y - 3z &= 5 \end{aligned}$

6. $\begin{aligned} -x + y + 2z &= 5 \\ x + y + z &= 3 \\ 2x + 3y + z &= 5 \end{aligned}$

In Problems 7 to 10 the solution of the corresponding system with variables x, y, and z can be determined directly from the given augmented matrix. Find the solution by thinking, not writing.

7. $\begin{bmatrix} 1 & 0 & 0 & | & 3 \\ 0 & 1 & 0 & | & -1 \\ 0 & 0 & 1 & | & 6 \end{bmatrix}$

8. $\begin{bmatrix} 1 & 0 & 0 & | & -2 \\ 0 & 0 & 1 & | & 3 \\ 0 & 1 & 0 & | & 0 \end{bmatrix}$

9. $\begin{bmatrix} 1 & 0 & 0 & | & 6 \\ 0 & 0 & 1 & | & 4 \\ 0 & 0 & 0 & | & 0 \end{bmatrix}$

10. $\begin{bmatrix} 1 & 0 & -2 & | & 1 \\ 0 & 1 & 17 & | & 2 \\ 0 & 0 & 1 & | & 0 \end{bmatrix}$

In Problems 11 to 14 find the general solution to the corresponding system whose variables are x, y, and z. List two particular solutions. Note that Problem 11 completes Example 2.

11. $\begin{bmatrix} 1 & 0 & 3 & | & -2 \\ 0 & 1 & 1 & | & -2 \\ 0 & 0 & 0 & | & 0 \end{bmatrix}$

12. $\begin{bmatrix} 1 & 2 & 0 & | & 5 \\ 0 & 0 & 1 & | & 4 \\ 0 & 0 & 0 & | & 0 \end{bmatrix}$

13. $\begin{bmatrix} 1 & 0 & 0 & | & 4 \\ 0 & 1 & -1 & | & 2 \\ 0 & 0 & 0 & | & 0 \end{bmatrix}$

14. $\begin{bmatrix} 1 & 2 & 3 & | & 12 \\ 0 & 0 & 0 & | & 0 \\ 0 & 0 & 0 & | & 0 \end{bmatrix}$

In Problems 15 to 18 find the general solution to the corresponding system whose variables are x, y, u, and v. Determine which variables are basic and which are free.

15. $\begin{bmatrix} 1 & 1 & 0 & 2 & | & 1 \\ 0 & 0 & 1 & 3 & | & 0 \end{bmatrix}$

16. $\begin{bmatrix} 1 & 1 & 0 & 0 & | & 4 \\ 0 & 0 & 1 & 0 & | & 3 \\ 0 & 0 & 0 & 1 & | & 2 \end{bmatrix}$

17. $\begin{bmatrix} 1 & 0 & 0 & 1 & | & 0 \\ 0 & 1 & 0 & 2 & | & 5 \\ 0 & 0 & 1 & 3 & | & 4 \end{bmatrix}$

18. $\begin{bmatrix} 1 & 2 & 3 & 0 & | & 0 \\ 0 & 0 & 0 & 1 & | & 5 \end{bmatrix}$

In Problems 19 to 22 determine whether each system has either no solution or infinitely many solutions. If there are infinitely many solutions, derive the general form and list the basic variables.

19. $\begin{aligned} x - y + 2z &= 4 \\ 2x - y - 2z &= -1 \\ 8x - 6y + 4z &= 14 \end{aligned}$

20. $\begin{aligned} x - y + 2z &= 4 \\ 2x - y - 2z &= -1 \\ 8x - 6y + 4z &= 0 \end{aligned}$

21. $\begin{aligned} -x - 2y + 2z &= 4 \\ 2x + 3y - 4z &= 7 \\ -x - 3y + 2z &= 19 \end{aligned}$

22. $\begin{aligned} -x - 2y + 2z &= 4 \\ 2x + 3y - 4z &= 7 \\ -x - 3y + 2z &= 0 \end{aligned}$

23. A hospital nutritionist prepares a menu of chicken and rice for patients on a low-fat, high-protein diet. It must contain exactly 12 grams (g) of fat, 9 milligrams (mg) of iron, and 1100 calories (cal). Each 3-ounce serving of chicken contains 3 g of fat, 2 mg of iron, and 200 cal. Each 1-cup serving of rice contains 6 g of fat, 5 mg of iron, and 700 cal. How many servings of chicken and rice should be included in the menu?

24. A hospital nutritionist prepares a menu of chicken and rice for patients on a low-fat, high-protein diet. It must contain exactly 16 g of fat, 44 mg of iron, and 1000 cal. Each 3-ounce serving of chicken contains 2 g of fat, 5 mg of iron, and 100 cal. Each 1-cup serving of rice contains 6 g of fat, 17 mg of iron, and 400 cal. How many servings of chicken and rice should be included in the menu?

•• Solve the systems of equations in Problems 25 to 34 by the Gauss–Jordan method. If there are infinitely many solutions, find the general form and two particular solutions.

25. $\begin{aligned} 2x + y - 2z &= 10 \\ 3x + 2y + 2z &= 1 \\ 5x + 4y + 3z &= 4 \end{aligned}$

26. $2x + y - 2z = 0$
    $3x + 2y + 2z = 0$
    $5x + 4y + 3z = 0$

27. $x - 3y + 5z = -17$
    $3x - 8y + 13z = -47$
    $-2x + 7y - 12z = 38$

28. $2x + 5y - z = 8$
    $x + 2y - z = 3$
    $-3x - 5y + 4z = -7$

29. $2x + y - 3z = 5$
    $3x - 2y + 2z = 5$
    $5x - 3y - z = 16$

30. $3x - 3y - 5z = 9$
    $x - y - 2z = 4$
    $-2x + 2y + z = 1$

31. $4x - 3y + 8z = 8$
    $x - y + 2z = 1$
    $4x - 5y + 8z = 0$

32. $2x + 4y + 6z = 2$
    $x - 2y - 5z = 4$
    $x + 2y + 3z = 0$

33. $3x - 6y - 9z = 12$
    $x + 2y - 5z = 4$
    $x - 2y - 3z = 0$

34. $-2x - 3y + 11z = -8$
    $-x - 4y + 18z = -4$
    $x + 2y - 8z = 4$

35. A hospital nutritionist prepares a menu of chicken and rice for patients on a low-fat, high-protein diet. It must contain exactly 5 g of fat, 11 mg of iron, and 1400 cal. Each 3-ounce serving of chicken contains 1 g of fat, 2 mg of iron, and 200 cal. Each 1-cup serving of rice contains 3 g of fat, 7 mg of iron, and 1000 cal. How many servings of chicken and rice should be included in the menu?

36. A hospital nutritionist prepares a menu of chicken and rice for patients on a low-fat, high-protein diet. It must contain exactly 7 g of fat, 30 mg of iron, and 1500 cal. Each 3-ounce serving of chicken contains 1 g of fat, 4 mg of iron, and 100 cal. Each 1-cup serving of rice contains 2 g of fat, 9 mg of iron, and 600 cal. How many servings of chicken and rice should be included in the menu?

37. A computer company makes three kinds of disk drives: hard, 1.2MB floppy, and 360K floppy. Each drive must be manufactured and tested. The hard drives require 3 hours of manufacturing and 1 hour of testing, 1.2MB drives require 6 hours of manufacturing and 2 hours of testing, and the 360K drives require 7 hours of manufacturing and 2 hours of testing. Manufacturing is possible exactly 67 hours per week and testing 20 hours per week. How many drives can be made per week?

38. An automobile plant makes cars in three price ranges: low (L), medium (M), and high (H). L-cars require 1 minute on the assembly line and 2 minutes of finishing; M-cars require 2 minutes on the assembly line and 3 minutes of finishing, and H-cars require 3 minutes on the assembly line and 8 minutes of finishing. The assembly line is available for exactly 63 minutes and finishing is available for 130 minutes. How many cars in each price range can be made?

●●● 39. Let a, b, c, and d be real numbers with $ad - bc \neq 0$. Reduce this matrix to echelon form.

$$\begin{bmatrix} a & b & 1 & 0 \\ c & d & 0 & 1 \end{bmatrix}$$

40. In Problem 37, what is the largest number of drives that can be made per week? What is the smallest number? (Assume that fractional parts of a drive cannot be made.)

41. Cost accounting is concerned with determining the cost of each department in a firm so that the firm can carry out a policy of setting prices, hiring or firing personnel, adding or eliminating departments, and preparing external reports.* Consider a firm with four departments: $A =$

*Adapted from the article by Roland Minch and Enrico Petri, "Matrix Models of Reciprocal Service Cost Allocation," *The Accounting Review*, July 1972, pp. 576–580.

accounting, M = maintenance, P = production, and F = finishing. Find the total costs x, y, z, w for each department, respectively, where

$$
\begin{aligned}
x &= 0.1y + 9{,}000 \\
y &= 0.2x + 6{,}000 \\
z &= 0.4x + 0.4y + 20{,}000 \\
w &= 0.4x + 0.5y + 16{,}000
\end{aligned}
$$

42. If a bettor in a pari-mutuel wagering system (such as horse racing or dog racing) seeks to win \$$w$ by betting on five horses whose odds of winning are a:1, b:1, c:1, d:1, and e:1,

then the amounts of money which should be bet on each horse are solutions to this system of equations.†

$$
\begin{aligned}
ax - y - z - r - s &= w \\
-x + by - z - r - s &= w \\
-x - y + cz - r - s &= w \\
-x - y - z + dr - s &= w \\
-x - y - z - r + es &= w
\end{aligned}
$$

If the odds on five particular horses are 2:1, 4:1, 7:1, 10:1, and 12:1, how much money should be bet on each horse in order to win \$100?

†The terminology and appropriate reference were given in Problem 41 of Section 1–4.

2–4 THE INVERSE

If you drive from Los Angeles to San Francisco and then return by retracing your route, the return trip is called the inverse route. More generally, any time you perform a procedure and then reverse it in such a way that you retrace the steps and end where you had begun, the reverse procedure is called the inverse. This section introduces the inverse of a matrix and describes a procedure for computing it. *All matrices in this section will be square.*

Invertible Matrices

Consider these 3×3 matrices.

$$
A = \begin{bmatrix} 1 & 0 & 8 \\ 1 & 2 & 3 \\ 2 & 5 & 3 \end{bmatrix}
\qquad
B = \begin{bmatrix} 9 & -40 & 16 \\ -3 & 13 & -5 \\ -1 & 5 & -2 \end{bmatrix}
$$

They satisfy $AB = I_3$ and $BA = I_3$, so B is called the *inverse* of A. Also, A is the inverse of B. In short, A and B are inverses of each other. The display contains the basic definitions.

Definitions

An $n \times n$ matrix A is **invertible** (or **nonsingular**) if there is an $n \times n$ matrix B such that

$$ AB = I_n \qquad \text{and} \qquad BA = I_n $$

B is called the **inverse** of A and is denoted by A^{-1} (read "A inverse"). Thus

$$ AA^{-1} = A^{-1}A = I_n $$

If A does not have an inverse, then A is called **noninvertible** (or **singular**).

An invertible matrix has only one inverse. Equivalently, *if a matrix has an inverse, then the inverse is unique.* The proof of this assertion is left to the exercises. The preceding 3×3 matrix A is invertible and its inverse is $A^{-1} = B$.

EXAMPLE 1 | Consider the matrix

$$A = \begin{bmatrix} 1 & 2 \\ 3 & 4 \end{bmatrix}$$

Problem

Which of these matrices is the inverse of A?

$$R = \begin{bmatrix} 4 & -2 \\ -3 & 1 \end{bmatrix} \qquad S = \begin{bmatrix} 1 & -2 \\ 0 & 1 \end{bmatrix} \qquad T = \begin{bmatrix} -2 & 1 \\ \frac{3}{2} & -\frac{1}{2} \end{bmatrix}$$

SOLUTION Form the matrix product AR.

$$AR = \begin{bmatrix} -2 & 0 \\ 0 & -2 \end{bmatrix}$$

Thus R is not the inverse of A.

The initial computations for AS seem to indicate that S is the inverse of A because

$$AS = \begin{bmatrix} 1 & 0 \\ * & * \end{bmatrix}$$

However, the $(2, 1)$ entry is equal to 3, not 0, so S is not the inverse of A.

The matrix T is the inverse of A because $AT = I$ and $TA = I$.

In Example 1 two equalities were needed to verify that T is the inverse of A: $AT = I$ and $TA = I$. However, only one of these equalities needs to hold, because if X and Y are $n \times n$ matrices with $XY = I_n$, then it follows that $YX = I_n$. The proof of this fact is beyond the scope of the book.

Gauss–Jordan Elimination

The definition of the inverse of a matrix does not indicate how to construct it. The procedure can be seen by examining a specific matrix A.

$$A = \begin{bmatrix} 1 & 0 & 8 \\ 1 & 2 & 3 \\ 2 & 5 & 3 \end{bmatrix}$$

Let the inverse of A be

$$A^{-1} = \begin{bmatrix} a & b & c \\ d & e & f \\ g & h & i \end{bmatrix}$$

By definition $AA^{-1} = I$, so

$$\begin{bmatrix} 1 & 0 & 8 \\ 1 & 2 & 3 \\ 2 & 5 & 3 \end{bmatrix} \begin{bmatrix} a & b & c \\ d & e & f \\ g & h & i \end{bmatrix} = \begin{bmatrix} 1 & 0 & 0 \\ 0 & 1 & 0 \\ 0 & 0 & 1 \end{bmatrix}$$

This equality leads to three systems of three linear equations in three variables.

$$\begin{array}{lll}
a \qquad\quad + 8g = 1 & b \qquad\quad + 8h = 0 & c \qquad\quad + 8i = 0 \\
a + 2d + 3g = 0 & b + 2e + 3h = 1 & c + 2f + 3i = 0 \\
2a + 5d + 3g = 0 & 2b + 5e + 3h = 0 & 2c + 5f + 3i = 1
\end{array}$$

Since all three systems have the same coefficient matrix, they can be solved simultaneously by augmenting the coefficient matrix with the identity matrix I_3 to form a 3×6 matrix. We state the procedure for finding the inverse of a matrix, then complete the details in Example 2.

Procedure for Finding the Inverse of an $n \times n$ Matrix M

1. Form the $n \times 2n$ matrix $[M \mid I]$.
2. Reduce it to echelon form $[X \mid Y]$, making sure to perform the same row operations on the augmented matrix I.
3. If X is the identity matrix, then Y is the inverse of M. If X is not the identity matrix, then M is not invertible.

EXAMPLE 2

Problem

Find the inverse of the matrix A, where

$$A = \begin{bmatrix} 1 & 0 & 8 \\ 1 & 2 & 3 \\ 2 & 5 & 3 \end{bmatrix}$$

SOLUTION First form the 3×6 matrix

$$\left[\begin{array}{ccc|ccc} 1 & 0 & 8 & 1 & 0 & 0 \\ 1 & 2 & 3 & 0 & 1 & 0 \\ 2 & 5 & 3 & 0 & 0 & 1 \end{array} \right]$$

The vertical bar separates the entries of A from the identity matrix. Next perform a series of row operations to reduce A to the identity matrix while carrying out the same operations on the right-hand side.

$$\left[\begin{array}{ccc|ccc} 1 & 0 & 8 & 1 & 0 & 0 \\ 0 & 2 & -5 & -1 & 1 & 0 \\ 0 & 5 & -13 & -2 & 0 & 1 \end{array} \right] \qquad \begin{array}{l} (-1)(\text{R}1) + (\text{R}2) \\ (-2)(\text{R}1) + (\text{R}3) \end{array}$$

$$\left[\begin{array}{ccc|ccc} 1 & 0 & 8 & 1 & 0 & 0 \\ 0 & 1 & -\frac{5}{2} & -\frac{1}{2} & \frac{1}{2} & 0 \\ 0 & 5 & -13 & -2 & 0 & 1 \end{array} \right] \qquad (\tfrac{1}{2})(\text{R}2)$$

$$\left[\begin{array}{ccc|ccc} 1 & 0 & 8 & 1 & 0 & 0 \\ 0 & 1 & -\frac{5}{2} & -\frac{1}{2} & \frac{1}{2} & 0 \\ 0 & 0 & -\frac{1}{2} & \frac{1}{2} & -\frac{5}{2} & 1 \end{array}\right] \qquad (-5)(R2) + (R3)$$

$$\left[\begin{array}{ccc|ccc} 1 & 0 & 8 & 1 & 0 & 0 \\ 0 & 1 & -\frac{5}{2} & -\frac{1}{2} & \frac{1}{2} & 0 \\ 0 & 0 & 1 & -1 & 5 & -2 \end{array}\right] \qquad (-2)(R3)$$

$$\left[\begin{array}{ccc|ccc} 1 & 0 & 0 & 9 & -40 & 16 \\ 0 & 1 & 0 & -3 & 13 & -5 \\ 0 & 0 & 1 & -1 & 5 & -2 \end{array}\right] \qquad \begin{array}{l}(-8)(R3) + (R1) \\ (\frac{5}{2})(R3) + (R2)\end{array}$$

The inverse of A is

$$A^{-1} = \begin{bmatrix} 9 & -40 & 16 \\ -3 & 13 & -5 \\ -1 & 5 & -2 \end{bmatrix}$$

Example 3 shows that the Gauss–Jordan method of constructing an inverse also detects noninvertible matrices. There are several other tests for invertibility, but this one produces the inverse at the same time, if it exists.

EXAMPLE 3

Problem

Find the inverse, if it exists, of the matrix M.

$$M = \begin{bmatrix} 1 & 2 & 3 \\ 4 & 5 & 6 \\ 7 & 8 & 9 \end{bmatrix}$$

SOLUTION

$$\left[\begin{array}{ccc|ccc} 1 & 2 & 3 & 1 & 0 & 0 \\ 4 & 5 & 6 & 0 & 1 & 0 \\ 7 & 8 & 9 & 0 & 0 & 1 \end{array}\right]$$

$$\left[\begin{array}{ccc|ccc} 1 & 2 & 3 & 1 & 0 & 0 \\ 0 & -3 & -6 & -4 & 1 & 0 \\ 0 & -6 & -12 & -7 & 0 & 1 \end{array}\right] \qquad \begin{array}{l}(-4)(R1) + (R2) \\ (-7)(R1) + (R3)\end{array}$$

$$\left[\begin{array}{ccc|ccc} 1 & 2 & 3 & 1 & 0 & 0 \\ 0 & -3 & -6 & -4 & 1 & 0 \\ 0 & 0 & 0 & 1 & -2 & 1 \end{array}\right] \qquad (-2)(R2) + (R3)$$

The matrix M cannot be reduced to I on the left-hand side of the vertical bar because of the row of 0's. Consequently M does not have an inverse.

The 2 × 2 Case

There is a separate procedure for finding the inverse of a 2 × 2 matrix that is much easier than the Gauss–Jordan method. Its verification is left for the exercises. Because the inverse of a 2 × 2 matrix is frequently needed, we single out the construction. Unfortunately the procedures for larger matrices are much more complicated.

Consider the matrix A from Example 1.

$$A = \begin{bmatrix} 1 & 2 \\ 3 & 4 \end{bmatrix}$$

The number $D = 1 \cdot 4 - 3 \cdot 2 = -2$ is called the **determinant** of A.

Procedure for Forming the Inverse of $M = \begin{bmatrix} a & b \\ c & d \end{bmatrix}$

1. Let $D = ad - bc$. If $D = 0$, then M is not invertible. If $D \neq 0$, then perform steps 2, 3, and 4.
2. Interchange a and d and negate b and c.
3. Multiply this matrix by the number $1/D$.
4. The inverse of M is

$$M^{-1} = \frac{1}{D} \begin{bmatrix} d & -b \\ -c & a \end{bmatrix}$$

EXAMPLE 4

Problem

What is the inverse of $M = \begin{bmatrix} -2 & -4 \\ 3 & 5 \end{bmatrix}$?

SOLUTION

$$M^{-1} = \frac{1}{(-2)(5) - (3)(-4)} \begin{bmatrix} 5 & 4 \\ -3 & -2 \end{bmatrix} = \frac{1}{2} \begin{bmatrix} 5 & 4 \\ -3 & -2 \end{bmatrix}$$

$$= \begin{bmatrix} \frac{5}{2} & 2 \\ -\frac{3}{2} & -1 \end{bmatrix}$$

The Inverse of a Product

If A and B are $n \times n$ matrices, then so is the product AB. Suppose that A and B are invertible. Then $(AB)(B^{-1}A^{-1}) = A(BB^{-1})A^{-1} = AI_nA^{-1} = AA^{-1} = I_n$. This means that AB is invertible. Since the inverse is unique, the inverse of AB is given by the formula

$$(AB)^{-1} = B^{-1}A^{-1}$$

Example 5 shows how to verify this formula. It will serve as a model for verifying other properties of the inverse.

EXAMPLE 5

Problem

Verify that $(AB)^{-1} = B^{-1}A^{-1}$, where

$$A = \begin{bmatrix} 1 & 2 \\ 3 & 4 \end{bmatrix} \qquad \text{and} \qquad B = \begin{bmatrix} -2 & -4 \\ 3 & 5 \end{bmatrix}$$

SOLUTION Form AB by matrix multiplication.

$$AB = \begin{bmatrix} 4 & 6 \\ 6 & 8 \end{bmatrix}$$

The inverse of AB is

$$(AB)^{-1} = -\tfrac{1}{4} \begin{bmatrix} 8 & -6 \\ -6 & 4 \end{bmatrix}$$

A^{-1} and B^{-1} were computed in Examples 1 and 4.

$$B^{-1} = \begin{bmatrix} \tfrac{5}{2} & 2 \\ -\tfrac{3}{2} & -1 \end{bmatrix} \quad \text{and} \quad A^{-1} = \begin{bmatrix} -2 & 1 \\ \tfrac{3}{2} & -\tfrac{1}{2} \end{bmatrix}$$

Multiply B^{-1} times A^{-1} in that order.

$$B^{-1}A^{-1} = \begin{bmatrix} -2 & \tfrac{3}{2} \\ \tfrac{3}{2} & -1 \end{bmatrix}$$

Note that $(AB)^{-1} = B^{-1}A^{-1}$.

A common mistake is to write $(AB)^{-1} = A^{-1}B^{-1}$. From Example 5 we can see that this equality does not always hold, since

$$A^{-1}B^{-1} = \begin{bmatrix} -\tfrac{13}{2} & -5 \\ \tfrac{9}{2} & \tfrac{7}{2} \end{bmatrix}$$

Note that $(AB)^{-1} \neq A^{-1}B^{-1}$.

EXERCISE SET 2–4

• In Problems 1 to 4 a matrix A is given. Determine which of the matrices X, Y, or Z is the inverse of A.

1. $A = \begin{bmatrix} 1 & 2 \\ 3 & 4 \end{bmatrix} \quad X = \begin{bmatrix} 1 & -2 \\ 0 & 1 \end{bmatrix}$

$Y = \begin{bmatrix} 4 & -2 \\ -3 & 1 \end{bmatrix} \quad Z = \begin{bmatrix} -2 & 1 \\ \tfrac{3}{2} & -\tfrac{1}{2} \end{bmatrix}$

2. $A = \begin{bmatrix} 4 & 3 \\ 3 & 2 \end{bmatrix} \quad X = \begin{bmatrix} 1 & 3 \\ -1 & -4 \end{bmatrix}$

$Y = \begin{bmatrix} 2 & -3 \\ -3 & 4 \end{bmatrix} \quad Z = \begin{bmatrix} -2 & 3 \\ 3 & -4 \end{bmatrix}$

3. $A = \begin{bmatrix} 1 & 0 & 0 \\ 0 & 2 & 0 \\ 0 & 0 & 3 \end{bmatrix} \quad X = \begin{bmatrix} 1 & 1 & 2 \\ 0 & -1 & 0 \\ 0 & 0 & -1 \end{bmatrix}$

$Y = \begin{bmatrix} 1 & 0 & 0 \\ 0 & \tfrac{1}{2} & 0 \\ 0 & 0 & \tfrac{1}{3} \end{bmatrix} \quad Z = \begin{bmatrix} 1 & 0 & 0 \\ 0 & \tfrac{1}{2} & 0 \\ 1 & 0 & \tfrac{1}{3} \end{bmatrix}$

4. $A = \begin{bmatrix} 1 & 0 & 1 \\ 0 & 1 & 0 \\ 0 & 0 & 1 \end{bmatrix} \quad X = \begin{bmatrix} 1 & 0 & 1 \\ 0 & 1 & 0 \\ 0 & 0 & -1 \end{bmatrix}$

$Y = \begin{bmatrix} 1 & 0 & -1 \\ 0 & 1 & 0 \\ 0 & 0 & 1 \end{bmatrix} \quad Z = \begin{bmatrix} 1 & 0 & 1 \\ 0 & 1 & 0 \\ -1 & 0 & -1 \end{bmatrix}$

In Problems 5 to 10 determine whether the given matrix is invertible and find its inverse if it is.

5. $\begin{bmatrix} 3 & 1 \\ 7 & 2 \end{bmatrix}$

6. $\begin{bmatrix} 8 & 4 \\ 4 & 2 \end{bmatrix}$

7. $\begin{bmatrix} 2 & 3 \\ 4 & 6 \end{bmatrix}$

8. $\begin{bmatrix} 5 & 4 \\ 4 & 3 \end{bmatrix}$

9. $\begin{bmatrix} 1 & -2 \\ -2 & 8 \end{bmatrix}$ 10. $\begin{bmatrix} 1 & -2 \\ 3 & 8 \end{bmatrix}$ $\begin{bmatrix} a & 0 & 0 \\ 0 & b & 0 \\ 0 & 0 & c \end{bmatrix}$

In Problems 11 to 18 determine whether the given matrix is invertible. If it is, compute its inverse.

11. $\begin{bmatrix} 1 & 1 & 1 \\ 2 & 1 & 1 \\ 2 & 2 & 3 \end{bmatrix}$ 12. $\begin{bmatrix} 1 & -1 & 0 \\ 0 & 1 & 1 \\ 2 & 0 & 1 \end{bmatrix}$

13. $\begin{bmatrix} 1 & 3 & 2 \\ 1 & 3 & 3 \\ 2 & 7 & 8 \end{bmatrix}$

14. $\begin{bmatrix} 1 & 4 & 2 \\ -2 & -8 & -3 \\ 0 & 1 & 1 \end{bmatrix}$

15. $\begin{bmatrix} 1 & -2 & -3 \\ 0 & 1 & 4 \\ 0 & 0 & 1 \end{bmatrix}$ 16. $\begin{bmatrix} 1 & -5 & 12 \\ 0 & 1 & -2 \\ 0 & 0 & 1 \end{bmatrix}$

17. $\begin{bmatrix} 0 & -3 & 0 \\ 2 & 0 & 0 \\ 0 & 0 & 4 \end{bmatrix}$ 18. $\begin{bmatrix} 0 & 0 & 2 \\ 0 & 1 & 0 \\ -1 & 0 & 1 \end{bmatrix}$

In Problems 19 and 20 two matrices A and B are given. Verify that $(AB)^{-1} = B^{-1}A^{-1}$.

19. $A = \begin{bmatrix} 5 & -2 \\ -3 & 1 \end{bmatrix}$ $B = \begin{bmatrix} 3 & 1 \\ 7 & 3 \end{bmatrix}$

20. $A = \begin{bmatrix} 7 & 3 \\ 5 & 2 \end{bmatrix}$ $B = \begin{bmatrix} 5 & -2 \\ -4 & 2 \end{bmatrix}$

•• In Problems 21 to 23 determine whether the given matrix is invertible, and find its inverse if it is.

21. $\begin{bmatrix} 0 & -1 \\ 1 & 4 \end{bmatrix}$ 22. $\begin{bmatrix} 2 & 0 \\ 0 & -4 \end{bmatrix}$ 23. $\begin{bmatrix} 1 & 3 \\ 0 & 0 \end{bmatrix}$

24. Let A be a matrix that has a row with all entries equal to zero. Can A be invertible? (*Hint:* See Problem 23.)

25. Compute the inverse of this matrix, if possible.

$$\begin{bmatrix} 1 & 0 & 0 \\ 0 & 3 & 0 \\ 0 & 0 & -1 \end{bmatrix}$$

26. Use the result of Problem 25 to compute the inverse of the following matrix, where $a \neq 0$, $b \neq 0$, and $c \neq 0$:

In Problems 27 and 28 find the inverse of the 4×4 matrix by the method for 3×3 matrices shown in Example 2.

27. $\begin{bmatrix} 1 & -2 & 3 & 1 \\ 0 & 1 & 2 & -4 \\ 0 & 0 & 1 & 0 \\ 0 & 0 & 0 & 1 \end{bmatrix}$ 28. $\begin{bmatrix} 1 & 2 & 5 & -8 \\ 0 & 1 & 3 & 0 \\ 0 & 0 & 1 & 2 \\ 0 & 0 & 0 & 1 \end{bmatrix}$

Let A be an invertible matrix. Then the inverse of A^{-1} is denoted by $(A^{-1})^{-1}$. In Problems 29 to 32 find A^{-1} and $(A^{-1})^{-1}$.

29. $A = \begin{bmatrix} 3 & 1 \\ 7 & 2 \end{bmatrix}$ 30. $A = \begin{bmatrix} 5 & 4 \\ 4 & 3 \end{bmatrix}$

31. $A = \begin{bmatrix} 1 & -2 \\ -2 & 8 \end{bmatrix}$ 32. $A = \begin{bmatrix} 1 & -2 \\ 3 & 8 \end{bmatrix}$

33. Use Problems 29 to 32 to discover a formula for $(A^{-1})^{-1}$.

34. Compute (a) $(A^{-1}B)^{-1}$ (b) $(AB^{-1}C)^{-1}$.

35. Find $(AB)^{-1}$ and AB, where

$$A^{-1} = \begin{bmatrix} 5 & -2 \\ -3 & 1 \end{bmatrix} \text{ and } B^{-1} = \begin{bmatrix} 3 & 1 \\ 7 & 3 \end{bmatrix}$$

36. Find A where

$$A^{-1} = \begin{bmatrix} 1 & 0 & 3 \\ 0 & 1 & 0 \\ 0 & 0 & 1 \end{bmatrix}$$

37. Let A and B be the following matrices. (a) Is B the inverse of A? (b) If not, what is?

$$A = \begin{bmatrix} -3 & 1 \\ 2 & 0 \end{bmatrix} \quad B = \begin{bmatrix} 0 & 1 \\ 2 & 3 \end{bmatrix}$$

38. Let A and B be the following matrices. (a) Is B the inverse of A? (b) If not, what is?

$$A = \begin{bmatrix} 4 & -6 & 2 \\ 6 & 3 & -3 \\ -14 & 3 & 5 \end{bmatrix}$$

$$B = \begin{bmatrix} 2 & 3 & 1 \\ 1 & 4 & 2 \\ 5 & 6 & 4 \end{bmatrix}$$

●●● 39. Prove that an invertible matrix A has exactly one inverse. (*Hint:* Assume there are two inverses, B and C. Substitute $CA = I$ into $B = IB$ and substitute $AB = I$ into $C = CI$ to show that $B = C$.)

40. Use the Gauss–Jordan method to compute the inverse of an arbitrary 2×2 matrix

$$A = \begin{bmatrix} a & b \\ c & d \end{bmatrix}$$

where $a \neq 0$ and $ad - bc \neq 0$.

41. Let A be the task of "putting on a sock" and B be the task of "putting on a shoe."
    (a) Interpret the meanings of A^{-1} and B^{-1}.
    (b) Interpret the meanings of AB and $(AB)^{-1}$.

(c) Explain why the right-hand side of the equation $(AB)^{-1} = B^{-1}A^{-1}$ is in the proper order.

42. In accounting, matrices are used to describe the financial interdependence of the departments within a firm. The inverse of a matrix is needed to compute the total cost of each department. Compute the inverse of the matrix M, which appeared in a classic paper in accounting.*

$$M = \begin{bmatrix} 1 & 0 & -0.05 & -0.10 & -0.20 \\ 0 & 1 & -0.10 & -0.05 & -0.20 \\ -0.10 & -0.10 & 1 & -0.05 & -0.20 \\ -0.05 & 0 & -0.10 & 1 & -0.20 \\ -0.10 & -0.10 & -0.05 & 0 & 1 \end{bmatrix}$$

*Thomas H. Williams and Charles H. Griffin, "Matrix Theory and Cost Allocation," *The Accounting Review*, July 1964, pp. 671–678.

2–5 SYSTEMS OF EQUATIONS

We have discussed two procedures for solving systems of linear equations, Gaussian elimination and the Gauss–Jordan method. This section describes an entirely different procedure, called the "inverse matrix method." Unlike the two elimination methods, the inverse matrix method applies only to square matrices. Therefore *all matrices in this section will be square matrices.*

The Matrix Equation

Let us return to the Pressman-Gutman example to show how matrix equations arise and how they can be solved. We need only consider the Fall River and Passaic finishing plants. Suppose the company determines that to break even the Fall River plant must produce $15 (= \$15,000,000)$ worth of material and the Passaic plant must produce $15.5 (= \$15,500,000)$. The problem is to find the prices to charge for cotton and synthetic material in order to break even. Let x be the price per yard of cotton and y the price per yard of synthetic material. To break even, x and y must satisfy the linear equations

$$2.0x + 4.5y = 15.0$$
$$2.5x + 4.0y = 15.5$$

Let A be the coefficient matrix, X the matrix of variables, and C the matrix of constants.

$$A = \begin{bmatrix} 2.0 & 4.5 \\ 2.5 & 4.0 \end{bmatrix} \qquad X = \begin{bmatrix} x \\ y \end{bmatrix} \qquad C = \begin{bmatrix} 15.0 \\ 15.5 \end{bmatrix}$$

Then the system of linear equations can be written as the **matrix equation**

$$AX = C$$

This matrix equation is similar to the numerical equation $ax = c$, where a and c are specified numbers and x is the number that is sought. Take a particular case.

$$2x = 5$$

It is solved by multiplying both sides by $\frac{1}{2}$, because $\frac{1}{2} = 2^{-1}$ is the multiplicative inverse of 2.

$$(\tfrac{1}{2})\cdot 2x = (\tfrac{1}{2})\cdot 5 = 2^{-1}\cdot 5$$
$$x = \tfrac{5}{2}$$

The solution to the matrix equation $AX = C$ is obtained in a similar way when the coefficient matrix A is invertible. The reasons are supplied in parentheses.

$$AX = C$$
$$A^{-1}(AX) = A^{-1}C \qquad \text{(multiply on the left by } A^{-1})$$
$$(A^{-1}A)X = A^{-1}C \qquad \text{(associative law)}$$
$$I_nX = A^{-1}C \qquad \text{(definition of } A^{-1})$$
$$X = A^{-1}C \qquad \text{(definition of } I_n)$$

If A is an invertible matrix, then the solution to the matrix equation

$$AX = C$$

is

$$X = A^{-1}C$$

If the matrix A is not invertible, then the equation $AX = C$ has either no solution or infinitely many solutions.

The solution by the matrix method involves finding the inverse and performing matrix multiplication. The multiplication is straightforward, but the procedure for computing the inverse is cumbersome for larger systems. Example 1 illustrates the matrix method when A^{-1} is given. Example 2 demonstrates the entire procedure from beginning to end.

EXAMPLE 1

Problem

Solve the Pressman-Gutman system of equations $AX = C$, where the inverse of the coefficient matrix A is

$$A^{-1} = \frac{-1}{3.25} \begin{bmatrix} 4.0 & -4.5 \\ -2.5 & 2.0 \end{bmatrix}$$

SOLUTION The solution to the matrix equation $AX = C$ is

$$X = A^{-1}C$$

$$\begin{bmatrix} x \\ y \end{bmatrix} = \frac{-1}{3.25} \begin{bmatrix} 4.0 & -4.5 \\ -2.5 & 2.0 \end{bmatrix} \begin{bmatrix} 15.0 \\ 15.5 \end{bmatrix}$$

$$\begin{bmatrix} x \\ y \end{bmatrix} = \begin{bmatrix} 3 \\ 2 \end{bmatrix}$$

Then $x = 3$ and $y = 2$. This solution can be checked by substituting directly into the original system of equations.

EXAMPLE 2

Problem

Solve the system of equations

$$x + 2y = 8$$
$$3x - 4y = -26$$

SOLUTION Write the system as a matrix equation.

$$\begin{bmatrix} 1 & 2 \\ 3 & -4 \end{bmatrix} \begin{bmatrix} x \\ y \end{bmatrix} = \begin{bmatrix} 8 \\ -26 \end{bmatrix}$$

The solution is

$$\begin{bmatrix} x \\ y \end{bmatrix} = \begin{bmatrix} 1 & 2 \\ 3 & -4 \end{bmatrix}^{-1} \begin{bmatrix} 8 \\ -26 \end{bmatrix}$$

$$\begin{bmatrix} x \\ y \end{bmatrix} = \left(\frac{1}{-10}\right) \begin{bmatrix} -4 & -2 \\ -3 & 1 \end{bmatrix} \begin{bmatrix} 8 \\ -26 \end{bmatrix}$$

$$\begin{bmatrix} x \\ y \end{bmatrix} = \left(\frac{1}{-10}\right) \begin{bmatrix} 20 \\ -50 \end{bmatrix}$$

$$\begin{bmatrix} x \\ y \end{bmatrix} = \begin{bmatrix} -2 \\ 5 \end{bmatrix}$$

Therefore $x = -2$ and $y = 5$.

The display assembles the four steps of the **inverse matrix method.**

Matrix Method for Solving a System of Equations

1. Translate the system into $AX = C$, where A is the coefficient matrix, X the matrix of variables, and C the matrix of constants.
2. Compute A^{-1} if A is invertible.
3. The solution is $X = A^{-1}C$.

Example 3 shows how to solve a system of three equations in three variables by the matrix inverse method.

EXAMPLE 3

Problem

Solve this system by the matrix method.

$$x + y + 2z = 4$$
$$2x + y + 2z = 3$$
$$2x \qquad + z = 2$$

SOLUTION Translate the system into $AX = C$, where

$$A = \begin{bmatrix} 1 & 1 & 2 \\ 2 & 1 & 2 \\ 2 & 0 & 1 \end{bmatrix} \qquad X = \begin{bmatrix} x \\ y \\ z \end{bmatrix} \qquad \text{and} \qquad C = \begin{bmatrix} 4 \\ 3 \\ 2 \end{bmatrix}$$

Compute A^{-1} by forming the 3×6 matrix $[A \mid I]$ and reducing A to the identity matrix while carrying out the same operations on the right-hand side.

$$\begin{bmatrix} 1 & 1 & 2 & 1 & 0 & 0 \\ 2 & 1 & 2 & 0 & 1 & 0 \\ 2 & 0 & 1 & 0 & 0 & 1 \end{bmatrix}$$

$$\begin{bmatrix} 1 & 1 & 2 & 1 & 0 & 0 \\ 0 & -1 & -2 & -2 & 1 & 0 \\ 0 & -2 & -3 & -2 & 0 & 1 \end{bmatrix} \qquad \begin{matrix} (-2)(R1) + (R2) \\ (-2)(R1) + (R3) \end{matrix}$$

$$\begin{bmatrix} 1 & 0 & 0 & -1 & 1 & 0 \\ 0 & 1 & 2 & 2 & -1 & 0 \\ 0 & 0 & 1 & 2 & -2 & 1 \end{bmatrix} \qquad \begin{matrix} (R1) + (R2) \\ (-1)(R2) \\ (-2)(R2) + (R3) \end{matrix}$$

$$\begin{bmatrix} 1 & 0 & 0 & -1 & 1 & 0 \\ 0 & 1 & 0 & -2 & 3 & -2 \\ 0 & 0 & 1 & 2 & -2 & 1 \end{bmatrix} \qquad (-2)(R3) + (R2)$$

The 3×3 matrix on the right-hand side is A^{-1}.

The solution to the original system of equations is $X = A^{-1}C$, so

$$\begin{bmatrix} x \\ y \\ z \end{bmatrix} = \begin{bmatrix} -1 & 1 & 0 \\ -2 & 3 & -2 \\ 2 & -2 & 1 \end{bmatrix} \begin{bmatrix} 4 \\ 3 \\ 2 \end{bmatrix} = \begin{bmatrix} -1 \\ -3 \\ 4 \end{bmatrix}$$

The solution is $x = -1$, $y = -3$, and $z = 4$.

Contrast in Methods

The two major ways to solve a system of n linear equations in n variables are elimination (Gaussian or Gauss–Jordan) and the inverse matrix method. Elimination can be applied to all systems, but the inverse matrix method applies only to square systems with an invertible coefficient matrix. Nonetheless there are times when the inverse matrix method is preferable.

Example 4 contrasts Gauss–Jordan elimination with the inverse matrix method. The coefficient matrix is very important in quantum mechanics and heat transfer.*

EXAMPLE 4

Problem

Solve the system of equations

$$
\begin{aligned}
2x - y &= -7 \\
-x + 2y - z &= 9 \\
-y + 2z &= -5
\end{aligned}
$$

SOLUTION (1) *Elimination*. Apply the Gauss–Jordan method by forming the augmented matrix and reducing it to echelon form.

$$
\left[\begin{array}{ccc|c}
2 & -1 & 0 & -7 \\
-1 & 2 & -1 & 9 \\
0 & -1 & 2 & -5
\end{array}\right]
$$

$$
\left[\begin{array}{ccc|c}
-1 & 2 & -1 & 9 \\
2 & -1 & 0 & -7 \\
0 & -1 & 2 & -5
\end{array}\right] \quad \text{(R1)} \leftrightarrow \text{(R2)}
$$

$$
\left[\begin{array}{ccc|c}
1 & -2 & 1 & -9 \\
0 & 3 & -2 & 11 \\
0 & -1 & 2 & -5
\end{array}\right] \quad \begin{array}{l} (-1)(\text{R1}) \\ 2(\text{R1}) + (\text{R2}) \end{array}
$$

$$
\left[\begin{array}{ccc|c}
1 & -2 & 1 & -9 \\
0 & -1 & 2 & -5 \\
0 & 3 & -2 & 11
\end{array}\right] \quad \text{(R2)} \leftrightarrow \text{(R3)}
$$

$$
\left[\begin{array}{ccc|c}
1 & 0 & -3 & 1 \\
0 & 1 & -2 & 5 \\
0 & 0 & 4 & -4
\end{array}\right] \quad \begin{array}{l} (-2)(\text{R2}) + (\text{R1}) \\ (-1)(\text{R2}) \\ 3(\text{R2}) + (\text{R3}) \end{array}
$$

$$
\left[\begin{array}{ccc|c}
1 & 0 & -3 & 1 \\
0 & 1 & -2 & 5 \\
0 & 0 & 1 & -1
\end{array}\right] \quad (\tfrac{1}{4})(\text{R3})
$$

$$
\left[\begin{array}{ccc|c}
1 & 0 & 0 & -2 \\
0 & 1 & 0 & 3 \\
0 & 0 & 1 & -1
\end{array}\right] \quad \begin{array}{l} 3(\text{R3}) + (\text{R1}) \\ 2(\text{R3}) + (\text{R2}) \end{array}
$$

The solution is $x = -2$, $y = 3$, and $z = -1$.

(2) *Inverse Matrix Method*. The first step is

$$
\begin{bmatrix} x \\ y \\ z \end{bmatrix} = \begin{bmatrix} 2 & -1 & 0 \\ -1 & 2 & -1 \\ 0 & -1 & 2 \end{bmatrix}^{-1} \begin{bmatrix} -7 \\ 9 \\ -5 \end{bmatrix}
$$

---

*See the article "What Is Quantum Field Theory?" by David C. Brydges in the journal *Bulletin of the American Mathematical Society*, January 1983, pp. 31–40. The augmented matrix is described on page 34.

Next perform steps similar to those in part (1) to find the inverse of the coefficient matrix. The solution is

$$\begin{bmatrix} x \\ y \\ z \end{bmatrix} = (\tfrac{1}{4}) \begin{bmatrix} 3 & 2 & 1 \\ 2 & 4 & 2 \\ 1 & 2 & 3 \end{bmatrix} \begin{bmatrix} -7 \\ 9 \\ -5 \end{bmatrix}$$

$$= (\tfrac{1}{4}) \begin{bmatrix} -8 \\ 12 \\ -4 \end{bmatrix} = \begin{bmatrix} -2 \\ 3 \\ -1 \end{bmatrix}$$

Once again, the solution is $x = -2$, $y = 3$, and $z = -1$.

In general, if one system is being solved it is better to use elimination. However, the inverse matrix method is preferable when there are many systems of equations having the same coefficients but different constants (on the right-hand side of the equations). For instance, the breakeven point for the Fall River and Passaic finishing plants is the solution to this system.

$$\begin{bmatrix} 2.0 & 4.5 \\ 2.5 & 4.0 \end{bmatrix} \begin{bmatrix} x \\ y \end{bmatrix} = \begin{bmatrix} 15.0 \\ 15.5 \end{bmatrix}$$

The numbers 15 and 15.5 represent the amounts of material that the two plants must produce. If a different amount of material has to be produced during the same month, then another matrix equation with the same coefficient matrix has to be solved. Example 5 illustrates how to handle this situation.

EXAMPLE 5 | **Problem**

Solve the system of equations

$$3x - 2y = a$$
$$-5x + 4y = b$$

for each of these pairs of values of a and b.

(1) $a = -1$ (2) $a = -7$ (3) $a = 12$
$\quad\; b = 3$ $b = 13$ $b = -20$

SOLUTION Rather than solve the three systems separately, apply the matrix method once. The solution to the general system is

$$\begin{bmatrix} x \\ y \end{bmatrix} = \begin{bmatrix} 3 & -2 \\ -5 & 4 \end{bmatrix}^{-1} \begin{bmatrix} a \\ b \end{bmatrix}$$

$$\begin{bmatrix} x \\ y \end{bmatrix} = \begin{bmatrix} 2 & 1 \\ \tfrac{5}{2} & \tfrac{3}{2} \end{bmatrix} \begin{bmatrix} a \\ b \end{bmatrix}$$

$$\begin{bmatrix} x \\ y \end{bmatrix} = \begin{bmatrix} 2a + b \\ (\tfrac{5}{2})a + (\tfrac{3}{2})b \end{bmatrix}$$

To obtain the solutions to the three specific systems, substitute their given values for a and b and set the corresponding entries equal. (1) In the first system $a = -1$ and $b = 3$, so

$$\begin{bmatrix} x \\ y \end{bmatrix} = \begin{bmatrix} 2(-1) + 3 \\ (\frac{5}{2})(-1) + (\frac{3}{2}) \cdot 3 \end{bmatrix} = \begin{bmatrix} 1 \\ 2 \end{bmatrix}$$

The solution is $x = 1$ and $y = 2$. (2) Similarly, the pair $a = -7$ and $b = 13$ yields $x = -1$ and $y = 2$, while (3) $a = 12$ and $b = -20$ yields $x = 4$ and $y = 0$.

EXERCISE SET 2–5

• In Problems 1 to 4 solve the system of linear equations by the inverse matrix method. Use the fact that the inverse of the coefficient matrix is given in Example 2 of Section 2–4.

1. $\begin{aligned} x + \quad\quad 8z &= -6 \\ x + 2y + 3z &= -3 \\ 2x + 5y + 3z &= -6 \end{aligned}$

2. $\begin{aligned} x + \quad\quad 8z &= 20 \\ x + 2y + 3z &= 9 \\ 2x + 5y + 3z &= 11 \end{aligned}$

3. $\begin{aligned} x + \quad\quad 8z &= 27 \\ x + 2y + 3z &= 9 \\ 2x + 5y + 3z &= 7 \end{aligned}$

4. $\begin{aligned} x + \quad\quad 8z &= 6 \\ x + 2y + 3z &= 3 \\ 2x + 5y + 3z &= 6 \end{aligned}$

In Problems 5 to 10 solve the system of linear equations by the inverse matrix method. Make use of the rule for computing the inverse of a 2×2 matrix described in Section 2–4.

5. $\begin{aligned} 2x + \ y &= 3 \\ 3x + 4y &= 7 \end{aligned}$ 6. $\begin{aligned} 2x + 5y &= 2 \\ 3x - 2y &= 3 \end{aligned}$

7. $\begin{aligned} 5x + 2y &= 1 \\ 2x - 3y &= 8 \end{aligned}$ 8. $\begin{aligned} 8x - 3y &= 1 \\ 3x + 4y &= 9 \end{aligned}$

9. $\begin{aligned} 2x - 3y &= 1 \\ 3x - 2y &= 3 \end{aligned}$ 10. $\begin{aligned} -4x + 2y &= -3 \\ 3x - 2y &= \ \ 3 \end{aligned}$

In Problems 11 to 14 follow the method shown in Example 3 to solve the system of linear equations.

11. $\begin{aligned} x + \ y + \ z &= -2 \\ 2x + \ y + \ z &= \ \ 3 \\ 2x + 2y + 3z &= -8 \end{aligned}$

12. $\begin{aligned} x - y \quad\quad &= \ \ 8 \\ y + z &= -7 \\ 2x + \quad z &= \ \ 6 \end{aligned}$

13. $\begin{aligned} x + 3y + 2z &= 9 \\ x + 3y + 3z &= 5 \\ 2x + 7y + 8z &= 7 \end{aligned}$

14. $\begin{aligned} x + 4y + 2z &= \ \ 19 \\ -2x - 8y - 3z &= -29 \\ y + \ z &= \ \ 10 \end{aligned}$

In Problems 15 to 18 solve the systems of linear equations for each pair of values of a and b.

15. $\begin{aligned} 5x + 2y &= a \\ 7x + 3y &= b \end{aligned}$
(a) $a = 13$ (b) $a = 5$ (c) $a = 6$
$\quad\ b = 19$ $\quad\quad b = 6$ $\quad\quad b = 9$

16. $\begin{aligned} 7x + 2y &= a \\ 3x + \ y &= b \end{aligned}$
(a) $a = 1$ (b) $a = 8$ (c) $a = -4$
$\quad\ b = 0$ $\quad\quad b = 4$ $\quad\quad b = -1$

17. $\begin{aligned} 5x + 3y &= a \\ 4x + 2y &= b \end{aligned}$
(a) $a = -7$ (b) $a = 17$ (c) $a = 21$
$\quad\ b = -6$ $\quad\quad b = 12$ $\quad\quad b = 12$

18. $\begin{aligned} 3x + 5y &= a \\ 2x + 2y &= b \end{aligned}$
(a) $a = -11$ (b) $a = -10$ (c) $a = 11$
$\quad\ b = \ \ -2$ $\quad\quad b = \ \ \ 0$ $\quad\quad b = 10$

•• In Problems 19 and 20 solve the systems of linear equations for each set of values of a, b, and c.

$$2x - y \quad\quad := a$$
$$-x + 2y - z = b$$
$$-y + 2z = c$$

19. (a) $a = 6, b = 3, c = -4$
    (b) $a = 11, b = 4, c = -17$
    (c) $a = 2, b = 0, c = -2$

20. (a) $a = 8, b = -9, c = 10$
    (b) $a = -11, b = 0, c = 13$
    (c) $a = -1, b = 0, c = -1$

In Problems 21 to 28 solve the system by the inverse matrix method if there is a unique solution.

21. $x + y + z = 3$
    $-x - 2y - 2z = -4$
    $4x + 3y + 4z = 9$

22. $2x \quad\quad + 4z = 0$
    $3x - 4y \quad\quad = 1$
    $-2y - 3z = 0$

23. $x + 3y + 2z = 20$
    $2x + 6y + 5z = 44$
    $x + 4y + 5z = 37$

24. $x - y \quad\quad = 12$
    $x \quad\quad + z = 13$
    $2x \quad\quad + z = 18$

25. $x - y \quad\quad = -3$
    $y + z = 7$
    $2x \quad\quad - z = 11$

26. $x + 4y + 2z = 11$
    $2x + 9y + 4z = 23$
    $-x - 4y - z = -6$

27. $2x \quad\quad + 4z = 1$
    $3x - 4y \quad\quad = 0$
    $-2y - 3z = 0$

28. $x + y + z = 1$
    $-2x - y - z = 0$
    $y + z = 2$

In Problems 29 and 30 use the results of Problems 27 and 28, respectively, of Section 2–4 to solve the system of four equations in four unknowns.

29. $x - 2y + 3z + t = -42$
    $y + 2z - 4t = 0$
    $z = -4$
    $t = 2$

30. $x + 2y + 5z - 8t = 8$
    $y + 3z = 9$
    $z + 2t = 1$
    $t = -1$

31. A company makes 120 suits of two types, wool and blend. It costs the company $100 to make each wool suit and $40 to make each blend suit. If the company allots exactly $6000 to make the suits, how many can it make of each type?

32. A company makes 450 pairs of denim and corduroy slacks. It costs the company $10 to make each pair of denim slacks and $25 to make each pair of corduroy slacks. If the company allots exactly $6750 to make the slacks, how many can it make of each type?

33. A person takes two daily vitamin supplements, C and E, via colored pills. Each red pill yields 45 units of C and 10 units of E. Each blue pill yields 9 units of C and 6 units of E. How many pills of each color must be taken daily to achieve 72 units of C and 28 units of E?

34. A person takes two daily vitamin supplements, C and E, via colored pills. Each red pill yields 90 units of C and 20 units of E. Each blue pill yields 36 units of C and 24 units of E. How many pills of each color must be taken daily to achieve 126 units of C and 44 units of E?

35. A transit company offers a discount on the first class fare, called an "executive" fare, to those passengers on a train who have no luggage. The train holds 190 passengers. The company is limited to a cost of $2250 per train, with the cost being $15 for each first class passenger, $15 for each executive passenger, and $5 for each economy class passenger. The company must reserve 10 cubic feet of storage for luggage for each first class passenger and 5 cubic feet of storage for each economy class passenger; 1350 cubic feet of storage must be used on each train. How many tickets of each type should be sold?

36. Repeat Problem 35 for a bus in which the costs and storage space remain the same, but the bus holds 45 passengers, the cost is $425 per bus, and the storage is 125 cubic feet.

37. Problem 41 in Section 2–3 introduced a cost accounting procedure for determining the total costs for each department in a firm with several interdependent departments. Solve that problem for x, y, z, w using matrix methods:*

$$\begin{bmatrix} x \\ y \end{bmatrix} = \begin{bmatrix} 0 & 0.1 \\ 0.2 & 0 \end{bmatrix} \begin{bmatrix} x \\ y \end{bmatrix} + \begin{bmatrix} 9000 \\ 6000 \end{bmatrix}$$

$$\begin{bmatrix} z \\ w \end{bmatrix} = \begin{bmatrix} 0.4 & 0.4 \\ 0.4 & 0.5 \end{bmatrix} \begin{bmatrix} x \\ y \end{bmatrix} + \begin{bmatrix} 20,000 \\ 16,000 \end{bmatrix}$$

38. Problem 42 in Section 2–4 asked for the inverse of the coefficient matrix of a system of linear equations derived from the system below. Use that result to solve the system, where each variable represents the total cost of one of the five service departments in a firm.†

$$
\begin{aligned}
x &= & 0.05z &+ 0.1u &+ 0.2v &+ & 8000 \\
y &= & 0.1z &+ 0.05u &+ 0.2v &+ & 12000 \\
z &= 0.1x + 0.1y & &+ 0.05u &+ 0.2v &+ & 6000 \\
u &= 0.05x & + 0.1z & &+ 0.2v &+ & 11000 \\
v &= 0.1x + 0.1y &+ 0.05z & & &+ & 13000
\end{aligned}
$$

39. A paper in accounting proposed an approximation method to the traditional reciprocal method of determining cost.‡ This new technique is much faster, and hence cheaper, to carry out. The matrices A and b are taken from that paper.

$$A = \begin{bmatrix} 0.05 & 0.15 & 0.15 \\ 0.10 & 0.10 & 0.15 \\ 0.20 & 0.20 & 0.05 \end{bmatrix} \quad b = \begin{bmatrix} 30,000 \\ 60,000 \\ 40,000 \end{bmatrix}$$

(a) Solve the matrix equation $(I_3 - A)x = b$ by the matrix inverse method (called the reciprocal method in accounting).

(b) Solve the matrix equation $(I_3 - A)x = b$ by the approximation method: $\hat{x} = 70{,}270y + b$, where the number $70{,}270$ is derived from the entries in A and b, and y is a column matrix whose entries are the sums of the rows of A.

(c) Find the relative error in each component of $\hat{x}$ as a percentage of each component of x.

40. Suppose that a bettor in a pari-mutuel wagering system seeks to win $\$w$ by dutching on two horses A and B, whose odds of winning are $a:1$ and $b:1$, respectively.§ Let $\$x$ be the amount of money which the bettor should bet on A and $\$y$ be the amount of money which the bettor should bet on B. Then x and y can be found by solving the matrix equation

$$\begin{bmatrix} a & -1 \\ -1 & b \end{bmatrix} \begin{bmatrix} x \\ y \end{bmatrix} = \begin{bmatrix} w \\ w \end{bmatrix}$$

If the odds on two particular horses are $2:1$ and $8:1$, how much money should be bet on each horse in order to win each of these amounts?
(a) $50   (b) $25   (c) $1000   (d) $5000

---

*Roland Minch and Enrico Petri, "Matrix Models of Reciprocal Service Cost Allocation," *The Accounting Review*, July 1972, pp. 576–580.

†Thomas H. Williams and Charles H. Griffin, "Matrix Theory and Cost Allocation," *The Accounting Review*, July 1964, pp. 671–678.

‡Frederick H. Jacobs and Ronald M. Marshall, "A Reciprocal Service Cost Approximation," *The Accounting Review*, January 1987, pp. 67–78.

§The terminology and appropriate references were given in Problem 41 of Section 1–4.

2–6 INPUT-OUTPUT ANALYSIS

Shortly after World War II broke out President Franklin Delano Roosevelt submitted an order for 50,000 airplanes, which required the country to produce more aluminum. But the potlines used to make aluminum required massive bus bars through which copper conducts electricity, and an unforeseen copper shortage threatened the entire production schedule. The government averted the crisis by substituting silver for copper. Where did they get all the silver? They borrowed it from Fort Knox.

The immediate result was that the 50,000 airplanes were built, but the end result was much more pervasive. It led to the construction of detailed tables analyzing the interdependence of various industries within the economy and to a new area of economics called "input-output analysis."

Input-output analysis has become one of the cornerstones of modern economics because it provides a picture of various interlocking parts of a given economy. The general pattern of VCRs, stereos, cars, soft drinks, insurance, and other items produced and consumed by Americans can be followed across the page of an input-output table. Such tables are of great use in economic forecasting. An input-output table constructed by the Department of Labor during World War II predicted correctly that, contrary to widespread opinion, there would not be a depression and massive unemployment at war's end. Today, industrial giants like ITT and General Motors, many smaller businesses, and some universities base their corporate policy on input-output analysis.

Input-Output Tables

At the heart of input-output analysis is an **input-output table,** which measures how much each industry depends on other industries for its production requirements. It is like a road map that allows you to read off distances from one city to another. In place of cities you have industries. And in place of miles between cities you have numbers giving the share of one industry's sales that are purchased by the other industries.

Table 1 describes a simplified economy consisting of a utility company that produces electricity (E) and water (W). The basic unit is of primary importance. Here we make it one dollar ($1). Thus the number 0.8 corresponds to $0.80, since $0.8 \times \$1 = \$0.80$.

Table 1 is like an economic crossword puzzle in which each entry can be read horizontally as the output of one industry or vertically as the input of another. The input columns explain the meaning of the entries. Consider the column under E. To produce each dollar worth of electricity, the utility company must spend $0.20 on electricity and $0.40 on water.

The rows of an input-output table tell what amounts of goods must be produced to meet **internal demands.** For Table 1 let

x = dollar amount of electricity produced
y = dollar amount of water produced

TABLE 1

		Input (purchase)	
		E	W
Output	E	0.2	0.3
(sale)	W	0.4	0.1

where each amount represents the production per household per month. The first row shows that in order to produce x dollars of electricity, the utility company must use $0.2x$ dollars of electricity and $0.3y$ dollars of water. Thus the internal demand for electricity can be expressed as

$$0.2x + 0.3y$$

The internal demand for water is determined from the second row of Table 1.

$$0.4x + 0.1y$$

In addition to internal demands there are also **external demands** by customers outside the company. Suppose the utility company determines that it must produce $20 of electricity and $5 of water per household each month. The total demands, both internal and external, for electricity and water become

electricity $0.2x + 0.3y + 20$
water $0.4x + 0.1y + 5$
            $\underbrace{}_{\text{internal}}$ $\underbrace{}_{\text{external}}$

The amounts produced, x for electricity and y for water, must equal the demand. This leads to two equations

$$x = 0.2x + 0.3y + 20$$
$$y = 0.4x + 0.1y + 5$$

This system can easily be solved by elimination. But in practice the numbers in an input-output table remain the same while only the external demand changes. Matrix methods present the most efficient means of solving such systems.

The Leontief Formula

Let X be the matrix of unknowns and D be the constant matrix.

$$X = \begin{bmatrix} x \\ y \end{bmatrix} \qquad D = \begin{bmatrix} 20 \\ 5 \end{bmatrix}$$

The constant matrix D is called the **bill of demands.** Let T be the coefficient matrix of the right-hand side of the system of equations. It is precisely Table 1 and is called the **technology matrix** (or the *consumption matrix* or the *input-output matrix*). The **Leontief Formula** is the equation

$$X = TX + D$$

It corresponds to the equality

$$\text{production} = (\text{internal demand}) + (\text{external demand})$$

The solution to the Leontief Formula is obtained as follows:

$$X = TX + D$$
$$IX - TX = D$$
$$(I - T)X = D$$
$$X = (I - T)^{-1}D$$

Procedure for Solving the Leontief Formula

Given
T and D.

Problem
Solve for X, where $X = TX + D$.

Steps
1. Form $I - T$.
2. Compute $(I - T)^{-1}$.
3. Multiply to get $X = (I - T)^{-1}D$.
4. Interpret the entries of X.

EXAMPLE 1

Consider an input-output analysis of a utility company that produces electricity (E) and water (W), where the technology matrix and the bill of demands are given by

$$T = \begin{matrix} \text{E} \\ \text{W} \end{matrix} \begin{array}{cc} E & W \\ \begin{bmatrix} 0.2 & 0.3 \\ 0.4 & 0.1 \end{bmatrix} \end{array} \qquad D = \begin{bmatrix} 20 \\ 5 \end{bmatrix}$$

Problem

Find the production schedule X that meets this demand.

SOLUTION Let x be the production schedule for E and let y be the production schedule for W. The unknown production schedule is

$$X = \begin{bmatrix} x \\ y \end{bmatrix}$$

The aim is to solve for X, where $X = TX + D$. By step 1

$$I - T = \begin{bmatrix} 1 & 0 \\ 0 & 1 \end{bmatrix} - \begin{bmatrix} 0.2 & 0.3 \\ 0.4 & 0.1 \end{bmatrix} = \begin{bmatrix} 0.8 & -0.3 \\ -0.4 & 0.9 \end{bmatrix}$$

By step 2 the inverse of $I - T$ is

$$(I - T)^{-1} = \begin{bmatrix} \frac{3}{2} & \frac{1}{2} \\ \frac{2}{3} & \frac{4}{3} \end{bmatrix}$$

Step 3 is to multiply the inverse by D.

$$X = (I - T)^{-1}D = \begin{bmatrix} \frac{3}{2} & \frac{1}{2} \\ \frac{2}{3} & \frac{4}{3} \end{bmatrix} \begin{bmatrix} 20 \\ 5 \end{bmatrix} = \begin{bmatrix} 32.5 \\ 20 \end{bmatrix}$$

By step 4 the utility company must produce \$32.50 worth of electricity and \$20 worth of water to meet the internal and external demands.

The solution to the system of equations in Example 1 is

$$32.5 = (0.2)(32.5) + (0.3)(20) + 20$$
$$20 = (0.4)(32.5) + (0.1)(20) + 5$$

The first equation means that each month the utility company should produce

$32.50 worth of electricity per household, with $6.50 going for the internal demand for electricity (since $6.50 = 0.2 × $32.50), $6 going for water (since $6 = 0.3 × $20), and the remaining $20 going to meet external demands. The utility company should produce $20 worth of water, with $13 going for electricity, $2 for the internal use of water, and $5 to meet external demands.

Example 2 shows why matrix methods are preferred to elimination for solving input-output problems.

EXAMPLE 2 | **Problem**

Let T be the technology matrix from Example 1. Find the production schedules for each of these demands.

(a) $D = \begin{bmatrix} 6 \\ 12 \end{bmatrix}$ (b) $D = \begin{bmatrix} 12 \\ 6 \end{bmatrix}$

SOLUTION Example 1 carried out steps 1 and 2 of the Leontief procedure. Therefore finding the production schedule involves only matrix multiplication by the inverse of $I - T$, which was computed in Example 1.

(a) $X = (I - T)^{-1}D = \begin{bmatrix} \frac{3}{2} & \frac{1}{2} \\ \frac{2}{3} & \frac{4}{3} \end{bmatrix} \begin{bmatrix} 6 \\ 12 \end{bmatrix} = \begin{bmatrix} 15 \\ 20 \end{bmatrix}$

(b) $X = (I - T)^{-1}D = \begin{bmatrix} \frac{3}{2} & \frac{1}{2} \\ \frac{2}{3} & \frac{4}{3} \end{bmatrix} \begin{bmatrix} 12 \\ 6 \end{bmatrix} = \begin{bmatrix} 21 \\ 16 \end{bmatrix}$

The matrix method is used to solve input-output problems because the bill of demands changes frequently but the technology matrix remains the same. Example 2 shows that production schedules are easy to compute once one production schedule has been computed.

Example 3 examines a three-sector economy.

EXAMPLE 3 | **Problem**

Find the production schedule for an economy with three industries, A, B, and C, where the technology matrix T and bill of demands D are given by

$$
\begin{array}{c} \\ \\ T = \begin{matrix} A \\ B \\ C \end{matrix} \end{array}
\begin{matrix} A & B & C \\ \begin{bmatrix} 0.8 & 0.1 & 0 \\ 0.1 & 0.8 & 0.1 \\ 0 & 0.1 & 0.8 \end{bmatrix} \end{matrix}
\qquad D = \begin{bmatrix} 8 \\ 6 \\ 12 \end{bmatrix}
$$

SOLUTION By step 1

$$
I - T = \begin{bmatrix} 1 & 0 & 0 \\ 0 & 1 & 0 \\ 0 & 0 & 1 \end{bmatrix} - \begin{bmatrix} 0.8 & 0.1 & 0 \\ 0.1 & 0.8 & 0.1 \\ 0 & 0.1 & 0.8 \end{bmatrix} = \begin{bmatrix} 0.2 & -0.1 & 0 \\ -0.1 & 0.2 & -0.1 \\ 0 & -0.1 & 0.2 \end{bmatrix}
$$

The inverse of $I - T$ is

$$
(I - T)^{-1} = 10 \begin{bmatrix} \frac{3}{4} & \frac{1}{2} & \frac{1}{4} \\ \frac{1}{2} & 1 & \frac{1}{2} \\ \frac{1}{4} & \frac{1}{2} & \frac{3}{4} \end{bmatrix}
$$

The solution is

$$X = (I - T)^{-1}D = 10 \begin{bmatrix} \frac{3}{4} & \frac{1}{2} & \frac{1}{4} \\ \frac{1}{2} & 1 & \frac{1}{2} \\ \frac{1}{4} & \frac{1}{2} & \frac{3}{4} \end{bmatrix} \begin{bmatrix} 8 \\ 6 \\ 12 \end{bmatrix}$$

$$= 10 \begin{bmatrix} 12 \\ 16 \\ 14 \end{bmatrix} = \begin{bmatrix} 120 \\ 160 \\ 140 \end{bmatrix}$$

This means that A must produce 120 units, B 160 units, and C 140 units to meet the internal and external demands placed on them.

Input-output analysis did not really blossom until after World War II, when the advent of computers made the computation of matrix inverses quick and reliable. The inverse of the matrix $I - T$ in Example 3 is computed in Exercise 13.

EXERCISE SET 2-6

• In Problems 1 to 4 use the technology matrix from Example 1 to compute the production schedule for the given bill of demands.

1. $D = \begin{bmatrix} 40 \\ 10 \end{bmatrix}$　　　　2. $D = \begin{bmatrix} 4 \\ 16 \end{bmatrix}$

3. $D = \begin{bmatrix} 20 \\ 25 \end{bmatrix}$　　　　4. $D = \begin{bmatrix} 12 \\ 5 \end{bmatrix}$

In Problems 5 to 8 let the technology matrix be

$$T = \begin{bmatrix} 0.5 & 0.4 \\ 0.75 & 0.2 \end{bmatrix}$$

Compute the production schedule to meet the given bill of demands.

5. $D = \begin{bmatrix} 20 \\ 10 \end{bmatrix}$　　　　6. $D = \begin{bmatrix} 4 \\ 2 \end{bmatrix}$

7. $D = \begin{bmatrix} 4 \\ 6 \end{bmatrix}$　　　　8. $D = \begin{bmatrix} 40 \\ 60 \end{bmatrix}$

In Problems 9 and 10 use the technology matrix from Example 3 to compute the production schedule for the given bill of demands.

9. $D = \begin{bmatrix} 12 \\ 6 \\ 8 \end{bmatrix}$　　　　10. $D = \begin{bmatrix} 20 \\ 2 \\ 28 \end{bmatrix}$

•• 11. A company makes hard and soft contact lenses. For each $1 output of hard lenses, the company must spend $0.60 on hard lenses and $0.10 on soft lenses. For each $1 of output on soft lenses, the company must spend $0.20 on hard lenses and $0.70 on soft lenses. Find the production schedule that satisfies a bill of demands of $60,000 for hard lenses and $50,000 for soft lenses.

12. Repeat Problem 11 for a bill of demands of $100,000 for hard lenses and $750,000 for soft lenses.

13. Verify that the matrices A and B are inverses, where

$$A = \begin{bmatrix} 0.2 & -0.1 & 0 \\ -0.1 & 0.2 & -0.1 \\ 0 & -0.1 & 0.2 \end{bmatrix}$$

$$B = \begin{bmatrix} 7.5 & 5 & 2.5 \\ 5 & 10 & 5 \\ 2.5 & 5 & 7.5 \end{bmatrix}$$

••• 14. Show that if $A = (0.1)B$, then $A^{-1} = (10)B^{-1}$.

15. If $A = kB$ for some number $k \neq 0$, how are A^{-1} and B^{-1} related?

Economics and mathematics have not always shared mutual interests. In fact Adam Smith's classic work on economics, *Wealth of Nations*, written in 1776, contains almost no mathematics. The real intrusion of mathematical methods into economics occurred a century later in the work of the 19th-century economist Alfred Marshall, who has been dubbed a "closet mathematician" because of his habit of removing all traces of the underlying mathematical principles from his published work in favor of real-life examples that illustrated his theory. However, one of Marshall's students at Cambridge, John Maynard Keynes, gained notoriety and world renown by applying mathematical methods to economic principles. Keynes not only took his degree in mathematics, but he published a book entitled *A Treatise on Probability*.

Of course not all economists today share the Keynesian view of mathematics. Harvard Professor John Kenneth Galbraith, himself a living disciple of Keynes, regards mathematics with a considerable degree of skepticism. In his *Economics, Peace, and Laughter* he wrote, "There can be no question that prolonged commitment to mathematical exercises in economics can be damaging. It leads to the atrophy of judgment and intuition."

In spite of Galbraith's viewpoint, a brief look at the Nobel prize laureates in economics attests to the close relationship that economics and mathematics have enjoyed.* These prizes were first awarded in 1969 by the National Bank of Sweden to commemorate its 300th anniversary. The first winners, Ragnar Frisch and Jan Tinbergen, were cited for developing "the mathematical expression of economic theory." In 1975 the mathematician Leonid Kantorovich (1912–1986) shared the prize for developing the mathematical theory of linear programming and for applying it to economic problems in optimizing the allocation of resources. (Chapters 3 and 4 are devoted to linear programming.) Altogether 12 of the 19 prizes in the years 1969–1987 have been awarded for work that is heavily mathematical.

This Case Study examines the work of the economist Wassily W. Leontief, who was awarded the 1973 Nobel prize in economics "for the development of the input-output method and for its application to important economic problems." Before describing his work, we will survey Leontief's accomplishments.†

Leontief was born in Leningrad in 1905. At the age of 15 he entered Leningrad University, graduating five years later with the title of "Learned Economist." It took him just two more years to obtain his Ph.D. from the University of Berlin. From there he traveled to China as an economic adviser

---

*Two excellent articles that describe the relationship between economics and mathematics are "Theoretical Models: Mathematical Form and Economic Content," by Gerard Debreu, the 1985 Nobel prize laureate in economics (the November 1986 issue of *Econometrica*, pp. 1259–1270), and "Mathematical Methods of Economics" by Joel Franklin, based on a lecture given at a meeting of the Society for Industrial and Applied Mathematics (printed in the April 1983 issue of the *American Mathematical Monthly*, pp. 229–246).

†See the July 1984 issue of *Omni* for an insightful interview with Leontief (pp. 78–82 and 110–118).

TABLE 1

Sector	Abbreviation	Industries
Final nonmetal	FN	Apparel, food, drugs, furniture
Final metal	FM	Motor vehicles, aircraft, communications
Basic metal	BM	Iron, steel, heating, plumbing
Basic nonmetal	BN	Livestock, agriculture products, fabrics, lumber
Energy	E	Electricity, gas, water, coal, petroleum
Services	S	Transportation, insurance, real estate, finance

and then to New York as a research associate at the National Bureau of Economic Research. In 1931 he began a 35-year tenure on the faculty of Harvard University. Since 1966 he has directed the Institute for Economic Analysis in New York. He lives in a Greenwich Village apartment with his wife Estelle, who is a poet.

Input-output analysis began in the 1920s when the federal government organized data to get a clearer picture of the workings of its economy. Leontief, then a young professor at Harvard, developed the primary methods of input-output analysis that later were instrumental in the policies of the important Controlled Materials Plan during World War II. After World War II the Air Force and the Department of Labor funded a project, called Project Scoop, which provided a detailed analysis of the economy. Since then input-output analysis has been regarded as one of the cornerstones of economic theory.

In 1965 Leontief published an account of the U.S. economy based on statistics from 1958.‡ His analysis examined the interdependence among 81 industrial sectors. We will present a more modest version in which the industries are lumped together into six sectors. Table 1 lists the sectors and some representative industries from each. Notice that it includes most of the articles involved in our daily life.

Table 2 gives Leontief's input-output table for the six-sector economy of the United States for 1958. The basic unit is one million dollars. Thus the number 0.004 in the (1, 2) entry of the table corresponds to $4000, since

$$0.004 \times \$1,000,000 = \$4000.$$

‡See Wassily W. Leontief, "The Structure of the U.S. Economy," *Scientific American*, April 1965, pp. 25–35.

TABLE 2

		Input					
		FN	FM	BM	BN	E	S
	FN	0.170	0.004	0.000	0.029	0.000	0.008
	FM	0.003	0.295	0.018	0.002	0.004	0.016
Output	BM	0.025	0.173	0.460	0.007	0.011	0.017
	BN	0.348	0.037	0.021	0.403	0.011	0.048
	E	0.007	0.001	0.039	0.025	0.358	0.025
	S	0.120	0.074	0.104	0.123	0.173	0.234

To solve the Leontief formula set

r = amount of FN goods produced in a year
s = amount of FM goods produced in a year
t = amount of BM goods produced in a year
x = amount of BN goods produced in a year
y = amount of E goods produced in a year
z = amount of S goods produced in a year

By including the external demands, Table 2 leads to the following system of six linear equations in six unknowns.

$$
\begin{aligned}
r &= 0.170r + 0.004s &&\; + 0.029x &&\; + 0.008z + &&99{,}640 \\
s &= 0.003r + 0.295s + 0.018t + 0.002x + 0.004y + 0.016z + &&75{,}548 \\
t &= 0.025r + 0.173s + 0.460t + 0.007x + 0.011y + 0.007z + &&14{,}444 \\
x &= 0.348r + 0.037s + 0.021t + 0.403x + 0.011y + 0.048z + &&33{,}501 \\
y &= 0.007r + 0.001s + 0.039t + 0.025x + 0.358y + 0.025z + &&23{,}527 \\
z &= 0.120r + 0.074s + 0.104t + 0.123x + 0.173y + 0.234z + &&263{,}985
\end{aligned}
$$

Example 1 solves this system and interprets its meaning for the U.S. economy.

EXAMPLE 1 | **Problem**

Find the production schedule to run the 1958 U.S. economy.

SOLUTION Let X be the unknown vector consisting of the production schedules for the six sectors of the economy and let D be the bill of demands.

$$
X = \begin{bmatrix} r \\ s \\ t \\ x \\ y \\ z \end{bmatrix}
\qquad
D = \begin{bmatrix} 99{,}640 \\ 75{,}548 \\ 14{,}444 \\ 33{,}501 \\ 23{,}527 \\ 263{,}000 \end{bmatrix}
$$

The technology matrix T is precisely Table 2. The Leontief Formula is $X = TX + D$, and its solution is $X = (I - T)^{-1}D$.

The difficult part of the solution is computing the inverse of $I - T$, which is this 6×6 matrix (rounded to three decimal places).

$$
\begin{bmatrix}
1.234 & 0.014 & 0.007 & 0.064 & 0.006 & 0.017 \\
0.017 & 1.436 & 0.056 & 0.014 & 0.019 & 0.032 \\
0.078 & 0.467 & 1.878 & 0.036 & 0.044 & 0.031 \\
0.752 & 0.133 & 0.101 & 1.741 & 0.065 & 0.123 \\
0.061 & 0.045 & 0.130 & 0.083 & 1.578 & 0.059 \\
0.340 & 0.236 & 0.307 & 0.315 & 0.375 & 1.348
\end{bmatrix}
$$

The solution vector X is obtained by multiplying this matrix by the bill of demands D. The result is

$$X = \begin{bmatrix} 131{,}033 \\ 120{,}459 \\ 80{,}681 \\ 178{,}732 \\ 431{,}562 \\ 66{,}929 \end{bmatrix}$$

This means that in order to run the economy, the FN sector should produce $131,033 worth of goods, the FM sector $120,459, the BM sector $80,681, the BN sector $178,732, the E sector $431,562, and the S sector $66,929.

Case Study Exercises

In Problems 1 to 4 find the production schedule to run the 1958 U.S. economy for the given bill of demands.

1.
$$D = \begin{bmatrix} 100{,}000 \\ 600{,}000 \\ 500{,}000 \\ 800{,}000 \\ 200{,}000 \\ 100{,}000 \end{bmatrix}$$

2.
$$D = \begin{bmatrix} 30{,}000 \\ 20{,}000 \\ 40{,}000 \\ 10{,}000 \\ 10{,}000 \\ 90{,}000 \end{bmatrix}$$

3.
$$D = \begin{bmatrix} 300{,}000 \\ 200{,}000 \\ 500{,}000 \\ 100{,}000 \\ 100{,}000 \\ 900{,}000 \end{bmatrix}$$

4.
$$D = \begin{bmatrix} 10{,}000 \\ 60{,}000 \\ 50{,}000 \\ 80{,}000 \\ 20{,}000 \\ 10{,}000 \end{bmatrix}$$

Problems 5 and 6 refer to the technology matrix T for the 1947 U.S. economy, where the three sectors are agriculture, manufacturing, and household.

$$T = \begin{bmatrix} 0.245 & 0.102 & 0.051 \\ 0.099 & 0.291 & 0.279 \\ 0.433 & 0.376 & 0.011 \end{bmatrix}$$

Find the production schedule to run the 1947 U.S. economy to meet the given bill of demands, using the fact that

$$(I - T)^{-1} = \begin{bmatrix} 1.454 & 0.291 & 0.157 \\ 0.533 & 1.763 & 0.525 \\ 0.837 & 0.791 & 1.278 \end{bmatrix}$$

5.
$$D = \begin{bmatrix} 2{,}000{,}000{,}000 \\ 1{,}000{,}000{,}000 \\ 4{,}000{,}000{,}000 \end{bmatrix}$$

6.
$$D = \begin{bmatrix} 1{,}000{,}000{,}000 \\ 3{,}000{,}000{,}000 \\ 100{,}000{,}000 \end{bmatrix}$$

CHAPTER REVIEW

Key Terms

2–1 **Addition and Scalar Multiplication**

Matrix	Addition of Matrices
Entries	Scalar Product
Size	Closed
Square Matrix	Associative
Row Matrix	Commutative
Column Matrix	Zero Matrix
Equal Matrices	Additive Inverse
Sum	

2–2 **Matrix Multiplication**

Matrix Multiplication	Constant Matrix
Product	Cancellation Law
Product Undefined	Main Diagonal
Coefficient Matrix	Identity Matrix
Variable Matrix	Distributive Laws

2–3 **The Gauss–Jordan Method**

Coefficient Matrix	Echelon Form
Augmented Matrix	Leading 1
Elementary Row Operations	Free Variable
General Solution	Basic Variable
Particular Solution	

2–4 **The Inverse**

Invertible Matrix	Noninvertible Matrix
Nonsingular Matrix	Singular Matrix
Inverse	Determinant

2–5 **Systems of Equations**

Matrix Equation
Inverse Matrix Method

2–6 **Input-Output Analysis**

Input-Output Table	Bill of Demands
Internal Demands	Leontief Formula
External Demands	Technology Matrix

REVIEW PROBLEMS

1. (a) What is the size of M?
   (b) Find m_{42}.
   (c) What notation describes the entry 7?

$$M = \begin{bmatrix} 1 & -6 & 9 \\ -2 & 4 & 3 \\ 1 & 2 & 3 \\ 4 & 5 & 6 \\ 2 & 7 & 8 \end{bmatrix}$$

2. Compute $3A + 2B - C$, where

$$A = \begin{bmatrix} 1 & -1 & 4 \\ 3 & -3 & 7 \end{bmatrix} \quad B = \begin{bmatrix} -3 & 6 & 5 \\ 4 & 0 & 1 \end{bmatrix}$$

$$C = \begin{bmatrix} 5 & 3 & 7 \\ 1 & 2 & 6 \end{bmatrix}$$

3. Determine whether it is possible to find values of x and y that satisfy the equality. If so, find x and y.

$$x \begin{bmatrix} 2 \\ 5 \end{bmatrix} + y \begin{bmatrix} 3 \\ 7 \end{bmatrix} = \begin{bmatrix} 7 \\ 16 \end{bmatrix}$$

4. Compute AB, where

$$A = \begin{bmatrix} 1 & 4 & 5 \\ 0 & 0 & 0 \\ 3 & 11 & 2 \end{bmatrix}$$

$$B = \begin{bmatrix} -2 & 1 & 0 \\ 0 & 6 & 3 \\ 0 & 0 & -2 \end{bmatrix}$$

5. Compute XY, where

$$X = \begin{bmatrix} 1 & 4 & 5 \\ 0 & 0 & 0 \\ 3 & 11 & 2 \end{bmatrix}$$

$$Y = \begin{bmatrix} 1 & 0 & -1 & 6 \\ 2 & 5 & 2 & 3 \\ 4 & 1 & 3 & 1 \end{bmatrix}$$

Problems 6 to 8 refer to the following matrices.

$$A = \begin{bmatrix} 0 & -1 \\ 0 & 3 \end{bmatrix} \quad B = \begin{bmatrix} 4 & -5 \\ 6 & -3 \end{bmatrix}$$

$$C = \begin{bmatrix} 5 & -4 \\ 6 & -3 \end{bmatrix}$$

6. Compute AB.

7. Compute AC.

8. What do you conclude from Problems 6 and 7?

Problems 9 and 10 deal with powers of a matrix A. By definition, $A^2 = A \cdot A$, $A^3 = A \cdot A^2$, and so on.

9. Compute A^3 for the matrix $A = \begin{bmatrix} 1 & 2 \\ 3 & 4 \end{bmatrix}$

10. Compute A^{10} for the matrix $A = \begin{bmatrix} 1 & 0 \\ -1 & 2 \end{bmatrix}$

11. Write the coefficient matrix A, the variable matrix X, and the constant matrix C for the system of equations

$$\begin{aligned} w + x - 5y &= -1 \\ 3x + y + 7z &= 8 \end{aligned}$$

12. Use the Gauss–Jordan method to find the unique solution to this system of equations.

$$\begin{aligned} x + y - z &= 4 \\ 2x - y + 3z &= 5 \\ 5x - 3y - 5z &= 12 \end{aligned}$$

13. Find the general solution to the system. List two particular solutions. The variables are x, y, and z.

$$\begin{bmatrix} 1 & 0 & 1 & -2 \\ 0 & 1 & -1 & 4 \\ 0 & 0 & 0 & 0 \end{bmatrix}$$

14. Find the general solution to the system. List two particular solutions. The variables are x, y, and z.

$$\begin{bmatrix} 1 & -2 & 0 & -2 \\ 0 & 0 & 1 & 4 \\ 0 & 0 & 0 & 0 \end{bmatrix}$$

15. Determine whether this system has either no solution or infinitely many solutions. If there are infinitely many solutions, derive the general form and list the basic variables.

$$\begin{aligned} -x + 2y - 3z &= 19 \\ -x + 2y - 2z &= 4 \\ 2x - 4y + 3z &= 7 \end{aligned}$$

16. Compute the inverse of this matrix if possible.

$$\begin{bmatrix} 1 & 1 & 1 \\ 2 & 1 & 1 \\ 0 & 1 & 1 \end{bmatrix}$$

17. Compute the inverse of this matrix if possible.

$$\begin{bmatrix} 1 & -1 & 0 \\ 0 & 1 & 1 \\ 2 & 0 & -1 \end{bmatrix}$$

18. The matrix $A = \begin{bmatrix} 1 & -2 \\ -2 & 8 \end{bmatrix}$ is invertible. The square of A^{-1} is denoted by A^{-2}.
(a) Compute A^{-2}.
(b) Compute A^2 and $(A^2)^{-1}$.
(c) How are A^{-2} and $(A^2)^{-1}$ related?

19. Compute (a) $(AB^{-1})^{-1}$ (b) $(ABC^{-1})^{-1}$.

20. Find A where $A^{-1} = \begin{bmatrix} 1 & 3 \\ 2 & 5 \end{bmatrix}$.

21. Derive A^{-1} by forming AB for the following matrices.

$$A = \begin{bmatrix} 3 & 2 & -1 \\ 1 & -1 & 0 \\ 2 & 0 & -1 \end{bmatrix}$$

$$B = \begin{bmatrix} 1 & 2 & -1 \\ 1 & -1 & -1 \\ 2 & 4 & -5 \end{bmatrix}$$

In Problems 22 and 23 solve the system by the inverse matrix method if possible.

22. $x + y + z = 1$
   $-2x - y - z = 0$
        $y + z = 0$

23. $2x \quad\quad + 4z = 16$
   $- 2y - 3z = 1$
   $3x - 4y + z = 29$

24. Solve the systems of linear equations for each pair of values of a and b.

$2x + 3y = a$
$5x + 7y = b$

(a) $a = 3$ (b) $a = 4$ (c) $a = -6$
   $b = 9$ $b = 0$ $b = 3$

25. Compute the production schedule where the technology matrix and the bill of demands are

$$T = \begin{bmatrix} 0.4 & 0.5 \\ 0.2 & 0.75 \end{bmatrix} \quad D = \begin{bmatrix} 100 \\ 80 \end{bmatrix}$$

3 Linear Programming: Geometric Approach

CHAPTER OVERVIEW Linear programming is a method of solving problems in which a particular quantity is to be minimized or maximized. The problem could be to maximize profit, minimize cost, or minimize caloric intake, or some similar optimization problem. The quantity to be optimized is limited in some ways. Profit is dependent on production capacity while cost depends on personnel. The solution of these problems involves linear inequalities and linear equations, and in Chapter 4 we will use matrices to solve more complicated problems.

Section 3–1 introduces linear inequalities. In Section 3–2 linear programming problems are translated into mathematics. Then in Section 3–3 we use the skills developed in Section 3–1 to solve the linear programming problems using a geometric approach.

CASE STUDY PREVIEW

In 1947 George Dantzig was a young mathematician looking for a method of solving very complicated optimization problems. The problems he considered were real-world problems originating in the military, business, and economics. He was employed by the military but it was his friends who were economists and mathematicians, not to mention the engineers who were working on building a new type of machine called a ''computer,'' who led him on the path to success.

Many of the problems facing Dantzig involved systems of equations with 50 or even 100 equations and as many unknowns. The method he created, called the *simplex method,* discussed in Chapter 4, became an instant hit. One of the first problems that he applied the method to was a transportation problem involving the large steel industry. These companies had built a virtual monopoly by claiming that the transportation cost of steel was too complicated to compute, so all steel companies, large and small, had to use the same simplified formula that favored the large companies. The simplex method showed them how to compute the cost, thereby allowing smaller companies to use their proximity to the user to cut cost and remain competitive.

After Section 3–2, we present a Case Study that shows how modern businesses use linear programming to solve major problems. It tells of the decision that Federal Express Corporation was faced with concerning whether to shift their operations from small aircraft to larger capacity ones.

The second Case Study, after Section 3–3, is based on the Berlin airlift, one of the first and most significant applications of linear programming. The problem was to supply Berlin and break the Russian blockade after World War II.

3–1 LINEAR INEQUALITIES

The general form of a **linear inequality** in two variables is

$$ax + by < c$$

where a, b, and c are constants. The inequality sign $<$ can be replaced by $\leq$, $>$, or $\geq$, so there are actually four versions of the general form of a linear inequality. Recall that $a \leq b$ means that $a < b$ or $a = b$. To review the notation for inequalities, remember that $6 < 8$, $6 \leq 6$, $-1 > -3$, and $-3 \geq -3$. Some elementary properties of inequalities are

Property 1. If $a < b$ and c is any number, then $a + c < b + c$.
Property 2. If $a < b$ and $c > 0$, then $ac < bc$.
Property 3. If $a < b$ and $c < 0$, then $ac > bc$.

A **solution** to a linear inequality is a pair of numbers that when substituted into the linear inequality produces a true statement. For instance, a solution to $3x + 5y < 10$ is $(0, 1)$ because $3 \cdot 0 + 5 \cdot 1 = 5$, which is less than 10. But $(2, 1)$ is not a solution because $3 \cdot 2 + 5 \cdot 1 = 11$, which is not less than 10.

Inequalities in two variables are graphed in the plane. The set of points that solve the inequality are **half-planes.** For example, to graph $y > 2x + 1$, first graph the set of points on the line $y = 2x + 1$. (See Figure 1.) They are not part of the solution to the inequality because y is not strictly greater than $2x + 1$. Select a point on the line, say, $(1, 3)$. Consider any point directly above $(1, 3)$, say, $(1, 5)$. Since $5 > 3$ and 3 equals $2x + 1$ when $x = 1$, we have $5 > 2x + 1$ when $x = 1$. This says that $(1, 5)$ is a solution to the inequality because it lies directly above a point on the line. Therefore all points that lie above the line are solutions. Let us summarize.

1. The set of points above the line satisfy $y > 2x + 1$.
2. The set of points below the line satisfy $y < 2x + 1$.
3. The set of points on the line satisfy $y = 2x + 1$.

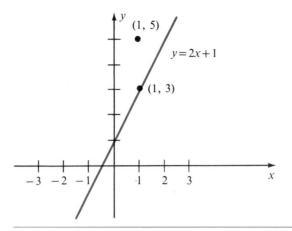

FIGURE 1

In general the solution to a linear inequality is determined by the corresponding equality. The solution to the inequality is one of the two half-planes into which the line divides the plane. All that is necessary is to test one of the points in the plane that is not on the line to determine if it is a solution to the inequality. If it is a solution, then the half-plane containing the point is the graph of the inequality. If it is not a solution, then the other half-plane is the graph.

Usually the easiest point to choose is (0, 0) because it is the simplest to substitute into the inequality. As long as (0, 0) does not lie on the line we will select it as the test point.

Procedure for Graphing Linear Inequalities

1. Replace the inequality sign by an equals sign and graph the corresponding straight line.
2. If the inequality sign is $\leq$ or $\geq$, then the line itself is part of the graph. Otherwise the line is not part of the graph.
3. Choose a test point that is not on the line and substitute the coordinates of the point into the inequality to determine if it satisfies the inequality. [Choose (0, 0) if it does not lie on the line.]
4. If the test point satisfies the inequality, then the half-plane containing the point is the graph of the inequality. If the test point does not satisfy the inequality, then the half-plane not containing the point is the graph.

We adopt the following two conventions when graphing inequalities.

1. If the line is part of the graph, it will be drawn as a solid line. If it is not part of the graph, it will be drawn as a dotted line.
2. The graph of the inequality will be the unshaded half-plane. In other words,

think of the shaded part as being "thrown away" and the unshaded part, the graph of the inequality, as kept.

The second convention might seem strange at first, but it will become clear why it is done this way when systems of inequalities are solved.

The next two examples illustrate the procedure.

EXAMPLE 1

Problem

Graph the inequalities.

(a) $2x + 3y \leq 6$ (b) $2x + 3y > 6$

SOLUTION (a) Replace the inequality sign with an equality sign. Graph the equation $2x + 3y = 6$. (See Figure 2.) Choose a convenient point that is not on the line; $(0, 0)$ is usually the easiest choice. Test $(0, 0)$ and get $2 \cdot 0 + 3 \cdot 0 = 0 \leq 6$, which is true. Thus $(0, 0)$ is a solution. From Figure 2, $(0, 0)$ lies below the line so that region is the graph. The region above the line is then shaded. Since the "equals" part of the inequality is included, the line is drawn as a solid line.

(b) The only difference between parts (a) and (b) is the different inequality sign. The solutions differ in two respects. Since the equals part of the inequality sign is not included, the line is graphed as a dotted line. Also, when we test $(0, 0)$ we get $2 \cdot 0 + 3 \cdot 0 = 0 > 6$, which is not true. Hence $(0, 0)$ is not a solution and the graph is the region not containing $(0, 0)$, the half-plane above the line. (See Figure 3.)

FIGURE 2

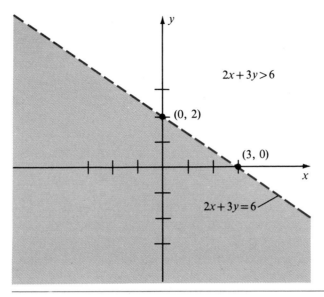

FIGURE 3

Thus far when the inequality was $>$ the region above the line was the solution. It might seem that this will always be the case; that $>$ means above the line and $<$ means below the line. It is not true. The next example illustrates this.

EXAMPLE 2

Problem

Graph the inequality $2x - 5y \leq 10$.

SOLUTION Follow the steps in the procedure.

1. Graph the line $2x - 5y = 10$. (See Figure 4.)

FIGURE 4

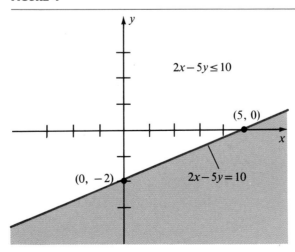

2. Use a solid line because the line is part of the graph.
3. Test $(0, 0)$; $2 \cdot 0 - 5 \cdot 0 = 0 \leq 10$, which is true so $(0, 0)$ is a solution and is part of the graph.
4. Since $(0, 0)$ is in the half-plane above the line, that region is the graph. Shade the region below the line.

Notice that the inequality in Example 2 has a $\leq$ sign but the graph is the half-plane above the line.

Systems of Linear Inequalities

Often in applications there is more than one inequality. A set with more than one inequality is called a **system of inequalities.** A *solution* to the system of inequalities is the set of points that simultaneously satisfy all the inequalities. For instance, in Figure 5 we graph the linear inequalities from Examples 1(b) and 2. They comprise the system of linear inequalities

$$2x + 3y > 6$$
$$2x - 5y \leq 10$$

The graph of the system is simply the two graphs superimposed on the same coordinate system. The two lines separate the plane into four regions; the region above both lines is the solution to the system. Because we have shaded the half-planes that are not the solutions, the unshaded region in Figure 5 is immediately seen to be the graph of the system.

FIGURE 5

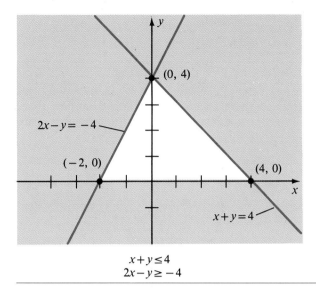

$$x+y \le 4$$
$$2x-y \ge -4$$

FIGURE 6

EXAMPLE 3 | **Problem**

Graph the system of inequalities.

$$
\begin{aligned}
x + y &\le 4 \\
2x - y &\ge -4 \\
y &\ge 0
\end{aligned}
$$

SOLUTION Graph each line: $x + y = 4$, $2x - y = -4$, and $y = 0$. (See Figure 6.) Test $(0, 0)$ in the first two inequalities; it is a solution to both. Since $(0, 0)$ lies below each line, the partial solution is the region below both lines. Since $(0, 0)$ is on $y = 0$, choose another point to test, say, $(0, 1)$; it is a solution and lies above the line, so shade the region below the line. The solution to the system is the unshaded triangular region.

The next example shows how many real-world problems can be expressed with linear inequalities. It is typical of the type of problem in the next two sections.

EXAMPLE 4 | **Problem**

A sausage manufacturer makes two types of sausages, regular and deluxe. Each regular sausage contains 2 pounds (lb) of filler and 1 lb of beef while each deluxe sausage contains 1 lb of filler and 2 lb of beef. How many sausages of each kind can be made with no more than 8 lb of filler and no more than 10 lb of beef?

SOLUTION Express the data in a table where the rows are labeled "filler" and "beef" and the columns are labeled "regular" and "deluxe."

	Regular	**Deluxe**	**Limiting Value**
Filler	2	1	8
Beef	1	2	10

Let x represent the number of regular sausages and y the number of deluxe sausages. Then $2x$ represents the amount of filler used in regular sausages and y represents the number of pounds of filler used in deluxe sausages. Since no more than 8 lb of filler can be used, we get

$$2x + y \leq 8$$

Similarly, x is the amount of beef used in regular sausages and $2y$ is the amount of beef used in deluxe sausages. Since no more than 10 lb of beef can be used, we get

$$x + 2y \leq 10$$

It makes no sense to let x and y take on negative values, so we get two additional inequalities:

$$x \geq 0 \qquad \text{and} \qquad y \geq 0$$

FIGURE 7

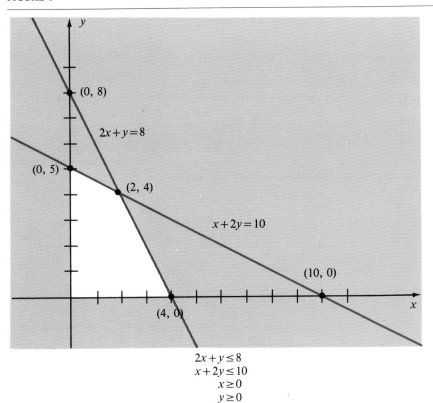

$$2x + y \leq 8$$
$$x + 2y \leq 10$$
$$x \geq 0$$
$$y \geq 0$$

The system of inequalities that represents the problem is

$$2x + y \leq 8$$
$$x + 2y \leq 10$$
$$x \geq 0, y \geq 0$$

The solution is obtained by graphing the lines $2x + y = 8$, $x + 2y = 10$, $x = 0$, and $y = 0$ and shading the appropriate regions. The last two inequalities limit the graph to the first quadrant. The solution to each of the other regions is the region below the lines, because $(0, 0)$ is a solution to each inequality. (See Figure 7.)

EXERCISE SET 3–1

• In Problems 1 to 4 determine whether the points satisfy the inequalities.

1. $x + y > 4$, $(1, 2), (2, 2)$

2. $2x - y < 0$, $(3, -1), (-2, 0)$

3. $-x + 7y \geq 8$, $(-1, 1), (0, 0)$

4. $3x - y \leq 3$, $(1, 0), (0, 1)$

In Problems 5 to 8 determine whether the points lie above, below, or on the line.

5. $x + y = 4$, $(0, 0), (1, 7)$

6. $2x - y = 0$, $(0, 0), (2, -1)$

7. $x - 3y = 1$, $(1, -1), (-1, -1)$

8. $2x + 3y = 6$, $(3, 0), (0, 0)$

In Problems 9 to 18 sketch the graph of the linear inequality by shading those points that do not satisfy the inequality.

9. $x + y > 2$ 10. $x + y < -3$

11. $2x + 3y \leq 12$ 12. $2x + 3y \geq 12$

13. $x + 3y > -6$ 14. $-x - 3y \leq 6$

15. $4x + 3y \geq 12$ 16. $4x - 3y \geq 12$

17. $y \leq 0$ 18. $x \leq 0$

In Problems 19 to 24 graph the system of linear inequalities.

19. $x + 3y \leq 6$ 20. $x + 2y \leq 4$
$\quad\ x - 2y > 4$ $\quad\ x - 3y \leq 6$

21. $2x + y \geq 4$ 22. $\quad\ x + y < -2$
$\quad\ x - y < 5$ $\quad 2x - 5y \geq 10$

23. $\quad x + y \leq 8$ 24. $x + y \leq 4$
$\quad 2x + y \leq 14$ $\quad x + 2y \leq 6$

•• In Problems 25 to 30 graph the system of linear inequalities.

25. $x + 3y \leq 6$ 26. $x + 2y \leq 4$
$\quad x - 2y > 4$ $\quad x - 3y \leq 6$
$\quad x \geq 0, y \geq 0$ $\quad x \geq 0, y \geq 0$

27. $2x + y \geq 4$ 28. $\quad x + y < -2$
$\quad x + y < 5$ $\quad 2x + 5y \geq 10$
$\quad x \geq 0, y \geq 0$ $\quad x \geq 0, y \geq 0$

29. $2x + y \leq 6$ 30. $x + 3y \leq 6$
$\quad x + y \leq 4$ $\quad x + 2y \leq 4$
$\quad x \geq 0, y \geq 0$ $\quad x \geq 0, y \geq 0$

In Problems 31 to 34 express the word problem as an inequality.

31. A manufacturer makes two types of jackets, a wool jacket and a cotton jacket. The wool jacket costs \$200 each to produce and the cotton jacket costs \$130 each. The manufacturer can budget no more than \$2600 per week to produce the jackets.

32. The manufacturer in Problem 31 can make no more than 100 jackets a week.

33. A travel agent has sold 400 tickets for a tour of Mexico. The agent can use two types of airplanes for the charter flights. Type 1 can carry at most 40 passengers and type 2 can carry at most 50 passengers.

34. The travel agent in Problem 33 is limited to a total air cost of $145,000. Each flight of a type 1 aircraft will cost $9000 and each type 2 flight will cost $15,000.

In Problems 35 to 38 express the word problem as a system of inequalities and graph its solution.

35. A manufacturer makes two types of toasters, large and small. The small toaster sells for $20 and the large toaster sells for $40. The manufacturer can budget no more than $2400 per week to produce the toasters and can make no more than 100 toasters per week.

36. A garden store sells two types of lawnmowers, gas and electric. The store wants to sell at least 45 mowers per week. The profit on each gas mower is $50 and the profit on each electric mower is $40. The store wants to earn at least $2000 per week on the sale of lawnmowers.

37. A hospital nutritionist prepares a menu consisting of beef and rice that must contain at least 45 grams (g) of protein and at least 15 milligrams (mg) of iron. Each serving of beef contains 45 g of protein and 9 mg of iron. Each serving of rice contains 9 g of protein, and 6 mg of iron.

38. A meat manufacturer mixes beef and pork in sausage links. Two types of links are made, regular and spicy. Each pound of regular sausage meat contains 0.3 pound (lb) of beef and 0.1 lb of pork, while each pound of spicy sausage meat contains 0.2 lb of beef and 0.4 lb of pork. There are 120 lb of beef and 60 lb of pork in stock.

●●● In Problems 39 to 45 graph the system of linear inequalities.

39. $x + 3y \leq 6$
    $x + 4y > 4$
    $x \geq 1, y \geq 1$

40. $x + 3y \leq 3$
    $x + 2y \leq 6$
    $x \geq 1, y \geq 2$

41. $4x + y \geq 8$
    $x + y \leq 5$
    $6x + y \leq 6$
    $x \geq 0, y \geq 0$

42. $x + y \leq 8$
    $x + 2y \leq 10$
    $2x + 5y \geq 10$
    $x \geq 0, y \geq 0$

43. $x + y \leq 10$
    $x + 2y \leq 12$
    $2x + y \leq 12$
    $x \geq 0, y \geq 0$

44. $x + y \leq 10$
    $x + 2y \geq 12$
    $2x + y \leq 12$
    $x \geq 0, y \geq 0$

45. $x + y \leq 10$
    $x + 2y \geq 12$
    $2x + y \geq 12$
    $x \geq 0, y \geq 0$

In Problems 46 and 47 express each constraint as an inequality.

46. In an article studying the effects of learning on resource allocation and scheduling, Joyce Chen used as an example a firm producing two products, type 1 and type 2.* To make each type 1 product requires 19 units of material and to make each type 2 product requires 20 units of material. No more than 14,440 units of material are to be used, and the total number of products made is not to exceed 750.

47. The transportation problem was one of the first types of practical problems to which linear programming was applied in the early 1950s. In a more recent application, Choypeng applied the procedure to problems in Thailand.† In one part of the study, two types of ships were considered, carriers and transport ships. The total number of ships available was 100. Each carrier required 30 administrative personnel and each transport required 10 administrative personnel. There was a total of 2500 administrative personnel available.

48. In an article describing how to apply linear programming to credit investigation and credit

*Chen, Joyce, "Modeling Learning Curve and Learning Complementarity for Resource Allocation and Production Scheduling," *Decision Sciences*, 14:170–186, 1983.

†Choypeng, P., et al., "Optimal Ship Routing and Personnel Assignment for Naval Recruitment in Thailand," *Interfaces*, 16:47–52, 1986.

granting procedures, John Stowe used the following analysis for customers with different types of credit ratings.‡ The net present value N of the loan, assuming credit was granted to the customer, was related to the size of the credit request x. For two types of customers, the two variables were related by

$$N \geq -6.1 + .26x$$
$$N \geq -2.4 + .23x$$

Sketch the graph of the solution set of the system of inequalities.

‡Stowe, John D., "An Integer Programming Solution for the Optimal Credit Investigation/Credit Granting Sequence," *Financial Management,* 14:66–76, 1985.

3–2 LINEAR PROGRAMMING PROBLEMS

A linear programming problem involves finding the largest or smallest value of a variable given that certain restrictions are met. For instance, a firm wants to maximize profit subject to limitations on production capacity and cost.

In this section we present three examples of typical linear programming problems. They are analogous to the ones that originated George Dantzig's research in this area. The aim is to develop a method for translating a given problem into mathematical terms. In the next section we will show how to solve the problems once the translation has been carried out.

We first state a typical problem and then use it to illustrate the first few steps in expressing linear programming problems in mathematical terms. It is a type of problem that is frequently found in the business world.

An Example: The Manufacturing Problem

Manufacturing problem   A clothing manufacturer makes two types of suits, wool and a blend of wool and polyester. The wool suit sells to the retail stores for $160 each and the blend for $90 each. The cost for manufacturing each wool suit is $100 and the cost of each blend suit is $40. If the manufacturer makes no more than 120 suits a week and budgets no more than $6000 per week, how many of each type should be made to maximize profit?

To express a linear programming problem in mathematical language, we must identify certain parts of the problem.

1. The variables, especially the one that is to be maximized or minimized, must be labeled. Assign a letter to each quantity and be careful about the units that are used.
2. Determine the equation that governs the quantity to be maximized or minimized. This equation is called the **objective function.**
3. Each problem contains restrictions that are called **constraints.** They are translated into linear inequalities. For each constraint the terms must have the same units.

Begin by rereading the problem once or twice. Often it is convenient to organize the data in a table. Let us demonstrate the technique by translating the manufacturing problem.

EXAMPLE 1 | There are two types of suits, wool and blend. The problem entails finding the number of each type of suit that maximizes profit. Therefore the two variables are seen to be

> x = the number of wool suits manufactured per week
> y = the number of blend suits manufactured per week

One constraint is that the total number of suits manufactured cannot exceed 120. This translates into the inequality

> $x + y \leq 120$

Generally the coefficients of the variables are entered into the table, but for this constraint it seems to be clearer if the variables themselves, x and y, are put into the table. Hence x and y are entered in the first row together with the limiting value, 120. The next constraint is cost, so we enter the cost of each type of suit, $100 and $40, respectively, together with the limiting value, $6000. From the last sentence in the problem the objective function entails maximizing profit, so one of the rows in the table consists of profit. The cost of each wool suit is $100 and the revenue is $160, so the profit of each wool suit is $60. The profit on each blend suit is $90 − $40 = $50. These values are entered in the last row of the table. The problem can be summarized in Table 1.

TABLE 1

	Wool Suits	Blend Suits	Limiting Value
Number manufactured	x	y	120
Cost ($)	100	40	6000
Profit ($)	60	50	

In the third row there is no limiting value because we are to find the largest value that profit can be under the given constraints. Let P represent profit and then the third row is interpreted as the equation

> $P = 60x + 50y$

Notice that each term represents dollars. P is dollars of profit and, since 60 represents dollars per wool suit and x is the number of wool suits, $60x$ is

> (dollars/wool suit)(number of wool suits) = dollars

Similarly, $50y$ is

> (dollars/blend suit)(number of blend suits) = dollars

The second row of Table 1 represents cost, and since no more than $6000 is to be spent per week, it is interpreted as

> $100x + 40y \leq 6000$

Two additional constraints are implicitly included in the problem. It makes no sense to allow x or y to take on negative values, and thus we must include the two inequalities $x \geq 0$ and $y \geq 0$ in the list of constraints. These constraints are easy to overlook because the problem does not explicitly mention them, but they are a part of almost every linear programming problem.

The problem is stated in mathematical language by the following:

$$\text{Maximize} \quad P = 60x + 50y \qquad \text{objective function}$$

$$\text{Subject to} \quad \left. \begin{array}{r} 100x + 40y \le 6000 \\ x + y \le 120 \\ x \ge 0,\, y \ge 0 \end{array} \right\} \text{constraints}$$

The objective function is separated from the constraints because they play different roles in the solution to the linear programming problem.

Definition of a Linear Programming Problem

A linear programming problem originates as a word problem. When it is translated into mathematics it is expressed in a similar way to the problem in Example 1. While each problem has specific characteristics, it is possible to describe in general the formulation of linear programming problems in two variables. In the next chapter we will study linear programming problems in more than two variables.

_____ **Definition** _____

A **linear programming problem** in two variables is a problem that entails maximizing or minimizing a linear expression $z = ax + by$, called the objective function, subject to several constraints. The constraints are expressed as linear inequalities such as $ax + by \ge c$ (or $\le c$). The object is to find values of x and y that satisfy the constraints and yield the largest (or smallest) value of z.

Key Steps in Translating the Problem

Example 1 illustrates the key steps in translating a linear programming problem into mathematical language. We record them for easy reference.

Key Steps in Translating a Linear Programming Problem into Mathematical Language

1. Read and reread the problem to assess what is known and what is to be determined.
2. Organize the data, preferably into a table.
3. Identify unknown quantities and assign variables to them.
4. Determine the objective function.
5. Translate the constraints into linear inequalities.

Steps 2, 3, and 4 were described earlier in more detail. Notice that their order was different. This is to emphasize that the objective function can be determined before or after the constraints; just be sure to keep it separate and express it as an equation rather than an inequality like the constraints.

Remember that most linear programming problems are complicated and require a second or third reading to get started. Also, when organizing the data in a table, check to ensure that the numbers in each row and column agree with its description.

In the remainder of this section we present two more examples of linear programming problems. One is from the health field and the other is a transportation problem.

EXAMPLE 2

A hospital nutritionist prepares a menu for patients on low-fat diets. The primary ingredients are chicken and corn. The meal must contain enough iron and protein to meet at least the U.S. recommended daily allowance (RDA) for each nutrient. The RDA for protein is 45 grams (g) and for iron it is 20 milligrams (mg). Each 3-ounce serving of chicken contains 45 g of protein, 10 mg of iron, and 4 g of fat. Each 1-cup serving of corn contains 9 g of protein, 6 mg of iron, and 2 g of fat. How much chicken and corn should be included in the meal in order to minimize the amount of fat?

Problem

Translate the problem into mathematical language.

SOLUTION There are two types of food: chicken and corn. The limiting values are at least 45 g of protein and at least 20 mg of iron. Hence the constraints are protein and iron, respectively. They become the rows of the table while chicken and corn are the columns. The objective function is to minimize the amount of fat. It becomes the third row. (See Table 2.)

TABLE 2

	Chicken (3-oz serving)	Corn (1-cup serving)	Limiting Value
Protein (g)	45	9	45
Iron (mg)	10	6	20
Fat (g)	4	2	

The variables are

x = the number of servings of chicken
y = the number of servings of corn

To determine the objective function, let F represent the amount of fat in the meal (in grams). From the third row it becomes

$F = 4x + 2y$

The first two rows are the constraints. In each case the limiting value is prefaced by the term "at least," which translates into $\geq$. Thus the two constraints are $45x + 9y \geq 45$ and $10x + 6y \geq 20$. As before, it makes no sense to let x and

y take on negative values, so the two additional constraints, $x \geq 0$ and $y \geq 0$ are included. The problem can be stated as follows:

$$\text{Minimize} \quad F = 4x + 2y \qquad \text{objective function}$$

$$\text{Subject to} \quad \left. \begin{array}{l} 45x + 9y \geq 45 \\ 10x + 6y \geq 20 \\ x \geq 0, \ y \geq 0 \end{array} \right\} \quad \text{constraints}$$

The next example is a common type of problem facing many businesses. It is called a transportation problem. We have chosen to illustrate this type of problem by maximizing profit subject to time and cost. In actual practice sometimes a firm needs to minimize time or cost subject to other limitations.

EXAMPLE 3

A bicycle manufacturer owns a retail outlet and, in addition, sells some bikes to a large chain store. One truck makes all deliveries from the manufacturing warehouse to the retail outlet and to the chain store distribution center. It takes 2 hours to make a round trip to the retail outlet and 4 hours to make a round trip to the distribution center. The cost of a round trip to the retail outlet is $50 and the cost of a round trip to the distribution center is $200. The profit of each truckload sold through the retail outlet is $210, and the profit is $400 for each truckload sold to the chain store. How many shipments per week should be scheduled to each store in order to maximize profit if the manufacturer is limited to no more than 40 hours of travel time per week and a cost of $1600?

Problem

Translate the problem into mathematical language.

SOLUTION From the last sentence of the problem the variables can be determined. They are

x = the number of trips per week to the retail outlet
y = the number of trips per week to the chain store
P = profit

Since the constraints involve time and cost, they are the labels of the rows. The data can then be put into the table. (See Table 3.)

TABLE 3

	Retail Outlet	Chain Store	Limiting Value
Time for a round trip (hours)	2	4	40
Cost of a round trip ($)	50	200	1600
Profit ($)	210	400	

The first two rows represent the constraints on time and cost. Since each uses the phrase "no more than," the inequalities involve $\leq$. They are $2x + 4y \leq 40$ and $50x + 200y \leq 1600$. The third row, representing profit, is the objective function, $P = 210x + 400y$. The problem can be stated as follows:

$$\text{Maximize} \quad P = 210x + 400y \qquad \text{objective function}$$

$$\text{Subject to} \quad \left. \begin{array}{r} 2x + 4y \leq 40 \\ 50x + 200y \leq 1600 \\ x \geq 0, \, y \geq 0 \end{array} \right\} \quad \text{constraints}$$

The method of solving linear programming problems involves two distinct operations. First the word problem must be expressed in mathematics and then the mathematical formulation must be solved. This section concentrates on the first skill, translating the word problem into mathematics. In the next section the method of solving the mathematical formulation of the linear programming problem will be studied.

When doing the exercises it is important to distinguish the objective function from the constraints. This is because the solution in the next section treats them separately. The first step in the solution is to graph the system of linear inequalities formed by the constraints. This graph produces a region in the plane. It is called the region of feasible solutions because the points in the region are the only ones that are allowed to be substituted into the objective function. Any point that is not in this region is not a possible solution. Then the points of intersection of the boundaries of the region are located. These points are called corner points. The corner points are then substituted into the objective function to find the solution. In the next section we will see why this method produces the solution.

EXERCISE SET 3-2

- In all the problems translate the linear programming problem into mathematical language.

1. A clothing manufacturer makes two types of outerwear, a wool coat and a polyester all-weather coat. The wool coat sells to the retail stores for $190 each and the polyester coat for $130 each. The cost for manufacturing each wool coat is $150 and the cost of each polyester coat is $100. If the manufacturer makes no more than 100 coats a week and budgets no more than $12,000 per week, how many of each type should be made to maximize profit?

2. A clothing manufacturer makes two types of jogging attire, regular and designer. The attire sells to the retail stores for $10 each for the regular suits and $40 each for the designer suits. The cost for manufacturing each regular suit is

$5 and it is $10 for a designer suit. If the manufacturer makes no more than 200 suits a week and budgets no more than $1500 per week, how many of each type should be made to maximize profit?

3. A hospital nutritionist prepares a menu for patients on low-fat diets. The primary ingredients are chicken and corn. The meal must contain at least 45 grams (g) of protein and at least 15 mg of iron. Each 3-ounce serving of chicken contains 45 g of protein, 9 mg of iron, and 4 g of fat. Each 1-cup serving of corn contains 9 g of protein, 6 mg of iron, and 2 g of fat. How much chicken and corn should be included in the meal in order to minimize the amount of fat?

4. Work Problem 3 with the constraint that the nutritionist wants the meal to contain at least

the full U.S. RDA for protein but requires only that at least one-half the U.S. RDA for iron be met.

5. The computer center director at a large university wants to staff consulting stations with two types of shifts: type 1 will contain 2 senior programmers and 1 student worker and type 2 will consist of 1 senior programmer and 5 student workers. The director wants to use no more than 36 individuals. There will be at least 24 hours to be filled during the week, with a type 1 shift serving for 3 hours and a type 2 shift serving for 2 hours. The cost of a type 1 shift is $44 per hour and $40 per hour for a type 2 shift. Determine the number of shifts of each type in order to minimize cost.

6. Work Problem 5 if a type 2 shift is to consist of 1 senior programmer and 7 student workers, so that the cost of a type 2 shift is $48, and all other figures remain the same.

7. A bicycle manufacturer has a retail outlet and, in addition, sells some bikes to a large chain store. One truck makes all deliveries, and it takes 3 hours to make a round trip to the retail store and 4 hours to make a round trip to the chain store. The cost of a round trip to the retail store is $100 and the cost of a round trip to the chain store is $200. The profit of each truckload sold through the retail store is $500; it is $800 for each truckload sold to the chain store. How many shipments per week should be scheduled to each store in order to maximize profit if the manufacturer is limited to no more than 48 hours per week and a cost of $2000?

8. Work Problem 7 if the profit on each truckload sold through the retail store is $600 and it is $800 for a truckload sold to the chain store.

9. A meat manufacturer mixes beef and pork in sausage links. Two types of links are made, regular and "all-beef." Each pound of regular sausage meat contains 0.3 pound (lb) of beef and 0.2 lb of pork, while each pound of all-beef sausage meat contains 0.4 lb of beef and 0.1 lb of pork. There are 120 lb of beef and 60 lb of pork in stock. If the profit for regular links is 50 cents per pound and the profit for all-beef links is 70 cents per pound, how many pounds of each type link should be made to maximize profit?

10. A consortium of travel agents has sold 1200 tickets to the Super Bowl. The weekend package includes air fare, and they have a choice of two types of airplanes for the charter flights. Type 1 can carry 100 passengers and type 2 can carry 150 passengers. Each flight of a type 1 aircraft will cost $9000 and each type 2 flight will cost $15,000. The consortium is allowed to lease no more than ten planes. How many airplanes of each type should be leased in order to minimize cost?

•• 11. The administration of a money market fund wants to invest up to $15 million in two types of investments. The funds are to be divided between short-term bank notes and Treasury notes. The current yield for bank notes is 15% and for Treasury notes it is 12%. Because the yield for bank notes fluctuates unpredictably, it has been decided that at least twice as much money is to be invested in Treasury notes as in bank notes. At least $2 million must be allocated in each type of investment. How much money should be invested in each type of investment to produce the largest gain?

12. A window manufacturer produces windows in two styles, regular and thermopane. It costs $100 to make each regular window, which the manufacturer sells for $150. It costs $120 to make each thermopane window and each sells for $175. The daily production capacity is 110 windows and the daily cost cannot exceed $12,000. How many windows of each type should be made per day to maximize profit?

13. An automobile leasing company buys two types of cars, compacts and midsized. The company wants to purchase at most 1000 cars. The cost of each compact is $8000 and of each midsized car is $10,000. The company has decided to allocate no more than $9 million for the pur-

chase of the cars. It must purchase at least 200 of each type of car. If the anticipated return from the sale of the cars at the end of the company's three-year leasing period is $1000 for a compact and $1200 for a midsized car, how many of each type should be purchased in order to maximize return?

14. Officials of the state high school basketball tournament must plan accommodations for at least 660 players. They have two types of rooms available, dormitory rooms that sleep three people and motel rooms that sleep two people. For meals they have budgeted $10 daily per person for those in dorms and $20 daily per person for those in motels, and they must not spend more than $12,000 in meal money per day. They have available at most 200 dormitory rooms and 150 motel rooms. If the daily cost of a room is $20 for a dormitory room and $30 for a motel room, how many rooms of each kind must they schedule in order to minimize cost?

15. A heating oil delivery firm has two processing plants, plant A and plant B. Plant A processes daily 150 barrels of high-grade oil for commercial use and 50 barrels of low-grade for residential use. Plant B processes daily 100 barrels of high-grade oil and 50 barrels of low-grade oil. The daily cost of operation is $20,000 for plant A and $15,000 for plant B. An order is placed for 1000 barrels of high-grade oil and 450 barrels for low-grade oil. Find the number of days that each plant should operate to fill the order and minimize the cost?

●●● 16. An automobile leasing company has warehouses in Los Angeles and San Francisco. A company in Las Vegas sends an order to lease 25 cars and a firm in Fresno orders 15 cars. The Los Angeles warehouse has 30 cars and the San Francisco warehouse has 20 cars. It costs $100 to ship a car from Los Angeles to Las Vegas, $90 to ship a car from Los Angeles to Fresno, $150 to ship a car from San Francisco to Las

Vegas, and $80 to ship a car from San Francisco to Fresno. Find the number of cars to be shipped from each warehouse to each city in order to minimize cost.

17. A consortium of travel agents has sold 1800 tickets to the Super Bowl. The weekend package includes air fare, and there is a choice of two types of airplanes for the charter flights. Type 1 can carry 100 passengers and type 2 can carry 300 passengers. Each flight of a type 1 aircraft will cost $9000 and each type 2 flight will cost $15,000. The consortium is allowed to lease no more than 12 planes. Each type 1 flight requires 2 stewards and each type 2 flight requires 3 stewards. No more than 30 stewards are available. How many airplanes of each type should be leased in order to minimize cost?

18. A manufacturer makes two types of sausage meat. Each pound of type 1 sausage meat contains 3 ounces (oz) of beef, 1 oz of pork, and 2 oz of spices. Each pound of type 2 sausage meat contains 6 oz of beef, 3 oz of pork, and 1 oz of spices. There is 225 lb of beef, 100 lb of pork, and 200 lb of spices in stock. If the profit for type 1 sausage meat is 90 cents per pound and the profit for type 2 sausage meat is 70 cents per pound, how many pounds of each type should be made to maximize profit?

19. In 1967 the Missouri Air Conservation Commission adopted air pollution regulations for metropolitan St. Louis that created heated controversy. Coal burning accounted for more than 30% of the pollutants, and a major coal producer sought an injunction to prohibit enforcement of the regulations, claiming, among other things, that it was economically infeasible. A great deal of study ensued, thereby establishing precedent on the legality of pollution standards. A study by Robert Kohn applied linear programming to determine the economic impact of the regulations on the coal industry.* One problem investigated by Kohn dealt with two methods of

*Kohn, Robert, "Application of Linear Programming to a Controversy on Air Pollution Control," *Management Science*, 17:609–621, 1971.

controlling the pollutant ammonia nitrate. The two control methods were the use of a wet scrubber and a cyclone collector. The regulation required that there be a reduction of at least 165 million lb of sulfur dioxide. Another constraint was that if the two controls operated at peak efficiency, the largest reduction of particulates would be 230 million lb. For each ton of ammonia nitrate that was controlled by the wet scrubber, there was a reduction of sulfur dioxide by 9.2 lb and particulates by 7.6 lb. For each ton of ammonia nitrate that was controlled by the cyclone collector, there was a reduction of sulfur dioxide by 4.6 lb and particulates by 3.1 lb. If the cost per ton was $.07 to operate the wet scrubber and $.03 to operate the cyclone collector, find the number of tons of ammonia nitrate production controlled by each type of control method in order to minimize cost.

20. The allocation of space is often a complicated problem for large enterprises. Large school systems and universities especially have a major task assigning classroom space. Gosselin and Truchon devised a linear programming model to help solve this type of allocation of space problem. The procedure has been applied at various locals, including the Université Laval.[†] As an example of their procedure, consider a seg- ment of a university that uses rooms with 20 seats, called preferred rooms, and rooms with 40 seats, called regular rooms. The total number of these rooms available is 50. There are 1600 students to accommodate. The objective is to minimize the number of times a request for a particular type of room is denied. This is called a penalty. Suppose 10% of requests for preferred rooms are usually denied and 5% of requests for regular rooms are usually denied. Find the number of each type of room that should be allocated to this segment of the university to minimize the penalty function.

21. In his book on linear programming, G. H. Symonds describes one of the first practical applications of the subject, which centered on the solution of oil refinery production problems.[‡] The following problem is adapted from this book. An oil refinery produces at most 2000 gallons of fuel oil per day. Two types are produced, commercial fuel oil and residential fuel oil. Blending requirements are that the number of gallons of commercial fuel oil be at least three times as many as the number of gallons of residential fuel oil. If the profit per gallon is $.34 for commercial fuel oil and $.22 for residential fuel oil, how many gallons of each type should be produced to maximize profit?

[†]Gosselin, K., and Truchon, M., "Allocation of Classrooms by Linear Programming," *Journal Operational Research Society*, 37:561–569, 1986.

[‡]Symonds, G. H., *Linear Programming: The Solution of Refinery Problems*, Esso Standard Oil Co., New York, 1955.

CASE STUDY   Federal Express Corporation*

Seldom does a research paper in a college political science course amount to more than simply a grade for the writer. But Mr. Frederick W. Smith's 1964 paper, written while he was a student at Yale University, describing a new type of business venture, was put into practice in 1972. It was then that Mr. Smith founded a company he called Federal Express Corporation. His concept was to airlift small packages overnight with door-to-door pickup and delivery.

*Adapted from John C. Camillus, "Federal Express Corporation: History and Concept of Business," in *Business Policy and Strategy Management*, by William Glueck and Laurence Jauck (eds.), New York, McGraw-Hill Book Co., 1980.

Just to open for business Mr. Smith needed a great deal of capital and planning. All airplanes, delivery trucks, offices, personnel, and government licenses had to be in place. In the eight years since writing his political science paper Mr. Smith firmed up his ideas while running a very successful airplane company. After leaving the Marines he purchased a company that serviced airplanes and gradually changed the business to buying and selling used aircraft. Over a two-year period Mr. Smith increased sales from $1 million to $9 million and turned its profit situation from a $40,000 loss to a $250,000 profit.

To verify the profit potential of his overnight delivery concept, Mr. Smith hired two business consultant firms. Both were enthusiastic. In fact the president of one firm, Mr. Arthur Bass, joined Mr. Smith as a Senior Vice-President for Planning in 1972. It was Mr. Bass who decided to use analytical methods to help in making the complex decisions Federal Express would have to face in the near future. According to Mr. Bass, it was his job to translate Mr. Smith's far-reaching ideas into reality, or to turn "Mr. Smith's view of Federal Express in 1990 into a profit in the 1970s." It was this view to the future, their ability to visualize and anticipate future trends, that provided them with almost limitless potential and distinguished them from the competition. In 1978, reflecting on the meteoric growth of Federal Express, Mr. Smith said:

> In five years we have gone through phases of growth which other companies normally experience over a period of twelve to fifteen years. From June 1977 to June 1978 alone our revenues have jumped from $108 million to $160 million. In order to understand where we are, we have to look not only at the past but also at the future.

The past achievements of Federal Express and potential growth in a market they defined were summarized in an article that appeared in *The South Magazine,* March–April 1975, which stated that:

> If anyone wants a refresher course in the free enterprise system, in the basic premise that creativity, dedication and sweat adds up to better, in the assumption that a work force can be inspired by the challenge of its mission, and that free give-and-take competition equals progress, then go to Federal Express. It is undoubtedly one of America's great adventures.

The key ingredient in Federal Express's strategy was that in a developed economy such as in the United States, with advanced technology and high capital investment, the need for swift transportation of parts and supplies is inevitable and frequent. While the needs of any individual customer might be unpredictable, from a statistical point of view, the timing and transportation requirements between large population centers would be fairly constant.

Federal Express's management believed that they could meet these demands and win 100% of the market if they could satisfy six primary requirements of their customers. They were: (1) coverage—they had to service all major population centers, (2) timing—they had to deliver before noon of the following day, (3) reliability—they would guarantee delivery, (4) cost—equitable rates were needed, (5) billing—it must be timely and accurate, (6) service—door-to-door pickup and delivery were essential.

Federal Express's most important consideration was the type of aircraft to use. The initial decision was guided by the Civil Aeronautics Board (CAB) ruling that distinguished between "air taxi" operators and cargo firms. The former could fly anywhere in the United States without the need to obtain a certificate of "public necessity and convenience." Operators not falling into this classification had to obtain a certificate for each city-pair route, of which Federal Express had 5000. No new certificate had been granted in over 20 years, as existing carriers' objections had to be overruled, with the burden of proof resting on the applicant. Federal Express's strategy to start up overnight with a multitude of city-pair routes made it impossible to classify their business as air cargo. They had to fall into the air taxi classification. This meant that the capacity of their planes was limited to 7500 pounds (lb). They chose the French-built Dassault Fanjet Falcon 20, and later added some Learjets.

These served as their only aircraft until 1978 when Federal Express's Washington lobbying resulted in an amendment to the CAB ruling, which essentially exempted cargo carriers from CAB control. This meant Federal Express could purchase larger aircraft. The obstacle was that their delivery system, based on the swift economy of small aircraft, was working efficiently and profitably. Would even a partial shift to large planes enhance operating efficiency? And if so, what mix would be most desirable?

United Airlines was retiring 20 Boeing 727s and Federal Express's management decided to analyze the feasibility of purchasing some or all of them. A comparison of the two types of planes yielded the following data:

	Boeing 727	**Falcon 20**
Direct operating cost	$1400/hr	$500/hr
Payload	42,000 lb	6000 lb

Hourly operating cost was limited to $35,000 and total payload had to be at most 672,000 lb. These facts translate into two constraints, but a third constraint was that only twenty 727s were available. The objective was to maximize the number of aircraft given these constraints.

To solve the problem, let

x = the number of 727s
y = the number of Falcon 20s

Then the problem becomes

Maximize $N = x + y$ subject to

$$1,400x + 500y \leq 35,000$$
$$42,000x + 6,000y \leq 672,000$$
$$x \geq 0, y \geq 0, x \leq 20$$

In the next section we will see how to find the solution to this type of problem. The solution is (10, 42). That is, to maximize the number of aircraft under the given constraints, Federal Express should purchase ten 727s and 42 Falcon 20s. (See Figure 1.)

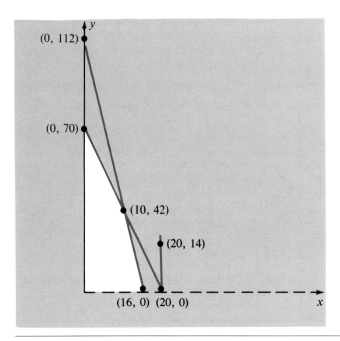

FIGURE 1

In 1978, following the removal of the CAB regulation, Federal Express purchased 10 Boeing 727s from United Airlines. The management of Federal Express was sure that the choice of the Fanjet Falcons and the Learjets played a vital role in the firm's early success. But as competition grew, both from new private companies and the U.S. Post Office, the larger aircraft would provide more flexibility and a wider range of cargo capability. In fact, as they were purchasing the 727s, Federal Express management was studying the feasibility of using 747s.

By the way, Mr. Smith received a grade of C on his paper.

3–3 GRAPHICAL SOLUTIONS TO LINEAR PROGRAMMING PROBLEMS

Once a linear programming problem in two variables has been translated into mathematical language, then the graphical solution can be applied. In this section we demonstrate this geometric approach to the solution. It is effective in handling problems in two unknowns, like those in Section 3–2. For problems with more variables the simplex method is used, which is covered in Chapter 4.

The Manufacturing Problem (Revisited)

Let us demonstrate the steps of the graphical method by looking at the manufacturing problem of Section 3–2. A clothing manufacturer made two types of suits. The problem was to maximize profit subject to constraints of cost and capacity. We start with the mathematical formulation of the problem.

Maximize $P = 60x + 50y$ subject to

$$100x + 40y \leq 6000$$
$$x + y \leq 120$$
$$x \geq 0, y \geq 0$$

The task is to find the values of x and y that simultaneously satisfy the inequalities and yield the largest value of P. The method is separated into three steps: The first step is to graph the inequalities. The second is to find the **corner points,** which are the points of intersection of the boundaries of the region of solutions of the inequalities. These skills have already been covered in Sections 3–1 and 1–3, respectively. The only new idea is the definition of "corner point." The third step is to find which corner point yields the largest number when substituted into the objective function. We will solve the manufacturing problem and then show why the method works.

1. *Graph the region defined by the inequalities.* As in Section 3–1, replace the inequality signs with equal signs and then graph the lines. Plot the intercepts and draw a line through the two points. For each line shade the region that is not part of the solution. The unshaded region is the solution of the inequalities. (See Figure 1.)

FIGURE 1

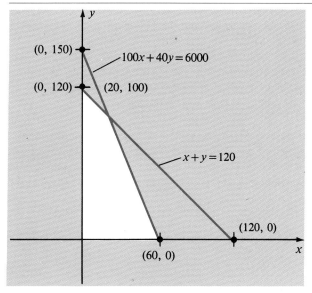

$$x + y \leq 120$$
$$100x + 40y \leq 6000$$
$$x \geq 0$$
$$y \geq 0$$

2. *Find the corner points.* The boundary of the solution of the unshaded region consists of the line segments of the four lines $x + y = 120$, $100x + 40y = 6000$, $x = 0$, and $y = 0$. From Figure 1 the boundary consists of four line segments, and thus there are four points of intersection. Three of them were determined when we graphed the lines: $(0, 0)$, $(60, 0)$, and $(0, 120)$. The fourth is the point of intersection of $x + y = 120$ and $100x + 40y = 6000$. By the methods of Section 1–3, multiply the first equation by -40 and add the two equations. The point of intersection is $(20, 100)$. Hence the four corner points are $(0, 0)$, $(60, 0)$, $(0, 120)$, and $(20, 100)$.

3. *Evaluate the objective function at the corner points.* Put the computations in the following table:

Corner Point	Value of $60x + 50y = P$
$(0, 0)$	$60 \cdot 0 + 50 \cdot 0 = 0$
$(60, 0)$	$60 \cdot 60 + 50 \cdot 0 = 3600$
$(0, 120)$	$60 \cdot 0 + 50 \cdot 120 = 6000$
$(20, 100)$	$60 \cdot 20 + 50 \cdot 100 = 6200$

The largest value of P is 6200. Therefore the solution is $(20, 100)$; it occurs when $x = 20$ and $y = 100$. Referring to the meaning of the description of the manufacturing problem, the largest profit is \$6200 when the manufacturer makes $x = 20$ wool suits and $y = 100$ blend suits.

The Graphical Method of Solution

Before presenting the general method we need some terminology. The set of points that are solutions of the constraints, the unshaded region in Figure 1, is called the **region of feasible solutions.** The feasible solutions are the only points in the plane that have a chance to be solutions of the linear programming problem. The vertices of the boundaries are called *corner points* and the **optimal solution** is the point (or points) that yields the largest or smallest value of the objective function, depending on whether it is to be maximized or minimized.

It is possible that a linear programming problem does not have a solution. For instance, if the region of feasible solutions is unbounded, then the objective function $z = x + y$ cannot be maximized. This means that if the region is bounded, the solution occurs at a corner point. If the region is unbounded, first check whether a solution exists. If it does, it will occur at a corner point.

The Graphical Method for Solving a Linear Programming Problem Expressed in Mathematical Language

1. Graph the constraints and find the region of feasible solutions.
2. Find the corner points of the region of feasible solutions. They are the points of intersection of the boundary line segments.

3. Evaluate the objective function at each corner point. The optimum solution, if it exists, is that point that yields
   (a) the largest value if the objective function is to be maximized.
   (b) the smallest value if the objective function is to be minimized.

Justification of the Method

The natural question to ask is why does the method work? There are an infinite number of points in the region of feasible solutions, each of which could possibly be a point that maximizes P. How can we be sure that the correct point is among the corner points? Let us answer this question for the manufacturing problem; the essence of the argument can be used to show that the method works for any linear programming problem.

The problem is to find values of x and y that are in the region of feasible solutions and yield the largest value of $P = 60x + 50y$. If P takes on a value, say, $P = 600$, then this equation becomes $600 = 60x + 50y$, which is a linear equation in x and y. The slope-intercept form of this equation is found by solving for y; so $50y = 600 - 60x$ and dividing by 50 yields the slope-intercept form

$$y = -(\tfrac{6}{5})x + 12$$

This line has slope $-\tfrac{6}{5}$ and y-intercept 12. Now let P take on another value, say, $P = 6200$. This is another linear equation in x and y. Its slope-intercept form is

$$y = -(\tfrac{6}{5})x + 124$$

This line has the same slope as the previous line. In general, if P takes on a value, say, $P = P_0$, then this expression is a linear equation in x and y whose slope is $-\tfrac{6}{5}$ and whose y-intercept is $P_0/50$. This means that all of these equations are parallel and they differ only in their y-intercept. Figure 2 gives some examples of the lines $60x + 50y = P_0$.

The optimal solution is a point in the region of feasible solutions that is also substituted into the objective function, $P = 60x + 50y$. Graphically this means we are looking for a point in the region of feasible solutions that is also on one of the lines $60x + 50y = P_0$. We need the point that yields the largest value of P_0. Making P_0 as large as possible means choosing the line with the largest y-intercept. Therefore we are searching for the line with the largest y-intercept that also touches the region of feasible solutions. From Figure 2 that line is the line that goes through (20, 100). Any line below it has a smaller y-intercept and any line above it does not touch the region of feasible solutions. This is why (20, 100) is the optimal solution.

This argument shows that the line with the largest y-intercept (or smallest if the objective function is to be minimized) that also touches the region of feasible solutions must be a corner point. Thus the corner points are the only ones that must be evaluated; the optimal solution must be one of them.

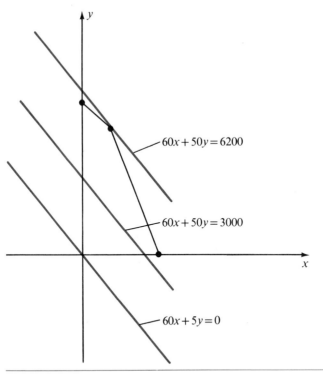

FIGURE 2   Some examples of the lines $60x + 50y = P_0$, the lines with slope $-\frac{6}{5}$.

We have implied that there may be more than one optimum solution. This can happen if the objective function has the same slope as one of the boundaries. In Figure 3 not only are points A and B optimum solutions, but all points on the line segment between A and B are also optimum solutions. Since they lie on the same line they yield the same value of the objective function.

Applying the Procedure

We can now apply the graphical method to solve the problems that were translated into mathematical language in Section 3–2.

EXAMPLE 1 | **Problem**

Find the optimum solution to the diet problem in Example 2 of Section 3–2.

Minimize $F = 4x + 2y$
Subject to $45x + 9y \geq 45$
$10x + 6y \geq 20$
$x \geq 0, \, y \geq 0$

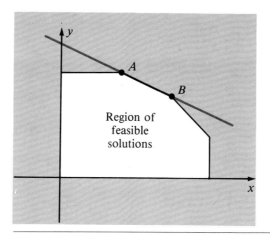

FIGURE 3

SOLUTION

1. Graph the constraints and find the region of feasible solutions. The inequality signs are replaced by equality signs and the intercepts are plotted. The graph is in Figure 4; it is the infinite region above the line segments.
2. The corner points are (2, 0), (0, 5), and (0.5, 2.5). The latter is the point of intersection of the two lines $45x + 9y = 45$ and $10x + 6y = 20$.

FIGURE 4

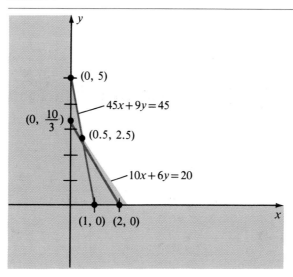

$$45x + 9y \geq 45$$
$$10x + 6y \geq 20$$
$$x \geq 0$$
$$y \geq 0$$

3. Evaluate the objective function at each corner point. The computations are in the following table:

Corner Point	Value of $4x + 2y$	$= F$
(2, 0)	$4 \cdot 2 + 2 \cdot 0$	$= 8$
(0, 5)	$4 \cdot 0 + 2 \cdot 5$	$= 10$
(0.5, 2.5)	$4 \cdot 0.5 + 2 \cdot 2.5$	$= 7$

The smallest value of F is 7. Therefore the solution is (0.5, 2.5); it occurs when $x = 0.5$ and $y = 2.5$. Referring to the meaning of the description of the diet problem, the optimum meal with the smallest amount of fat has 0.5 serving of chicken (1.5 ounces) and 2.5 servings of corn (2.5 cups).

If the problem in Example 1 called for maximizing the objective function, there would be no solution because the region of feasible solutions is unbounded. The objective function, $F = 4x + 2y$, can be made as large as can be by choosing large enough values of x and y. Since the region is unbounded, such values can always be found in the region. This would be an example of the situation where you must check to see if the problem has a solution before applying the method.

EXAMPLE 2 | **Problem**

Find the optimum solution to the transportation problem in Example 3 of Section 3–2.

$$\text{Maximize} \quad P = 210x + 400y$$
$$\text{Subject to} \quad 2x + 4y \le 40$$
$$50x + 200y \le 1600$$
$$x \ge 0, y \ge 0$$

SOLUTION

1. Graph the constraints and find the region of feasible solutions. (See Figure 5.)
2. The corner points are (20, 0), (0, 8), and (8, 6).
3. Evaluate the objective function at each corner point.

Corner Point	Value of $210x + 400y = P$
(20, 0)	$210 \cdot 20 + 400 \cdot 0 = 4200$
(0, 8)	$210 \cdot 0 + 400 \cdot 8 = 3200$
(8, 6)	$210 \cdot 8 + 400 \cdot 6 = 4080$

The optimum solution is (20, 0). The manufacturer should schedule 20 trips to the retail outlet and no trips to the chain store each week to maximize profit.

$$2x + 4y \leq 40$$
$$50x + 200y \geq 1600$$
$$x \geq 0$$
$$y \geq 0$$

FIGURE 5

EXERCISE SET 3–3

• In Problems 1 to 6 find the graph of the set of feasible solutions and the corner points.

1. $x + 4y \leq 12$
   $x + 3y \leq 10$
   $x \geq 0, y \geq 0$

2. $x + y \leq 10$
   $3x + 2y \leq 12$
   $x \geq 0, y \geq 0$

3. $x + 6y \leq 18$
   $2x + 3y \geq 12$
   $x \geq 0, y \geq 0$

4. $x + 3y \leq 12$
   $2x + 5y \leq 10$
   $x \geq 0, y \geq 0$

5. $2x + 3y \leq 10$
   $4x + 7y \geq 28$
   $x \geq 0, y \geq 0$

6. $3x + 2y \geq 18$
   $x + 4y \leq 12$
   $x \geq 0, y \geq 0$

In Problems 7 to 12 solve the linear programming problem with the given objective function and the constraints from the previous problem in this section's exercises.

7. Maximize $z = 4x + 10y$ subject to the constraints in Problem 1.

8. Minimize $z = 3x + 4y$ subject to the constraints in Problem 1.

9. Minimize $z = 20x + 18y$ subject to the constraints in Problem 3.

10. Maximize $z = 10x + 12y$ subject to the constraints in Problem 2.

11. Minimize $z = 20x + 18y$ subject to the constraints in Problem 5.

12. Maximize $z = 10x + 12y$ subject to the constraints in Problem 6.

In Problems 13 to 22 solve the indicated problem from the exercise set in Section 3–2.

13. Problem 1

14. Problem 2

15. Problem 3

16. Problem 4

17. Problem 5

18. Problem 6

19. Problem 7

20. Problem 8

21. Problem 9

22. Problem 10

•• In Problems 23 to 30 solve the indicated problem from the exercise set in Section 3–2.

23. Problem 11

24. Problem 12

25. Problem 13

26. Problem 14

27. Problem 15

••• 28. Problem 16

29. Problem 17

30. Problem 18

In Problems 31 to 38 find the values of x and y that maximize $P = 2x + 3y$ subject to the given constraints.

31. $x + 3y \le 6$
     $x + 4y \ge 4$
     $x \ge 1, y \ge 1$

32. $x + 3y \le 3$
     $x + 2y \le 6$
     $x \ge 1, y \ge 2$

33. $4x + y \le 8$
     $x + y \ge 5$
     $x + 3y \ge 9$
     $x \ge 0, y \ge 0$

34. $x + y \le 8$
     $x + 2y \le 10$
     $2x + 5y \ge 10$
     $x \ge 0, y \ge 0$

35. $x + y \le 10$
     $x + 2y \le 12$
     $2x + y \le 12$
     $x \ge 0, y \ge 0$

36. $x + y \le 10$
     $x + 2y \ge 12$
     $2x + y \le 12$
     $x \ge 0, y \ge 0$

37. $x + y \le 10$
     $x + 2y \ge 12$
     $2x + y \ge 12$
     $x \ge 0, y \ge 0$

38. $x + y \le 20$
     $x + 3y \ge 30$
     $y \le 25$
     $x \ge 5, y \ge 0$

39. A firm manufactures two types of lamps, gold-plated and chrome-plated. Each gold-plated lamp requires 2 hours of plating, 1 hour of finishing, and 3 hours of assembling. Each chrome-plated lamp requires 2 hours of plating, 2 hours of finishing, and $1\frac{1}{2}$ hours of assembling. The capacity of the plating department is 200 hours, the capacity of the finishing department is 120 hours, and the capacity of the assembly department is 180 hours. If the profit of each gold-plated lamp is $50 and the profit of each chrome-plated lamp is $30, find the number of each type of lamp that should be manufactured to maximize profit.

40. A hospital nutritionist prepares a menu consisting of chicken and rice that must contain at least 40 units of protein, at least 90 units of carbohydrates, and at least 50 units of iron. Each serving of chicken contains 4 units of protein, 1 unit of carbohydrates, 1 unit of iron, and 4 units of fat. Each serving of rice contains 1 unit of protein, 1 unit of carbohydrates, 3 units of iron, and 5 units of fat. How many servings of chicken and rice should the menu contain to minimize the amount of fat?

41. A meat manufacturer mixes beef and pork in sausage links. Two types of links are made, regular and spicy. Each pound of regular sausage meat contains 0.1 pound (lb) of beef, 0.1 lb of pork, and 0.1 lb of spices. Each pound of spicy sausage meat contains 0.3 lb of beef, 0.1 lb of pork, and 0.2 lb of spices. The manufacturer can use no more than 120 lb of beef and 80 lb of pork, but at least 100 lb of spices must be used. The profit of each pound of regular sausage meat is $.40, while the profit of each pound of spicy sausage meat is $.60. Find the number of pounds of each type of sausage meat that should be made to maximize profit.

42. Solve Problem 41 if the manufacturer is also required to manufacture at least 100 lb of regular sausage meat but no more than 700 lb of spicy sausage meat.

43. Many farming strategies require crop rotations. In a model that was tested on a northeastern Oregon farm, researchers studied crop rotations of four crops: corn, potatoes, wheat and al-

falfa.* In one part of the study, the number of acres planted with corn and the number of acres planted with potatoes were investigated. The total number of acres planted was limited to at most 100. Each acre of corn required 1 lb of nitrogen while each acre of potatoes required 3 lb of nitrogen. No more than 270 lb of nitrogen was to be used. Find the number of acres of each type of crop that should be planted to maximize profit if the profit per acre was $126 for corn and $1600 for potatoes.

44. In an article exploring new techniques in the application of linear programming to various aspects of planning and management, Robert Bland studied a small brewery whose ale and beer are in high demand but whose production is limited by materials that are in short supply.† Ale and beer are made from different proportions of corn, hops and malt. Each barrel of ale requires 5 lb of corn, 4 oz of hops and 35 lb of malt. Each barrel of beer requires 15 lb of corn, 4 oz of hops and 20 lb of malt. The brewery has on hand 480 lb of corn, 160 oz of hops and 1190 lb of malt. If the profit per barrel is $13 for ale and $23 for beer, how many barrels of each type should be produced to maximize profit?

45. An integrated crop and intensive beef production firm is a farm enterprise that grows much of its own foodstuffs for its cattle. To operate with a minimum cost goal requires detailed long-term planning. In an article applying linear programming to the solution of this type of problem, J. J. Glen considered various foodstuff rations for the livestock.‡ In one part of the study, two variables were considered, the number of cattle fed primarily enterprise-grown wheat and the number of cattle fed primarily enterprise-grown hay. The number of cattle in the study was at least 500. Each animal fed primarily enterprise-grown wheat consumed 0.2 hectare of foodstuff while each animal fed primarily enterprise-grown hay consumed 0.7 hectare of foodstuff. The number of hectares for these animals was limited to 140. Find the number of animals that should be given each type of ration to minimize cost if the cost is $35 to feed an animal the wheat ration diet and the cost is $30 to feed an animal the hay ration diet.

*El-Nazer, Talaat, and McCarl, Bruce, "The Choice of Crop Rotation: A Modeling Approach and Case Study," *American Journal of Agricultural Economics*, 68:127–136, 1986.

†Bland, Robert G., "The Allocation of Resources by Linear Programming," *Scientific American*, June 1983, pp. 126–144.

‡Glen J. J., "A Linear Programming Model for an Integrated Crop and Intensive Beef Production Enterprise," *Journal Operational Research Society*, 37:487–494, 1986.

CASE STUDY   The Berlin Airlift

One of the first concrete problems that George Dantzig's linear programming helped solve was the massive supply to West Berlin shortly after World War II.

Berlin surrendered to the Russian army on May 2, 1945. During the following weeks the Russians shipped most of the city's industrial goods to Russia. The American, British, and French troops did not arrive until July 1945. Even before the war ended the Allies decided to divide the city into four sectors, each country occupying one. Berlin lay deep in the Russian sector of the country, but the Western powers assumed that the Soviets would allow them free access to the city. However, on June 24, 1948 the Soviets blocked all land and water routes through East Germany to Berlin. They hoped to drive the Western powers from East Germany. The problem facing the Western Allies was how to keep West Berlin supplied during the Russian blockade.

The problem was turned over to the Planning Research Division of the U.S. Air Force. Their staff had been working on similar programming problems, most of which were theoretical. But now they were asked to apply their new methods to a very practical problem. Their solution helped shatter Soviet hopes of using the blockade to win total occupation of Berlin. A gigantic airlift was organized to supply more than 2 million people in West Berlin. It was a large-scale program that required intricate planning. To break the blockade, hundreds of American and British planes delivered massive quantities of food, clothing, coal, petroleum, and other supplies. At its peak a plane landed in West Berlin every 45 seconds.

The number of variables in the formulation of the problem exceeded 50. They included the number of planes, crew capacity, runways, supplies in Berlin, supplies in West Germany, and money. A simplified version of the problem is to maximize the cargo capacity of the planes subject to the restrictions: (1) The number of planes is limited to 44; (2) the American planes are larger than the British planes and therefore need twice the number of personnel per flight, so if we let one "crew" represent the number of flight personnel required for a British plane, and thus two "crews" are required for an American flight, then the restriction is that the number of crews available is 64; and (3) the cost of an American flight is $9000 and the cost of a British flight is $5000 and the total cost per week is limited to $300,000. The cargo capacity of an American plane is 30,000 cubic feet and 20,000 cubic feet for a British plane.

These data can be arranged in the following table:

	American Planes	British Planes	Limiting Value
Number of planes	x	y	44
Crew capacity	2	1	64
Cost of flights	9,000	5,000	300,000
Capacity	30,000	20,000	

The corresponding linear programming problem is

Maximize $C = 30,000x + 20,000y$ subject to

$$x + y \le 44$$
$$2x + y \le 64$$
$$9,000x + 5,000y \le 300,000$$

The solution is $x = 20$ and $y = 24$. Thus 20 American and 24 British planes should be used to maximize cargo capacity.

CHAPTER REVIEW

Key Terms

3–1 Linear Inequalities

Linear Inequality   Half-Plane

Solution   System of Linear Inequalities

3–2 Linear Programming Problems

Linear Programming Problems   Constraints

Objective Function

3–3 Graphical Solutions to Linear Programming Problems

Region of Feasible Solutions   Optimal Solution

Corner Point

REVIEW PROBLEMS

In Problems 1 to 3 sketch the graph of the linear inequality by shading those points that do not satisfy the inequality.

1. $2x + 3y \geq 12$ **2.** $3x - 5y \leq 30$

3. $5x + 2y < 20$

In Problems 4 to 8 graph the system of linear inequalities.

4. $x + 2y \geq 6$ **5.** $x + 3y \leq 6$
   $x + y \geq 4$ $x - 2y > 4$

6. $5x + 3y \leq 30$ **7.** $x + 3y \leq 6$
   $y \geq 5$ $x - 2y > 6$
                               $x \geq 0, y \geq 0$

8. $2x + 3y \leq 12$
   $x + 3y \geq 9$
   $x \geq 0, y \geq 0$

9. Express the word problem as an inequality and graph its solution. A manufacturer makes two types of stoves, gas and electric. The cost of a gas stove is $280 and the cost of an electric stove is $330. The manufacturer can budget no more than $4800 per week to produce the stoves.

10. Express the word problem as a system of inequalities and graph its solution. A manufacturer makes two types of lamps, large and small. The small lamp sells for $20 and the large lamp sells

for $40. The manufacturer can budget no more than $2400 per week to produce the lamps and can make no more than 100 lamps per week.

In Problems 11 to 14 translate the linear programming problem into mathematical language.

11. A consortium of travel agents has sold 2500 tickets to the world's fair. The weekend package includes air fare, and there is a choice of two types of airplanes for the charter flights. Type 1 can carry 150 passengers and type 2 can carry 200 passengers. Each flight of a type 1 aircraft will cost $12,000 and each type 2 flight will cost $15,000. The consortium is allowed to lease no more than 15 planes. How many airplanes of each type should be leased in order to minimize cost?

12. A door manufacturer produces doors in two styles, regular and deluxe. It costs $100 to make each regular door, which the manufacturer sells for $175. It costs $120 to make each deluxe door and each sells for $250. The daily production capacity is 110 doors and the daily cost cannot exceed $12,000. How many doors of each type should be made per day to maximize profit?

13. Officials of a convention must plan accommodations for at least 800 participants. They have two types of rooms available, motel rooms that sleep three people and hotel rooms that sleep two people. For meals, they have budgeted $20 daily per person for those in motel rooms and $40 daily per person for those in hotels, and they must not spend more than $24,000 in meal money per day. If the daily cost of a room is $40 for a motel room and $60 for a hotel room, how many rooms of each kind must they schedule in order to minimize cost?

14. A manufacturer makes two types of hot dog meat. Each pound of type 1 hot dog meat contains 3 ounces (oz) of beef, 1 oz of pork, and 1 oz of ham. Each pound of type 2 hot dog meat contains 4 oz of beef, 2 oz of pork, and 1 oz of ham. There is 120 pounds (lb) of beef, 50 lb of pork, and 30 lb of ham in stock. If the profit for type 1 hot dog meat is $.30 per pound and the profit for type 2 hot dog meat is $.40 per pound, how many pounds of each type should be made to maximize profit?

15. Find the region of feasible solutions, the corner points, and the point that maximizes $z = 4x + 5y$ subject to

$$2x + 3y \geq 12$$
$$3x + y \leq 11$$
$$x, y \geq 0$$

16. Find the region of feasible solutions, the corner points, and the point that minimizes $z = x + 8y$ subject to

$$x + 3y \geq 24$$
$$x + 2y \leq 20$$
$$x, y \geq 0$$

In Problems 17 to 20 solve the linear programming problem for the given problem in this set of exercises.

17. Problem 11

18. Problem 12

19. Problem 13

20. Problem 14

4 Linear Programming: The Simplex Method

CHAPTER OVERVIEW

In the previous chapter we considered linear programming problems with two variables. The approach was geometric and we could "see" the solution emerging in a figure because it could be viewed graphically. Unfortunately this method does not extend to problems with more than two variables because it is much more difficult, if not impossible, to get a graphical depiction.

The simplex method can handle problems with many variables so it is applied to a wide variety of settings. However, unlike the geometric approach there is little intuition to guide you; thus it will appear to be a "cookbook" approach by contrast.

The first section lays the groundwork by defining an appropriate matrix and a special solution associated with it. The next section shows how to move to the next matrix if the present one does not yield the final solution. Section 4–3 finalizes the method for standard problems and Section 4–4 handles nonstandard problems. Section 4–5 studies a general type of problem.

CASE STUDY PREVIEW

The problems facing George Dantzig and his contemporaries in the late 1940s were monumental. They arose in the military, management, shipping, and economics. The most perplexing part was that most of the problems they faced could be expressed only in systems of equations and inequalities that contained far too many variables—often one hundred or more. Linear programming was created in 1947 when Dantzig invented the simplex method—a way to skirt most of the seemingly necessary calculations.

Today most large corporations have a department called something similar to "Operations Research" or "Industrial Management." They use mathematical analysis to solve problems facing the firm. One of the primary tools of these departments is linear programming. They solve problems in various areas, such as transportation cost, task assignment, production levels, and needs analysis of personnel and machinery.

The Case Study deals with the Shell Oil Company's method of handling the distribution of their oil. Since 20% of Shell's revenue is spent on transportation of oil, an efficient distribution is essential. Shell uses linear programming to determine how many thousands of barrels of oil to ship from their refineries to their storage and reshipment terminals. The Case Study investigates the solution of Shell's oil transportation problem in the Chicago region.

4–1 SLACK VARIABLES AND THE SIMPLEX TABLEAU

At first we solve linear programming problems that are expressed in a specific form. In Section 4–4 we consider other types of linear programming problems and express them in this form. This shows that the method applies in all cases.

Standard Form

Consider the following two linear programming problems. The first is in the proper form but the second is not.

Problem 1	**Problem 2**

Maximize $m = 6x + 8y$ subject to

$$x + y \leq 10$$
$$2x + 3y \leq 12$$
$$x \geq 0$$
$$y \geq 0$$

Minimize $m = 2x + 3y$ subject to

$$x + 2y \geq 15$$
$$4x + 3y \leq 12$$
$$x > -10$$
$$y \geq 0$$

We say that Problem 1 is in *standard form* because it has three special properties.

───────────────────────── **Definition** ─────────────────────────

A linear programming problem is in **standard form** if

i. The objective function is to be maximized.

ii. The variables are all nonnegative; that is, they all are greater than or equal to 0.

iii. In the other constraints the linear expressions involving the variables are less than or equal to a nonnegative constant; that is, they are all in the form

(linear polynomial) $\leq a$

where a is a positive number or 0.

Problem 2 is not in standard form for three reasons: because the objective function is to be minimized, violating condition (i); because x can be negative, violating condition (ii); and because the first constraint is in the form greater than or equal 15, which violates condition (iii).

The First Step: Slack Variables

From this point on until Section 4–4 we assume that each linear programming problem is in standard form. In particular, throughout the entire chapter we assume that all variables are greater than or equal to 0, so we will not continue to write this stipulation.

The first step in the simplex method is to translate each inequality into an equation. To illustrate, consider the inequality

$$2x + y \leq 10$$

The point $(1, 3)$ satisfies the inequality because

$$2(1) + (3) = 5 \leq 10$$

If $2x + y$ is less than 10, then let s_1 represent the difference between these two numbers. Hence

$$2x + y + s_1 = 10$$

Note that $s_1 > 0$; it is a positive number if $2x + y < 10$ or it is 0 if $2x + y = 10$. For the point $(1, 3)$, $s_1 = 5$ because

$$2(1) + 3 + 5 = 10$$
$$\quad\;\diagdown \qquad \diagdown \quad\; \diagdown$$
$$\qquad x \qquad y \quad\; s_1$$

Another way of saying that s_1 is the difference between $2x + y$ and 10 is that s_1 "takes up the slack" between them. For this reason s_1 is called a **slack variable.** Slack variables are used in this way to transform the inequalities of a linear programming problem into equalities. The subscripts indicate the inequality into which the slack variable was introduced. Here is an example to help clarify these ideas.

EXAMPLE 1 | **Problem**

Use slack variables to transform the linear programming problem with inequalities into one with equalities.

Maximize $m = 6x + 8y$ subject to
$$x + \;\; y \leq 10$$
$$2x + 3y \leq 12$$

SOLUTION Use s_1 with the first inequality to get the equation $x + y + s_1 = 10$. For the second inequality we use s_2 and get $2x + 3y + s_2 = 12$. Hence the modified linear programming problem is

Maximize $m = 6x + 8y$ subject to
$$x + \;\; y + s_1 \qquad\quad = 10$$
$$2x + 3y + \qquad + s_2 = 12$$

In Example 1 two new variables, s_1 and s_2, were introduced so the modified system has four variables. Since x and y are greater than or equal to 0 so must s_1 and s_2 be greater than or equal to 0. Thus all four variables satisfy condition (ii) of the definition of standard form.

As in Chapter 2, each variable will be kept in a single column. The constraints are expressed as equations to take advantage of the techniques for solving systems of equations and matrices that were covered in Chapters 1 and 2. Another advantage is that the objective function and the constraints can be treated together in the same system of equations rather than separately.

The Proper Form of the Objective Function

The objective function is not in the same form of an equation as the constraints. It does not have a nonnegative constant to the right of the equals sign. By rearranging terms the objective function can be expressed in proper form. For example, the objective function in Problem 1 can be written as

$$-6x - 8y + m = 0$$

Using this form of the objective function, we can express Problem 1 in terms of linear equations as follows:

Maximize m such that
$$\left. \begin{array}{l} x + y + s_1 \qquad\qquad = 10 \\ 2x + 3y \qquad + s_2 \qquad = 12 \end{array} \right\} \text{ constraints}$$
$$-6x - 8y \qquad\qquad + m = 0 \qquad \text{objective function}$$

Hence the constraints and the objective function form a system of linear equations. The augmented matrix of the system is the matrix A.

$$A = \begin{bmatrix} x & y & s_1 & s_2 & m & \\ 1 & 1 & 1 & 0 & 0 & 10 \\ 2 & 3 & 0 & 1 & 0 & 12 \\ \hline -6 & -8 & 0 & 0 & 1 & 0 \end{bmatrix}$$

The columns are labeled to emphasize the variables to which each corresponds. We also separate the last column corresponding to the constant terms. In addition, the last row corresponding to the objective function must be distinguished from the others.

The Simplex Tableau

The matrix of the linear programming problem (with the constraints and the objective function as the corresponding equations), is called the **initial simplex tableau.** The term "tableau" is synonymous here with "matrix." The term "simplex" refers to the method of solution employed. "Initial" implies that we will derive other tableaus in the simplex method.

A simplex tableau has more columns than rows so there is an infinite number of solutions to the system. Each is a feasible solution to the linear programming problem.

Which feasible solution is best? The key is the form of the initial simplex tableau. In matrix A the third, fourth, and fifth columns each have exactly one entry equal to 1 and all others equal to 0. This type of column is in **basic form** and the variable corresponding to it is called a **basic variable.** Hence the basic variables for matrix A are s_1, s_2, and m. Columns 1 and 2 are not in basic form, so their variables, x and y, are called **nonbasic variables.**

Basic Feasible Solutions

The particular solution derived from the simplex tableau is found by setting the nonbasic variables equal to 0. Then the remainder of the solution is apparent. For example, in matrix A we set the nonbasic variables x and y equal to 0 so the corresponding equations become

$$
\begin{aligned}
s_1 \quad &= 10 \\
s_2 \quad &= 12 \\
m &= 0
\end{aligned}
$$

This type of solution is called a **basic feasible solution** of the linear programming problem. Hence the basic feasible solution for matrix A is $x = y = 0$, $s_1 = 10$, $s_2 = 12$, $m = 0$. We first encountered solutions to systems like this in Section 2–3. The general procedure can easily be extended to linear programming problems with more variables and constraints.

Procedure for Determining a Basic Feasible Solution

Let matrix M be a simplex tableau having some columns in basic form and some not in basic form.
1. Set each nonbasic variable of M equal to 0.
2. The resulting system of equations has one and only one basic variable in each equation. Each equation yields a value in the last column for a basic variable.

The basic feasible solution of a simplex tableau is a corner point of the region of feasible solutions. For the initial tableau it is the origin, which is almost never the optimum solution. The simplex method consists of constructing a sequence of tableaus and basic feasible solutions, which are also corner points of the region of feasible solutions. This sequence ultimately yields an optimal solution to the original linear programming problem. Section 4–2 will describe the construction of this sequence.

Example 2 shows how to apply the procedure for determining a basic feasible solution to tableaus that are not initial ones.

EXAMPLE 2 | Consider the following two examples of simplex tableaus:

(a)
$$\left[\begin{array}{ccccc|c} x & y & s_1 & s_2 & m & \\ -1 & 0 & 1 & -2 & 0 & 2 \\ 1 & 1 & 0 & 1 & 0 & 5 \\ \hline 1 & 0 & 0 & 3 & 1 & 15 \end{array}\right]$$

(b)
$$\left[\begin{array}{ccccccc|c} x & y & z & s_1 & s_2 & s_3 & m & \\ 1 & 1 & 0 & 0 & 1 & 0 & 0 & 2 \\ 0 & 1 & 0 & 1 & -1 & 1 & 0 & 5 \\ 0 & -1 & 1 & -2 & 3 & 0 & 0 & 10 \\ \hline 0 & 2 & 0 & 0 & 2 & 0 & 1 & 15 \end{array}\right]$$

Problem

Identify the basic and nonbasic variables and then compute the basic feasible solution for each matrix.

SOLUTION (a) The columns in basic form are the second, third, and fifth columns, so the basic variables are y, s_1, and m. Thus the nonbasic variables are x and s_2. To compute the basic feasible solution, let $x = s_2 = 0$ and so the equations corresponding to the matrix are

$$\begin{aligned} s_1 &= 2 \\ y &= 5 \\ m &= 15 \end{aligned}$$

The basic feasible solution is $s_1 = 2$, $y = 5$, $m = 15$, and $x = s_2 = 0$.
(b) The basic variables are x, z, s_3, and m and the nonbasic variables are y, s_1, and s_2. Let $y = s_1 = s_2 = 0$, so the equations corresponding to the matrix are

$$\begin{aligned} x &= 2 \\ s_3 &= 5 \\ z &= 10 \\ m &= 15 \end{aligned}$$

The basic feasible solution is $x = 2$, $z = 10$, $s_3 = 5$, $m = 15$, $y = s_1 = s_2 = 0$.

In this section we showed both how to express a linear programming problem so it can be handled by the simplex method and how to compute the basic feasible solution. Example 3 combines these procedures.

EXAMPLE 3 | **Problem**

Find the initial simplex tableau for the linear programming problem and compute its basic feasible solution.

Maximize $m = 2x + 3y + 4z$ subject to
$$x + 2y + 3z \le 12$$
$$2x \quad\quad + 5z \le 10$$
$$3x + y + z \le 6$$

SOLUTION Add slack variables and put the objective function in proper form. The equations are

$$
\begin{aligned}
x + 2y + 3z + s_1 \quad\quad\quad\quad\quad\quad &= 12 \\
2x \quad\quad + 5z \quad\quad + s_2 \quad\quad\quad\quad &= 10 \\
3x + y + z \quad\quad\quad\quad + s_3 \quad\quad &= 6 \\
-2x - 3y - 4z \quad\quad\quad\quad\quad\quad + m &= 0
\end{aligned}
$$

The initial simplex tableau is

$$
\begin{array}{ccccccc}
x & y & z & s_1 & s_2 & s_3 & m \\
\end{array}
$$
$$
\left[
\begin{array}{ccccccc|c}
1 & 2 & 3 & 1 & 0 & 0 & 0 & 12 \\
2 & 0 & 5 & 0 & 1 & 0 & 0 & 10 \\
3 & 1 & 1 & 0 & 0 & 1 & 0 & 6 \\
\hline
-2 & -3 & -4 & 0 & 0 & 0 & 1 & 0
\end{array}
\right]
$$

The basic variables are s_1, s_2, s_3, and m. The nonbasic variables are x, y, and z. To determine the basic feasible solution we set $x = y = z = 0$. The resulting equations are

$$
\begin{aligned}
s_1 \quad\quad\quad\quad &= 12 \\
s_2 \quad\quad\quad &= 10 \\
s_3 \quad\quad &= 6 \\
m &= 0
\end{aligned}
$$

Hence the basic feasible solution is $x = y = z = 0$, $s_1 = 12$, $s_2 = 10$, $s_3 = 6$, $m = 0$.

By way of summary, we give the procedure for translating a linear programming problem in standard form into the mathematical form for solving it by the simplex method.

1. Add a slack variable to each constraint that is expressed as an inequality in order to transform it into an equation.
2. Put the objective function in proper form and include it with the constraints as a system of linear equations.
3. Form the augmented matrix of the system. Label the columns with the corresponding variables. Separate the last row corresponding to the objective function and the last column corresponding to the constant terms.

EXERCISE SET 4–1

• In Problems 1 to 4 determine whether the linear programming problem is in standard form.

1. Maximize $m = 3x + 4y$ subject to
$$x + y \leq 15$$
$$2x - y \leq 6$$
$$x \geq 0, y \geq 0$$

2. Minimize $m = 3x + 5y$ subject to
$$2x + 3y \leq 10$$
$$x + 2y \geq 5$$
$$x \geq 0, y \geq 0$$

3. Maximize $m = 5x - y$ subject to
$$x + 3y \leq 0$$
$$3x + y \leq 8$$
$$x \geq 0, y \geq 0$$

4. Maximize $m = 2x + y + z$ subject to
$$x + 3z \leq 5$$
$$x + y \leq 8$$
$$x + 2y \leq 5$$
$$x \geq 0, y \geq 0, z \geq 0$$

In Problems 5 and 6 add slack variables to the inequalities to translate them into equations.

5. $x + 2y \leq 4$
$x - y \leq 5$

6. $x - y + z \leq 8$
$x + y - z \leq -2$

In Problems 7 and 8 put the objective function in proper form.

7. $m = 3x + 4y$

8. $m = 2x + 8y + 3z$

Throughout the rest of the exercises assume the problem is in standard form.

In Problems 9 to 14 find the initial simplex tableau for the linear programming problem.

9. Maximize $m = 3x + 4y$ subject to
$$x + 2y \leq 10$$
$$3x + y \leq 8$$

10. Maximize $m = 4x + 9y$ subject to
$$3x + 4y \leq 12$$
$$2x + 5y \leq 10$$

11. Maximize $m = x + 5y$ subject to
$$3x + 8y \leq 1$$
$$4x - 5y \leq 4$$
$$2x + 7y \leq 6$$

12. Maximize $m = x + 2y + 4z$ subject to
$$x + 2y + 8z \leq 16$$
$$x + y + 3z \leq 12$$
$$x + z \leq 5$$

13. Maximize $m = 3x + y + 5z$ subject to
$$x + z \leq 15$$
$$x + 2y \leq 8$$
$$y + 5z \leq 10$$

14. Maximize $m = x - z$ subject to
$$x + 4y + 5z \leq 10$$
$$5x + 2y - z \leq 20$$
$$x + 3y + 6z \leq 12$$

In Problems 15 to 20 determine the basic variables and the nonbasic variables corresponding to the simplex tableau.

15.
$$\begin{array}{ccccc|c} x & y & s_1 & s_2 & m & \\ 2 & 1 & 0 & 2 & 0 & 3 \\ 3 & 0 & 1 & 1 & 0 & 1 \\ \hline -1 & 0 & 0 & 3 & 1 & 6 \end{array}$$

16.
$$\begin{array}{ccccc|c} x & y & s_1 & s_2 & m & \\ 0 & 3 & 2 & 1 & 0 & -1 \\ 1 & 1 & 0 & 0 & 0 & 2 \\ \hline 0 & -2 & 0 & 0 & 1 & 0 \end{array}$$

17.
$$\begin{array}{ccccc|c} x & y & s_1 & s_2 & m & \\ 1 & 0 & 2 & 0 & 0 & 3 \\ 0 & 1 & 3 & 1 & 0 & 4 \\ \hline 0 & -1 & -6 & 0 & 1 & 8 \end{array}$$

18.
$$\begin{array}{ccccc|c} x & y & s_1 & s_2 & m & \\ -2 & 0 & 1 & 0 & 0 & 3 \\ 0 & 1 & 0 & -1 & 0 & 4 \\ \hline 0 & 0 & 0 & 5 & 1 & 8 \end{array}$$

19.
$$\begin{array}{ccccccc|c} x & y & z & s_1 & s_2 & s_3 & m & \\ 1 & 0 & 1 & 0 & 0 & 1 & 0 & 1 \\ 2 & 0 & 0 & 0 & 1 & 2 & 0 & 3 \\ 0 & 1 & 0 & 3 & 0 & 1 & 0 & 4 \\ \hline 0 & 0 & 0 & -4 & 0 & 3 & 1 & 6 \end{array}$$

20.
$$\begin{array}{ccccccc|c} x & y & z & s_1 & s_2 & s_3 & m & \\ 0 & 0 & 0 & 2 & 0 & 1 & 1 & 2 \\ 1 & 7 & 0 & -1 & 1 & 2 & 0 & 5 \\ 0 & 1 & 1 & 0 & 2 & 1 & 0 & 14 \\ \hline 0 & 0 & 0 & 0 & -2 & 3 & 0 & 22 \end{array}$$

In Problems 21 to 26 compute the basic feasible solution to the simplex tableaus.

21. Problem 15.

22. Problem 16.

23. Problem 17.

24. Problem 18.

25. Problem 19.

26. Problem 20.

•• In Problems 27 to 30 construct the original linear programming problem (without slack variables) assuming that the matrix is the initial simplex tableau.

27. $$\begin{bmatrix} 2 & 3 & 1 & 0 & 0 & | & 3 \\ 4 & 1 & 0 & 1 & 0 & | & 6 \\ -1 & -3 & 0 & 0 & 1 & | & 0 \end{bmatrix}$$

28. $$\begin{bmatrix} 3 & 3 & 1 & 0 & 0 & | & 3 \\ 4 & 1 & 0 & 1 & 0 & | & 6 \\ -1 & 0 & 0 & 0 & 1 & | & 10 \end{bmatrix}$$

29. $$\begin{bmatrix} 0 & 2 & 1 & 1 & 0 & 0 & 0 & | & 3 \\ 0 & 3 & 2 & 0 & 1 & 0 & 0 & | & 0 \\ 1 & 4 & 0 & 0 & 0 & 1 & 0 & | & 1 \\ -5 & -4 & -7 & 0 & 0 & 0 & 1 & | & 40 \end{bmatrix}$$

30. $$\begin{bmatrix} 1 & 3 & 0 & 1 & 0 & 0 & 0 & | & 3 \\ 1 & 4 & 1 & 0 & 1 & 0 & 0 & | & 0 \\ 0 & 4 & 2 & 0 & 0 & 1 & 0 & | & 1 \\ -9 & -2 & 0 & 0 & 0 & 0 & 1 & | & 8 \end{bmatrix}$$

In Problems 31 to 34 consider the two inequalities.

$x + 2y \le 10$
$2x + 5y \le 20$

Let s_1 and s_2 be slack variables that turn these inequalities into equalities so that

$x + 2y + s_1 = 10$
$2x + 5y + s_2 = 20$

Find the values of s_1 and s_2 that satisfy the equations for the given values of x and y.

31. $x = 1, y = 1$ 32. $x = 2, y = 1$

33. $x = 0, y = 3$ 34. $x = 1, y = 3$

In Problems 35 and 36 find the initial simplex tableau for the linear programming problem. Assume the problem is in standard form.

35. Maximize $m = 3x + y + 5z + 6w$ subject to
$x + z + 2w \le 13$
$x + 2y + 3w \le 8$
$y + 3z + w \le 10$
$x + 3z + 2w \le 12$

36. Maximize $m = x + z + 2w$ subject to
$x + 4y + 5z \le 18$
$5x + 2y + 3w \le 29$
$x + 3y + 6w \le 22$
$2y + 3z + w \le 20$

In Problems 37 and 38 determine the basic variables, the nonbasic variables, and the basic feasible solution of the simplex tableau.

37. $$\begin{bmatrix} x & y & z & s_1 & s_2 & s_3 & m & \\ 0 & 0 & 0 & 2 & 1 & 1 & 0 & | & 5 \\ 1 & 0 & 0 & 1 & 0 & 0 & 0 & | & 8 \\ 0 & 6 & 1 & 3 & 0 & 0 & 0 & | & 2 \\ 0 & -1 & 0 & -5 & 3 & 0 & 1 & | & 9 \end{bmatrix}$$

38. $$\begin{bmatrix} x & y & z & s_1 & s_2 & s_3 & m & \\ 1 & 1 & 0 & 0 & 0 & 0 & 0 & | & 1 \\ 0 & 0 & 0 & 1 & 1 & 0 & | & 4 \\ 0 & 0 & 1 & 1 & 0 & 0 & 0 & | & 6 \\ 0 & -1 & 0 & 2 & 1 & 0 & 1 & | & 2 \end{bmatrix}$$

••• 39. Show that the number of variables of a linear programming problem with slack variables is greater than the number of equations.

In Problems 40 to 42 find the initial simplex tableau for the linear programming problem.

40. A firm manufactures three types of tents, small, medium, and large. Each small tent takes 2 hours to cut and 4 hours to sew. Each medium tent takes 3 hours to cut and 5 hours to sew. Each large tent takes 4 hours to cut and 7 hours to sew. There is a total of 310 hours available to cut and 500 hours available to sew the tents. No more than 100 tents are to be made. How many of each type of tent should be made to maximize profit if the profit is $30 on each small tent, $50 on each medium tent, and $70 on each large tent?

41. A soft-drink company makes three types of fruit juice drinks by mixing lemonade, limeade, and orange juice. Each gallon of lemon-lime drink

uses 0.5 gallon (gal) of lemonade and 0.5 gal of limeade. Each gallon of orangeade drink uses 0.5 gal of orange juice, 0.25 gal of lemonade, and 0.25 gal of limeade. Each gallon of fruit punch drink uses 0.10 gal of orange juice, 0.5 gal of lemonade, and 0.4 gal of limeade. The company will use no more than 30 gal of orange juice, 40 gal of lemonade, and 25 gal of limeade. If the profit per gallon is $2 for lemonade, $1 for orangeade, and $2.50 for fruit punch drink, how many gallons of each type of drink should be made to maximize profit?

42. A farmer has 100 acres of land and wants to plant corn, beans, and tomatoes. Each acre of corn requires 4 days of labor, $140 in seed and fertilizer, and $50 to harvest. Each acre of beans requires 5 days of labor, $100 in seed and fertilizer, and $80 to harvest. Each acre of tomatoes requires 2 days of labor, $130 in seed and fertilizer, and $90 to harvest. The farmer has available 120 total days of labor, $10,500 to invest in seed and fertilizer, and $6800 to spend in harvesting. If the profit per acre is $320 for corn, $400 for beans, and $500 for tomatoes, how many acres of each plant should the farmer plant to maximize profit?

In Problems 43 to 45 a linear programming problem adapted from a research study is given. Express the problem mathematically, find the initial simplex tableau, and determine whether it is in standard form.

43. The following problem is from an article studying the dairy industry of New Zealand.* The production of three types of products is studied: whole milk, cream and buttermilk. The total number of liters of the three products produced per day is no more than 12 million. The number of liters of cream and buttermilk produced is less than $\frac{1}{2}$ the number of liters of whole milk produced. If the profit per liter is $.54 for whole milk, $.44 for cream and $.37 for buttermilk, how many millions of liters of each type of product should be produced to maximize profit?

44. This problem is from an article studying stock location problems, with a specific example of a library with limited storage capacity.† A new library is divided into three general storage areas: literature, science, and fiction. The total number of books stored is no more than 1 million. The number of nonfiction books is no more than 700,000. The number of literature books is at least 200,000. Find how many of each type of book should be stored to minimize cost if the cost is $25 for each literature book, $30 for each science book, and $22 for each fiction book.

45. Cleanup capability is a necessary evil in the oil industry. In an article studying various aspects of cleanup of oil spills, researchers set up the following type of problem.‡ Three types of cleanup equipment were considered. The transportation cost (in dollars per gallon of oil spilled per mile to the site) to a particular potential spill site is $.0015 for type 1 equipment, $.0015 for type 2 equipment and $.00015 for type 3 equipment. The response time (in hours to reach the site) is 3 hours for type 1 equipment, 4 hours for type 2 equipment and 5 hours for type 3 equipment. The operating cost (in dollars per gallon of oil spilled) is $5 for type 1 equipment, $7 for type 2 equipment, and $1 for type 3 equipment. If the total transportation cost is limited to $.012/gal/mi and the total response time is limited to 30 hours, find the number of each type of equipment that should be used to minimize the operating cost.

*Benseman, B. R., "Production Planning in the New Zealand Dairy Industry," *Journal Operational Research Society*, 37:747–754, 1986.

†Evans, James R., "The Factored Transportation Problem," *Management Science*, 30:1021–24, 1984.

‡Psaraftis, Harilaos N., et al., "Optimal Response to Oil Spills: The Strategic Decision Case," *Operations Research*, 34:203–217, 1986.

4–2 THE MAXIMALITY TEST AND PIVOTING

The simplex method begins by expressing the linear programming problem as a system of equations. Then from the initial simplex tableau a sequence of simplex tableaus is computed to obtain a succession of solutions. The procedure ensures that the value of the objective function increases at each step. We continue the process until an optimal solution is found or we find that there is no solution.

Test for a Maximal Solution

For each new tableau in the procedure we must test its basic feasible solution to determine if it is the final optimal solution of the linear programming problem. Since the problem is in standard form, the objective function must be maximized. The test itself is easy to apply, but understanding why it works is another matter. Suppose the last row of a simplex tableau is

$$\begin{matrix} x & y & s_1 & s_2 & m \\ |2 & -1 & 0 & 0 & 1 & 70| \end{matrix}$$

This row corresponds to the objective function. Its corresponding equation is

$$2x - y + m = 70$$

The numbers in the row are the coefficients of the variables in the equation. The object of the problem is to increase m as much as possible. This is done by increasing the other variables, in this case x and y. The central idea is the sign of the coefficients: x has a positive coefficient and y has a negative one. If $x = y = 0$, then $m = 70$. If x increases to 10 and $y = 0$, then m decreases to 50. If y increases to 10 and $x = 0$, then m increases to 80. Thus positive coefficients cause a decrease in m while negative coefficients cause an increase in m. This means that m cannot be increased further when there are no negative coefficents in the objective function. Since these coefficients are the numbers in the last row, we get the following test:

Maximality Test

A simplex tableau will yield an optimal solution if and only if the last row, corresponding to the objective function, contains no negative entries.

Remember that this test is applied to each tableau. If there are no negative entries in the last row, the solution is maximal. If the test fails, we compute another simple tableau using a method that will be described shortly. Example 1 demonstrates how to use the maximality test.

EXAMPLE 1 Suppose A and B are the last rows of two simplex tableaus.

$$\begin{matrix} & x & y & z & s_1 & s_2 & s_3 & m \\ A & [1 & 0 & 2 & -2 & 1 & 0 & 0 & 6] \\ B & [0 & 0 & 1 & 2 & 12 & 0 & 1 & 32] \end{matrix}$$

Problem

Apply the maximality test to each row.

SOLUTION There is a negative number in row A so the basic feasible solution is not maximal. Another simplex tableau must be computed. In row B there are no negative numbers so the basic feasible solution is maximal. Therefore the maximum value of m is 32.

Suppose the Test Fails: The Entering Variable

If there are negative entries in the last row of the simplex tableau, then the maximality test says that the basic feasible solution is not maximal, so we must compute another tableau. This is done by **pivoting,** that is, by selecting one entry to be the pivot entry and using elementry row operations to change it to 1 and the other entries in its column to 0. This will be done in such a way that m increases by the largest possible amount. In the remainder of this section the pivot operation will be explained. In the next section the selection of the pivot entry will be discussed.

After the pivot operation is completed the variable corresponding to its column becomes a basic variable because its column will be in basic form. Thus pivoting turns a nonbasic variable into a basic one. It is easy to determine the proper variable, called the *entering variable* (because it ''enters'' the set of basic variables). If there is only one negative entry in the last row of the tableau, then the variable corresponding to the column containing that negative entry is the entering variable and that column is called the *pivot column.* Suppose, however, there are two or more negative entries. Then the one that is most negative will produce the largest increase in m and its variable is chosen as the entering variable. This is the next step in the procedure.

Selecting the Pivot Column

The column that has the most negative entry in the last row is the **pivot column.** The variable corresponding to its column is the **entering variable.**

If there are two or more numbers that are equal and the most negative, any one of them can be chosen for the pivot column. Example 2 will help clarify how to select the pivot column.

EXAMPLE 2   Suppose A and B are the last rows of two simplex tableaus.

$$\begin{array}{ccccccccc} & x & y & z & s_1 & s_2 & s_3 & m & \\ A\big[& 3 & 0 & -4 & 0 & 0 & -1 & 1 & 84 \big] \\ B\big[& -8 & -1 & 0 & -8 & 0 & 0 & 1 & 12 \big] \end{array}$$

Problem

Identify the pivot column and the entering variable for each tableau.

SOLUTION For matrix A the most negative entry is -4, so the pivot column is column 3 and the entering variable is z. For matrix B the most negative entry is -8. Since it appears twice, the pivot column is either column 1 or column 4. The entering variable is either x or s_1.

Once the pivot column has been selected, elementary row operations are used to transform it into basic form. That is, make one entry 1 and the others 0. This is what is meant by the pivot process. The next example shows how to pivot.

EXAMPLE 3   Consider tableau A.

$$A = \begin{array}{c} \\ \\ \\ \end{array} \begin{array}{ccccc} x & y & s_1 & s_2 & m \\ \end{array}$$
$$A = \left[\begin{array}{ccccc|c} 1 & 2 & 0 & 3 & 0 & 26 \\ 0 & ① & 1 & 1 & 0 & 10 \\ 0 & -2 & 0 & 10 & 1 & 12 \end{array}\right]$$
$$\uparrow$$

The maximality test fails because there is a negative entry in the last row. There is only one negative entry in the last row, so the second column is the pivot column and y is the entering variable. The $(2, 2)$ entry (the row 2, column 2 entry), is circled because it is the pivot entry (we will explain how it was chosen in the next section). To transform the second column into basic form, insert a 0 in the first row (in place of 2) and 0 in the third row (in place of -2). This is done by the two operations: replace the first row by -2 times row 2 plus row 1, R1 + (-2)R2, and replace row 3 by 2 times row 2 plus row 3, R3 + 2R2. This yields the new tableau B.

$$B = \begin{array}{ccccc} x & y & s_1 & s_2 & m \\ \end{array}$$
$$B = \left[\begin{array}{ccccc|c} 1 & 0 & -2 & 1 & 0 & 6 \\ 0 & 1 & 1 & 1 & 0 & 10 \\ 0 & 0 & 2 & 12 & 1 & 32 \end{array}\right] \quad \begin{array}{l} \text{R1} + (-2)\text{R2} \\ \\ \text{R3} + 2(\text{R2}) \end{array}$$

We have pivoted about the $(2, 2)$ entry, thus letting y "enter" the set of basic variables.

There are two important observations about Example 3. First, elementary row operations are used in the pivot process to compute tableau B from tableau A. Hence the two matrices are *equivalent,* meaning that they have identical solution sets. Therefore the solutions to tableau B are the solutions of the original linear programming problem. This ensures that when we compute a new simplex tableau using the pivot process, the same solution set is retained. Also, the test for a maximal solution is applied whenever a new simplex tableau is computed. The matrix B

passes the maximality test so its basic feasible solution is the maximal solution for the original linear programming problem.

When a new tableau is computed, the set of basic variables changes: the variable corresponding to the pivot column becomes a new basic variable and so one of the former basic variables becomes a nonbasic variable. The next example illustrates the pivot process. It begins with a linear programming problem and uses three pivots to obtain the maximal solution. Recall that we are assuming that each linear programming problem in this section and the next is in standard form, so, in particular, all variables are greater than or equal to 0.

EXAMPLE 4 | **Problem**

Maximize $m = 3x + 2y + z$ subject to

$$
\begin{aligned}
x + \quad z &\leq 200 \\
y + 2z &\leq 400 \\
2x + \quad y &\leq 600
\end{aligned}
$$

SOLUTION First use slack variables to express the problem as a system of equations while expressing the objective function in proper form.

$$
\begin{aligned}
x \qquad + \quad z + s_1 \qquad\qquad\qquad &= 200 \\
y + 2z \qquad + s_2 \qquad\qquad &= 400 \\
2x + \quad y \qquad\qquad\quad + s_3 \qquad &= 600 \\
-3x - 2y - \quad z \qquad\qquad\qquad + m &= \quad 0
\end{aligned}
$$

The initial simplex tableau is

$$
\begin{bmatrix}
x & y & z & s_1 & s_2 & s_3 & m & \\
\textcircled{1} & 0 & 1 & 1 & 0 & 0 & 0 & 200 \\
0 & 1 & 2 & 0 & 1 & 0 & 0 & 400 \\
2 & 1 & 0 & 0 & 0 & 1 & 0 & 600 \\
-3 & -2 & -1 & 0 & 0 & 0 & 1 & 0
\end{bmatrix}
$$

The maximality test fails because there are negative numbers in the last row. The pivot column is the first column since -3 is the most negative number in the last row. Suppose the pivot entry is the $(1, 1)$ entry. We get

$$
\begin{bmatrix}
x & y & z & s_1 & s_2 & s_3 & m & \\
1 & 0 & 1 & 1 & 0 & 0 & 0 & 200 \\
0 & 1 & 2 & 0 & 1 & 0 & 0 & 400 \\
0 & 1 & -2 & -2 & 0 & 1 & 0 & 200 \\
0 & -2 & 2 & 3 & 0 & 0 & 1 & 600
\end{bmatrix}
\qquad
\begin{aligned}
& \\
& \\
& \text{R3} + (-2)\text{R1} \\
& \text{R4} + 3(\text{R1})
\end{aligned}
$$

The maximality test fails because there is a negative entry in the last row. The pivot column is column 2. Let us pivot about the $(3, 2)$ entry. We get

$$\begin{array}{ccccccc}x & y & z & s_1 & s_2 & s_3 & m\end{array}$$

$$\begin{bmatrix} 1 & 0 & 1 & 1 & 0 & 0 & 0 & | & 200 \\ 0 & 0 & 4 & 2 & 1 & -1 & 0 & | & 200 \\ 0 & 1 & -2 & -2 & 0 & 1 & 0 & | & 200 \\ 0 & 0 & -2 & -1 & 0 & 2 & 1 & | & 1000 \end{bmatrix}$$

R2 + (−1)R3

R4 + 2(R3)

The maximality test fails again. The pivot column is column 3. Let us pivot about the (2, 3) entry. First perform $\frac{1}{4}$(R2) to make the pivot entry 1.

$$\begin{array}{ccccccc}x & y & z & s_1 & s_2 & s_3 & m\end{array}$$

$$\begin{bmatrix} 1 & 0 & 0 & \frac{1}{2} & -\frac{1}{4} & \frac{1}{4} & 0 & | & 150 \\ 0 & 0 & 1 & \frac{1}{2} & \frac{1}{4} & -\frac{1}{4} & 0 & | & 50 \\ 0 & 1 & 0 & -1 & \frac{1}{2} & \frac{1}{2} & 0 & | & 300 \\ 0 & 0 & 0 & 0 & \frac{1}{2} & \frac{3}{2} & 1 & | & 1100 \end{bmatrix}$$

R1 + (−1)(R2)

$\frac{1}{4}$(R2)

R3 + 2(R2)

R4 + 2(R2)

The maximality test passes and the basic feasible solution is maximal. It is

$$x = 150, \quad y = 300, \quad z = 50, \quad s_1 = s_2 = s_3 = 0, \quad m = 1100$$

Hence the solution to the original problem is $m = 1100$ when $x = 150$, $y = 300$, and $z = 50$.

EXERCISE SET 4–2

- In Problems 1 to 6 assume that each row matrix is the last row of a simplex tableau. Determine whether it passes the maximality test.

$$\begin{array}{cccccc}x & y & s_1 & s_2 & m\end{array}$$
1. $\begin{bmatrix} 2 & 0 & 0 & -1 & 1 & 10 \end{bmatrix}$

$$\begin{array}{cccccc}x & y & s_1 & s_2 & m\end{array}$$
2. $\begin{bmatrix} 0 & 3 & -8 & 0 & 1 & 0 \end{bmatrix}$

$$\begin{array}{cccccc}x & y & s_1 & s_2 & m\end{array}$$
3. $\begin{bmatrix} 0 & 0 & 3 & 0 & 1 & 0 \end{bmatrix}$

$$\begin{array}{ccccccc}x & y & z & s_1 & s_2 & s_3 & m\end{array}$$
4. $\begin{bmatrix} 2 & 1 & 0 & 0 & 3 & 0 & 1 & 10 \end{bmatrix}$

$$\begin{array}{ccccccc}x & y & z & s_1 & s_2 & s_3 & m\end{array}$$
5. $\begin{bmatrix} -2 & 1 & 0 & 0 & -5 & 1 & 6 & 8 \end{bmatrix}$

$$\begin{array}{ccccccc}x & y & z & s_1 & s_2 & s_3 & m\end{array}$$
6. $\begin{bmatrix} 3 & 0 & 1 & 0 & 8 & 0 & 0 & 0 \end{bmatrix}$

In Problems 7 to 10 find the basic feasible solution of the simplex tableau and then apply the maximality test.

$$\begin{array}{ccccc}x & y & s_1 & s_2 & m\end{array}$$
7. $\begin{bmatrix} 0 & 3 & 1 & 2 & 0 & | & 10 \\ 1 & -2 & 0 & 1 & 0 & | & 15 \\ 0 & -1 & 0 & 0 & 1 & | & 20 \end{bmatrix}$

$$\begin{array}{ccccc}x & y & s_1 & s_2 & m\end{array}$$
8. $\begin{bmatrix} 1 & 3 & 0 & 2 & 0 & | & 12 \\ 0 & -2 & 1 & 1 & 0 & | & 25 \\ 0 & 1 & 0 & 0 & 1 & | & 50 \end{bmatrix}$

$$\begin{array}{ccccccc}x & y & z & s_1 & s_2 & s_3 & m\end{array}$$
9. $\begin{bmatrix} 0 & 0 & 1 & 1 & 1 & 1 & 0 & | & 10 \\ 1 & 0 & 0 & 2 & 3 & -7 & 0 & | & 28 \\ 0 & 1 & 0 & 4 & 4 & 4 & 0 & | & 34 \\ 0 & 0 & 0 & 0 & 0 & 31 & 1 & | & 22 \end{bmatrix}$

$$\begin{array}{ccccccc}x & y & z & s_1 & s_2 & s_3 & m\end{array}$$
10. $\begin{bmatrix} 2 & 0 & 1 & 1 & 0 & 3 & 0 & | & 10 \\ 1 & 0 & 0 & 2 & 1 & -1 & 0 & | & 28 \\ 8 & 1 & 0 & 4 & 0 & 2 & 0 & | & 34 \\ -3 & 0 & 1 & 0 & 0 & -1 & 1 & | & 22 \end{bmatrix}$

In Problems 11 to 16 select the pivot column and the entering variable.

11.
$$\begin{array}{ccccc} x & y & s_1 & s_2 & m \\ \end{array}$$
$$\left[\begin{array}{ccccc|c} 0 & 0 & 0 & 1 & 0 & 1 \\ 1 & 1 & 1 & 0 & 0 & 3 \\ \hline 0 & -3 & -1 & 0 & 1 & 8 \end{array}\right]$$

12.
$$\begin{array}{ccccc} x & y & s_1 & s_2 & m \\ \end{array}$$
$$\left[\begin{array}{ccccc|c} 2 & 1 & 0 & 1 & 0 & 1 \\ 1 & 0 & 1 & 0 & 0 & 7 \\ \hline -8 & 0 & 0 & -3 & 1 & 5 \end{array}\right]$$

13.
$$\begin{array}{ccccccc} x & y & z & s_1 & s_2 & s_3 & m \\ \end{array}$$
$$\left[\begin{array}{ccccccc|c} 1 & 1 & 0 & 0 & 2 & 0 & 0 & 3 \\ 0 & 0 & 0 & 1 & 1 & 1 & 0 & 8 \\ 0 & 0 & 1 & 3 & 0 & 0 & 0 & 9 \\ \hline 0 & 2 & 0 & -1 & -2 & 0 & 1 & 3 \end{array}\right]$$

14.
$$\begin{array}{ccccccc} x & y & z & s_1 & s_2 & s_3 & m \\ \end{array}$$
$$\left[\begin{array}{ccccccc|c} 0 & 0 & 1 & 0 & 1 & 0 & 0 & 0 \\ 1 & 0 & 0 & 1 & 2 & 1 & 0 & 4 \\ 2 & 1 & 0 & 0 & 0 & 1 & 0 & 2 \\ \hline -3 & 0 & 0 & 0 & -1 & 1 & 1 & 5 \end{array}\right]$$

15.
$$\begin{array}{ccccccc} x & y & z & s_1 & s_2 & s_3 & m \\ \end{array}$$
$$\left[\begin{array}{ccccccc|c} 0 & 0 & 1 & 0 & -4 & 0 & 0 & 10 \\ 1 & 0 & 3 & 1 & 2 & 1 & 0 & 9 \\ 0 & 1 & -1 & 0 & 1 & 1 & 0 & 2 \\ \hline 0 & 0 & -9 & 0 & -5 & 3 & 1 & 45 \end{array}\right]$$

16.
$$\begin{array}{ccccccc} x & y & z & s_1 & s_2 & s_3 & m \\ \end{array}$$
$$\left[\begin{array}{ccccccc|c} 2 & -1 & 1 & 0 & -4 & 0 & 0 & 10 \\ 1 & 0 & 0 & 1 & 2 & 0 & 0 & 18 \\ 0 & 1 & 0 & 0 & 0 & 1 & 0 & 2 \\ \hline 4 & -1 & 0 & 0 & -2 & 0 & 1 & 45 \end{array}\right]$$

In Problems 17 to 28 perform the pivot process on the simplex tableau in the indicated problem using the given entry as the pivot entry.

17. Problem 11, the (2, 2) entry.

18. Problem 12, the (1, 1) entry.

19. Problem 13, the (1, 5) entry.

20. Problem 14, the (3, 1) entry.

21. Problem 15, the (2, 3) entry.

22. Problem 16, the (2, 5) entry.

23. Problem 15, the (1, 3) entry.

24. Problem 16, the (1, 5) entry.

25. Problem 15, the (3, 5) entry.

26. Problem 16, the (3, 2) entry.

27. Problem 15, the (3, 2) entry.

28. Problem 16, the (1, 1) entry.

In Problems 29 to 35 find the initial simplex tableau and form a new tableau by pivoting about the indicated entry.

29. Maximize $m = 5x + 7y$ subject to
$$x + 3y \le 12$$
$$2x + 3y \le 18$$
Pivot about the (1, 2) entry.

30. Maximize $m = 8x + 5y$ subject to
$$4x + 3y \le 24$$
$$9x + 2y \le 18$$
Pivot about the (2, 1) entry.

31. Maximize $m = 4x + 7y$ subject to
$$3x + 5y \le 15$$
$$x + 5y \le 10$$
Pivot about the (2, 2) entry.

32. Maximize $m = x + 2y + 3z$ subject to
$$x + 2y + z \le 40$$
$$x + y + z \le 30$$
Pivot about the (2, 3) entry.

33. Maximize $m = x + 3y + 4z$ subject to
$$x + 3y \le 300$$
$$x + z \le 150$$
$$x + 2y + z \le 200$$
Pivot about the (2, 3) entry.

34. Maximize $m = 4x + 6y + 5z$ subject to
$$2x + 3y + z \le 900$$
$$3x + y + z \le 350$$
$$4x + 2y + z \le 400$$
Pivot about the (3, 2) entry.

•• In Problems 35 to 40 translate the linear programming problem into mathematics. Then find the initial simplex tableau and form a new tableau by pivoting about the (1, 1) entry.

35. A clothing manufacturer makes three types of jogging attire, discount, regular, and designer.

The attire sells to the retail stores for $5 each for the discount suits, $10 each for the regular suits, and $40 each for the designer suits. The cost for manufacturing each discount suit is $3, for each regular suit $5, and for each designer suit $10. The number of designer suits produced must be less than the sum of discount and regular suits made. If the manufacturer makes no more than 200 suits a week and budgets no more than $1500 per week, how many of each type should be made to maximize profit?

36. A hospital nutritionist prepares a menu for patients on low-fat, high-protein diets. The primary ingredients are chicken, corn, and rice. The meal must contain no more than 1000 calories (cal), 5 grams (g) of fat, and 15 milligrams (mg) of iron. Each 3-ounce serving of chicken contains 45 g of protein, 500 cal, 9 mg of iron, and 4 g of fat. Each 1-cup serving of corn contains 9 g of protein, 150 cal, 6 mg of iron, and 2 g of fat. Each 1-cup serving of rice contains 7 g of protein, 100 cal, 5 mg of iron, and 2 g of fat. How much chicken, corn, and rice should be included in the meal in order to maximize the amount of protein?

37. The computer center director at a large university wants to staff consulting sections with three types of shifts: type 1 will contain 2 senior programmers and 1 student worker, type 2 will consist of 1 senior programmer and 5 student workers, and type 3 will consist of 1 senior programmer and 3 student workers. The director wants to use no more than 50 individuals. The number of senior programmers is limited to 20. There will be at least 24 hours to be filled during the week, with a type 1 shift serving for 3 hours, a type 2 shift serving for 2 hours, and a type 3 shift serving for 1 hour. The cost of a type 1 shift is $44 per hour, of a type 2 shift $40 per hour, and of a type 3 shift $30. Determine the number of shifts of each type in order to minimize cost.

38. A bicycle manufacturer has two retail outlets and, in addition, sells some bikes to a large

chain store. One truck makes all deliveries, and it takes 3 hours to make a round trip to the first retail store, 2 hours to the second retail store, and 4 hours to the chain store. The cost of a round trip to the first retail store is $100, to the second retail store $75, and to the chain store $200. The profit of each truckload sold through the first retail store is $500, for the second retail store $400, and for each truckload sold to the chain store, $800. How many shipments per week should be scheduled to each store in order to maximize profit if the manufacturer is limited to no more than 48 hours per week and a cost of $3000?

●●● 39. An investment manager wants to invest at most $25,000 in four types of mutual funds. From past experience it is assumed that in the short run type A fund will yield 8%, type B fund will yield 10%, type C fund will yield 12%, and type D fund will yield 11%. The manager wants to invest no more than $10,000 in type A and type B funds together; no more than $15,000 in type A, type B, and type C funds together; and no more than $20,000 in type A and type D funds together. How much should be invested in each type of fund to maximize profit?

40. A firm manufactures four types of tents, small, medium, large, and deluxe. Each small tent takes 2 hours to cut and 4 hours to sew; each medium tent takes 3 hours to cut and 5 hours to sew; each large tent takes 4 hours to cut and 7 hours to sew; each deluxe tent takes 6 hours to cut and 9 hours to sew. There is a total of 310 hours available to cut and 500 hours available to sew the tents. No more than 100 tents are to be made. How many of each type of tent should be made to maximize profit if the profit is $90 on each small tent, $50 on each medium tent, $70 on each large tent, and $80 on each deluxe tent?

41. Under Title III of the 1973 Older Americans Act over 500 Area Agencies on Aging (AAAs) have been established to coordinate services to the elderly. In an article analyzing how AAAs can most efficiently operate their centers, re-

searchers based their recommendations on a preference index.* Persons served by the centers were divided into one of three categories: type 1 consisted of persons with no impairment, type 2 persons had one impairment, and type 3 persons had more than one impairment. Each center could serve all types of persons, but some were equipped to serve a larger proportion of one type than another. The preference index measured the proportion of each type of person a center was best equipped to serve. For example, one center had the preference index $P = 5x + 6y + 3z$, where x is the number of type 1 persons it could serve, y is the number of type 2 persons and z is the number of type 3 persons. The object was to maximize the preference index. This means the center could best serve twice as many type 2 persons as type 3 persons, and just a few less type 1 than type 2. For this center, the number of persons it could serve was 60. Its rehabilitation department, which served type 2 and 3 persons, could serve at most 40 persons. The remedial training department, which served type 1 and 2 persons, could serve at most 50 persons. Find the number of each type of person the center could serve to maximize its preference index.

42. The management of school districts is big business. Many large districts have budgets exceeding $25 million. Most corporations in the private sector with budgets this high have large accounting departments to manage their fiscal operations. In an article analyzing financial management for school districts, the primary technique is linear programming.† The most important consideration is cash flow. School districts have large revenues that are usually paid in lump sums, such as property taxes and state aid. Most expenditures are also made in several lump sums. Revenues must be invested wisely before an expenditure is due to optimize financial management. Suppose a district keeps its capital in three types of investments: liquid accounts (such as savings accounts and money markets), one-month certificates of deposit (CDs), and six-month CDs. The liquid accounts earn 6% interest, the one-month CDs earn 8% interest, and the six-month CDs earn 9%. The sum of the amounts in the three accounts is limited to $2.5 million. At least $1 million must be kept in the liquid accounts, meaning that the sum of the amounts in the CDs can be no more than $1.5 million. State law requires that the sum of 80% of the liquid accounts and 30% of the one-month CDs must be no more than the district's short term debt, which is $1.8 million. Find the amount of each account that maximizes interest.

43. The following problem is adapted from an article studying how to maximize profit from a large scale assembly line.‡ The problem is in standard form. The initial simplex tableau is given. Express the problem as a word problem, that is, describe the objective function and the constraints. The variables represent the capacity of the four machines in the assembly line; that is, they are the number of items that the machine can treat per day. The numbers in the last row represent the profit per item per machine.

$$\begin{bmatrix} x & y & z & w & s_1 & s_2 & s_3 & m & \\ 1 & 1 & 1 & 1 & 1 & 0 & 0 & 0 & 300 \\ 1 & 1 & 0 & 0 & 0 & 1 & 0 & 0 & 100 \\ 0 & 1 & 1 & 1 & 0 & 0 & 1 & 0 & 230 \\ -2 & -3 & -4 & 0 & 0 & 0 & 0 & 1 & 0 \end{bmatrix}$$

*Burton, Richard M., and Dellinger, David C., "Making the Area Agencies on Aging Work: The Role of Information," *Socio-Economic Planning Science,* 14:1–9, 1980.

†Dembowski, Frederick L., "An Integer Programming Approach to School District Financial Management," *Socio-Economic Planning Science,* 14:147–153, 1980.

‡Chandrasekaran, R., "Production Planning in Assembly Line Systems," *Management Science,* 30:713–719, 1984.

4–3 THE SIMPLEX METHOD FOR PROBLEMS IN STANDARD FORM

There is one more skill in the simplex method. In Section 4–2 we saw that the simplex method uses elementary row operations to compute a sequence of tableaus that ends when the maximality test passes. Example 4 illustrated such a sequence, but at each step the pivot entry was given. In this section the rule to select the pivot entry is presented. Then all the steps in the simplex method are reviewed.

Choosing the Pivot Row and the Pivot Entry

Once the pivot column is identified, the **pivot entry** is chosen by selecting the correct row—the **pivot row.** In keeping with the goal of finding a maximal solution, we choose the pivot entry so that, after pivoting, the new basic feasible solution will give the largest increase in m. We must also avoid making any variable negative since the problem must remain in standard form.

The key to choosing the correct pivot entry is the last column because that is where the values of the basic variables in the basic feasible solution are found. The main objective is to avoid generating negative numbers in the last column when pivoting because that would violate the stipulation that the variables be nonnegative. So let us explain how a negative number could be produced.

Suppose a tableau has the following two rows where the first column is the pivot column and all its entries are 1.

$$\begin{array}{cc} \text{pivot column} & \text{constant} \end{array}$$

$$\begin{array}{c} \text{row 1} \\ \text{row 2} \end{array} \begin{bmatrix} 1 & \cdots & 3 \\ 1 & \cdots & 1 \\ \vdots & & \\ & & \end{bmatrix}$$

Look at the consequence of choosing row 1 as the pivot row. To pivot, perform R2 + (−1)(R1) to put a 0 in the (2, 1) entry. In the last column of the second row this results in $1 + (-1)3 = -2$, a negative number—exactly what we must avoid. It is produced because the entry in the pivot row, 3, was larger than the entry in the other row, 1. This means that if all the entries in the pivot column are 1, then the pivot row must be that row having the smallest nonnegative number in the last column.

Suppose the numbers in the pivot column are not all 1. Then the pivot row is selected by forming quotients. Each entry in the pivot column is the denominator of a quotient, unless that entry is 0 or a negative number, in which case no quotient is formed. The numerator of each quotient is the entry from the constant column of the corresponding row. No quotient is formed for the objective function, the bottom row. Hence quotients are formed only for rows above the bottom row with positive numbers in the pivot column. The pivot row is then the row with the smallest quotient. This is the criterion.

Choosing the Pivot Row

For each row, except the last one, with a positive entry in the pivot column, compute the ratio of the number in the last column to the number in the pivot column. The pivot row is the row with the smallest nonnegative ratio.

If there are two or more rows that have the same ratio, any of these rows can be selected. The next example shows how to apply the criterion.

EXAMPLE 1 | **Problem**

For each tableau find the pivot column and the pivot row. Then identify the pivot entry.

$$A = \begin{bmatrix} x & y & s_1 & s_2 & m & \\ 3 & -1 & 1 & 0 & 0 & 6 \\ 5 & 2 & 0 & 1 & 0 & 15 \\ \hline -4 & 1 & 0 & 0 & 1 & 10 \end{bmatrix}$$

$$B = \begin{bmatrix} x & y & z & s_1 & s_2 & s_3 & m & \\ 1 & 3 & 0 & -10 & 0 & 3 & 0 & 10 \\ 0 & -1 & 0 & 5 & 1 & 8 & 0 & 50 \\ 0 & 2 & 1 & 6 & 0 & -2 & 0 & 30 \\ \hline 0 & -3 & 0 & -5 & 0 & 6 & 1 & 10 \end{bmatrix}$$

SOLUTION For tableau A the first column is the pivot column because -4 is the only negative number in the last row. The two ratios are

for row 1 the ratio is $\frac{6}{3} = 2$
for row 2 the ratio is $\frac{15}{5} = 3$

Row 1 has the smaller nonnegative ratio and is therefore the pivot row. Thus the (1, 1) entry, 3, is the pivot entry.
     For tableau B, column 4 is the pivot column because -5 is the most negative. Do not consider the ratio for row 1 because its entry in the pivot column, -10, is negative. The other ratios are

for row 2 the ratio is $\frac{50}{5} = 10$
for row 3 the ratio is $\frac{30}{6} = 5$

Row 3 is the pivot row. Thus the pivot entry is the (3, 4) entry, 6.

     There is a reason for stipulating that only nonnegative ratios be considered. It might seem superfluous because all of the constants are nonnegative so the ratios must be nonnegative. We inserted the stipulation because in the next section constraints whose constant term is negative are considered and we want the same procedure to hold.

All the major steps in the simplex method have been covered. They are assembled in the following procedure.

The Simplex Method for Problems in Standard Form

1. State the problem in terms of a system of linear equations by inserting slack variables into the constraints and expressing the objective function in proper form.
2. Construct the initial simplex tableau.
3. Apply the maximality test. If the basic feasible solution is maximal, the problem is solved. If not, go to step 4.
4. Compute a new simplex tableau using the following procedure:
   (a) Choose the pivot column.
   (b) Choose the pivot row.
   (c) Pivot about the pivot entry.
5. Go to step 3.

It might be necessary to use steps 3, 4, and 5 several times before the test for a maximal solution passes. It is possible to determine when a linear programming problem does not have a solution by looking at the ratios used in step 4, part (b). If all of the ratios are negative numbers, the problem has no solution. Example 2 illustrates the procedure.

EXAMPLE 2

Problem

Maximize $m = 15x + 30y$ subject to
$$x + 2y \leq 40$$
$$5x + 2y \leq 80$$

SOLUTION

1. State the problem in terms of linear equations.

$$
\begin{aligned}
x + 2y + s_1 & & & = 40 \\
5x + 2y & + s_2 & & = 80 \\
-15x - 30y & & + m & = 0
\end{aligned}
$$

2. Construct the initial simplex tableau.

$$
\begin{array}{ccccc}
x & y & s_1 & s_2 & m \\
\end{array}
$$

$$
\left[
\begin{array}{ccccc|c}
1 & 2 & 1 & 0 & 0 & 40 \\
5 & 2 & 0 & 1 & 0 & 80 \\
-15 & -30 & 0 & 0 & 1 & 0
\end{array}
\right]
$$

3. The maximality test fails because the last row has negative numbers.
4. The pivot column is column 2 and the ratios are $\frac{40}{2} = 20$ for row 1 and $\frac{80}{2} = 40$ for row 2. Hence row 1 is the pivot row so the pivot entry is the $(1, 2)$ entry. Pivoting produces the tableau

$$
\begin{array}{ccccc|c}
x & y & s_1 & s_2 & m & \\
\frac{1}{2} & 1 & \frac{1}{2} & 0 & 0 & 20 \\
4 & 0 & -1 & 1 & 0 & 40 \\
\hline
0 & 0 & 15 & 0 & 1 & 600
\end{array}
\qquad
\begin{array}{l}
(\frac{1}{2})\text{R}1 \\
\text{R}2 + (-2)\text{R}1 \\
\text{R}3 + (30)\text{R}1
\end{array}
$$

5. The maximality test passes and therefore the basic feasible solution is the maximal solution: it is $x = s_1 = 0$, $y = 20$, $s_2 = 40$, $m = 600$.

Hence the solution to the original problem is $m = 600$ when $x = 0$ and $y = 20$.

The next example requires three pivots to obtain the optimal solution.

EXAMPLE 3 | **Problem**

Maximize $m = 3x + 5y + 4z$ subject to

$$
\begin{aligned}
x + y &\leq 200 \\
2x + z &\leq 350 \\
2y + z &\leq 650
\end{aligned}
$$

SOLUTION First use slack variables to express the problem as a system of equations while expressing the objective function in proper form.

$$
\begin{array}{rcrcrcrcrcrcl}
x &+& y & & & +& s_1 & & & & & & &=& 200 \\
2x &+& & & z & & &+& s_2 & & & & &=& 350 \\
& & 2y &+& z & & & & &+& s_3 & & &=& 650 \\
-3x &-& 5y &-& 4z & & & & & & &+& m &=& 0
\end{array}
$$

Then find the initial simplex tableau. It is

$$
\begin{array}{ccccccc|c}
x & y & z & s_1 & s_2 & s_3 & m & \\
1 & 1 & 0 & 1 & 0 & 0 & 0 & 200 \\
2 & 0 & 1 & 0 & 1 & 0 & 0 & 350 \\
0 & 2 & 1 & 0 & 0 & 1 & 0 & 650 \\
\hline
-3 & -5 & -4 & 0 & 0 & 0 & 1 & 0
\end{array}
$$

The maximality test fails because there are negative numbers in the last row. The pivot column is the second column since -5 is the most negative number in the last row. There are two positive numbers in the second column so we compute the two ratios: for row 1 the ratio is $200/1 = 200$; for row 3 the ratio is $650/2 = 325$. So row 1 is the pivot row. Hence the pivot entry is the $(1, 2)$ entry, 1. Pivoting yields the new tableau

$$\begin{array}{ccccccc} x & y & z & s_1 & s_2 & s_3 & m \\ \left[\begin{array}{ccccccc|c} 1 & 1 & 0 & 1 & 0 & 0 & 0 & 200 \\ 2 & 0 & 1 & 0 & 1 & 0 & 0 & 350 \\ -2 & 0 & 1 & -2 & 0 & 1 & 0 & 250 \\ \hline 2 & 0 & -4 & 5 & 0 & 0 & 1 & 1000 \end{array}\right] \end{array}$$

$$\text{R3} + (-2)\text{R1}$$
$$\text{R4} + 5\text{R1}$$

The maximality test again fails because there is still a negative entry in the last row. The pivot column is column 3. There are two positive numbers in column 3 so we compute the two ratios: for row 2 the ratio is $350/1 = 350$; for row 3 the ratio is $250/1 = 250$. So row 3 is the pivot row. Hence the pivot entry is the $(3, 3)$ entry, 1. Pivoting yields the new tableau

$$\begin{array}{ccccccc} x & y & z & s_1 & s_2 & s_3 & m \\ \left[\begin{array}{ccccccc|c} 1 & 1 & 0 & 1 & 0 & 0 & 0 & 200 \\ 4 & 0 & 0 & 2 & 1 & -1 & 0 & 100 \\ -2 & 0 & 1 & -2 & 0 & 1 & 0 & 250 \\ \hline -6 & 0 & 0 & -3 & 0 & 4 & 1 & 2000 \end{array}\right] \end{array}$$

$$\text{R2} + (-1)\text{R3}$$
$$\text{R4} + 4\text{R3}$$

The maximality test fails again. The pivot column is column 1. There are two positive numbers in the pivot column so we compute the two ratios: for row 1 the ratio is $200/1 = 200$; for row 2 the ratio is $100/4 = 25$. So the pivot row is row 2. Hence the pivot entry is the $(2, 1)$ entry, 4. First perform $(\frac{1}{4})\text{R2}$ and then the remaining operations. The new tableau is

$$\begin{array}{ccccccc} x & y & z & s_1 & s_2 & s_3 & m \\ \left[\begin{array}{ccccccc|c} 0 & 1 & 0 & \frac{1}{2} & -\frac{1}{4} & \frac{1}{4} & 0 & 175 \\ 1 & 0 & 0 & \frac{1}{2} & \frac{1}{4} & -\frac{1}{4} & 0 & 25 \\ 0 & 0 & 1 & -1 & \frac{1}{2} & \frac{1}{2} & 0 & 300 \\ \hline 0 & 0 & 0 & 0 & \frac{3}{2} & \frac{5}{2} & 1 & 2150 \end{array}\right] \end{array}$$

$$\text{R1} + (-1)\text{R2}$$
$$(\tfrac{1}{4})\text{R2}$$
$$\text{R3} + 2\text{R2}$$
$$\text{R4} + 6\text{R2}$$

The maximality test passes and the basic feasible solution is maximal. It is

$$x = 25, \quad y = 175, \quad z = 300, \quad s_1 = s_2 = s_3 = 0, \quad m = 2150$$

Hence the solution to the original problem is $m = 2150$ when $x = 25$, $y = 175$, and $z = 300$.

The next example shows how to determine when a linear programming problem has no solution.

EXAMPLE 4 | **Problem**

Maximize $m = x + 2y$ subject to
$$3x - 2y \le 18$$
$$3x - y \le 24$$

SOLUTION Construct the initial simplex tableau

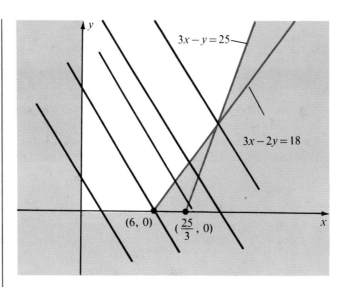

FIGURE 1   Various lines $m = x + 2y$. There is no largest value for $m = x + 2y$.

$$\begin{bmatrix} x & y & s_1 & s_2 & m & \\ 3 & -2 & 1 & 0 & 0 & 18 \\ 3 & -1 & 0 & 1 & 0 & 24 \\ \hline -1 & -2 & 0 & 0 & 1 & 0 \end{bmatrix}$$

The most negative number in the last row is -2 in the second column. But both entries above -2 are negative, which means there is no number that qualifies as the pivot. Hence there is no solution. The graph of the region of feasible solutions is an unbounded region. (See Figure 1.) The objective function, $m = x + 2y$, can be made as large as desired by choosing large enough values of x and y in the region of feasible solutions. Therefore m does not have a maximum value.

EXERCISE SET 4–3

• Assume all problems are in standard form.

In Problems 1 to 6 two columns of a simplex tableau are given. Assume the one on the left is the pivot column and the other is the last column. Compute the appropriate ratios and select the pivot entry.

1.
$$\begin{bmatrix} 2 & \cdots & 10 \\ 1 & \cdots & 6 \\ \hline -2 & \cdots & 10 \end{bmatrix}$$

2.
$$\begin{bmatrix} 3 & \cdots & 8 \\ 6 & \cdots & 12 \\ \hline -1 & \cdots & 10 \end{bmatrix}$$

3.
$$\begin{bmatrix} -1 & \cdots & 4 \\ 2 & \cdots & 10 \\ \hline -5 & \cdots & 10 \end{bmatrix}$$

4.
$$\begin{bmatrix} 3 & \cdots & 9 \\ 3 & \cdots & 8 \\ \hline -1 & \cdots & 1 \end{bmatrix}$$

5.
$$\begin{bmatrix} 3 & \cdots & 11 \\ -2 & \cdots & 2 \\ 6 & \cdots & 12 \\ \hline -3 & \cdots & 8 \end{bmatrix}$$

6.
$$\begin{bmatrix} 0 & \cdots & 1 \\ -3 & \cdots & 3 \\ 1 & \cdots & 10 \\ \hline -8 & \cdots & 10 \end{bmatrix}$$

In Problems 7 to 12 find the pivot entry and perform the pivot operation.

7.
$$\begin{array}{c}\begin{array}{ccccc}x & y & s_1 & s_2 & m\end{array}\\\left[\begin{array}{ccccc|c}0 & 4 & 1 & 2 & 0 & 70\\1 & 1 & 0 & 1 & 0 & 15\\0 & -1 & 0 & 0 & 1 & 20\end{array}\right]\end{array}$$

8.
$$\begin{array}{c}\begin{array}{ccccc}x & y & s_1 & s_2 & m\end{array}\\\left[\begin{array}{ccccc|c}0 & 2 & 0 & 1 & 0 & 2\\1 & 1 & 1 & 0 & 0 & 3\\0 & -4 & -1 & 0 & 1 & 8\end{array}\right]\end{array}$$

9.
$$\begin{array}{c}\begin{array}{ccccc}x & y & s_1 & s_2 & m\end{array}\\\left[\begin{array}{ccccc|c}2 & 1 & 0 & 1 & 0 & 10\\1 & 0 & 1 & 0 & 0 & 8\\\hline-8 & 0 & 0 & -4 & 1 & 20\end{array}\right]\end{array}$$

10.
$$\begin{array}{c}\begin{array}{ccccccc}x & y & z & s_1 & s_2 & s_3 & m\end{array}\\\left[\begin{array}{ccccccc|c}2 & 0 & 1 & 1 & 0 & 3 & 0 & 18\\2 & 0 & 0 & 2 & 1 & -1 & 0 & 28\\4 & 1 & 0 & 5 & 0 & 1 & 0 & 34\\-4 & 0 & 2 & 0 & 0 & -3 & 1 & 22\end{array}\right]\end{array}$$

11.
$$\begin{array}{c}\begin{array}{ccccccc}x & y & z & s_1 & s_2 & s_3 & m\end{array}\\\left[\begin{array}{ccccccc|c}1 & 3 & 0 & 0 & 2 & 0 & 0 & 20\\0 & 1 & 0 & 1 & 0 & 1 & 0 & 2\\0 & 1 & 1 & 3 & 1 & 0 & 0 & 9\\0 & 1 & 0 & -1 & -3 & 0 & 1 & 3\end{array}\right]\end{array}$$

12.
$$\begin{array}{c}\begin{array}{ccccccc}x & y & z & s_1 & s_2 & s_3 & m\end{array}\\\left[\begin{array}{ccccccc|c}2 & 0 & 1 & 0 & 1 & 0 & 0 & 20\\1 & 0 & 0 & 1 & 2 & 1 & 0 & 30\\2 & 1 & 0 & 0 & 0 & 1 & 0 & 10\\-3 & 0 & 0 & 0 & -1 & 1 & 1 & 5\end{array}\right]\end{array}$$

In Problems 13 to 32 solve the linear programming problem by the simplex method.

13. Maximize $m = 5x + 7y$ subject to
$$x + 3y \le 12$$
$$2x + 3y \le 18$$

14. Maximize $m = 8x + 5y$ subject to
$$4x + 3y \le 24$$
$$9x + 2y \le 18$$

15. Maximize $m = 4x + 7y$ subject to
$$3x + 5y \le 15$$
$$x + 5y \le 10$$

16. Maximize $m = 3x + 7y$ subject to
$$3x + y \le 15$$
$$4x + 5y \le 20$$

17. Maximize $m = x + 2y + 3z$ subject to
$$x + 2y + z \le 40$$
$$x + y + z \le 30$$

18. Maximize $m = 3x + 7y + 2z$ subject to
$$3x + 5y + 3z \le 15$$
$$4x + 5y + 2z \le 20$$

19. Maximize $m = 3x + 4y + 2z$ subject to
$$x + 5y + 3z \le 15$$
$$x + 2y + 2z \le 20$$

20. Maximize $m = 4x + 7y + z$ subject to
$$3x + y + z \le 30$$
$$4x + y + 2z \le 45$$

21. Maximize $m = 5x + 7y$ subject to
$$3x + 5y \le 30$$
$$x + 5y \le 20$$
$$x \le 8$$

22. Maximize $m = 12x + 7y$ subject to
$$3x + y \le 30$$
$$2x + 3y \le 55$$
$$x \le 8$$

23. Maximize $m = 10x + 7y$ subject to
$$3x + y \le 48$$
$$2x + 5y \le 84$$
$$y \le 14$$

24. Maximize $m = 12x + 7y$ subject to
$$3x + y \le 30$$
$$2x + 3y \le 55$$
$$y \le 16$$

25. Maximize $m = 6x + 8y$ subject to
$$x + y \le 10$$
$$2x + 3y \le 24$$
$$3x + 2y \le 24$$

26. Maximize $m = 7x + 5y$ subject to
$$x + y \le 20$$
$$x + 2y \le 20$$
$$2x + 3y \le 20$$

27. Maximize $m = x + 3y + 4z$ subject to
$$x + 3y \le 300$$
$$x + z \le 150$$
$$x + 2y + z \le 200$$

28. Maximize $m = 4x + 6y + 5z$ subject to
$$2x + 3y + z \leq 550$$
$$3x + y + z \leq 350$$
$$4x + 2y + z \leq 400$$

•• 29. Maximize $m = 2x + 3y + 5z + 2w$ subject to
$$x + 4y + 2z \leq 60$$
$$x + y + w \leq 20$$
$$x + 3y + 2w \leq 80$$
$$y + 3w \leq 90$$

30. Maximize $m = 20x + 35y + 45z + 30w$ subject to
$$x + 4y + z \leq 350$$
$$x + 5y \leq 650$$
$$x + 3y + w \leq 200$$
$$x + w \leq 90$$

31. Maximize $m = 8x + 7y + 6w$ subject to
$$x + 2y + 2w \leq 30$$
$$x + z + w \leq 40$$
$$y + z \leq 20$$
$$x + z + 2w \leq 30$$

32. Maximize $m = x + 2y + 3w$ subject to
$$x + 3y + z + 4w \leq 50$$
$$x + z + w \leq 60$$
$$y + z + 2w \leq 80$$
$$x + 2y + z \leq 30$$

••• 33. An airline offers a discount on their first class fare to those who have no luggage to check in. The new class is called an "executive" fare. The plane on which it is offered has 190 seats available. The price of a first class ticket is $310, the price of an executive ticket is $270, and the price of an economy class ticket is $140. The airline's cost is $150 for a first class ticket, $150 for an executive class ticket, and $50 for an economy class ticket. The airline must reserve 10 cubic feet (cu ft) of storage for luggage for each first class passenger and 5 cu ft of storage for each economy class passenger. It is decided that the cost of each flight will not exceed $22,500 and the maximum storage space available is 1350 cu ft. How many tickets of each type should be sold to maximize profit?

34. In one part of the farm a farmer plants three crops, tomatoes, potatoes, and corn. The profit per acre for tomatoes is $1200, for potatoes $700, and for corn $800. The cost per acre for fertilizer is $8 for tomatoes, $3 for potatoes, and $6 for corn. The cost per acre for maintenance is $8 for tomatoes, $9 for potatoes, and $6 for corn. Time spent per acre per week for maintenance is 2 hours for tomatoes, 1 hour for potatoes, and 1 hour for corn. No more than $780 is to be spent for fertilizer and no more than $804 is to be spent for maintenance. Time spent for maintenance is to be limited to no more than 150 hours per week. How many acres of each crop should be planted to maximize profit?

35. A lamp manufacturer makes three types of lamps: a deluxe model that is sold in the firm's own store, a regular model that is shipped to a nearby department store, and an economy model that is shipped to a discount store across town. The cost of manufacturing each type of lamp is $80 for a deluxe model, $60 for a regular model, and $30 for an economy model. It costs $20 to ship each regular model and $50 to ship each economy model. The maximum weekly costs are $7000 for manufacturing and $2200 for shipping. It takes 40 minutes to assemble each deluxe model and 20 minutes to assemble each regular and each economy model. The maximum amount of time available for assembly each week is 500 machine-hours; that is, ten machines are available for 50 hours per week. How many lamps of each type should be made to maximize profit if the profit is $140 on each deluxe model, $110 on each regular model, and $90 on each economy model?

36. A city planning/consulting firm has three types of consulting teams to handle different types of business: type I consists of 1 engineer, 1 architect, and 3 salespersons; type II consists of 3 engineers, 1 architect, and 3 salespersons; and type III consists of 4 engineers, 2 architects, and 2 salespersons. There are 100 engineers available, 40 architects available, and 100 salespersons available to make up the teams. If the

average weekly revenue from each type of team is $4000 for type I, $7000 from type II, and $8000 from type III, how many teams of each type should be made to maximize profit?

37. A firm manufactures three types of tool sheds, small, medium, and large. Each small shed takes 2 hours to cut and 2 hours to assemble; each medium shed takes 1 hour to cut and 3 hours to assemble, and each large shed takes 1 hour to cut and 4 hours to assemble. There is a total of 320 hours available to cut and 640 hours available to assemble the sheds. No more than 200 sheds are to be made. How many of each type of shed should be made to maximize profit if the profit is $90 on each small shed, $60 on each medium shed, and $30 on each large shed?

38. A paint manufacturer runs a sale on three colonial colors of indoor trim paint. The color is mixed by blending yellow, red, and blue pigment in base white paint. Each gallon of gunmetal blue has 0.1 ounce (oz) of yellow, 0.2 oz of red, and 0.4 oz of blue. Each gallon of Williamsburg orange has 0.3 oz of yellow, 0.2 oz of red, and 0.1 oz of blue. Each gallon of Saratoga rust has 0.2 oz of yellow, 0.4 oz of red, and 0.1 oz of blue. The company will use no more than 32 oz of yellow, 30 oz of red, and 10 oz of blue. If the profit per gallon is $1 for gun-metal blue, $2 for Williamsburg orange, and $3 for Saratoga rust, how many gallons of each type of paint should be made to maximize profit?

39. In 1983 the Dominican Republic began planning an emergency medical service for the capital city of Santo Domingo. One component of the plan used the simplex method to solve a linear programming problem that described how to allocate ambulance locations efficiently.* The following problem is adapted from this study. The city is divided into zones, each covered by an ambulance location. There are three types of ambulances: type 1 is designed to cover 2 zones, type 2 covers 3 zones and type 3 covers 4 zones. The total number of ambulances available in this part of the study is 72. The average response time is 5 minutes for a type 1 ambulance, 7 minutes for type a 2 ambulance, and 10 minutes for a type 3 ambulance. The total response time is to be no greater than 800 minutes. Find the number of each type of ambulance needed to maximize the number of zones covered.

40. Large meat processing firms usually blend sausage meats using computer-based linear programming models.† A typical problem would entail maximizing profit while the constraints include the number of pounds of meat and spice available, the amount of protein content, and the amount of fat in the blend. Suppose the amount of the blend of beef, pork and spice is to be no more than 400 lb. Each pound of beef contains 90 g (grams) of protein and each pound of pork contains 60 g of protein while the spice contains no protein. The total amount of protein in the blend is limited to 15,000 g. Each pound of beef contains 10 g of fat, each pound of pork contains 5 g of fat and each pound of spice contains 2.5 g of fat. The total amount of fat in the blend is limited to 1400 g. The profit per pound is $1 for beef, $2 for pork and $.50 for spice. Find the number of pounds of each ingredient that maximizes profit.

41. Banks for cooperatives (BCs) are specialty banks that service specific types of firms, such as farm cooperatives. Their loan policies differ from traditional lending institutions in that they tailor payment and interest schedules to the individual customer. They are also governed by different tax and accounting regulations. In an article analyzing lending policies of BCs, linear programming was used to maximize the profit, which is called the *present value of the loan,* subject to

*Eaton, David J., et al., "Determining Ambulance Deployment in Santo Domingo, Dominican Republic," *Journal Operational Research Society,* 37:113–126, 1986.

†Steuer, Ralph E., "Sausage Blending Using Multiple Objective Linear Programming," *Management Science,* 30:1376–84, 1984.

constraints, including variable interest rates.‡ For example, consider a loan that is to be paid back in three variable payments, p_1, p_2, and p_3. The sum of the payments is to be no more than $50,000. The variable interest rate is 9% on the first payment, 8% on the second payment, and 6% on the third payment. The total interest is to be no more than $4000. The first payment is to be no more than $20,000. The present value of the loan is $.9p_1 + .8p_2 + .7p_3$. Find the amount of each payment that maximizes the present value.

‡Ahmad, Ismail, et al., "Analyzing Equity Capital Programs of Banks for Cooperatives," *American Journal of Agricultural Economics,* 68:849–856, 1986.

4–4 NONSTANDARD PROBLEMS

Thus far we have restricted ourselves to linear programming problems in standard form. In this section we apply the simplex method to other types of problems. Recall that a problem in standard form satisfies three properties.

  i. The objective function is to be maximized.
  ii. The variables are all nonnegative; that is, they are greater than or equal to 0.
  iii. The other constraints must be of the form

    (linear polynomial) $\leq a$

    where a is nonnegative.

In this section we consider problems on which these restrictions are relaxed. Property (ii) is still assumed, so we still assume that all the variables are greater than or equal to 0. Since property (ii) still holds, a problem that is not in standard form requires that the objective function either be minimized or have constraints with greater than (>) instead of less than (<). We first study the type of nonstandard problem with the objective function to be minimized and then handle the second type.

Minimizing the Objective Function

Suppose the problem requires the objective function to be minimized, so property (i) is not satisfied. We need a fact about numbers. Let S be a finite set of numbers and let $-S$ be the set of negatives of all the numbers in S. If p is the smallest number in S, then $-p$ is the largest number in $-S$. This is true because $p \leq x$ for all x in S implies that $-p \geq -x$ for all $-x$ in $-S$. This means that if the problem is to minimize m, then we can solve the problem that maximizes $-m$. Example 1 makes use of this property.

EXAMPLE 1 Suppose a linear programming problem is to minimize

$$m = 10x + 30y$$

Suppose there are four feasible solutions: (2, 2), (3, 1), (4, 3), and (0, 4). Let us compute m and $-m$ for each point.

Feasible Solutions	$m = 10x + 30y$	$-m = -10x - 30y$
(2, 2)	$20 + 60 = 80$	$-20 - 60 = -80$
(3, 1)	$30 + 30 = 60$	$-30 - 30 = -60$
(4, 3)	$40 + 90 = 130$	$-40 - 90 = -130$
(0, 4)	$0 + 120 = 120$	$0 - 120 = -120$

The smallest value of m is $m = 60$, which occurs at (3, 1). The largest value of $-m$ is $-m = -60$, also at (3, 1). Hence the same point that minimizes m also maximizes $-m$.

Example 1 shows that minimum problems can be solved via the simplex method by multiplying the objective function m by -1 and maximizing $-m$. The problem then satisfies property (i). The optimal solution that maximizes $-m$ will also minimize m.

For example, suppose the problem is to

Minimize $m = 5x + 7y$ subject to
$x + 5y \leq 10$
$x + 3y \leq 12$

This problem is not in standard form because the objective function is to be minimized. To put it into standard form express it as

Maximize $-m = -5x - 7y$ subject to
$x + 5y \leq 10$
$x + 3y \leq 12$

The preceding discussion shows that the optimal solution of this problem is also the optimal solution of the original problem.

Inequalities in Nonstandard Form

Now we deal with the second type of nonstandard problem, one that violates property (iii), that is, one in which an inequality has greater than or equal to ($\geq$) instead of less than or equal to ($\leq$). For instance, consider

$x - 3y \geq 5$

To write the inequality using $\leq$ instead of $\geq$, multiply it by -1 and get

$-x + 3y \leq -5$

However, this violates the part of property (iii) that requires the constant term to be nonnegative. To alleviate that difficulty, we form the simplex tableau and perform an appropriate pivot.

In order to see how to handle a negative constant, consider the following row of a simplex tableau and assume that column 4 is in basic form.

$$\begin{array}{ccccc} x & y & s_1 & s_2 & m \\ [\,-1 & 5 & 0 & 1 & 0 & -50\,] \end{array}$$

The difficulty is the negative number in the last column. When we calculate the basic feasible solution corresponding to this tableau, we get $s_2 = -50$. This violates property (ii), which requires all variables to be nonnegative. In order to apply the simplex method, we must change the sign of any number in the last column that is negative, except the last one. This is done by **pivoting by a negative number** that is in the same row as the negative number in the last column. In the preceding tableau -50 appears in the last column. Since all variables are greater than or equal to 0, the only way a number in the last column can be negative is for at least one of the numbers in its row to be negative. Otherwise the equation corresponding to that row would have all positive coefficients and variables on the left-hand side of the equation set equal to a negative constant, which is the number in the last column. This cannot happen since it implies a negative number is set equal to a positive number.

Refer again to the preceding row with -50 in the last column. Locate a negative number in that row. There is only one, -1, in the first column. Use it as the pivot column. Pivoting in this manner will remove the negative number from that entry, as is described in Example 2. This procedure might have to be repeated several times because another negative number might appear or may have been in the last column to begin with.

Remember that we are not concerned with negative numbers in the bottom entry of the last row. It corresponds to the objective function and a negative number in that spot does not violate property (iii).

EXAMPLE 2   Consider the simplex tableau

$$\begin{bmatrix} 1 & 2 & 1 & 0 & 0 & | & 40 \\ -1 & 3 & 0 & 1 & 0 & | & -20 \\ \hline 1 & -4 & 0 & 0 & 1 & | & 0 \end{bmatrix}$$

Problem

Remove the negative number in the last column.

SOLUTION Look for a negative number in the same row as the negative number in the last column. The (2, 1) entry is negative so choose its column as the pivot column. As before, compute the ratios and select the pivot entry by identifying the smaller positive ratio. In this case the ratios are

for the first row the ratio is $\frac{40}{1} = 40$
for the second row the ratio is $\frac{-20}{-1} = 20$

The second row has the smaller ratio, and thus it is the pivot row. Hence the pivot entry is the (2, 1) entry, -1. When we pivot about it we get the new tableau

$$\begin{bmatrix} 0 & 5 & 1 & 1 & 0 & | & 20 \\ 1 & -3 & 0 & -1 & 0 & | & 20 \\ \hline 0 & -1 & 0 & 1 & 1 & | & -20 \end{bmatrix}$$

Notice that all of the entries in the last column (except the last one) are nonnegative. (It might be necessary to pivot more than once to achieve this.) Thus the basic feasible solution has all variables nonnegative. Now the simplex method can be applied since the tableau is in standard form.

We now summarize the simplex method for problems that are not in standard form.

The Simplex Method for Problems in Nonstandard Form

1. If the problem is to minimize m, then maximize $= -m$.
2. If any constraint is in the form

   (linear polynomial) $\geq a$

   then multiply the inequality by -1 to get it in the form

   (linear polynomial) $\leq -a$

   Then form the initial simplex tableau.
3. If no negative number appears in the last column (except the last entry), go to step 6.
4. If a negative number appears in the last column (except the last one), change it to a positive number by pivoting as follows:
   (a) Select a negative entry in the same row as the negative number in the last column. Its column becomes the pivot column.
   (b) Compute all ratios, including those corresponding to negative entries in the pivot column. The pivot row is the one with the smallest positive ratio.
   (c) Perform the pivot above the pivot entry.
5. Go to step 3.
6. Now the tableau is in standard form. Use the simplex method for problems in standard form.

In the next example the method is applied twice, and each time a negative entry appears in the last column. It also shows how to handle the case when two negative numbers appear.

EXAMPLE 3 | **Problem**

Minimize $m = x + y$ subject to
$$2x - y \geq 30$$
$$-x + y \geq 50$$

SOLUTION

1. Multiply the objective function by -1, which yields

   Maximize $-m = -x - y$

2. Both constraints are in the form (linear polynomial) greater than or equal to a, so multiply each by -1, which yields

   $$-2x + y \leq -30$$
   $$x - y \leq -50$$

   The initial simplex tableau is

   $$\begin{bmatrix} -2 & 1 & 1 & 0 & 0 & | & -30 \\ 1 & -1 & 0 & 1 & 0 & | & -50 \\ 1 & 1 & 0 & 0 & 1 & | & 0 \end{bmatrix}$$

3. There are two negative numbers in the last column so we go to step 4.
4. Either number can be chosen. We arbitrarily choose -50 in the second row. There is one other negative entry in the second row, -1, in the second column. The two ratios are $-30/1 = -30$ and $-50/-1 = 50$. Since 50 is the only positive ratio, choose the (2, 2) entry as the pivot entry. The resulting simplex tableau is

   $$\begin{bmatrix} -1 & 0 & 1 & 1 & 0 & | & -80 \\ -1 & 1 & 0 & -1 & 0 & | & 50 \\ 2 & 0 & 0 & 1 & 1 & | & -50 \end{bmatrix}$$

   The number -80 is the only negative number in the last column above the last entry. The only other negative entry in the first row is -1 in the first column. The two ratios are $-80/-1 = 80$ and $50/-1 = -50$. The only positive ratio is 80 so the (1, 1) entry, -1, is the pivot entry. The resulting simplex tableau is

   $$\begin{bmatrix} 1 & 0 & -1 & -1 & 0 & | & 80 \\ 0 & 1 & -1 & -2 & 0 & | & 130 \\ 0 & 0 & 2 & 3 & 1 & | & -210 \end{bmatrix}$$

   There is no negative entry in the last row to the left of the last entry so the maximality test passes, and hence the basic feasible solution for this tableau yields the optimal solution. The basic feasible solution is $x = 80$, $y = 130$, $s_1 = s_2 = 0$, $-m = -210$. Hence the values $x = 80$ and $y = 130$ yield the minimum value of $m = 210$.

Dantzig's Breakthrough

The simplex method for solving linear programming problems is a fairly complex procedure as compared to the geometric approach used in Chapter 3. There are two reasons that the simplex method is superior. First it can handle many equations and variables while the geometric method can deal with only two variables. Second, it is very mechanical and does not rely on visual or geometric interpretations of data, and so it is much easier to program a computer to do the calculations.

In fact in 1947 when George Dantzig was searching for a method to solve these problems, he was able to consider the simplex method only because the invention of the computer was imminent. His simplex method reduced the number of calculations drastically, but there were still too many for the method to be useful. Once the computer was invented, the simplex method became a way of life for economists and big business. An example is described in some detail at the end of this chapter.

EXERCISE SET 4–4

• In Problems 1 to 6 determine which of the three defining properties of a linear programming problem in standard form is (or are) violated.

1. Maximize $m = x + y$ subject to
$$2x - y \leq 3$$
$$x + y \geq 4$$

2. Maximize $m = -x - y$ subject to
$$3x + y \geq -1$$
$$2x - y \geq 4$$

3. Minimize $m = x + y - 3z$
$$x - z \leq 2$$
$$y + z \geq 4$$

4. Minimize $m = x + 3z$ subject to
$$x + y \geq 1$$
$$x - z \leq 2$$

5. Minimize $m = x + 3y + 3z$
$$x + y - z \leq 12$$
$$x - y + z \geq 4$$
$$x + y + z \geq 3$$

6. Minimize $m = x + y + z$ subject to
$$x + y \geq -1$$
$$ y - z \leq 0$$

In Problems 7 to 12 convert the objective function to one in standard form and then write the last row of the initial simplex tableau, assuming that there are the given number of variables and constraints in the problem.

7. Minimize $m = x + y$, two variables, two constraints.

8. Minimize $m = 2x + 5y$, two variables, two constraints.

9. Minimize $m = x + 3y$, two variables, three constraints.

10. Minimize $m = 2x + 3y$, two variables, three constraints.

11. Minimize $m = x - y + 2z$, three variables, three constraints.

12. Minimize $m = 2x + z$, three variables, four constraints.

In Problems 13 to 18 convert the linear programming problem into one in standard form.

13. Minimize $m = x + y$ subject to
$$x + 2y \leq 4$$
$$x + 3y \geq 1$$

14. Minimize $m = 3x + y$ subject to
$$3x + 2y \leq 14$$
$$4x + 3y \geq 11$$

15. Minimize $m = 2x - y$ subject to
$$2x + y \geq 2$$
$$x - y \geq 4$$

16. Minimize $m = x - 5y$ subject to
$$2x + 3y \geq 42$$
$$5x - 3y \geq 54$$

17. Minimize $m = 3x + y + z$ subject to
$$x - z \leq -2$$
$$x + y \geq 4$$

18. Minimize $m = x - 3y$ subject to
$$x + y - 2z \geq 0$$
$$x \qquad - 4z \leq -2$$
$$2x + 3y \qquad \geq -4$$

In Problems 19 and 20 suppose the solution of a minimum problem yields the given matrix as the final simplex tableau. What is the minimum value that is obtained by the optimal solution to the original problem?

19. $\begin{bmatrix} 1 & 0 & -1 & 0 & 0 & | & 12 \\ 0 & 1 & 2 & 1 & 0 & | & 15 \\ 0 & 0 & 1 & 1 & 1 & | & -60 \end{bmatrix}$

20. $\begin{bmatrix} 1 & 2 & 0 & 3 & 0 & | & 10 \\ 0 & 1 & 1 & 1 & 0 & | & 6 \\ 0 & 1 & 0 & 0 & 1 & | & -110 \end{bmatrix}$

In Problems 21 to 33 solve the linear programming problems by the simplex method.

21. Maximize $m = -x + 3y$ subject to
$$5x + 2y \leq 100$$
$$y \geq 10$$
$$x \geq 0$$

22. Maximize $m = x + 4y$ subject to
$$6x + 9y \leq 18$$
$$x \geq 2$$
$$y \geq 0$$

23. Minimize $m = 4x + 13y$ subject to
$$x + 2y \geq 11$$
$$x + y \geq 10$$

24. Minimize $m = 2x + 7y$ subject to
$$x - y \geq 10$$
$$x + 2y \geq 6$$

25. Minimize $m = 2x - 4y$ subject to
$$2x + 6y \leq 18$$
$$-2x + 3y \geq 6$$

26. Minimize $m = x - 2y$ subject to
$$3x + y \geq 18$$
$$x + y \leq 9$$

•• 27. Minimize $m = x + z$ subject to
$$x + 2y \qquad \leq 8$$
$$x - \qquad z \leq 4$$
$$y + 4z \geq 4$$

28. Minimize $m = x + y$ subject to
$$x + y \qquad \geq 10$$
$$x \qquad - z \leq 14$$
$$y + 2z \geq 6$$

29. Minimize $m = x + 2y + z$ subject to
$$2x + y \qquad \leq 10$$
$$3x + y \qquad \leq 12$$
$$y + 2z \geq 8$$

30. Minimize $m = 6x + 7y + 3z$ subject to
$$2x + 5y \qquad \leq 20$$
$$2x + y \qquad \leq 8$$
$$y + z \leq 8$$

31. Minimize $m = x + 2y + z$ subject to
$$3x + 2y + z \leq 12$$
$$x \qquad + z \leq 8$$
$$2x + 5y \qquad \leq 10$$

32. Minimize $m = x + 3y + 2z$ subject to
$$x + 5y + z \leq 20$$
$$x \qquad + z \leq 10$$
$$2x + 5y + z \leq 15$$

••• 33. Minimize $m = 2x + y + 3z + u$ subject to
$$x + y \qquad \leq 100$$
$$x + y + z \qquad \leq 150$$
$$y + z \qquad \leq 100$$
$$x \qquad + u \leq 200$$

34. Show that if S is a finite set of numbers and p is the smallest number in S, then $-p$ is the largest number in $-S$, the set of all negatives of numbers in S.

35. A television service store manager must purchase at least 200 VCR tapes. There are two types of tapes to choose from, high-quality and regular. The high-quality tapes play for 2 hours and cost $8 each and the regular tapes play for 1 hour and cost $6. The tapes must produce at least 100 hours of playing time. How many of each type should the manager purchase to minimize cost?

36. A fuel oil manufacturing firm has two processing plants, plant I and plant II. Plant I processes daily 250 barrels of high-grade oil for commercial use and 100 barrels of low-grade oil for residential use. Plant II processes daily 100 bar-

rels of high-grade oil and 100 barrels of low-grade oil. The daily cost of operation is $50,000 for plant I and $30,000 for plant II. An order is placed for 1500 barrels of high-grade oil and 1200 barrels for low-grade oil. Find the number of days that each plant should operate to fill the order and minimize the cost.

37. A candy manufacturer makes three types of 1-pound (lb) chocolate bars. The chocolate-nut bar contains $\frac{3}{4}$ lb chocolate and $\frac{1}{4}$ lb nuts, the chocolate-peanut butter bar contains $\frac{3}{4}$ lb chocolate and $\frac{1}{4}$ lb peanut butter, and the chocolate-cluster bar contains $\frac{1}{2}$ lb chocolate, $\frac{1}{4}$ lb nuts, and $\frac{1}{4}$ lb peanut butter. Each day the manufacturer must use at least 10 lb of nuts, no more than 40 lb of chocolate, and no more than $7\frac{1}{2}$ lb of peanut butter. If it costs $4 to make each chocolate-nut bar, $2 to make each chocolate-peanut butter bar, and $3 to make each chocolate-cluster bar, how many bars of each type should the manufacturer make each week to minimize cost?

38. An overnight delivery service has daily flights to primary depots in four cities, Boston, Chicago, Dallas, and Los Angeles. The total daily flights to Boston and Chicago must be at least 10; the total daily flights to Chicago and Dallas must be at least 12; there must be no more than 15 total daily flights to Boston and Los Angeles; there must be no more than 10 total daily flights to Dallas and Los Angeles. The cost of each flight is $8000 to Boston, $20,000 to Chicago, $10,000 to Dallas, and $6000 to Los Angeles. How many flights to each city should be scheduled to minimize cost?

39. Linear programming has been used in the petroleum industry for over 25 years to minimize the cost of blending fuel. In an article analyzing linear programming techniques used at Exxon, blending gasoline in processing tanks was studied.* As an example, suppose that at one point of the blending process there are three tanks. The total amount of gasoline that passes through the three tanks is no more than 5 million gallons, and no more than 1 million gallons passes through tank 3. One goal of the blending process is to extract lead from the gasoline. From each gallon in tank 1, .2 g (grams) of lead is extracted, from each gallon in tank 2, .5 g of lead is extracted, and from each gallon in tank 3, .1 g of lead is extracted. At least 1.5 million g of lead must be extracted. The cost per gallon is $.10 for tank 1, $.15 for tank 2, and $.20 for tank 3. Find the number of gallons that should be processed by each tank to minimize cost.

40. The cost of transportation is an important consideration for many enterprises. School systems especially feel the pressure to hold down busing costs while maintaining adequate service. Swersey and Ballard applied linear programming to a bus scheduling problem in the school system of New Haven, Connecticut.† A typical problem in the study involves a busing schedule utilizing three types of routes that service a school with 2000 students. The object is to minimize the number of buses. The first type of route uses two buses, and the second and third types of routes use three buses each. Each bus on a type 1 route has 50 seats while each bus on a type 2 or 3 route has 60 seats. The operating cost per year is $15,000 for each bus used on a type 1 route, $20,000 for each bus used on a type 2 route and $25,000 for each bus used on a type 3 route. If the cost per year to maintain the fleet is limited to $600,000, find the number of each type of route that should be used to minimize the number of buses.

*Baker, Thomas E., and Lasdon, Leon S., ''Successive Linear Programming at Exxon,'' *Management Science,* 31:264–274, 1985.

†Swersey, Arthur J., and Ballard, Wilson, ''Scheduling School Buses,'' *Management Science,* 30:844–853, 1984.

4–5 THE DUAL PROBLEM

This section describes the dual method for solving a nonstandard problem whose objective function is minimized and whose constraints are all in the form greater than or equal to ($\geq$). Such a problem can be solved by the methods of Section 4–4, but they require numerous pivots. The **dual method,** however, converts the original problem to a problem that is in standard form and whose solution contains the solution to the original problem.

Matrix Inequalities

Let M and N be two matrices. Recall that $M = N$ if they have the same size and their corresponding entries are equal. Similarly, we define $M \leq N$ if they have the same size and $m_{ij} \leq n_{ij}$ for all i,j.

Linear programming problems can be written in terms of matrices. Consider the diet problem from Section 3–2.

$$\begin{aligned} \text{Minimize} \quad & m = 4x + 2y \\ \text{Subject to} \quad & 5x + y \geq 5 \\ & 5x + 3y \geq 10 \\ & x \geq 0, y \geq 0 \end{aligned}$$

Let A be the coefficient matrix of the first two constraints, C be the constant matrix, R be the row matrix of coefficients of m, and X be the column matrix of variables.

$$A = \begin{bmatrix} 5 & 1 \\ 5 & 3 \end{bmatrix} \quad C = \begin{bmatrix} 5 \\ 10 \end{bmatrix} \quad R = [4 \quad 2] \quad X = \begin{bmatrix} x \\ y \end{bmatrix}$$

The diet problem becomes

$$\begin{aligned} \text{Minimize} \quad & m = RX \\ \text{Subject to} \quad & AX \geq C \\ & X \geq 0 \end{aligned}$$

In Example 1 another linear programming problem is converted to matrix notation.

EXAMPLE 1 | **Problem**

Write this linear programming problem in terms of matrices.

$$\begin{aligned} \text{Minimize} \quad & m = 7x + 5y \\ \text{Subject to} \quad & x + 3y \geq 2 \\ & 2x + y \geq 4 \\ & x + y \geq 6 \\ & x \geq 0, y \geq 0 \end{aligned}$$

SOLUTION Let A be the coefficient matrix of the first three constraints, C be the constant matrix, R be the row matrix of coefficients of m, and X be the column matrix of variables.

$$A = \begin{bmatrix} 1 & 3 \\ 2 & 1 \\ 1 & 1 \end{bmatrix} \quad C = \begin{bmatrix} 2 \\ 4 \\ 6 \end{bmatrix} \quad R = [7 \quad 5] \quad X = \begin{bmatrix} x \\ y \end{bmatrix}$$

The linear programming problem can be written in terms of matrices as

Minimize $m = RX$
Subject to $AX \geq C$
$X \geq 0$

Notice in Example 1 that A and C must have the same number of rows and that R and X must have the same number of entries.

The Transpose

The **transpose** of a matrix, A, denoted by A', is formed by switching the rows and columns of A so that the first row becomes the first column, the second row becomes the second column, and so on. For instance

$$\text{if} \quad A = \begin{bmatrix} 1 & 3 & 0 \\ 2 & 7 & 4 \end{bmatrix} \quad \text{then} \quad A' = \begin{bmatrix} 1 & 2 \\ 3 & 7 \\ 0 & 4 \end{bmatrix}$$

In general the transpose of an $m \times n$ matrix is an $n \times m$ matrix. Thus the transpose of a row matrix is a column matrix, and vice versa. The central role of the transpose in the dual problem is described in the display.

The Dual Problem

If a linear programming problem is

Minimize $m = RX$
Subject to $AX \geq C$
$X \geq 0$

then its *dual problem* is

Maximize $m = C'U$
Subject to $A'U \leq R'$
$U \geq 0$

where U is a column matrix whose size is the same size as C.

The dual problem is in standard form so it can be solved by standard means. Its objective function m is the same as the objective function m of the original problem. The proof of this fact is beyond the scope of this book.

There is a more efficient way to express the dual problem. Let us illustrate it on the diet problem.

$$\text{Minimize} \quad m = 4x + 2y$$
$$\text{Subject to} \quad 5x + y \geq 5$$
$$5x + 3y \geq 10$$
$$x \geq 0, \, y \geq 0$$

First form the matrix T, called the **primal matrix,** whose bottom row comes from the objective function and whose other rows come from the explicit constraints.

$$T = \begin{bmatrix} 5 & 1 & 5 \\ 5 & 3 & 10 \\ 4 & 2 & 0 \end{bmatrix}$$

Its transpose is

$$T' = \begin{bmatrix} 5 & 5 & 4 \\ 1 & 3 & 2 \\ 5 & 10 & 0 \end{bmatrix}$$

The dual problem can be read from T', with the objective function again coming from the bottom row.

$$\text{Maximize} \quad m = 5u + 10v$$
$$\text{Subject to} \quad 5u + 5v \leq 4$$
$$u + 3v \leq 2$$
$$u \geq 0, \, v \geq 0$$

EXAMPLE 2 | **Problem**

Write the dual problem of the linear programming problem in Example 1.

SOLUTION Form the primal matrix T.

$$T = \begin{bmatrix} 1 & 3 & 2 \\ 2 & 1 & 4 \\ 1 & 1 & 6 \\ 7 & 5 & 0 \end{bmatrix}$$

Then form the transpose T'.

$$T' = \begin{bmatrix} 1 & 2 & 1 & 7 \\ 3 & 1 & 1 & 5 \\ 2 & 4 & 6 & 0 \end{bmatrix}$$

The dual problem of the original problem is

$$\text{Maximize} \quad m = 2u + 4v + 6w$$
$$\text{Subject to} \quad u + 2v + w \leq 7$$
$$3u + v + w \leq 5$$
$$u \geq 0, \, v \geq 0, \, w \geq 0$$

The Solution

The dual problem is in standard form so it is solved by the usual means. The final tableau yields the solution to the original problem because the variable m in the dual problem is the same as the variable m in the original problem. The values of the variables x and y are obtained from the final tableau in a distinctive way. Consider the diet problem and its dual together.

<table>
<tr><td colspan="2" align="center">**Diet problem**</td><td colspan="2" align="center">**Dual problem**</td></tr>
<tr><td>Minimize</td><td>$m = 4x + 2y$</td><td>Maximize</td><td>$m = 5u + 10v$</td></tr>
<tr><td>Subject to</td><td>$5x + y \geq 5$</td><td>Subject to</td><td>$5u + 5v \leq 4$</td></tr>
<tr><td></td><td>$5x + 3y \geq 10$</td><td></td><td>$u + 3v \leq 2$</td></tr>
<tr><td></td><td>$x \geq 0, y \geq 0$</td><td></td><td>$u \geq 0, v \geq 0$</td></tr>
</table>

The initial tableau of the dual problem is

$$
\begin{array}{ccccc}
u & v & s_1 & s_2 & m \\
\end{array}
$$
$$
\left[
\begin{array}{ccccc|c}
5 & 5 & 1 & 0 & 0 & 4 \\
1 & 3 & 0 & 1 & 0 & 2 \\
\hline
-5 & -10 & 0 & 0 & 1 & 0 \\
\end{array}
\right]
$$

It reduces to the final tableau

$$
\begin{array}{ccccc}
u & v & s_1 & s_2 & m \\
\end{array}
$$
$$
\left[
\begin{array}{ccccc|c}
1 & 0 & \frac{3}{10} & -\frac{1}{2} & 0 & \frac{1}{5} \\
0 & 1 & -\frac{1}{10} & \frac{1}{2} & 0 & \frac{3}{5} \\
\hline
0 & 0 & \frac{1}{2} & \frac{5}{2} & 1 & 7 \\
\end{array}
\right]
$$

The solution to the dual problem is $u = \frac{1}{5}$, $v = \frac{3}{5}$, and $m = 7$. The value $m = 7$ is the maximum value of the objective function of the dual problem and the minimum value of the objective function of the original problem. The values for x and y lie in the bottom row of the final tableau in the columns of the slack variables, $x = \frac{1}{2}$ and $y = \frac{5}{2}$. (The proof of this fact is beyond the scope of this book.) This solution corresponds to the point $(\frac{1}{2}, \frac{5}{2})$ derived geometrically in Example 1 of Section 3–3.

Example 3 continues Examples 1 and 2.

EXAMPLE 3 | **Problem**

Solve the problem in Example 1.

SOLUTION From Example 2 the dual problem is

$$
\begin{aligned}
\text{Maximize} \quad & m = 2u + 4v + 6w \\
\text{Subject to} \quad & u + 2v + w \leq 7 \\
& 3u + v + w \leq 5 \\
& u \geq 0, v \geq 0, w \geq 0
\end{aligned}
$$

The initial tableau is

$$
\begin{array}{cccccc}
u & v & w & s_1 & s_2 & m \\
\end{array}
$$
$$
\left[
\begin{array}{cccccc|c}
1 & 2 & 1 & 1 & 0 & 0 & 7 \\
3 & 1 & 1 & 0 & 1 & 0 & 5 \\
\hline
-2 & -4 & -6 & 0 & 0 & 1 & 0 \\
\end{array}
\right]
$$

Pivot about the (2, 3) entry:

$$
\begin{array}{cccccc}
u & v & w & s_1 & s_2 & m \\
\left[\begin{array}{cccccc|c}
-2 & 1 & 0 & 1 & -1 & 0 & 2 \\
3 & 1 & 1 & 0 & 1 & 0 & 5 \\
\hline
16 & 2 & 0 & 0 & 6 & 1 & 30
\end{array}\right]
\end{array}
$$

The values for x and y lie in the bottom row under the slack variables. The solution is $x = 0$, $y = 6$, $m = 30$.

The dual method consists of three steps: (1) convert the original problem to its dual problem, (2) solve the dual problem by standard techniques, and (3) interpret the final tableau of the dual problem to obtain the solution to the original problem. Example 4 illustrates the entire method.

EXAMPLE 4   **Problem**

Solve by the dual method.

Minimize $m = 12x + 3y + 4z$
Subject to $2x + 3y + 4z \geq 4$
             $3x + y + z \geq 6$
             $x \geq 0, y \geq 0, z \geq 0$

SOLUTION To convert the original problem to its dual, form the primal matrix T and its transpose.

$$
T = \begin{bmatrix} 2 & 3 & 4 & 4 \\ 3 & 1 & 1 & 6 \\ 12 & 3 & 4 & 0 \end{bmatrix}
\qquad
T' = \begin{bmatrix} 2 & 3 & 12 \\ 3 & 1 & 3 \\ 4 & 1 & 4 \\ 4 & 6 & 0 \end{bmatrix}
$$

So the dual problem is

Maximize $m = 4u + 6v$
Subject to $2u + 3v \leq 12$
             $3u + v \leq 3$
             $4u + v \leq 4$
             $u \geq 0, v \geq 0$

The initial tableau is

$$
\begin{array}{cccccc}
u & v & s_1 & s_2 & s_3 & m \\
\left[\begin{array}{cccccc|c}
2 & 3 & 1 & 0 & 0 & 0 & 12 \\
3 & 1 & 0 & 1 & 0 & 0 & 3 \\
4 & 1 & 0 & 0 & 1 & 0 & 4 \\
\hline
-4 & -6 & 0 & 0 & 0 & 1 & 0
\end{array}\right]
\end{array}
$$

Pivot about the (2, 2) entry.

$$\begin{array}{cccccc} u & v & s_1 & s_2 & s_3 & m \\ \left[\begin{array}{cccccc|c} -7 & 0 & 1 & -3 & 0 & 0 & 3 \\ 3 & 1 & 0 & 1 & 0 & 0 & 3 \\ \hline 1 & 0 & 0 & -1 & 1 & 0 & 1 \\ 14 & 0 & 0 & 6 & 0 & 1 & 18 \end{array}\right] \end{array}$$

This passes the maximality test. The solution to the original problem is $x = 0$, $y = 6$, $z = 0$, $m = 18$.

EXERCISE SET 4–5

- In Problems 1 to 4 write the linear programming problem in terms of matrices. In all problems $x \geq 0$, $y \geq 0$, and $z \geq 0$.

1. Minimize $m = 4x + 2y$
   Subject to $3x + 2y \geq 4$
   $4x - 5y \geq 9$

2. Minimize $m = 4x + y + z$
   Subject to $x + 3y - 2z \geq 6$
   $x - 2y + 3z \geq 6$

3. Minimize $m = 2x + 4y$
   Subject to $3x + 5y \geq 8$
   $-x + 3y \geq 4$
   $2x - 6y \geq 7$

4. Minimize $m = 4x + y + z$
   Subject to $x + 3y - 2z \geq 6$
   $x - 2y + 3z \geq 6$
   $-x + 4y + 5z \geq 9$

In Problems 5 to 10 form the transpose of the given matrix.

5. $\begin{bmatrix} 5 & -1 \\ 2 & -3 \end{bmatrix}$

6. $\begin{bmatrix} 4 & 6 \\ -5 & 8 \\ 10 & 7 \end{bmatrix}$

7. $\begin{bmatrix} -4 \\ 2 \end{bmatrix}$

8. $\begin{bmatrix} 4 & 0 & -3 \\ 2 & -1 & 8 \end{bmatrix}$

9. $\begin{bmatrix} 1 & -4 & 11 \\ 6 & 15 & 0 \\ -2 & 1 & -2 \end{bmatrix}$

10. $[3 \quad 0 \quad -6]$

In Problems 11 to 14 write the dual problem of the given linear programming problem.

11. Problem 1.

12. Problem 2.

13. Problem 3.

14. Problem 4.

In Problems 15 to 18 solve each linear programming problem by the dual method. In all problems $x \geq 0$, $y \geq 0$, and $z \geq 0$.

15. Minimize $m = x + 2y$
    Subject to $2x + y \geq 10$
    $3x + y \geq 12$

16. Minimize $m = 6x + 7y$
    Subject to $2x + 5y \geq 20$
    $2x + y \geq 8$

17. Minimize $m = 4x + 13y$
    Subject to $x + 2y \geq 11$
    $x + y \geq 10$

18. Minimize $m = 2x + 7y$
    Subject to $x - y \geq 10$
    $x + 2y \geq 6$

19. A dietician is preparing a meal of chicken and rice so that each meal contains at least 5 grams (g) of protein and 5 milligrams (mg) of iron. Each serving of chicken contains 5 g of protein, 3 mg of iron, and 4 g of fat. Each serving of rice contains 1 g of protein, 2 mg of iron, and 2 g of fat. How many servings of chicken and rice should be prepared to minimize the amount of fat?

20. A heating oil firm has two processing plants. Plant A processes daily 3000 barrels of No. 2

fuel for residential use and 1000 barrels of No. 6 fuel for industrial use. Plant B processes daily 2000 barrels of No. 2 fuel and 1000 barrels of No. 6 fuel. The daily cost of operation is $20,000 for Plant A and $15,000 for Plant B. An order is placed for 20,000 barrels of No. 2 fuel and 9000 barrels of No. 6 fuel. Find the number of days that each plant should be operated to fill the order and minimize the cost.

•• In Problems 21 to 26 solve each linear programming problem by converting it to the dual problem. Assume that $x \geq 0$, $y \geq 0$, $z \geq 0$, and $w \geq 0$.

21. Minimize $m = x + 2y + z$
    Subject to $3x + 2y + z \geq 12$
    $x + z \geq 8$
    $2x + 5z \geq 10$

22. Minimize $m = 2x + y + 3z + w$
    Subject to $x + y \geq 100$
    $x + y + z \geq 150$
    $y + z \geq 100$
    $x + w \geq 0$

23. Minimize $m = 3x + 4y + 2z + 3w$
    Subject to $x + y + w \geq 8$
    $2x + z \geq 7$
    $y + z + w \geq 9$
    $2x + y + 2w \geq 6$

24. Minimize $m = 5x + 6y + 8z + 3w$
    Subject to $x + y + w \geq 1$
    $3x + z + 2w \geq 2$
    $x + y + z + w \geq 0$
    $4x + y + 2z \geq 3$

25. Minimize $m = x + 2y + z$
    Subject to $3x + 2y + z \geq 12$
    $x + z \geq 8$

26. Minimize $m = 2x + y + 3z + w$
    Subject to $x + y + z \geq 15$
    $y + z \geq 10$
    $x + w \geq 0$

27. A dietician is preparing a menu of chicken, rice, and spinach so that each meal contains at least 30 g of protein, 50 mg of iron, and 40 mg of starch. Each serving of chicken contains 1 g of protein, 1 mg of iron, and 20 g of fat. Each serving of rice contains 2 g of protein, 1 mg of starch, and 35 g of fat. Each serving of spinach contains 2 g of protein, 1 mg of starch, and 65 g of fat. How many servings of chicken, rice, and spinach should be prepared to minimize the amount of fat?

28. A heating oil firm has three processing plants. The amount of No. 2 fuel oil processed per hour is 1000 barrels at Plant A and 3000 barrels at Plant B. The amount of No. 6 fuel processed per hour is 1000 barrels at Plant A and 1000 barrels at Plant C. The amount of blended oil processed per hour is 1000 barrels at Plant A, 2000 barrels at Plant B, and 1000 barrels at Plant C. The daily cost of operation is $100,000 for Plant A, $300,000 for Plant B, and $400,000 for Plant C. An order is placed for 300,000 barrels of No. 2 fuel oil, 150,000 barrels of No. 6 fuel, and 200,000 barrels of blended oil. Find the number of hours that each plant should be operated to fill the order and minimize the cost.

••• 29. One of the most widely used applications of linear programming is in scheduling problems. In an article describing plant design and scheduling problems in the food industry, Regina Beneviste analyzed efficient ways to mix powdered ingredients in weighing machines, called hoppers.* A specific plant used three hoppers in mixing batches of several recipes. One type of problem involved minimizing the amount of time spent weighing the ingredients subject to the number of times certain recipes were required. Recipe A used hopper 1 and hopper 2. Recipe B used hopper 2 and hopper 3. Recipe C used hopper 1 and hopper 3. The batch requires recipe A be made at least 100 times, recipe B be made at least 240 times and recipe C be made at least 140 times. It takes .4 minutes to use hopper 1, .8 minutes to use hopper 2 and .7

*Beneviste, Regina, "An Integrated Plant Design and Scheduling Problem in the Food Industry," *Journal Operational Research Society*, 37:435–461, 1986.

minutes to use hopper 3. Find the number of times each hopper should be scheduled to minimize the amount of weighing time.

30. Dotmar Packaging Ltd. is a firm with a corrugated cardboard box factory that operates four finishing machines in the corrugation process. Boxes are made by cutting a continuous strip of corrugated board into customer-specified dimensions. Cost is affected by the amount of waste left from the cutting process. Dotmar hired consultants to develop a production schedule that would minimize cutting cost while satis-fying customer demand.[†] A typical problem involved four cutting patterns performed on the four finishing machines. Pattern A uses machines 1 and 2. Pattern B uses machines 3 and 4. Pattern C uses machines 1 and 3. Pattern D uses machines 2 and 4. The demand on a particular day requires 75 runs of pattern A, 180 runs of pattern B, 105 runs of pattern C, and 150 runs of pattern D. If the cost per run is $1.00 for machine 1, $1.80 for machine 2, $1.60 for machine 3, and $.90 for machine 4, find the number of times each machine should be used to minimize cost while satisfying demand.

[†]Bookbinder, James H., and Higginson, James K., "Customer Service vs Trim Waste in Corrugated Box Manufacture," *Journal Operational Research Society,* 37:1061–1071, 1986.

CASE STUDY   Shell Oil Company

One of the most important applications that George Dantzig's simplex method was used for was the solution of the transportation problem. Virtually the same methods are used today in industry to solve similar problems. All large firms, and many smaller ones, have very complex schemes and formulas to determine the cost of transporting their merchandise. Usually the procedure entails the use of a computer.

As an example we will look at the Shell Oil Company and their distribution problems.[*] Since 20% of Shell's revenue is spent on transportation of oil, an efficient distribution is critical. Shell divides the country into regions and each region is divided into subregions. We will center our attention on the Chicago area subregion. This area has two primary refineries where the oil is turned from crude oil to the many varied grades of commercial grade oil. They are located in East Chicago, Indiana and Hammond, Indiana. The two major storage and reshipment terminals are located in Des Plaines, Illinois, and Niles, Michigan. Figure 1 illustrates Shell's distribution scheme. In actual practice the problem is more complex than we can consider here. It involves over 1200 variables and 800 constraints. This is because there are more complex decisions to be made such as which mode of transportation to use (including pipelines, barges, trucks, and tankers). The typical problem faced each day is solved on a computer. It takes about $\frac{1}{2}$ hour to run and costs about $100. Such a report will usually generate around ten optional reports because there are various goals and different types of individuals using the same data.

Let us now concentrate on a typical problem encountered in the Chicago area subregion. On a given day the East Chicago refinery will produce at least

[*]T. K. Zierer, et al., "Practical Applications of Linear Programming to Shell's Distribution Problems," *Interfaces,* Vol. 6, No. 4, August 1976.

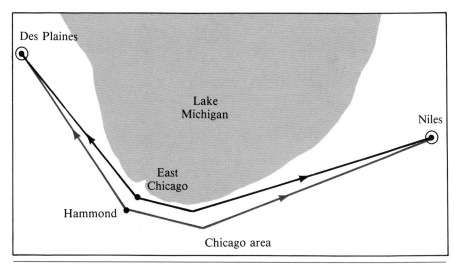

Des Plaines

Lake
Michigan

Niles

East
Chicago

Hammond

Chicago area

FIGURE 1

50 thousand barrels of commercial grade oil; the Hammond refinery will produce at least 120 thousand barrels; the Niles terminal requires at least 70 thousand barrels; and the Des Plaines terminal requires at least 100 thousand barrels. The average transportation cost of shipping from the East Chicago refinery is $8 (per mile per 1000 barrels) to Niles and $16 to Des Plaines, while the average cost from the Hammond refinery is $14 to Niles and $7 to Des Plaines. The problem is to determine how many barrels should be shipped from each refinery to each terminal in order to minimize the transportation cost subject to the given constraints. To solve the problem we can arrange the data in Table 1.

We let the variables represent the number of barrels (measured in 1000s) shipped per day from a refinery to a terminal, so that

x = the number from East Chicago to Niles
y = the number from East Chicago to Des Plaines
z = the number from Hammond to Niles
t = the number from Hammond to Des Plaines
c = the total cost

TABLE 1

From	East Chicago		Hammond		Limiting
To	Niles	Des Plaines	Niles	Des Plaines	Value
Produced (E	1	1	0	0	50
by (H	0	0	1	1	120
Received (N	1	0	1	0	70
by (D	0	1	0	1	100
Cost	8	16	14	7	

Thus the problem becomes

Minimize $c = 8x + 16y + 14z + 7t$ subject to

$$
\begin{aligned}
x + y &\geq 50 \\
z + t &\geq 120 \\
x + z &\geq 70 \\
y + t &\geq 100 \\
x, y, z, t &\geq 0
\end{aligned}
$$

Note that the problem is not in standard form because it is a minimum problem and each constraint is in the greater than or equal to ($\geq$) form. Therefore if we express the problem in matrix form, we can solve the dual problem. In matrix form the problem is

Minimize $c = RX$
Subject to $AX \geq C$

where

$$
R = \begin{bmatrix} 8 & 16 & 14 & 7 \end{bmatrix} \quad X = \begin{bmatrix} x \\ y \\ z \\ t \end{bmatrix} \quad A = \begin{bmatrix} 1 & 1 & 0 & 0 \\ 0 & 0 & 1 & 1 \\ 1 & 0 & 1 & 0 \\ 0 & 1 & 0 & 1 \end{bmatrix} \quad C = \begin{bmatrix} 50 \\ 120 \\ 70 \\ 100 \end{bmatrix}
$$

In order to form the dual problem, we compute the transpose of each matrix. We get

$$
R' = \begin{bmatrix} 8 \\ 16 \\ 14 \\ 7 \end{bmatrix}, \quad U = \begin{bmatrix} u \\ v \\ w \\ r \end{bmatrix}, \quad A' = \begin{bmatrix} 1 & 0 & 1 & 0 \\ 1 & 0 & 0 & 1 \\ 0 & 1 & 1 & 0 \\ 0 & 1 & 0 & 1 \end{bmatrix}
$$

$$
C' = \begin{bmatrix} 50 & 120 & 70 & 100 \end{bmatrix}
$$

Thus the dual problem is

Maximize $c = C'U$ subject to
$A'U \geq R'$

The corresponding simplex tableau is

$$
\begin{bmatrix}
1 & 0 & 1 & 0 & 1 & 0 & 0 & 0 & 0 & 8 \\
1 & 0 & 0 & 1 & 0 & 1 & 0 & 0 & 0 & 16 \\
0 & 1 & 1 & 0 & 0 & 0 & 1 & 0 & 0 & 14 \\
0 & 1 & 0 & 1 & 0 & 0 & 0 & 1 & 0 & 7 \\
-50 & -120 & -70 & -100 & 0 & 0 & 0 & 0 & 1 & 0
\end{bmatrix}
$$

The entry in the last row that is most negative is -120 in the second column. The two ratios are $\frac{7}{1} = 7$ and $\frac{14}{1} = 14$, so the pivot entry is 1, the (4, 2) entry. The new tableau after pivoting is

$$
\begin{bmatrix}
1 & 0 & 1 & 0 & 1 & 0 & 0 & 0 & 0 & 8 \\
1 & 0 & 0 & 1 & 0 & 1 & 0 & 0 & 0 & 16 \\
0 & 0 & 1 & -1 & 0 & 0 & 1 & -1 & 0 & 7 \\
0 & 1 & 0 & 1 & 0 & 0 & 0 & 1 & 0 & 7 \\
-50 & 0 & -70 & 20 & 0 & 0 & 0 & 120 & 1 & 840
\end{bmatrix}
$$

The most negative entry in the last row is -70 in the third column. The ratios are $\frac{7}{1} = 7$ and $\frac{8}{1} = 8$ so the pivot entry is 1, in the third row. The new tableau is

$$
\begin{bmatrix}
1 & 0 & 0 & 1 & 1 & 0 & -1 & 1 & 0 & 1 \\
1 & 0 & 0 & 1 & 0 & 1 & 0 & 0 & 0 & 16 \\
0 & 0 & 1 & -1 & 0 & 0 & 1 & -1 & 0 & 7 \\
0 & 1 & 0 & 1 & 0 & 0 & 0 & 1 & 0 & 7 \\
-50 & 0 & 0 & -50 & 0 & 0 & 70 & 50 & 1 & 1330
\end{bmatrix}
$$

The most negative entry in the last row is -50, and it appears in the first and fourth columns so we can use either one. We arbitrarily choose the first row. The ratios are $\frac{1}{1} = 1$ and $\frac{16}{1} = 16$, so the pivot entry is 1 in the first row. The new tableau is

$$
\begin{bmatrix}
1 & 0 & 0 & 1 & 1 & 0 & -1 & 1 & 0 & 1 \\
0 & 0 & 0 & 0 & -1 & 1 & 1 & -1 & 0 & 15 \\
0 & 0 & 1 & -1 & 0 & 0 & 1 & -1 & 0 & 7 \\
0 & 1 & 0 & 1 & 0 & 0 & 0 & 1 & 0 & 7 \\
0 & 0 & 0 & 0 & 50 & 0 & 20 & 100 & 1 & 1380
\end{bmatrix}
$$

The maximality test passes and thus the solution is $c = 1380$. Therefore the minimum cost for Shell under the given constraints is $1380 (per mile per 1000 barrels). The values of the variables are found in the last row in the columns headed by the slack variables. Hence we have $x = 50$, $y = 0$, $z = 20$, and $t = 100$. This means that Shell will achieve its minimum cost by shipping 50 thousand barrels from East Chicago to Niles, none from East Chicago to Des Plaines, 20 thousand barrels from Hammond to Niles, and 100 thousand barrels from Hammond to Des Plaines.

Case Study Exercises

1. Suppose the average transportation cost of shipping from the East Chicago refinery is $16 (per mile per 1000 barrels) to Niles and $8 to Des Plaines while the average cost from the Hammond refinery is $7 to Niles and $14 to Des Plaines. Determine how many barrels should be shipped from each refinery to each terminal in order to minimize the transportation cost.

2. Suppose all the transportation costs are $10. Determine how many barrels should be shipped from each refinery to each terminal in order to minimize the transportation cost.

CHAPTER REVIEW

Key Terms

4–1   Slack Variables and the Simplex Tableau

Standard Form	Basic Form
Slack Variables	Basic Variable
Simplex Tableau	Nonbasic Variable
Initial Simplex Tableau	Basic Feasible Solution

4–2   The Maximality Test and Pivoting

Maximality Test	Pivot Column
Pivoting	Entering Variable

4–3   The Simplex Method for Problems in Standard Form
Pivot Entry
Pivot Row

4–4   Nonstandard Problems
Pivoting by a Negative Number

4–5   Dual Problem

Dual Problem	Primal Matrix
Transpose	

REVIEW PROBLEMS

In Problems 1 and 2 determine whether the linear programming problem is in standard form.

1. Minimize $m = 3x + 5y$ subject to
$$2x + 3y \le 10$$
$$x + 2y \le 5$$
$$x \ge 0, y \ge 0$$

2. Maximize $m = 7x + 8y$ subject to
$$3x + 7y \le 50$$
$$2x + 5y \ge 45$$
$$x \ge 0, y \ge 0$$

3. Add slack variables to the inequalities and express them as equations.
$$4x + 3y \le 30$$
$$2x + 7y \le 65$$

In Problems 4 and 5 find the initial simplex tableau for the linear programming problem. Assume the problem is in standard form.

4. Maximize $m = 2x + 5y$ subject to
$$2x + 3y \le 10$$
$$3x + 4y \le 12$$

5. Maximize $m = x + 4y + 2z$ subject to
$$x + 2y + 5z \le 30$$
$$2x + 3y + z \le 40$$
$$4x + 3y \le 20$$

6. Determine the basic variables, the nonbasic variables, and the basic feasible solution of the simplex tableau.

$$
\begin{bmatrix}
x & y & z & s_1 & s_2 & s_3 & m & \\
0 & 0 & 0 & 1 & 1 & 1 & 0 & 15 \\
1 & 0 & 0 & 2 & 0 & 0 & 0 & 18 \\
0 & 4 & 1 & 4 & 1 & 0 & 0 & 22 \\
0 & -1 & 0 & -5 & 3 & 0 & 1 & 49
\end{bmatrix}
$$

7. Apply the maximality test to the simplex tableau.

$$\begin{array}{ccccccc} x & y & z & s_1 & s_2 & s_3 & m \\ \left[\begin{array}{ccccccc|c} 0 & 0 & 0 & 1 & 1 & 0 & 0 & 22 \\ 1 & 3 & -3 & 2 & 0 & 0 & 0 & 35 \\ 0 & 4 & 2 & 4 & 0 & 1 & 0 & 28 \\ 0 & 1 & 0 & -5 & 0 & 0 & 1 & 82 \end{array}\right] \end{array}$$

8. Select the pivot column and the entering variable.

$$\begin{array}{cccccc} x & y & z & s_1 & s_2 & s_3 & m \\ \left[\begin{array}{ccccccc|c} 2 & 0 & 1 & 0 & 4 & 0 & 0 & 20 \\ 1 & 0 & 0 & 1 & 2 & 1 & 0 & 10 \\ 1 & 1 & 0 & 0 & 0 & 1 & 0 & 12 \\ -3 & 0 & 0 & 0 & 3 & 1 & 1 & 55 \end{array}\right] \end{array}$$

9. For the simplex tableau in Problem 8 perform the pivot process using the (2, 1) entry as the pivot entry.

10. Find the initial simplex tableau and form a new tableau by pivoting about the indicated entry.
Maximize $m = 2x + 3y$ subject to
$$x + 3y \le 15$$
$$2x + 5y \le 10$$
$$x \ge 0, \ y \ge 0$$
Pivot about the (1, 2) entry.

11. Translate the linear programming problem into mathematics. Then find the initial simplex tableau and form a new tableau by pivoting about the (1, 1) entry.

A hospital nutritionist prepares a menu for patients on low-fat, high-protein diets. The primary ingredients are chicken, beans, and potatoes. The meal must contain no more than 1000 calories (cal), 5 grams (g) of fat, and 15 milligrams (mg) of iron. Each 3-ounce serving of chicken contains 45 g of protein, 500 cal, 9 mg of iron, and 4 g of fat. Each serving of beans contains 9 g of protein, 150 cal, 6 mg of iron, and 2 g of fat. Each serving of potatoes contains 7 g of protein, 100 cal, 5 mg of iron, and 2 g of fat. How much chicken, beans, and potatoes should be included in the meal in order to minimize the amount of fat?

In Problems 12 and 13 find the pivot entry and perform the pivot operation.

12.
$$\begin{array}{ccccc} x & y & s_1 & s_2 & m \\ \left[\begin{array}{ccccc|c} 2 & 0 & 1 & 1 & 0 & 20 \\ 1 & 1 & 0 & 0 & 0 & 16 \\ -8 & 0 & 0 & -4 & 1 & 40 \end{array}\right] \end{array}$$

13.
$$\begin{array}{ccccccc} x & y & z & s_1 & s_2 & s_3 & m \\ \left[\begin{array}{ccccccc|c} 1 & 0 & 1 & 1 & 0 & 2 & 0 & 12 \\ 5 & 0 & 0 & 2 & 1 & -1 & 0 & 24 \\ 4 & 1 & 0 & 0 & 1 & 0 & 0 & 25 \\ -4 & 0 & 3 & 0 & 0 & -2 & 1 & 42 \end{array}\right] \end{array}$$

14. Solve the linear programming problem by the simplex method.

Maximize $m = 8x + 7y$ subject to
$$2x + 3y \le 21$$
$$x + 3y \le 18$$
$$x \ge 0, \ y \ge 0$$

15. Solve the linear programming problem by the simplex method.

Maximize $m = 4x + 5y + 3z$ subject to
$$x + 5y + 3z \le 15$$
$$x + 2y + 2z \le 20$$
$$x \ge 0, \ y \ge 0, \ z \ge 0$$

16. Solve the linear programming problem by the simplex method. A firm manufactures three types of sails, small, medium, and large. Each small sail takes 3 hours to cut and 2 hours to assemble; each medium sail takes 4 hours to cut and 2 hours to assemble; each large sail takes 2 hours to cut and 4 hours to assemble. There is a total of 80 hours available to cut and 320 hours available to assemble the sails. No more than 50 sails are to be made. How many of each type of sail should be made to maximize profit if the profit is $100 on each small sail, $150 on each medium sail, and $50 on each large sail?

17. Convert the linear programming problem into one in standard form.

Minimize $m = 2x + 5y$ subject to
$$x + 4y \le 4$$
$$x + 2y \ge 7$$
$$x \ge 0, \ y \ge 0$$

18. Solve the linear programming problem by the simplex method.

Minimize $m = x + y + 2z$ subject to
$$x + 4y \qquad \leq 8$$
$$x - y \qquad \leq 6$$
$$\qquad y + 4z \geq 4$$
$$x \geq 0, y \geq 0, z \geq 0$$

In Problems 19 and 20 solve the linear programming problem by the dual method.

19. Minimize $m = 12x + 6y$ subject to
$$x + y \geq 2$$
$$4x + y \geq 3$$
$$x \geq 0, y \geq 0$$

20. Minimize $m = 40x + 30y$
$$x + y \geq 1$$
$$2x + y \geq 2$$
$$x + y \geq 3$$
$$x \geq 0, y \geq 0$$

5 Finance

CHAPTER OVERVIEW In this chapter we deal primarily with the mathematics of finance, but we also discuss inflation and population growth. In Section 5–1 we introduce simple interest on savings and loans and investigate why the "simple discount method" is really no discount at all. In Section 5–2 the interest is compounded over different periods, with the emphasis on the customary banking practice of interest compounded daily. We also show how much money should be invested at a given time to produce a desired amount at some later date. This kind of problem is a particular concern to college students who must plan tuition payments years ahead of time.

In Section 5–3 we apply the exponential function to inflation, cost of living, and population growth.

In Sections 5–4 and 5–5 we describe transactions in which the payments are made periodically. Both sections rely on the formula for the sum of a geometric series, which is derived in Section 5–4. In Section 5–5 we discuss the kinds of loans on cars and houses that most of us have to take out during our lifetime.

Since tables cannot possibly contain all of the necessary figures, a calculator is required. We introduce estimations before some of the solutions to encourage students to think about the reasonableness of answers. This is particularly important when a calculator is used.

CASE STUDY PREVIEW

"Tastes great!!!" "Less filling!!!!"

The commercials based on these two claims have become marketing classics. The Miller Brewing Company developed the advertisements for product identification in 1974. At the time, Miller held onto a distant fifth place share of the brewing market in the United States. Within five years, however, the advertisements vaulted Miller into a close second place behind Anheuser-Busch, and the two breweries have dominated the domestic market ever since. The Case Study examines Miller's rise to the top.

5–1 SIMPLE INTEREST AND DISCOUNT

In this section we describe simple interest and introduce the concept of present value. We also examine the so-called "discount interest method." A good working knowledge of percents is required. The solutions to all financial problems are expressed in a dollars-and-cents form, so a calculator might be necessary for some computations.

A Matter of Interest

Suppose that you invest $1000 at a 6% rate of interest. The amount of money invested is called the **principal,** denoted by P. Here $P = \$1000$. The **interest** I is the amount of money earned on the investment, and the **rate of interest** r is the percentage that is earned. Here $r = 6\%$. In all calculations we will write r in decimal form, so $r = 0.06$.

    The rate of interest can be a simple rate or a rate that is compounded over various periods. In **simple interest** the rate of interest applies only to the principal; in **compound interest** the rate of interest applies to the previously accumulated interest as well as to the principal. The formula for simple interest is given in the display.

Simple Interest

The simple interest I earned on a principal P invested at $r = p\%$ annual rate of interest for t years is

$$I = Prt$$

For the $1000 invested at a 6% rate of interest the simple interest in one year is $I = (\$1000)(0.06)(1) = \$60$. When interest accrues over 5 years, the simple interest is $I = (\$1000)(0.06)(5) = \$300$.

    Newspaper advertisements for banks frequently tout interest rates that grab our attention, but the fine print usually reveals conditions like "minimum deposit $10,000." Example 1 illustrates the simple interest formula on a $10,000 investment over fractional parts of a year. It assumes an **exact interest year** of 365 days (as opposed to an **ordinary interest year** of 360 days). In general we will *assume 365 days in a year* since most banks pay interest on savings for 365 days (but only

charge interest on loans for 360 days). When interest is calculated in terms of days,
$t = \dfrac{d}{365}$, where d is the number of days; when it is calculated in terms of months,

$t = \dfrac{m}{12}$, where m is the number of months.

EXAMPLE 1

Suppose that a $10,000 certificate is deposited at a 10.75% rate of interest.

Problem

(a) What is the simple interest for $2\frac{1}{2}$ years? (b) What is the simple interest from March 8 to April 15?

ESTIMATION 10% of $10,000 yields $1000 per year, so the interest will be about $2500 for $2\frac{1}{2}$ years and $100 for 5 weeks.

SOLUTION From the information given, $P = 10,000$ and $r = 10.75\%$. Write $r = 0.1075$. (a) Here $t = 2.5$, so $I = Prt = (10,000)(0.1075)(2.5) = 2687.5$. The simple interest is $2687.50. (b) There are 38 days from March 8 to April 15, so the time is $t = 38/365 = 0.10411$. Then $I = Prt = (\$10,000)(0.1075)(0.10411) = \$111.92$.

The examples so far have emphasized savings, but in reality most students are concerned with loans. Aside from a different viewpoint and a year based on 360 days, however, the basic concepts remain the same. Example 2 illustrates this fact.

EXAMPLE 2

Problem

What interest must be paid on a loan of $2500 for 90 days at 14% simple interest?

SOLUTION Set $P = 2500$, $r = 0.14$, and $t = 90/360 = 0.25$. Then $I = Prt = (2500)(0.14)(0.25) = 87.5$, so the interest on the loan is $87.50.

Almost all loans of a year or less are calculated as simple interest. However, most savings earn compound interest.

Present Value

We have seen that the simple interest on a principal $P = \$1000$ that is invested at $r = 6\%$ simple interest for $t = 5$ years is $I = \$300$. The sum of $1300 = \$1000 + \$300$ is called the **future amount.** It is denoted by A and given by the formula $A = P + I$. By substituting for I and factoring the common term P, we get $A = P + Prt = P(1 + rt)$. This is called the Future Amount Formula.

Future Amount Formula

$$A = P(1 + rt)$$

EXAMPLE 3 | **Problem**

If a $10,000 savings certificate is purchased at a $6\frac{1}{2}\%$ rate of simple interest, what is the future amount in (a) 60 days, (b) 30 months, (c) 20 years?

SOLUTION In all three cases $P = 10,000$ and $r = 0.065$. The remaining variable in the Future Amount Formula is t. (a) For 60 days the time is $t = 60/365 = 0.16438$. Then $A = P(1 + rt) = 10,000(1 + 0.065 \cdot 0.16438) = 10,106.85$. Thus the future amount is $10,106.85. (b) For 30 months $t = 30/12 = 2.5$. Then the future amount is $A = P(1 + rt) = \$10,000(1 + 0.065 \cdot 2.5) = \$11,625$. (c) For 20 years $t = 20$, so the future amount is $A = \$10,000(1 + 0.065 \cdot 20) = \$23,000$.

In the Future Amount Formula the principal P is called the **present value.** There is a reciprocal relationship between A and P: A is the future amount of P, and P is the present value of A. In Example 3 the present value P was given and the future amount A was solved for. The reverse case occurs when some future amount is needed and money can be invested at a given rate. The present value is obtained by dividing both sides of the Future Amount Formula by $(1 + rt)$.

Present Value Formula

$$P = \frac{A}{1 + rt}$$

Example 4 considers a resourceful parent who wants to invest a lump sum to help pay for a newborn child's future college education.

EXAMPLE 4 | **Problem**

How much money should be invested at 12% simple interest to yield $20,000 in 18 years?

ESTIMATION Since 12% of $10,000 yields $1200 per year, 18 years will yield more than $20,000. Therefore the answer must be less than $10,000.

SOLUTION The future amount is $A = 20,000$, the rate of interest is $r = 0.12$, and the time period is $t = 18$ years. By the Present Value Formula

$$P = \frac{20,000}{1 + (0.12)(18)} = \frac{20,000}{3.16} = 6,329.11$$

Therefore $6329.11 should be invested now in order to produce $20,000 in 18 years.

Simple Discount

Some lending institutions loan money by the **discount interest method.** This method involves deducting the interest on the loan before giving the balance to the borrower. The deducted money is called the **discount** and the amount the borrower actually receives is called the **proceeds.**

For instance, consider a loan of $1000 for 9 months at 8% simple interest using the discount interest method. Then $P = 1000$, $r = 0.08$, and $t = \frac{9}{12} = 0.75$. The discount corresponds to simple interest.

$$I = Prt = (1000)(0.08)(0.75) = 60$$

Thus the discount is $60. The proceeds PR are defined by the formula $PR = P - I$, so $PR = \$940$.

Here is how the discount interest method works. The $60 is deducted from the $1000 loan first. Then the borrower is handed the proceeds of $940. This is really the amount the borrower is borrowing. In effect the borrower is paying $60 interest on only $940. Example 5 will compute the real interest being charged.

There is another way to view discount interest. It involves the true annual interest rate, which is called the **effective rate** and is computed from the formula for simple interest in the form $I = (PR)(er)t$. Solving for the variable er yields

$$er = \frac{I}{(PR)t}$$

The effective rate allows the consumer to compare the discount rate with the rate that is actually being charged. Example 5 carries out such a comparison.

EXAMPLE 5 | Suppose that $1000 is borrowed for 9 months at 8% simple interest using the discount interest method.

Problem

What is the effective rate?

SOLUTION The term is 9 months, so $t = \frac{9}{12} = 0.75$. We have already calculated the discount $I = \$60$ and the proceeds $PR = \$940$. The effective rate is

$$er = \frac{I}{(PR)t} = \frac{60}{(940)(0.75)} = 0.0851$$

The effective rate is 8.51%. Compare this with the stated rate of 8%.

The "Truth in Lending Law" requires that the effective rate of interest appear on all contracts. Example 5 should convince you that it was in the consumer's interest to enact this law. After all the term "discount interest method" is a real misnomer!

EXERCISE SET 5–1

• In Problems 1 to 4 find the simple interest on each of the following savings. Assume 365 days in a year.

1. $1000 at $5\frac{1}{2}$% interest for
   (a) 7 years   (b) 7 months   (c) 1 week

2. $2500 at 6% interest for
   (a) 3 years   (b) 3 months   (c) 30 days

3. $1988 at 14.2% interest for
   (a) 13 years   (b) 21 months   (c) 58 days

4. $1988 at 14.1% interest for
   (a) 13 years   (b) 21 months   (c) 58 days

In Problems 5 and 6 find the simple interest on each of the following loans. Assume 360 days in a year.

5. $1580 at 11% interest for
   (a) 5 years   (b) 40 days

6. $1480 at 12% interest for
   (a) 4 years   (b) 50 days

In Problems 7 to 10 find the future amount of each of the following savings.

7. $1000 at $5\frac{1}{2}$% interest for
   (a) 7 years   (b) 7 months   (c) 1 week

8. $2500 at 6% interest for
   (a) 3 years   (b) 3 months   (c) 30 days

9. $1988 at 14.2% interest for
   (a) 13 years   (b) 21 months   (c) 58 days

10. $1988 at 14.1% interest for
    (a) 13 years   (b) 21 months   (c) 58 days

In Problems 11 and 12 find the total amount that must be repaid on each of the following loans. Assume 360 days in a year.

11. $1580 at 11% interest for
    (a) 4 years   (b) 4 months   (c) 4 days

12. $1480 at 12% interest for
    (a) 4 years   (b) 4 months   (c) 4 days

In Problems 13 and 14 find the present value of the given future amount in the time specified.

13. $5000 at 10% interest for
    (a) 10 years   (b) 10 months   (c) 10 days

14. $1000 at 5% interest for
    (a) 5 years   (b) 5 months   (c) 5 days

In Problems 15 to 18 find the discount, proceeds, and effective rate. Assume 360 days in a year.

15. $2000 at 10% interest for 3 years.

16. $1000 at 5% interest for 8 months.

17. $176 at 11.2% interest for 60 days.

18. $1776 at 15.1% interest for March 1 through April 15.

19. Suppose a loan at 9% simple interest is necessary to pay for a semester's tuition of $2998. How much interest will have to be paid in 4 years?

20. Suppose that a parent refuses to pay the senior year's tuition because of the liberal life-style the student has adopted. How much interest does the student have to pay on a loan to cover the $3200 tuition at $6\frac{1}{2}$% interest for 8 months?

•• In Problems 21 and 22 find the simple interest on each of the following loans. Assume 360 days in a year.

21. $10,000 at 12.3% interest made on September 10 and due on
    (a) February 10   (b) February 28

22. $12,345 at 10% interest made on April 3 and due on
    (a) November 3   (b) November 26

In Problems 23 and 24 find the total amount that must be repaid on these loans. Assume 360 days in a year.

23. $10,000 at 12.3% interest made on September 10 and due on
    (a) February 10   (b) February 28

24. $12,345 at 10% interest made on April 3 and due on
    (a) November 3   (b) November 26

25. A family wants to deposit a certain amount of money in an account that pays 14.2% simple interest so their first offspring can have $100,000 at age 18 for 4 years' tuition at an Ivy League school. How much should they deposit now?

26. A family wants to send its newborn child to a state university when the child turns 18 years old. How much should they invest now in an account that pays 13.4% simple interest to ensure that their child will have $40,000 for 4 years' tuition?

In Problems 27 to 30 the principal P, interest I, and time t of an investment are given. Find the simple interest rate.

27. $P = \$1000$, $I = \$48.75$, $t = 9$ months.

28. $P = \$2500$, $I = \$131.25$, $t = 1$ year.

29. $P = \$500$, $I = \$31.25$, $t = 1$ year.

30. $P = \$1000$, $I = \$28.75$, $t = 6$ months.

In Problems 31 to 34 the future amount A, principal P, and rate of simple interest r are given. Find the time t of the investment.

31. $A = \$2000$, $P = \$1000$, $r = 12\%$.

32. $A = \$1500$, $P = \$1000$, $r = 8\%$.

33. $A = \$500$, $P = \$375$, $r = 5.5\%$.

34. $A = \$500$, $P = \$250$, $r = 9.8\%$.

In Problems 35 to 38 the future amount A, principal P, and time t are given for an investment that earns simple interest. Find the interest rate r as a percent.

35. $A = \$2000$, $P = \$1000$, $t = 39$ months.

36. $A = \$1500$, $P = \$1000$, $t = 4$ years.

37. $A = \$500$, $P = \$375$, $t = 3$ years.

38. $A = \$500$, $P = \$425$, $t = 30$ months.

••• Problems 39 and 40 consider a surprising view of the historic tax bill that was passed by the Senate in 1986. John D. Paulus, the managing director of Morgan Stanley's economics department, contended that interest rates would drop when the cost of borrowing rose. This paradox ran counter to the prevailing theory of finance.*

39. The Senate tax bill lowered the tax rate for someone in the highest tax bracket from 50 to 27%. Suppose such a person borrowed $100,000.
    (a) What is the interest if the money is borrowed for 1 year at 10% simple interest?
    (b) What are the costs of borrowing the money for a person in the 50% tax bracket and a person in the 27% tax bracket?
    (c) What percent did the after-tax cost rise while the tax rate dropped from 50 to 27%?

40. The Senate tax bill also affected people in lower tax brackets. This problem concerns a person who borrowed $10,000 and whose tax bracket was reduced from 30 to 15%.
    (a) What is the interest if the money is borrowed for 1 year at 10% simple interest?
    (b) What are the costs of borrowing the money for a person in the 30% tax bracket and a person in the 15% tax bracket?
    (c) What percent did the after-tax cost rise while the tax rate dropped from 30 to 15%?

*J. A. Livingston, "Business Outlook: Taking Interest in a Paradox," *The Philadelphia Inquirer,* June 11, 1986.

5–2 COMPOUND INTEREST

In this section we develop the ''Compound Interest Formula'' for finding the principal on investments and loans whose interest rates are compounded annually, quarterly, monthly, and daily. The emphasis is on interest rates that are compounded daily because most banks use this method.

   The solutions to the problems are expressed in a dollars-and-cents form, which can be obtained by means of a calculator or from one of three tables. Some expressions, however, cannot be evaluated by the tables, so they will require a calculator with an exponential key. Because calculators round their answers in different ways, your results might differ slightly from ours. The tables were compiled on a microcomputer; thus they too are subject to slight rounding errors.

Annual Rates

When money is invested at a simple rate of interest the interest applies only to the principal. When money is invested at a compound rate of interest the interest applies to the principal and to the previously earned interest. The expression **interest compounded annually** refers to an interest rate that is applied once a year.

   We compare simple interest and compound interest on a principal of $P = 1000 that is deposited at an annual interest rate of $r = 6\%$ for $t = 3$ years. The simple interest is

$$I = Prt = (\$1000)(0.06)(3) = \$180$$

An amount of $60 is paid at the end of each year.

   For compound interest the $60 interest that is earned the first year is added to the $1000 principal at the end of the year. The second year's interest is earned on the new principal of $1060. We change the notation to reflect the change in the principal. Let P_t be the principal after t years. In particular, P_1 is the principal after 1 year.

$$P_1 = P + P(0.06) = P(1.06) = \$1000(1.06) = \$1060$$

The principal after 2 years is

$$P_2 = P_1 + P_1(0.06) = P_1(1.06)$$
$$= (\$1060)(1.06) = \$1123.60$$

The principal after 3 years is

$$P_3 = P_2(1.06) = (\$1123.60)(1.06) = \$1191.02$$

Therefore the compound interest after 3 years is $1191.02 − $1000 = $191.02, compared to the simple interest of $180.

   This method of calculating interest compounded annually leads to a derivation of a formula for computing the principal in any year directly without having to compute the interests for the intervening years. Recall that the Future Amount

Formula is $A = P(1 + rt)$. Therefore the principal after $t = 1$ year is

$$P_1 = P(1 + r)$$

When interest is compounded annually the principal after 2 years is

$$P_2 = P_1(1 + r) = P(1 + r)(1 + r)$$
$$= P(1 + r)^2$$

The principal after 3 years is

$$P_3 = P_2(1 + r) = P(1 + r)^2(1 + r)$$
$$= P(1 + r)^3$$

By continuing this process, we find the formula for the principal after t years is

$$P_t = P(1 + r)^t$$

This is the formula for the principal when interest is compounded annually. We refer to it as "the formula for interest compounded annually." It is the analog of the Future Amount Formula for simple interest. The term $(1 + r)^t$ is called the **growth factor.**

For the preceding example a principal of $1000 that is invested for 3 years at 6% interest compounded annually is obtained directly by substituting $P = 1000$, $r = 0.06$, and $t = 3$.

$$P_3 = \$1000(1 + 0.06)^3 = \$1000(1.06)^3$$

This expression can be converted into dollars and cents by evaluating the growth factor $(1.06)^3$ by a calculator or by the Compound Interest Table in the Appendix. If the calculator has an exponential key $\boxed{y^x}$, the usual method for evaluating $(1.06)^3$ is to use the key sequence

$$1.06 \boxed{y^x} \boxed{3} \boxed{=}$$

Otherwise ordinary multiplication can be used.

$$(1.06)^3 = (1.06)(1.06)(1.06) = 1.19102$$

The principal is

$$P_3 = \$1000(1.06)^3 = \$1000(1.19102) = \$1191.02$$

The growth factor $(1.06)^3$ is located in the Compound Interest Table at $n = 3$ and $r = 6\%$. The entry 1.19102 agrees with the figure obtained by calculator.

Example 1 applies the formula for interest compounded annually to a "loan problem."

EXAMPLE 1 | **Problem**

What is the interest on a $2500 loan that is borrowed at 10% interest compounded annually for 5 years?

ESTIMATION The answer must exceed $(5)(10\%)(\$250) = \$1250$.

SOLUTION It is given that $P = 2500$, $r = 0.1$, and $t = 5$. Therefore

$$P_5 = 2500(1 + 0.1)^5 = 2500(1.1)^5$$

By calculator or the Compound Interest Table (at $n = 5$ and $r = 10\%$)

$$P_5 = 2500(1.1)^5 = 2500(1.61051) = 4026.275$$

Thus the principal is $4026.28. The interest is equal to

$$P_5 - P = \$4026.28 - \$2500 = \$1526.28$$

The Compound Interest Formula

We introduce interest rates that are compounded over various periods by examining a principal of $1000 that is invested for 2 years at 8% interest compounded quarterly. The word "quarterly" means that the interest is added to the principal four times a year. Since $\dfrac{12 \text{ months}}{4} = 3$ months, the interest is added every 3 months. This is called the **conversion period** (or just the **period**). Moreover, the rate of interest that is paid every 3 months is $\dfrac{0.08}{4} = 0.02$. It is called the **periodic interest.** Thus the expression "8% compounded quarterly" means that 2% interest is earned every 3 months. The interest earned after the first quarter is 2% of $1000 (or $20), so the principal increases to $1020. It can be obtained from the formula

$$Q_1 = P + P(0.02) = P(1.02)$$

The principal after two quarters is

$$Q_2 = Q_1(1.02) = P(1.02)(1.02) = P(1.02)^2$$

The principal after three quarters is

$$Q_3 = Q_2(1.02) = P(1.02)^2(1.02) = P(1.02)^3$$

Similarly, the principal after four quarters, or 1 year, is

$$Q_4 = P(1.02)^4$$

and the principal after 2 years is

$$Q_8 = P(1.02)^8$$

This formula is a particular case of the **Compound Interest Formula,** which determines the principal on interest rates that are compounded over given time periods. Its derivation is analogous to this specific case.

By the Compound Interest Formula the principal after 2 years on $1000 invested at 8% interest compounded quarterly is obtained by substituting $r = 0.08$, $k = 4$, $t = 2$, and $P = 1000$. Then $i = 0.08/4 = 0.02$ and $n = 8$, so

$$A = 1000(1 + 0.02)^{2 \cdot 4} = 1000(1.02)^8$$

The growth factor $(1.02)^8$ can be obtained by calculator or from the Compound

> **Compound Interest Formula**
>
> $$A = P(1 + i)^n$$
>
> where
> A is the compound amount
> P is the principal
> $i = r/k$ is the periodic interest, where
> r is the annual rate of interest
> k is the number of periods per year the interest is compounded
> $n = tk$ is the total number of periods, where
> t is the number of years

Interest Table at $n = 8$ (not $n = 2$) and $r = 2\%$ (not $r = 8\%$) as 1.17116. Then

$$A = 1000(1.02)^8 = (1000)(1.17166) = 1171.66$$

Therefore the compound amount after 2 years is $1171.66.

EXAMPLE 2

Problem

Compute the compound amount on a $1000 certificate that is deposited for 5 years in an account that pays 16% interest compounded quarterly.

ESTIMATION The answer must exceed $1000 + (5)($160) = $1800.

SOLUTION We have $P = \$1000$, $r = 0.16$, and $t = 5$. Since the interest is compounded quarterly, $k = 4$. Then $i = 0.16/4 = 0.04$ and $n = 20$, so $A = 1000(1.04)^{20}$. The growth factor in the table at $n = 20$ and $r = 4\%$ is 2.19112, so $(1.04)^{20} = 2.19112$. The compound amount is $A = \$1000(2.19112) = \$2191.12$.

Example 3 considers a loan at an interest rate that is compounded monthly. It shows that a calculator must be used when the available tables do not contain the given rate.

EXAMPLE 3

Problem

How much interest is paid on a 3-year loan of $2500 at 10% interest compounded monthly?

SOLUTION First, we find the compound amount after 3 years. Set $P = 2500$, $r = 0.1$, $t = 3$, and $k = 12$. Then

$$i = \frac{0.1}{12} \text{ and } n = 36$$

so

$$A = 2500\left(1 + \frac{0.1}{12}\right)^{36}$$

The growth factor $\left(1 + \frac{0.1}{12}\right)^{36}$ cannot be found in the Compound Interest Table. To express the answer in dollars-and-cents form, use a calculator with an exponential key to get $\left(1 + \frac{0.1}{12}\right)^{36} = 1.3481818$. Then the compound amount after 3 years is

$$A = \$2500(1.3481818) = \$3370.45$$

To obtain the interest I, subtract the borrowed amount of $2500 from the compound amount A.

$$I = A - P = \$3370.45 - \$2500 = \$870.45$$

Annual Yield

The most common interest rates are those that are compounded daily. We assume a 365-day year.

EXAMPLE 4 | **Problem**

What is the interest on a $1000 investment for 1 year at 8% interest compounded daily?

SOLUTION We are given $P = \$1000$, $t = 1$, $r = 0.08$, and $k = 365$. Then $i = 0.08/365 = 0.0002191$ and $n = tk = 365$. Substitute these values into the Compound Interest Formula and use a calculator.

$$A = \$1000(1 + 0.0002191)^{365} = \$1083.28$$

The interest is $1083.28 - \$1000.00 = \$83.28$.

In Example 4 the stated rate of 8% is called the **nominal rate.** It is equivalent to a simple interest rate of 8.328%, which is called the **effective rate.** The effective rate provides a common unit for comparing various nominal rates. The Truth in Lending Law provides the consumer with this common unit of measure.

Most banks today use the term **annual yield** for the effective rate of a nominal rate that is compounded daily. Example 5 shows how to find the annual yield.

EXAMPLE 5 | **Problem**

What is the annual yield on an account that pays 7.7% compounded daily?

SOLUTION The nominal rate is 7.7%. The compound interest in 1 year from

HIGH YIELD CERTIFICATES

Annual Yield	Annual Rate	Term
8.00%	**7.70**%	*30 Months*
7.88%	**7.58**%	*18 Months*
7.63%	**7.35**%	*6 Months*
7.30%	**7.05**%	*91 Days*

Effective rate   Rate of compound interest

FIGURE 1

an initial investment of $1 is obtained by substituting $P = 1$, $i = 0.077/365 = 0.000210958$, and $n = 365$ into the Compound Interest Formula.

$$A = 1(1.000210958)^{365} = 1.08003$$

This means that every dollar will grow to $1.08003. The annual yield is obtained by converting $A - 1$ into a percent. Since $A - 1 = 0.08003$, the annual yield is 8.003%.

The annual yield of 8.003% in Example 5 is customarily stated as 8%. See the top row in Figure 1 for a newspaper advertisement listing these rates. The general method for converting from nominal rates to annual yields is to calculate A and then convert $A - 1$ to a percent.

Present Value

The Compound Interest Formula can also be used to calculate the amount that should be invested to produce a desired amount at some later date. For instance, suppose that $1000 will be needed in 3 years from an investment that earns 6% compounded annually. The problem is to determine the present value P from the Compound Interest Formula.

$$A = P(1 + i)^n$$
$$A = P(1 + 0.06)^3$$
$$1000 = P(1.06)^3$$

Multiply both sides by $(1.06)^{-3}$ to get

$$1000(1.06)^{-3} = P$$

The value of $(1.06)^{-3}$ can be found in the Present Value I table in the Appendix or by a calculator. The entry in the table at $n = 3$ and $r = 6\%$ is 0.839619, meaning that $(1.06)^{-3} = 0.839619$. For a calculator with an exponential key $\boxed{y^x}$, use the key sequence

$$1.06 \boxed{y^x} -3 \boxed{=}$$

Sometimes -3 is obtained by the sequence $\boxed{3}\ \boxed{\pm}$. The present value is $P = 1000(0.839619) = 839.619$. Therefore an investment of $839.62 will mature to $1000 if it is invested for 3 years at an interest rate of 6% compounded annually.

EXAMPLE 6 | **Problem**

How much money should be invested to produce $1500 in 2 years at 8% interest compounded annually?

ESTIMATION The answer must be less than $1500.

SOLUTION We are given that $t = 2$, $A = 1500$, $r = 0.08$, and $k = 1$. Then $i = r/k = 0.08$ and $n = tk = 2$. Substitute these values into the Compound Interest Formula.

$$A = P(1 + i)^n$$
$$1500 = P(1.08)^2$$

Solve for P.

$$P = 1500(1.08)^{-2}$$

By the Present Value I table, the entry for $n = 2$ and $r = 8\%$ is 0.857339. Thus $P = 1500(0.857339) = 1286.0085$. The present value is $1286.01.

The next example reconsiders an example from Section 5–1 at an interest rate that is compounded daily. Its solution makes use of the Present Value II table in the Appendix. The Present Value I table is used in problems whose interest rates are compounded annually while the Present Value II table is used in problems whose interest rates are compounded daily.

EXAMPLE 7 | A parent of a newborn child wants to buy a certificate that will mature to $20,000 when the child turns 18. A bank advertises "All-Savers" certificates that pay 12% interest compounded daily.

Problem

How much should be invested in the certificate?

SOLUTION From the information given $t = 18$, $A = 20,000$, $r = 0.12$, and $k = 365$. Substitute these values into the Compound Interest Formula.

$$A = P(1 + i)^n$$

$$20,000 = P\left(1 + \frac{0.12}{365}\right)^{18 \cdot 365}$$

Solve for P.

$$P = 20,000\left(1 + \frac{0.12}{365}\right)^{-18 \cdot 365}$$

This form can be evaluated by a calculator with an exponential key or from the Present Value II table. The entry at $r = 12\%$ and $n = 18$ is 0.115376, so

$$\left(1 + \frac{0.12}{365}\right)^{-18 \cdot 365} = 0.115376$$

The present value is

$$P = 20,000(0.115376) = 2307.52$$

Therefore an investment of \$2307.52 at 12% interest compounded daily will mature to \$20,000 in 18 years.

Contrast Example 7 with Example 4 of Section 5–1, which required \$6329.11 at 12% simple interest.

EXERCISE SET 5–2

● In Problems 1 to 6 compute the interest on each savings or loan, where the rate is compounded annually.

1. \$500 for 3 years at 6%.

2. \$1000 for 10 years at 10%.

3. \$1000 for 10 years at 8.15%.

4. \$1500 for 3 years at 5.75%.

5. \$1994 for 10 years at 13.056%.

6. \$1984 for 10 years at 13.678%.

In Problems 7 to 10 compute the compound amount on the following investments, where the interest is compounded annually for 30 months.

7. \$2500 at 10%   8. \$500 at 6%

9. \$2500 at 9.67%   10. \$2500 at 10.16%

In Problems 11 and 12 determine the compound amount on the savings, where the interest rates are compounded quarterly.

11. \$1000 for 10 years at 8%.

12. \$2500 for 3 years at 12%.

In Problems 13 and 14 determine the interest on loans at interest rates that are compounded quarterly.

13. $2500 for 3 years at 12%.

14. $5000 for 2 years at 16%.

In Problems 15 and 16 determine the interest on loans at interest rates that are compounded monthly.

15. $1000 for 5 years at 10.77%.

16. $500 for 10 years at 5.5%.

In Problems 17 and 18 determine the interest on savings accounts at interest rates that are compounded daily.

17. $1000 for 10 years at 8%.

18. $2500 for 3 years at 12%.

In Problems 19 and 20 determine the interest on loans at interest rates that are compounded daily.

19. $2500 for 3 years at 12%.

20. $5000 for 2 years at 16%.

In Problems 21 and 22 determine the annual yield on the following nominal rates.

21. 5.5%   22. 13.8%

In Problems 23 to 26 determine the amount of money that should be invested now to return $10,000 in the stated time at the given interest rate compounded annually.

23. 5 years, 6%   24. 10 years, 6%

25. 30 years, 10%   26. 50 years, 2%

In Problems 27 and 28 determine the amount of money that should be invested now to return $10,000 in the stated time and at the given interest rate compounded daily.

27. 5 years, 6%   28. 50 years, 2%

In Problems 29 and 30 determine what size certificate should be purchased now in order to return the stated dividend when the interest is compounded daily.

FIGURE 2

DIME 30-MONTH SAVINGS CERTIFICATES IN EFFECT THRU JAN. 18th GIVE YOU:

15.25% effective annual yield on   **14.00%** a year interest rate

MINIMUM DEPOSIT $500

GIFTS AROUND THE CIRCLE OR **$20 CASH**
When you renew or buy a $10,000 6-Month Certificate OR When you buy a new All Savers Tax-Free or a 30-Month Certificate for $5,000 or more.

DIME 6-MONTH SAVINGS CERTIFICATES IN EFFECT THRU JANUARY 18th GIVE YOU:

13.678% effective annual yield on   **13.056%** a year interest rate

ALL SAVERS 1-YEAR CERTIFICATES NOW GIVE YOU:

10.16% effective annual yield on   **9.67%** a year interest rate

UP TO $2,000 INTEREST TAX-EXEMPT
Any existing 6-Month Certificate can be converted to an All Savers Certificate without penalty.

18-MONTH IRA RATES

CURRENT VARIABLE RATE	GUARANTEED FIXED RATE
13.69%	**13.67**%
COMPOUNDED DAILY FOR AN EFFECTIVE ANNUAL YIELD OF	COMPOUNDED DAILY FOR AN EFFECTIVE ANNUAL YIELD OF
14.67%	**14.60**%

FIGURE 3

29. $4000 in 3 years at 12% interest.

30. $8000 in 4 years at 10% interest.

•• 31. How much interest is owed on a 3-year $2500 loan at 10% interest compounded in the following ways?
(a) quarterly   (b) monthly   (c) daily

32. How much interest is owed on a $2500 loan at 14% interest compounded daily for the following periods?
(a) 60 days   (b) 9 months   (c) 2 years

33. Verify that the annual yield of 7.3% corresponds to the nominal rate of 7.05% in Figure 1.

34. Would you prefer to have interest compounded daily at 5% for the first 3 years and at 6% for the next 3 years, or vice versa?

35. Use the Present Value I table to find the present value, in dollars-and-cents form, of $10,000 in 3 years at 8% compounded semiannually.

36. Use the Present Value I table to find the present value, in dollars-and-cents form, of $10,000 in 3 years at 8% compounded quarterly.

37. Which is worth more, a lump sum of $10,000 or a $5000 certificate that was purchased 9 years ago at an 8% rate of interest compounded annually?

38. Repeat Problem 37 for an interest rate that is compounded daily.

••• 39. Figure 2 displays an advertisement from the *New York Times* for the Dime Savings Bank offering various rates of interest on savings certificates. Verify that the three annual yields correspond to the stated nominal rates.

40. Figure 3 shows part of a newspaper advertisement for 18-month IRA rates. Verify that each annual yield in the bottom row corresponds to the nominal rate above it.

41. Find the annual yield on the money market funds listed in Figure 4, which displays the "Consumer Rates" chart from the *New York Times* of April 17, 1988.

42. Experiment with the Compound Interest Formula to determine about how long it takes a $1000 investment to double if the 8% interest is compounded daily.

Hershel Walker signed a $6 million, 4-year contract to play football in the fledgling USFL. Steve Young signed a $40 million, 43-year contract at the same time to play in the same league. Most sources feel that Hershel Walker got the better financial deal.* Problems 43 to 46 present a simplified analysis of the comparison using a guaranteed rate of interest of 10% compounded annually.

*B. Campsey and E. Brigham, "Reenlisting with the Generals," which appeared in Chapter 14 of the book *Introduction to Financial Management,* Hinsdale, Ill., The Dryden Press, 1985, pp. 355–378.

CONSUMER RATES
in percent

Tax-Exempt Bonds Bond Buyer 20-Bond Index	**7.81**
Money Market Funds Donoghue's Money Fund 7-Day Average Yield	**6.28**
Bank Money Market Accounts Bank Rate Monitor	**5.67**
Home Mortgage Federal Home Loan Bank, national average	**10.34**
Unsecured Personal Loans Minimum rate, 3-year maturity	
Chase Manhattan, New York	**13.90**
Continental Illinois, Chicago	**14.00**
Bank of America, San Francisco	**17.25**

FIGURE 4

43. Steve Young will be paid $9 million in 1990. What was the present value in 1984, the first year of the contract?

44. Steve Young will be paid $30 million in the year 2027. What will the present value be in 1990, when those payments begin?

45. Hershel Walker was paid $1.5 million in 1984, 1985, 1986, and 1987. What were the present values in 1984 of each of the last 3 years?

46. Whose contract contained the higher present value in 1984, Steve Young's or Hershel Walker's?

CASE STUDY   The Advertising Coup of the Miller Brewing Company*

The modern era of the brewing industry began after World War II when several firms decided to sell beer nationally instead of operating from a single plant. The number of breweries decreased from 404 in 1947 to 110 in 1970 as the large national firms cut into the sales of the smaller local companies. Even though some firms increased sales dramatically, the industry as a whole showed little growth. This was due to demographic factors. According to *Brewers Almanac 1976* (p. 82), the age group 21–45 accounts for better than 70% of the market.

*William Glueck and Lawrence Jauck, *Business and Policy Management,* 4th ed., New York, McGraw-Hill, 1984, pp. 622–640.

TABLE 1

	1974	
	Barrels (000)	Market Share
Anheuser-Busch	34.1	23.2%
Schlitz	22.7	15.4
Pabst	14.3	9.7
Coors	12.3	8.4
Miller	9.1	6.2

This age group had little growth from 1948 to 1959. While the total consumption of beer declined from 1959 to 1970, the number of firms with capacity of greater than 4 billion barrels increased from 2 to 11. By 1974 the top five breweries accounted for over 60% of the market. Their barrel shipments and market share in 1974 are given in Table 1.

In order to keep their market share, the national companies used advertising, especially television commercials, to increase product identification. While competition was fierce in the 1960s, in 1970 a new rivalry developed when Philip Morris Inc., a large multiproduct company, bought Miller Brewing. The assets and sales of Philip Morris were larger than those of the other top four breweries combined, and Philip Morris was an experienced, aggressive advertiser. In 1975 Miller launched two campaigns that changed the market dramatically. Miller devised a brilliant strategy to promote a low calorie beer, Lite, which they had bought from Meister Brau Inc. of Chicago in 1972. Low calorie beers had been advertised heavily in the past with little success. Miller reasoned that a large segment of the market was comprised of young sports fans who dreamed of athletic prowess. People were developing a passion for exercise. Miller used retired athletes in appealing commercials to spread the message that you could drink plenty of good-tasting Lite and still retain athletic skills. Lite was not just for weight-watchers anymore.

In an equally successful promotion Miller's ads for their premium beer were directed at young people having a good time at boisterous parties. The competition came out with comparatively bland messages. For instance, Schlitz commercials had a gritty cowboy pointing a gun at the screen and threatening the viewer to buy its beer.

In 1975 Miller's market share increased from 6.2% to 8.5%. In the same period Schlitz' market share increased from 15.4 to 15.5%. Example 1 determines the percentage increase for each company.

EXAMPLE 1 | **Problem**

What is the percentage increase when a firm's market share increases (a) from 6.2 to 8.5% (b) from 15.4 to 15.5%?

SOLUTION (a) The actual increase is $8.5\% - 6.2\% = 2.3\% = 0.023$. The percentage increase is $0.023/0.062 \approx 0.37$, or about 37%. (b) The actual increase is $15.5\% - 15.4\% = 0.1\% = 0.001$. The percentage increase is $0.001/0.154 \approx 0.006$, or about 0.6%.

In the next 3 years Miller's market share increased to 12.0% in 1976, 15.2% in 1977, and 18.9% in 1978. Meanwhile Schlitz' market share was 15.8% in 1976, 13.9% in 1977, and 11.8% in 1978. Example 2 determines the percentage increase (or decrease) for each year and each company.

EXAMPLE 2

Problem

Find the percentage increase (or decrease) of Miller's and Schlitz' market share in the years 1976 to 1978.

SOLUTION The percentages can be found by dividing the present year's share by the previous year's share and then subtracting 1 from that number. For Miller the dividends are $12.0/8.5 \approx 1.41$, $15.2/12.0 \approx 1.27$ and $18.9/15.2 \approx 1.24$. Therefore Miller's market share increased by 41, 27, and 24% in the period from 1976 to 1978. Schlitz' dividends are $15.8/15.5 \approx 1.02$, $13.9/15.8 \approx 0.88$, and $11.8/13.9 \approx 0.85$. Therefore Schlitz' market share increased by 2% in 1976, decreased by 12% in 1977 (because $0.88 - 1 = -0.12$), and decreased by 15% in 1978 (because $0.85 - 1 = -0.15$).

Compounded annual growth rate   Marketing analysts use *compounded annual growth rate* instead of simple growth rate as a measure of a firm's performance over a period of years. This is because compounded growth rate takes into account not only the actual annual percentage increase or decrease, but also the rate of increase on the previous year's increase. That is, just as compound interest includes interest on interest, compound growth rate includes the percent of growth on the previous year's growth rate.

In the brewing industry, marketing analysts use barrel shipments as a measure of market share. Table 2 gives the increased barrel shipments, the market share percentage change, and the compounded annual shipment growth in the period 1974–1978 for the five market leaders.

Miller's aggressive ad campaigns paid huge dividends. The company moved from a distant fifth in market share in 1974 to a close second in 1978, as its market share increased from 6.2 to 18.9%. The Anheuser-Busch market share went from 23.2% in 1974 to 25.1% in 1978. Schlitz went from 15.4% in 1974 to 11.8% in 1978.

TABLE 2

	1974–1978		
	Increased Barrels (000)	Change in Market Share (%)	Compounded Annual Shipment Growth Rate (%)
Anheuser-Busch	7.5	+ 1.9	5.1
Miller	22.2	+12.7	36.2
Schlitz	− 3.1	− 3.6	− 3.3
Pabst	1.1	− 0.4	1.9
Coors	0.3	− 0.8	0.4

While the change in market share is significant, marketing analysts point to the compounded annual shipment growth rate when assessing Miller's phenomenal success in this time span. Compare Miller's increase to that of Anheuser-Busch. The change in market share is about six times greater for Miller, but the compounded annual shipment growth rate is better than seven times larger.

To quote August Busch III, Chairman of the Board of Anheuser-Busch, "This business is now a two-horse race."

So who says advertising doesn't pay?

Case Study Exercises

1. Miller's market share increased from 6.2% in 1974 to 18.9% in 1978. Find the percentage increase.

2. Schlitz' market share decreased from 15.4% in 1974 to 11.8% in 1978. Find the percent of decrease.

3. The actual number of barrel shipments by Miller from 1974 to 1978 were (in millions) 9.1, 12.8, 18.4, 24.2, and 31.3. Find the percentage increase for each year and show that the compounded annual shipment growth was 36.2%.

4. The actual number of barrel shipments by all other breweries (all except the top five) from 1974 to 1978 were (in millions) 53.0, 49.7, 48.2, 45.2, and 41.7. Find the percentage increase for each year and show that the compounded annual shipment growth was -4.9%.

5. Show that the two methods for determining percentage increase in Examples 1 and 2 are equivalent; that is, let a and b represent the two percentages and show that $(b - a)/a = b/a - 1$.

5–3 INFLATION AND POPULATION GROWTH

The Compound Interest Formula can be applied to many areas outside banking. In this section we discuss inflation, cost of living, and population growth.

Inflation

The Compound Interest Formula provides a model of economic inflation when the compound amount A is interpreted as the cost in t years of an item that costs P dollars today. However, it must be assumed that the annual rate of inflation remains fixed from year to year. Then $k = 1$ so $n = tk = t$ and $i = r/k = r$.

Take, for instance, a record album that cost $9 in 1988. If the annual rate of inflation is 10%, how much will it cost in 1994? This is equivalent to finding the future amount of a $9 account that is deposited for 6 years at 10% interest compounded annually. Substitute $n = 6$ and $i = 0.1$ into the Compound Interest Formula.

$$A = P(1 + i)^n = 9(1.1)^6$$

From the Compound Interest Table or by calculator

$$A = 9(1.77156) = 15.944$$

Therefore the album will cost $15.94 in 1994.

Example 1 considers how salary increases affect base salaries.

EXAMPLE 1 | **Problem**

If a base salary was $15,600 in 1985 and a 6% pay raise is received in each of the next 32 years, what will the base salary be in 2017?

SOLUTION From the information given, $P = 15,600$, $i = 0.06$, and $n = 32$. The salary in the year 2017 will be the compound amount in the Compound Interest Formula.

$$A = \$15,600(1.06)^{32} = \$15,600(6.45338) = \$100,672.83$$

Thus the salary will be more than $100,000 a year.

The next example applies the Compound Interest Formula to a present value problem. It determines what today's millionaire was worth 32 years ago.

EXAMPLE 2 | **Problem**

If the rate of inflation remained a constant 4% over the past 32 years, what amount would have grown to $1,000,000 today?

SOLUTION Substitute $r = 0.04$, $n = 32$, and $A = 1,000,000$ into the Compound Interest Formula.

$$A = P(1 + i)^n$$
$$1,000,000 = P(1.04)^{32}$$

Solve for P.

$$P = 1,000,000(1.04)^{-32}$$

By calculator or the Present Value I table (with $n = 32$ and $r = 4\%$) the present value is

$$P = 1,000,000(0.285059) = 285,059$$

This means that someone who was worth $285,059 thirty-two years ago is equivalent to a millionaire today.

The effects of high inflation become more apparent when presented in practical terms. Take a look at Table 1. We have already calculated the cost of an album in

TABLE 1 Effects of a 10% Annual Rate of Inflation

Item	1988 Cost ($)	1994 Cost ($)	2020 Cost ($)	2088 Cost ($)
Nothing	.05	.09	1.06	689
Album	9	15.94	190	124,025
Bicycle	100	177	2,111	1,378,061
Automobile	8,000	14,172	168,910	110,244,899
House	50,000	88,578	1,055,689	689,030,617

1994. Notice that a bicycle that cost $100 in 1988 will cost over a million dollars a century later!

Cost of Living

Cost of living problems can also be solved by the Compound Interest Formula. Consider this news item.

> The cost of living rose eight-tenths of one percent in January. This amounts to an increase of over 10% for the year.

How can such a conclusion be warranted? After all a year has 12 months and $(12)(0.8\%) = 9.6\%$ is less than 10%.

Example 3 will answer the question by considering an item that costs $1 on January 1. It computes the cost 1 year later under the assumption that the price is determined solely by the cost of living.

EXAMPLE 3 | **Problem**

If the cost of living rises 0.8% each month for 1 year, what is the cost of living over the entire year?

SOLUTION Special care is needed to put this problem in a form that can be handled by the Compound Interest Formula. It is equivalent to finding A for $i = 0.8\%$ and $n = 12$. Let $P = \$1$ serve as a standard unit. Substitute into the Compound Interest Formula.

$$A = \$1(1.008)^{12} = \$1.10034$$

This means that a $1 item will be worth over $1.10 a year later. The rise in the cost of living over a year is $A - 1 = 0.10034$, which is 10.034%. This verifies the conclusion "over 10% for the year."

Population Growth

The Compound Interest Formula also serves as a model of population growth under certain assumptions. Example 4 demonstrates one aspect.

EXAMPLE 4 | **Problem**

The world's population has been increasing at a fairly constant rate of 2% per year. How much larger will the population be in 100 years?

SOLUTION Assume that the principal P is 1. Substitute $i = 0.02$ and $n = 100$ into the Compound Interest Formula.

$$A = 1(1.02)^{100} = 7.24$$

Thus the population will be 7.24 times as large in 100 years as it is today.

Example 4 contains a dire implication in spite of the seemingly small population growth. It indicates that the world population will increase sixfold in the next century. (Doubling is regarded as a onefold increase.) In different terms, it means that the 170×170 yards of land and water now occupied by each of us 5 billion earthlings will reduce to 24×24 yards in 100 years. The commandment "love thy neighbor" will, of necessity, attain even greater significance.

EXERCISE SET 5–3

• In Problems 1 to 4 assume a constant rate of inflation during the stated time period.

1. What will a $9 album cost in 20 years at 10% inflation?

2. What will an $8000 car cost in 10 years at 10% inflation?

3. What will a $20,000 salary be in 10 years with an 8% pay raise each year?

4. What will a $100,000 salary be in 5 years with a 12% pay raise each year?

In Problems 5 and 6 assume a constant 2% increase in the population over the given time period.

5. The population of the United States in 1980 was 226,504,825. What will it be in (a) 1990 (b) 2000?

6. In 1980 the population of China, including Tibet but excluding Taiwan, was approximately 971 million. Wow! What will it be in (a) 1990 (b) 2000?

7. If the cost of living rises 0.6% per month for a year, how much will it rise over the entire year?

8. If the cost of living rises 0.2% per month for a year, how much will it rise over the entire year?

9. Suppose the rate of inflation had remained a constant 4% for the past 50 years. What amount of money then would be worth $1,000,000 today?

10. Repeat Problem 9 for a period of 10 years.

In Problems 11 to 14 assume a constant rate of inflation of 6% from 1968 to 1988. How much would each item have cost in 1968 if its price in 1988 is the one stated?

11. $18 ticket to a concert

12. $100 bicycle

13. $12,000 automobile

14. $80,000 house

•• Problems 15 and 16 list the 1980 population for certain states and the average rate that the population changed per year from 1970 to 1980. Find each state's population in 1990, assuming the same change throughout the 1980s.

15.

State	1980 Population	Percent Change
(a) California	23,668,562	1.85
(b) Arizona	2,717,866	5.31
(c) Oklahoma	3,025,266	1.82
(d) North Carolina	5,874,429	1.55
(e) Illinois	11,418,461	0.28
(f) New York	17,557,288	−0.38

16.

State	1980 Population	Percent Change
(a) Alaska	400,481	3.24
(b) Colorado	2,888,834	3.07
(c) Texas	14,228,383	2.71
(d) South Carolina	3,119,208	2.04
(e) Ohio	10,797,419	0.13
(f) Rhode Island	947,154	−0.03

17. In 1985 the rate of inflation in Brazil was 63.7%. If this rate remains constant, what will a 2500-cruzeiro item (about $1) cost in 1990?

18. In 1980 the rate of inflation is Israel was 132.9%. If this incredible rate continued for a decade, what would a 25-cent bagel cost in 1990?

19. Use the Compound Interest Table to determine how many years it takes for prices to double if the annual rate of inflation is 6%.

20. In 1982 the rate of inflation in Great Britain was 14%. If this rate continues, what will a 120,000-pound Rolls Royce cost in 1990?

21. Rare coins minted before 1933 have increased in value at an annual appreciation rate of 28.7%. How much would a 1928 quarter be worth in 1988?

22. (a) Pieter Minuit, a Dutch colonist, bought Manhattan Island from the Indians in 1626 for about $24. If instead he had invested the $24 in a bank paying 5% interest per year, what would the principal have been when the United States celebrated its bicentennial in 1976? (b) If Minuit had put his $24 in 1626 in a savings account that paid 5% interest compounded daily, how much would have been in his account in 1976?

23. In January 1982 the cost of living in Argentina rose 9%. If this increase continued throughout the year, how much would a 100-peso item cost in January 1983?

24. If the cost of living rises 0.75% each month for a year, will the annual cost of living exceed 10%?

25. Gerald Baumann, a numismatist at Manfred, Tordella & Brooks, stated, ''People have become very aware of U.S. coins as an investment vehicle. They no longer have any qualms about paying more than $50,000 for a single coin.'' Rare coins increased their value from 1968 to 1986 at an average rate of 12.7% per year. If this growth continued, how much would a $50,000 coin purchased in 1986 be worth in the year 2000?

••• The facts stated in Problems 26 and 27 are based on a newspaper article which appeared in 1987, the year that the world's population reached 5 billion.*

26. The Population Crisis Committee predicted that the population of the world would grow by an additional billion by the end of the century. Substitute various values for i into the Compound Interest Formula to approximate the annual growth rate of the world's population between 1987 and the year 2000.

27. It is predicted that Mexico City's population will soar from 15 million in 1987 to 26 million by the year 2000. Substitute various values for i into the Compound Interest Formula to approximate the annual growth rate of Mexico City's population between 1987 and 2000.

*Dan Shannon, ''World Population at 5 Billion,'' *Los Angeles Times,* July 12, 1987.

5–4 ANNUITIES AND SINKING FUNDS

In this section each investment consists of a sequence of payments instead of one lump sum. We rely on the Compound Interest Formula from Section 5–2. As in earlier sections, all answers are written in a dollars-and-cents form that is obtained by either a calculator or a table.

Annuities

A family is planning a special vacation 2 years in advance by depositing $500 at the end of every 3 months in a savings account that earns 8% interest compounded quarterly. After 2 years (eight deposits) the amount deposited will total $4000 altogether. How much interest will accrue from the deposits?

This scenario is a typical annuity problem, where an **annuity** is a sequence of equal payments made over equal time periods. It is an **ordinary annuity** because the time periods correspond to the period of the compound rate of interest and the payments are made at the end of the time period. The sum of the payments and the interest is called the **future value** (or the **amount**). We will calculate the future value of the family's annuity.

The Compound Interest Formula $A = P(1 + i)^n$ gives the compound amount on each payment $P = \$500$. The first payment, which is deposited at the end of the first quarter, earns 8% interest compounded quarterly for seven quarters, so $r = 0.08$, $k = 4$, and $t = 7/4$. Then $i = r/k = 0.08/4 = 0.02$, and $n = tk = 7$; thus by the Compound Interest Formula the compound amount is $500(1.02)^7$. Figure 1 shows a time line with the compound amount of each payment. The top row lists the compound amount on the first payment. The other compound amounts are also listed. The future value A of the annuity is the sum of these compound amounts.

$$A = 500(1.02)^7 + 500(1.02)^6 + \cdots + 500(1.02) + 500$$

Each term can be evaluated by the Compound Interest Table or a calculator. However, it is more efficient to derive a formula that gives the future value without having to evaluate each of these expressions. Set $a = 1.02$. Then

$$\begin{aligned} A &= 500a^7 + 500a^6 + \cdots + 500a + 500 \\ &= 500(a^7 + a^6 + \cdots + a + 1) \end{aligned}$$

The sum in the parentheses is called a **geometric series.** A geometric series can be evaluated as follows:

FIGURE 1

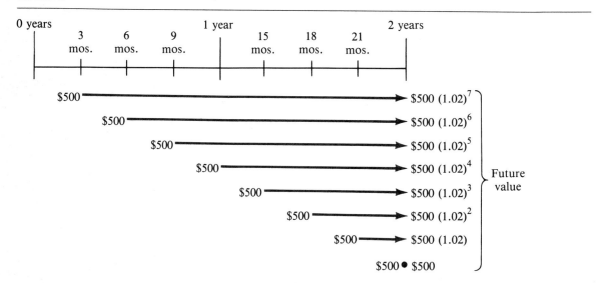

$$a^n + a^{n-1} + \cdots + a + 1 = \frac{(a^n + a^{n-1} + \cdots + a + 1)(a - 1)}{(a - 1)}$$

$$= \frac{a^{n+1} - 1}{a - 1}$$

The general result is stated in the display.

Sum of a Geometric Series

$$a^n + \cdots + a + 1 = \frac{a^{n+1} - 1}{a - 1}$$

In particular, $a^7 + \cdots + a + 1 = \dfrac{a^8 - 1}{a - 1}$, so

$$A = 500 \frac{a^8 - 1}{a - 1} = 500 \frac{(1.02)^8 - 1}{0.02}$$

The last expression is a particular case of the Annuity Formula, whose derivation is similar. It is stated in the display.

Annuity Formula

$$A = R \frac{(1 + i)^n - 1}{i}$$

where
  A is the future value (or amount)
  R is the periodic payment
  $i = r/k$ is the periodic interest rate
  $n = tk$ is the number of periods
(The payments are made at the end of each period.)

The fractional multiple of R in the Annuity Formula is commonly denoted by $s_{\overline{n}|i}$ (read "s angle n at i").

$$s_{\overline{n}|i} = \frac{(1 + i)^n - 1}{i}$$

The Annuity Table in the Appendix provides some values of $s_{\overline{n}|i}$. For $n = 8$ and $i = 2\%$, $s_{\overline{n}|i} = 8.58297$. Therefore the future value of the family's annuity is

$A = \$500(8.58297) = \$4291.48$. The value of $s_{\overline{8}|0.02}$ can also be obtained by a calculator with an exponential key $\boxed{y^x}$ using this key sequence.

$$1.02 \; \boxed{y^x} \; 8 \; \boxed{=} \; \boxed{-} \; 1 \; \boxed{=} \; \boxed{\div} \; .02 \; \boxed{=}$$

The display should approximate 8.58297.

     Examples 1 and 2 demonstrate the Annuity Formula.

EXAMPLE 1

Problem

An annuity consists of payments of $1000 made at the end of each year for 12 years at 6% compounded annually. What is the future value of the annuity?

SOLUTION We are given that $R = 1000$, $i = 0.06$, and $n = 12$. By the Annuity Formula, and either a calculator or the Annuity Table,

$$A = 1000 \, \frac{(1.06)^{12} - 1}{0.06} = 1000(16.86994) = 16869.94$$

Therefore the future value of the annuity is $16,869.94.

EXAMPLE 2

Problem

What is the future value of an annuity consisting of payments of $100 made at the end of each month for 3 years at 18% compounded monthly?

ESTIMATION The answer must exceed $(3)(12)(\$100) = \$3600$.

SOLUTION From the information given $R = 100$, $i = r/k = 0.18/12 = 0.015$, and $n = tk = 3(12) = 36$. The future value of the annuity is

$$A = \$100 \, \frac{(1.015)^{36} - 1}{0.015} = \$100(47.27597) = \$4727.60$$

A calculator has to be used to obtain the final form because the Annuity Table in the Appendix does not provide entries for $s_{\overline{n}|i}$ when $i = 1.5\%$.

     Suppose the family that was planning its vacation 2 years in advance decided to deposit the payments at the beginning of each quarter instead of at the end. The future value of this annuity is calculated in Example 3. When the payments on an annuity are paid at the beginning of the period it is called an **annuity due.**

EXAMPLE 3

Problem

What is the future value of an annuity consisting of payments of $500 made at the beginning of each quarter for 2 years at 8% compounded quarterly?

SOLUTION Refer to Figure 2. The time line divides the 2 years into eight quarters. The exponent in the first row under the line is 8 because the first

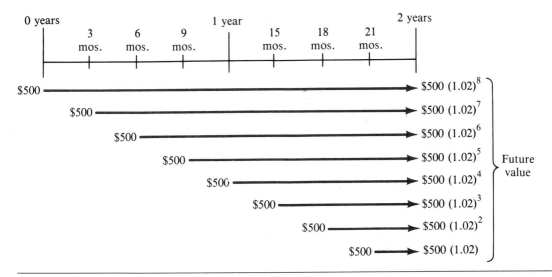

FIGURE 2

payment earns interest for eight periods (quarters). The future value of the annuity is the sum of all the compound amounts.

$$A = 500(1.02)^8 + \cdots + 500(1.02)$$

Notice that $A + 500$ corresponds to an annuity for nine periods. By the Annuity Formula

$$A + 500 = 500 \frac{(1.02)^9 - 1}{0.02} = 500(9.75463) = 4877.31$$

Then $A = 4877.31 - 500 = 4377.31$, so the future value is \$4377.31.

From Example 3 the future value of the family's annuity is \$4377.31 if payments are made at the beginning of each quarter. Compare this with the future value of the annuity (calculated earlier) if payments are made at the end of each quarter: \$4291.48.

Sinking Funds

The Annuity Formula expresses a relationship between the future value A and the payments R. So far we have been given R and asked to solve for A. Example 4 illustrates the reverse situation for a problem we encountered in Sections 5–1 and 5–2.

EXAMPLE 4 A parent wants to present a child with \$20,000 when the child turns 18. The parent wants to make a deposit on each birthday, from the first to the 18th, and present it on the 18th birthday.

Problem

How large should the deposits be if the account earns 6% compounded annually?

SOLUTION According to the Annuity Formula

$$20,000 = R\frac{(1.06)^{18} - 1}{0.06}$$

By calculator or the Annuity Table

$$20,000 = R(30.90564)$$

Solve for R.

$$R = \frac{20,000}{30.90564} = 647.13$$

Each payment will be $647.13.

In Example 4 the annual payment of $647.13 seems small compared to the amount of $20,000. It attests to the power of the exponential function y^x. The annuity in this example is called a **sinking fund,** which is an account established to receive periodic payments to build some future amount. Example 5 provides another illustration.

EXAMPLE 5 | **Problem**

Determine the quarterly payments of a sinking fund that will provide $4000 in 5 years from an account earning 12% compounded quarterly.

ESTIMATION Each payment must be less than ($4000/5)/4 = $200.

SOLUTION We have $A = 4000$, $k = 4$, $t = 5$, and $r = 0.12$. Then $i = 0.03$ and $n = 20$. By calculator,

$$4000 = R\frac{(1.03)^{20} - 1}{0.03} = R(26.87037)$$

Therefore the quarterly payments are

$$R = \frac{\$4000}{26.87037} = \$148.86$$

EXERCISE SET 5–4

• In Problems 1 to 4 determine the future value of the stated annuity, where the payments are made at the end of each year and the interest is compounded annually.

1. The payments are $1000 per year for 8 years at 6%.

2. The payments are $1000 per year for 12 years at 8%.

3. The payments are $2000 per year for 5 years at 12%.

4. The payments are $5000 per year for 3 years at 10%.

In Problems 5 to 10 determine the future value of the stated annuity where the payments are made at the end of each period.

5. The payments are $500 per quarter for 2 years at 8% compounded quarterly.

6. The payments are $1000 per quarter for 5 years at 16% compounded quarterly.

7. The payments are $100 per month for 2 years at 12% compounded monthly.

8. The payments are $100 per month for 3 years at 18% compounded monthly.

9. The payments are $300 per quarter for 3 years at 16% compounded quarterly.

10. The payments are $200 per quarter for 4 years at 6% compounded quarterly.

In Problems 11 to 14 determine the future value of the stated annuity, where the payments are made at the beginning of each period.

11. The payments are $500 per quarter for 2 years at 8% compounded quarterly.

12. The payments are $1000 per quarter for 5 years at 16% compounded quarterly.

13. The payments are $100 per month for 2 years at 12% compounded monthly.

14. The payments are $100 per month for 3 years at 18% compounded monthly.

In Problems 15 to 18 a parent wants to present a child with a cash gift by making periodic deposits in an account paying the stated interest compounded annually. How large should the deposits be?

15. $10,000 gift after 5 years at 6%.

16. $5000 gift after 3 years at 8%.

17. $2000 gift after 2 years at 10%.

18. $800 gift after 4 years at 12%.

In Problems 19 to 22 determine the quarterly payments of a sinking fund that is set up to satisfy the given conditions.

19. $3000 in 3 years at 8% compounded quarterly.

20. $3000 in 3 years at 16% compounded quarterly.

21. $4000 in 4 years at 6% compounded quarterly.

22. $4000 in 4 years at 10% compounded quarterly.

•• 23. Suppose that a professional athlete deposits $25,000 at the end of each year for 10 years in an account paying 8% compounded annually. How much money will be in the account at that time?

24. Suppose that a TV news anchor deposits $2000 every 3 months for the duration of a 3-year contract in an account paying 8% compounded quarterly. How much money will be in the account at the end of the contract?

25. Tax laws allow all individuals to deposit $2000 per year in IRAs (individual retirement accounts). If at age 30 someone deposits $2000 a year in an account paying 8% compounded annually, how much will be in the account upon retirement at age 67?

26. Tax laws allow self-employed individuals to open a Keogh retirement plan, which is not taxable until the money is withdrawn. If $5000 is deposited every year for 30 years at 10% compounded annually, how much will be in the account upon retirement?

27. For the first 5 years of their marriage, a couple deposits $400 at the end of each month in an account paying 8% compounded monthly. How much will be in the account after 5 years?

28. The couple in Problem 27 could have invested $5000 at the end of each year in an account paying 9% compounded annually. Did they make a wise choice?

29. A family wants to purchase a $1600 camcorder in 6 months. How much money should they deposit at the end of each month in an account paying 8% compounded monthly?

30. A 14-year-old boy dreams of saving his earnings from his newspaper route to buy a $22,000 Corvette when he turns 18. How much money should he deposit at the end of each month in an account paying 6% compounded monthly to realize his dream?

31. A 17-year-old dreamer wants to become a millionaire at the age of 40 by making monthly deposits in an account paying 12% compounded monthly. How big must the payments be for the dream to come true?

TABLE 1   The Nation's Top Markets (in millions of dollars)

Metro Area	1985	1990 Estimate	Percent Increase 1985–90
1. Los Angeles–Long Beach	$50,985	$76,557	50.2
2. New York	41,390	59,971	44.9
3. Chicago	35,628	52,260	46.7
4. Philadelphia	28,394	42,106	48.3
5. Boston-Lawrence-Salem-Lowell-Brockton	27,704	40,438	56.0
6. Washington, D.C.	25,220	38,781	53.8
7. Detroit	26,267	37,800	43.9
8. Houston	21,643	31,823	47.0
9. Nassau-Suffolk, N.Y.	19,423	28,819	48.4
10. Atlanta	17,419	27,593	58.4

Source: Sales and Marketing Management.

32. A newly married couple wants to be able to put down $30,000 on a new house in 3 years. How much money should they deposit at the end of each month in an account paying 9% compounded monthly?

33. If you deposit $10 a day in an account paying 6% compounded daily, how much will be in the account after 1 year?

34. If you deposit $1 a day in an account paying 8% compounded daily, how much will be in the account after 1 year?

35. A firm sets up a sinking fund to produce $400,000 in equipment in 5 years at 16% compounded quarterly. How big are the quarterly payments?

36. A city establishes a sinking fund to produce $4 million revenue in 20 years at 7% compounded annually. How big are the annual payments?

37. What daily deposits have to be made in an account paying 8% compounded daily to have $1000 in 1 year?

38. What daily deposits have to be made in an account paying 6% compounded daily to have $1000 in 1 year?

39. The physicist Albert Bartlett, concerned about the energy crisis, wrote, "Many people find it hard to believe that when the rate of consumption is growing a mere 7% per year, the consumption in one decade exceeds the total of all of the previous consumption." Use the assumption that the rate of consumption is 7% and the fact that 1000 trillion barrels of oil were consumed in 1986 to project how many trillion will be consumed in (a) 1987 (b) 1988 (c) 1996 (d) the sum of all of the years from 1987 through 1996.

••• Problems 40 and 41 are based on Table 1, which lists the ten metropolitan areas in the United States which had the biggest markets in 1985.

40. If the Los Angeles–Long Beach area experiences a 10% increase for every year from 1985 to 2000, what will the market be in the year 2000?

41. If, for every year from 1985 to 2000, Atlanta experiences an 11.6% increase and Detroit experiences an 8.7% increase, will Atlanta's market exceed Detroit's by the year 2000?

5–5 AMORTIZATION

In this section we make use of the present value concept from Section 5–2 to develop a formula for the payments on loans that many of us take out at one time or another. Here are two typical amortization problems that we will solve in this section.

Problem A. A new car costs $17,000. The dealer offers $5000 for a trade-in. The down payment is $2000. What will the monthly installments be if the $10,000 loan is taken out for 4 years at 18% annual interest compounded monthly?

Problem B. A house costs $90,000. The terms of sale are 10% down and the remainder to be paid in monthly payments for 25 years at an annual interest rate of 9% compounded monthly. What are the monthly payments?

Present Value

Consider an investment that pays 8% annual interest compounded quarterly. How much money should be invested as a lump sum today in order to remove $1500 every 3 months for 2 years? The answer must be less than $12,000. How much less? To answer these questions, regard each of the $1500 dividends separately. A certain amount of money A_1 must produce the first dividend of $1500 in 3 months. The quarterly rate of interest is $\dfrac{0.08}{4} = 0.02$, so by the Compound Interest Formula $1500 = A_1(1.02)$. Then $A_1 = \dfrac{1500}{1.02}$. Recall that A_1 is called the present value of $1500. Similarly, the present value A_2 of the $1500 dividend for two periods (6 months) is $A_2 = \dfrac{1500}{1.02^2}$. This process continues through eight periods (2 years), with the present value of the final dividend being $A_8 = \dfrac{1500}{1.02^8}$. The eight present values are shown below the time line in Figure 1.

Since the sequence of payments is an annuity, the total amount P that will produce the annuity is called the **present value of the annuity.** P is the sum of

FIGURE 1

the individual present values. Each one could be evaluated separately but, like the last section, it is more efficient to derive a general formula.

$$P = A_1 + A_2 + \cdots + A_8$$

$$= \frac{1500}{1.02} + \frac{1500}{1.02^2} + \cdots + \frac{1500}{1.02^8}$$

Set $a = 1.02$, factor the common term 1500, and simplify.

$$P = 1500 \left(\frac{1}{a} + \frac{1}{a^2} + \cdots + \frac{1}{a^8} \right)$$

$$= \frac{1500}{a^8} \left(\frac{a^8}{a} + \frac{a^8}{a^2} + \cdots + \frac{a^8}{a^8} \right)$$

$$= \frac{1500}{a^8} (a^7 + a^6 + \cdots + 1)$$

Now form the sum of the geometric series and simplify.

$$P = \frac{1500}{a^8} \cdot \frac{a^8 - 1}{a - 1}$$

$$= 1500 \cdot \frac{1 - \dfrac{1}{a^8}}{a - 1}$$

$$= 1500 \cdot \frac{1 - \dfrac{1}{1.02^8}}{0.02}$$

The last expression is a particular case of the formula for the present value of an annuity. Its derivation is similar to the one derived here. It is summarized in the display.

Present Value of an Annuity

$$P = R \cdot \frac{1 - \dfrac{1}{(1 + i)^n}}{i}$$

where
  P is the present value of all payments
  R is the periodic payment
  $i = r/k$ is the periodic interest rate
  $n = tk$ is the number of periods

The fractional multiplier of R is commonly denoted by $a_{\overline{n}|i}$ (read "a angle n at i").

$$a_{\overline{n}|i} = \frac{1 - \dfrac{1}{(1 + i)^n}}{i}$$

The Amortization Table in the Appendix provides some values of $a_{\overline{n}|i}$. For $n = 8$ and $i = 2\%$, $a_{\overline{n}|i} = 7.32548$. Therefore the present value for the stated investment is $P = (\$1500)(7.32548) = \$10{,}988.22$. The value of $a_{\overline{8}|0.02}$ can also be obtained by a calculator with an exponential key and a reciprocal key using this procedure: calculate $(1 + i)^n$, take its reciprocal, subtract from 1, and divide by i. The present value is the product of the number displayed (here 7.32548) and the periodic payment R (here \$1500).

EXAMPLE 1

Problem

What is the present value of an annuity that pays \$50 a month for 4 years at 12% annual interest compounded monthly?

ESTIMATION The answer must be less than $(4)(12)(\$50) = \$2400$.

SOLUTION From the information given, $r = 0.12$, $k = 12$, and $t = 4$, so the monthly interest is $i = 0.01$ and the number of periods is $n = 48$. By the formula for the present value of an annuity

$$P = 50 \cdot \frac{1 - \dfrac{1}{1.01^{48}}}{0.01}$$

By calculator, $a_{\overline{48}|0.01} = 37.97396$. Therefore the present value is $P = (\$50)(37.97396) = \$1898.70$.

Amortization

The root of the term "amortize" is the French word *mort* meaning death. Thus the expression "to **amortize** a debt" means to pay it off (literally to "kill" it). Problems A and B from the beginning of this section are amortization problems that ask for the monthly payments on loans. We will develop the formula for solving these problems by solving a similar problem whose figures are more tractable.

Consider a \$4000 loan that must be repaid in five yearly payments at an interest rate of 8% compounded annually. What will the size of the payments R be? This is an annuity problem in which the present value is known and the periodic payment must be solved for. Each payment must exceed $\$4000/5 = \$800$. To determine the exact payment, substitute $P = 4000$, $i = 0.08$, and $n = 5$ into the formula for the present value of an annuity.

$$4000 = R \cdot \frac{1 - \dfrac{1}{(1.08)^5}}{0.08} = R \cdot a_{\overline{5}|0.08}$$

By the Amortization Table or calculator $a_{\overline{5}|0.08} = 3.99271$, so

$$4000 = R \cdot 3.99271$$

Then

$$R = \frac{4000}{3.99271} = 1001.83$$

Therefore the payments that amortize the loan are $1,001.83.

In general suppose a loan of $P must be repaid in equal periodic payments of $R at an interest rate that is compounded over the same period of time. The Amortization Formula determines the size of the payments. Its derivation is similar to the preceding particular case.

Amortization Formula

$$R = P \cdot \frac{i}{1 - \dfrac{1}{(1 + i)^n}} = P \cdot \frac{1}{a_{\overline{n}|i}}$$

where
P is the present value of all payments
R is the periodic payment
$i = r/k$ is the periodic interest rate
$n = tk$ is the number of periods

The Amortization Formula allows us to solve the problems stated at the beginning of this section. Example 2 solves Problem A.

EXAMPLE 2

Suppose that a new car costs $17,000. The dealer offers $5000 for a trade-in with a down payment of $2000.

Problem

What will the monthly installments be if the $10,000 loan is for 4 years at 18% annual interest compounded monthly?

ESTIMATION The answer must exceed $10,000/48 \approx $200.

SOLUTION We have $P = 10,000$. There are 4 years of monthly payments, so $n = 48$. The monthly interest rate is $i = 0.18/12 = 0.015$. By the Amortization Formula

$$R = 10,000 \cdot \frac{1}{a_{\overline{48}|0.015}}$$

By calculator $a_{\overline{48}|0.015} = 34.04255$, so

$$R = \frac{10,000}{34.04255} = 293.75.$$

Therefore the monthly payments are $293.75.

Let us reflect on Example 2. The total amount of money paid over the life of the loan is ($293.75)(48) = $14,100. Therefore the amount of interest paid on the $10,000 loan is $4100. A *payment schedule* (also called an **amortization schedule**) lists all payments of a loan and breaks down each payment into the principal and interest. Let us illustrate this with an example having fewer periods.

Reconsider the $4000 loan at 8% interest compounded annually. It was repaid in five yearly payments of $R = $1001.83. For the first payment the interest applies to the entire loan, so the interest is 8% of $4000, or $320. The principal that is applied to the loan is the difference between the payment and the interest, or $1001.83 − $320 = $681.83. Then the remaining balance of the loan after one payment is $4000 − $681.83 = $3318.17. This is shown in the row for payment 1 in the accompanying payment schedule. The figures for payment 2 are obtained in a similar manner. The interest is 8% of the remaining balance, or (0.08)(3318.17) = 265.45. The principal that is applied to the loan is $1001.83 − $265.45 = $736.38, so the remaining balance of the loan after two payments is $3318.17 − $736.38 = $2581.79. The other payments are obtained similarly and are shown in the payment schedule.

Payment	Interest	Principal	Balance
0			$4000.00
1	$320.00	$681.83	3318.17
2	265.45	736.38	2581.79
3	206.54	795.29	1786.50
4	142.92	858.91	927.59
5	74.21	927.62	−0.03

Payment schedule for $4000 loan at 8% interest for 5 years

Sometimes an adjustment has to be made in the final payment. Here the final payment would be reduced by three cents.

Example 3 solves Problem B stated at the beginning of the section.

EXAMPLE 3

A house costs $90,000. The terms of sale are 10% down and the remainder to be paid in monthly payments for 25 years. The loan is taken at an annual interest rate of 9% compounded monthly.

Problem

What are the monthly payments? Construct the payment schedule for the first four payments.

SOLUTION Initially 10% of $90,000 (= $9000) must be paid, so the amount of the loan is $P = \$81,000$. There are $n = (25)(12) = 300$ monthly periods and the monthly interest rate is $i = 0.09/12 = 0.0075$. By the Amortization Formula the monthly payments are

$$R = P \cdot \frac{1}{a_{\overline{300}|0.0075}}$$

The value of $a_{\overline{300}|0.0075}$ is not provided in the Amortization Table. By calculator $a_{\overline{300}|0.0075} = 119.16162$, so $R = 81,000/119.16162 = 679.75$. Therefore the monthly payments are $679.75.

Here is the payment schedule for the first four payments.

Payment	Interest	Principal	Balance
0			$81,000.00
1	$607.50	$72.25	80,927.75
2	$606.96	$72.79	80,854.96
3	$606.41	$73.34	80,781.62
4	$605.86	$73.89	80,707.73

Payment schedule for $81,000 loan at 0.75% per month for 300 months

The total amount paid for the house in Example 3 is ($679.75)(300) = $203,925. Therefore the amount of interest paid is $203,925 − $81,000 = $122,925, which is one and one-half times the amount of the loan. The abbreviated payment schedule supplies some evidence for this fact, because after four payments of $679.75 each, the principal of the loan has been lowered from $81,000 to only $80,707.73. Most homeowners have experienced the anguish that accompanies this realization.

EXERCISE SET 5–5

• In Problems 1 to 4 determine the amount of money that should be invested as a lump sum today to receive the stated dividend.

1. A dividend of $400 every 3 months for 2 years in an account that pays 8% compounded quarterly.

2. A dividend of $400 every 3 months for 1 year in an account that pays 8% compounded quarterly.

3. A dividend of $100 every 3 months for 1 year in an account that pays 6% compounded quarterly.

4. A dividend of $100 every 3 months for 4 years in an account that pays 6% compounded quarterly.

In Problems 5 to 8 determine the present value of the stated annuity.

5. Pays $100 a month for 2 years at 24% annual interest compounded monthly?

6. Pays $150 a month for 1 year at 24% annual interest compounded monthly?

7. Pays $250 a month for 2 years at 12% annual interest compounded monthly?

8. Pays $20 a month for 4 years at 12% annual interest compounded monthly?

In Problems 9 and 10 determine the size of the payments for a $4000 loan that must be repaid in five yearly payments at the stated annual interest rate.

9. 6%   10. 10%

In Problems 11 and 12 determine the size of the payments for a $3500 loan that must be repaid in four yearly payments at the stated annual interest rate.

11. 4%   12. 12%

In Problems 13 to 16 determine the monthly car installments on the stated loan at the annual interest rate compounded monthly.

13. $2000 loan for 4 years at 18%.

14. $1000 loan for 2 years at 18%.

15. $2500 loan for 2 years at 6%.

16. $1200 loan for 3 years at 6%.

In Problems 17 to 20 determine the monthly house payments on the stated loan at the annual interest rate compounded monthly.

17. $60,000 loan for 20 years at 9%.

18. $40,000 loan for 25 years at 8%.

19. $120,000 loan for 30 years at 8%.

20. $75,000 loan for 25 years at 9%.

21. Construct the payment schedule for the loan in Problem 9.

22. Construct the payment schedule for the loan in Problem 11.

23. Construct the payment schedule for the first four payments of the loan in Problem 17.

24. Construct the payment schedule for the first four payments of the loan in Problem 18.

•• 25. Suppose that a new car costs $13,000. The dealer offers $4000 for a trade-in with a down payment of $1000. What will the monthly installments be if the loan is for 4 years at 12% annual interest compounded monthly?

26. Suppose that a new car costs $22,000. The dealer offers $8000 for a trade-in. What will the monthly

installments be if there is no down payment and the loan is for 4 years at 12% annual interest compounded monthly?

27. A house costs $800,000. The terms of sale are 25% down and the remainder to be paid in monthly payments for 25 years. The loan is taken at an annual interest rate of 9% compounded monthly.
    (a) What are the monthly payments?
    (b) Construct the payment schedule for the first four payments.

28. A house costs $65,000. The terms of sale are 20% down and the remainder to be paid in monthly payments for 30 years. The loan is taken at an annual interest rate of 9% compounded monthly.
    (a) What are the monthly payments?
    (b) Construct the payment schedule for the first four payments.

29. Construct the entire payment schedule for the loan in Problem 15.

30. Construct the entire payment schedule for the loan in Problem 14.

31. A movie star signs a "10 Million Dollar Contract," which pays $1,000,000 a year for 10 years. If the studio can invest its money at 10% interest compounded annually, what is the present value of the contract?

32. Two donors agree to give the same amount of money to their alma mater. One pledges $1000 a year for 10 years. The other prefers to donate one lump sum now. What size should the lump sum be if the money earns 10% interest compounded annually?

33. A student wants to have $200 each month for spending money for the next 9 months at school. How much money has to be set aside from summer earnings to do this if the money is deposited in an account that pays 6% interest compounded monthly?

34. If the monthly payments for a car loan were $150 for 4 years at 9% interest compounded monthly, what was the amount of the loan?

35. An art museum benefactor donated $500,000

for acquisitions, with the understanding that the money was to be withdrawn quarterly for 5 years. If the account paid 8% annual interest compounded quarterly, how much money could the art museum withdraw each quarter?

36. What are the monthly payments on a $3000 dining room set that is paid for in 24 monthly installments at a monthly interest rate of 1.5%?

37. A small business buys a local network of microcomputers costing $30,000. The business amortizes the purchase with monthly payments over 6 years at 12% interest compounded monthly. What is the size of each payment?

38. Determine the monthly payments on a $10,000 loan under each of these repayment conditions.
    (a) 5 years at 12% interest compounded monthly.
    (b) 3 years at 9% interest compounded monthly.

39. If a house was purchased with a down payment of $10,000 and a 30-year mortgage at 9.5% interest compounded monthly with monthly payments of $400, what was the price of the house?

CHAPTER REVIEW

Key Terms

5–1 Simple Interest and Discount

Principal
Interest
Rate of Interest
Simple Interest
Compound Interest
Exact Interest Year
Ordinary Interest Year

Future Amount
Present Value
Discount Interest Method
Discount
Proceeds
Effective Rate

5–2 Compound Interest

Interest Compounded:
　　Annually
　　Quarterly
　　Monthly
　　Daily
Growth Factor

Period
Periodic Interest
Compound Interest Formula
Nominal Rate
Effective Rate
Annual Yield

5–4 Annuities and Sinking Funds

Annuity
Ordinary Annuity
Future Value

Geometric Series
Annuity Due
Sinking Fund

5–5 Amortization

Present Value of an Annuity
Amortize

Amortization Schedule

Summary of Important Concepts

Formulas

Simple interest

$$I = Prt$$

Future amount

$$A = P(1 + rt)$$

Interest compounded annually

$$P_t = P(1 + r)^t$$

Compound Interest Formula

$$A = P(1 + i)^n$$

Sum of a geometric series

$$a^n + \cdots + a + 1 = \frac{a^{n+1} - 1}{a - 1}$$

Annuity Formula

$$A = R \cdot \frac{(1 + i)^n - 1}{i} = R \cdot s_{\overline{n}|i}$$

Present value of an annuity

$$P = R \cdot \frac{1 - \dfrac{1}{(1 + i)^n}}{i} = R \cdot a_{\overline{n}|i}$$

Amortization Formula

$$R = P \cdot \frac{i}{1 - \dfrac{1}{(1 + i)^n}} = P \cdot \frac{1}{a_{\overline{n}|i}}$$

REVIEW PROBLEMS

1. Find the simple interest on a $2500 savings account that earns $5\frac{1}{4}\%$ interest for 2 years.

2. Find the simple interest on a $4000 loan for 90 days at 18% interest.

3. Determine the interest on a $500 loan for 3 years at 15.5% interest compounded annually.

4. After 3 years, what is the future value of a savings account with $700 that pays 5.25% interest compounded daily?

5. Determine the present value of $10,000 over 3 years at 8% compounded (a) annually (b) daily.

6. What is the annual yield if the nominal rate is 7.58%?

7. If the cost of a ticket to a Broadway play rises 8% a year for the next 20 years, what will the price of a $40 ticket be in 20 years?

8. In 1985 Mexico's rate of inflation was 63.7%. If a burrito cost 1000 pesos in 1985, and that inflation continued for a decade, what would the burrito cost in 1995?

9. In May of 1986 Mexico's rate of inflation was 5.6%. If that rate continued for an entire year, what would the annual rate of inflation be?

10. If the world population increases 2% each year for the next 25 years, how much larger will the population be?

11. Determine the future value of an annuity in which a payment of $250 is made at the end of every quarter for 3 years at 8% compounded quarterly.

12. Determine the quarterly payments of a sinking fund that must produce $12,000 in 4 years at 8% compounded quarterly.

13. If an individual deposits $2000 a year in a retirement account paying 6% compounded annually, how much will be in the account after 25 years?

14. How much money should be invested as a lump sum today to receive a dividend of $500 every 3 months for 4 years in an account that pays 8% compounded quarterly?

15. What is the present value of an annuity that pays $250 a month for 4 years at 9% annual interest compounded monthly?

16. Construct the payment schedule for a $5000 loan that must be repaid in four yearly payments at an interest rate of 10% compounded annually.

17. A house costs $80,000. The terms of sale are 25% down and the remainder to be paid in monthly payments for 30 years. The loan is taken at an annual interest rate of 12% compounded monthly.

   (a) What are the monthly payments?
   (b) Construct the payment schedule for the first four payments.

6 Sets and Counting

CHAPTER OVERVIEW　This chapter lays the foundation for the next two chapters on probability and statistics. It begins with a review of the basic terms and notation from set theory, where various sets of real numbers are introduced. Since several meanings are attached to the verb "count," a discussion of the mathematical meaning is given. It is followed by a development of two fundamental counting techniques, the Inclusion-Exclusion Principle and the Multiplication Principle. Venn diagrams and tree diagrams are introduced as aids in visualizing certain types of situations, like the scheme concocted by a gang of crooks to rig the Pennsylvania lottery. The last concepts described, permutations and combinations, are applied to the binomial theorem, which itself has applications in probability, statistics, and calculus.

CASE STUDY PREVIEW

Write down the license plate number of your car (or a friend's car if you do not own one). Most likely it is not a "number" at all, but rather, consists of three letters followed by three digits. The reason for this assertion is that more than half of the states currently use this so-called standard format.

California has more passenger vehicles than any other state; so many that the standard format cannot account for all of the license plates. A second format was introduced, and the combined formats accounted for all of the state's motoring needs. But then California introduced yet a third format. Is this an instance of irrational behavior in the Golden State, or is there a mathematical justification? The Case Study will reveal the answer.

6–1 SETS

In mathematics the little word "set" plays a very big role. Set theory was developed to organize, synthesize, and unify all of mathematics.

This section introduces some basic aspects of set theory and describes various sets of real numbers. We have already used sets intuitively in Chapters 3 and 4, where sets of feasible solutions to linear programming problems were encountered.

Notation

There are several ways to describe a **set.** The most direct way is the *descriptive method,* in which the phrase "the set of all . . . ." precedes a description of the set. The set must be *well defined* in the sense that it is possible to tell whether or not a given object is in the set. For instance, the "set of all large numbers" is not well defined because the word "large" is ambiguous.

EXAMPLE 1 | **Problem**

Determine whether each of the following collections is a well-defined set:
(a) good senators; (b) nonnegative numbers.

SOLUTION (a) Each senator is regarded as good by some people and bad by others. Since the collection is not well defined, it is not a set.
(b) The nonnegative numbers consist of zero and all positive numbers. This collection is well defined, so it is a set.

Usually capital letters stand for sets. *Throughout this chapter the letter S will denote the set of all license plates in standard format,* where **standard format** means three letters followed by three digits.

The objects in a set are called **elements.** The symbol $\in$ means "is an element of" while the symbol $\notin$ means "is not an element of." Thus AOK 007 $\in$ S but 007 AOK $\notin$ S.

Another way to specify a set is to use *braces* to enclose the elements. The notation $A = \{4, 5, 6\}$ means that A is the name of the set whose elements are 4, 5, and 6. It is read "A is equal to the set consisting of 4, 5, and 6." Braces can also define sets implicitly. The set

$$F = \{1, 2, 3, \ldots, 50\}$$

refers to the numbers from 1 to 50. The sets A and F are finite. Two examples of infinite sets are the **natural numbers**

$$N = \{1, 2, 3, \ldots\}$$

and the **integers**

$$Z = \{\ldots, -3, -2, -1, 0, 1, 2, 3, \ldots\}$$

A third way of describing sets is to use *set-builder notation* in the form $\{\_:\_\_\_\}$. This is read "the set of all $\_$ such that $\_\_\_$." The set of **rational numbers** Q, which is sometimes called the set of fractions, can be defined by

$$Q = \left\{ \frac{a}{b} : a,b \in Z, \quad b \neq 0 \right\}$$

In words, Q consists of all quantities that can be expressed in the form a/b, where a and b are integers such that $b \neq 0$. Some examples of rational numbers are $\frac{8}{5}$, $4 \ (= \frac{4}{1})$, $0 \ (= \frac{0}{1})$, $1.25 \ (= \frac{5}{4})$, and $0.333 \cdots \ (= \frac{1}{3})$.

Subsets

Let X and Y be arbitrary sets. Then X is a **subset** of Y if every element in X also belongs to Y. Equivalently X is a subset of Y if and only if $x \in X$ implies that $x \in Y$. It is written $X \subset Y$. If $X \subset Y$, then Y *contains* X. If X is not a subset of Y, we write $X \not\subset Y$. This occurs when some element is in the set X but not in the set Y. Since every natural number is an integer, it follows that for the infinite sets N and Z described earlier, $N \subset Z$.

EXAMPLE 2 | **Problem**

Is the set $D = \{0, \frac{1}{2}, 1\}$ a subset of N? Is $D \subset Z$? Is $D \subset Q$?

SOLUTION The number 0 belongs to D but not to N, so $D \not\subset N$. Similarly, $D \not\subset Z$ since $\frac{1}{2} \in D$ but $\frac{1}{2} \notin Z$. Each element of D is a rational number, so $D \subset Q$.

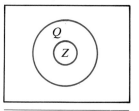

Figure 1

Since the set of rational numbers contains all of the natural numbers, all of their negatives, and 0, it follows that $Z \subset Q$. A **Venn diagram** depicts this in Figure 1. The innermost circle represents Z and is contained in the larger circle representing Q.

Two sets are **equal** if they contain the same elements. Thus $\{i, o, u\} = \{u, i, o\}$ but $\{i, o, u\} \neq \{i, o, x\}$. Also if $B = \{x \in Z : x > 0\}$, then $B = N$.

Figure 2

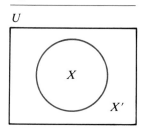

Operations on Sets

The **complement** of a set X is the set of elements that do not belong to X. It is written X', read "X prime." The definition raises the issue of how far to look outside X. Consider, for instance, the set Q of rational numbers. Is the kitchen sink "not in Q"? Yes, it is not in Q. However, we usually limit the search to a set called the **universal set,** denoted by U. The universal set for Q in this book will be the set R of real numbers.

The Venn diagram in Figure 2 shows a set and its complement. The universal set U is the area inside the rectangle, X is the area inside the circle, and X' is the area that is inside the rectangle U but outside the circle X.

Usually the universal set is understood from context but sometimes it must be specified. Example 3 addresses this issue.

EXAMPLE 3

Problem

Find A' for the following choices of a set A and a universal set U.
(a) $A = \{1, 2, 3\}$, $U = \{1, 2, 3, 4, 5, 6\}$
(b) $A = \{1, 2, 3\}$, $U = N$
(c) A is the set of even positive integers, $U = N$

SOLUTION (a) $A' = \{4, 5, 6\}$ (b) $A' = \{4, 5, 6, 7, \ldots\}$ (c) A' is the set of odd positive integers.

Two additional operations on sets are union and intersection. Let X and Y be arbitrary sets. The **intersection** of X and Y, written $X \cap Y$, is the set of all elements that are in X and Y simultaneously. The **union** of the set X with the set Y, written $X \cup Y$, is the set of all elements that are in X or in Y (or in their intersection).

EXAMPLE 4

Problem

Consider the sets $A = \{0, 1, 2, 3, 4\}$, $B = \{1, 3, 5\}$, and natural numbers N. Find (a) $A \cap B$, (b) $A \cap N$, (c) $A \cup B$, (d) $A \cup N$.

SOLUTION (a) $A \cap B = \{1, 3\}$ (b) $A \cap N = \{1, 2, 3, 4\}$
(c) $A \cup B = \{0, 1, 2, 3, 4, 5\}$ (d) $A \cup N = \{0, 1, 2, 3, \ldots\}$

The number $\sqrt{2}$ cannot be written as the ratio of any pair of natural numbers, so it is called an **irrational number.** (We will not prove this fact in this book.) The set Q' consists of all irrational numbers. Since no number can be both rational and irrational simultaneously, $Q \cap Q'$ has no elements. The set with no elements is called the **empty** (or **null**) **set,** denoted by $\emptyset$. Hence $Q \cap Q' = \emptyset$.

Two sets X and Y having no elements in common are called **disjoint.** Every set X and its complement X' are disjoint because $X \cap X' = \emptyset$. In particular Q and Q' are disjoint.

For any set X and universal set U we have $X \cup X' = U$. What about the union $Q \cup Q'$? It is equal to the set R of all **real numbers.** Therefore R is the union of the set of rational numbers and the set of irrational numbers so that $R = Q \cup Q'$, where Q and Q' are disjoint sets. The Venn diagram in Figure 3 shows the set inclusions

$$N \subset Z \subset Q \subset R$$

Example 5 applies set theory to a kind of problem that is encountered frequently.

Figure 3

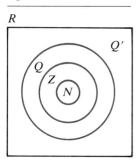

TABLE 1   Relative Costs of Living in Ten Major Cities[a]

City	All Items	Grocery Items	Housing	Utilities	Transportation	Health Care	Misc. Goods & Services
New York	136.3	107.6	179.8	172.9	117.6	164.3	108.9
San Francisco	127.2	109.9	189.9	59.5	119.9	151.6	113.4
San Diego	116.8	95.5	162.1	82.6	119.7	122.6	105.6
Philadelphia	116.4	106.8	117.8	152.4	113.1	125.2	107.2
Los Angeles	114.1	95.4	145.8	65.2	113.1	130.5	115.9
San Jose, Calif.	113.0	102.4	158.6	57.7	109.9	130.4	103.2
Dallas	112.0	110.5	119.6	120.7	111.3	108.8	105.2
Denver	107.6	105.8	134.9	75.1	110.1	111.7	98.7
Phoenix	106.3	99.2	118.6	102.5	106.9	119.0	99.6
Columbus, Ohio	105.2	106.8	98.2	120.3	107.7	104.5	103.1

Source: American Chamber of Commerce Researchers Association.
[a]Cost-of-living index for major items; national average = $100

EXAMPLE 5

In 1987, the American Chamber of Commerce Researchers Association conducted a study to see how much it costs to maintain a middle-management life-style in 242 cities. Table 1 shows some parts of the study for 10 major cities. Let G be the set of cities whose grocery items cost less than $100, and L be the set of cities whose utilities cost less than $100.

Problem

Determine the elements in the sets G, L, $G \cap L$, and $G \cup L$.

SOLUTION The sets G and L can be read from the third and fifth columns in Table 1.

$G = \{$San Diego, Los Angeles, Phoenix$\}$
$L = \{$San Francisco, San Diego, Los Angeles, San Jose, Denver$\}$

The definitions of intersection and union lead to

$G \cap L = \{$San Diego, Los Angeles$\}$
$G \cup L = \{$San Francisco, San Diego, Los Angeles, San Jose, Denver, Phoenix$\}$

EXERCISE SET 6–1

- In Problems 1 to 4 determine if the collection is well defined.

1. The set of all large dogs.

2. The set of all tall people.

3. The set of all positive integers.

4. The set of rational numbers between 1 and 2 inclusive.

In Problems 5 to 8 list the elements in the set.

5. The set of positive integer divisors of 12.

6. The set of integers between 0 and 5 inclusive.

7. The set of integers between -3 and 0 inclusive.

8. The set of even integers between -10 and -4 inclusive.

In Problems 9 to 12 determine whether A or B is a subset of X.

9. $X = \{1, 2, 3, 4\}, A = \{0, 1, 2\}, B = \{2, 3, 4\}$

10. $X = \{-1, 0, 1, 2, 3, 4\}, A = \{0, 1, 4\},$
    $B = \{1, 3, 4\}$

11. $X = Z, A = \{-2, 0, 2, 4\}, B = \{2, 4, 6, 8\}$

12. $X = Z, A = \{0\}, B = \{-\frac{1}{2}, 1, 2\}$

In Problems 13 to 16 find X' for the given choice of U.

13. $X = \{1, 2, 3\}, U = \{1, 2, 3, 4, 5, 6\}$

14. $X = \{1, 2, 3\}, U = \{0, 1, 2, 3, 4, 5, 6, 7\}$

15. $X = \{1, 2, 3\}, U = N$

16. $X = \{2, 4, 6, 8, 10\}, U = N$

In Problems 17 to 20 find $X \cup Y$.

17. $X = \{1, 2, 3\}, Y = \{2, 3, 4\}$

18. $X = \{-1, 0, 1\}, Y = \{-1, 1\}$

19. $X = Q, Y = \varnothing$

20. $X = Q, Y = Z$

In Problems 21 to 24 find $X \cap Y$.

21. $X = \{1, 2, 3, 4\}, Y = \{3, 4, 5\}$

22. $X = \{1, 2, 3, 4\}, Y = \{4, 5\}$

23. $X = \{-2, -1, 0, 1, 2\}, Y = N$

24. $X = \{\frac{2}{2}, \frac{3}{2}, \frac{4}{2}, \frac{5}{2}\}, Y = Z$

•• In Problems 25 to 28 determine whether A or B is a subset of X.

25. $X = Q, A = \{x : x^2 = 4\}, B = \{x : x^2 = 2\}$

26. $X = Q, A = \{x : x^2 = -1\}, B = \{x : x^2 = 1\}$

27. $X = Q, A = \{\ldots, -3, -2, -1\},$
    $B = \{x \in Z : x < 0\}$

28. $X = Q, A = \{5, 6, 7, \ldots\},$
    $B = \{x \in Z : x < -4\}$

In Problems 29 and 30 find X' for the given choice of U.

29. $X = \{1, 3, 5, \ldots\}, U = \{2, 4, 6, \ldots\}$

30. $X = \{4, 8, 12, \ldots\}, U = \{2, 4, 6, \ldots\}$

In Problems 31 to 34 find $X \cup Y$.

31. $X = \{-2, -1, 0, 1\}, Y = N$

32. $X = \{\sqrt{2}, 2, \sqrt{3}, 3\}, Y = Q$

33. $X = N, Y = \{x \in Z : x < 0\}$

34. $X = N, Y = \{x \in Z : x \geq 0\}$

In Problems 35 to 38 find $X \cap Y$.

35. $X = Q, Y = \{\sqrt{2}, \sqrt{3}, \sqrt{4}, \sqrt{5}, \ldots\}$

36. $X = Q', Y = \{\sqrt{2}, \sqrt{3}, \sqrt{4}, \sqrt{5}, \ldots\}$

37. $X = N, Y = \{x \in Z : x < 4\}$

38. $X = \{x \in Z : x < 7\}, Y = \{x \in Z : x < 3\}$

39. Describe $F \cap X$ where $F = \{4, 8, 12, 16, \ldots\}$ and $X = \{6, 12, 18, 24, \ldots\}$.

40. Describe $M \cap K$ where M consists of all multiples of m and K consists of all multiples of k where m and k are given positive integers.

••• Problems 41 to 44 refer to Table 1 in Example 5.

41. List the set of cities whose cost of housing or miscellaneous goods and services is less than $100.

42. List the set of cities whose cost of housing and utilities exceeds $100.

43. Let U = 10 major cities, H = cost of housing that is more than $100, and M = cost of miscellaneous goods and services that is more than $100. Describe in words the fact that $H \cup M = U$.

44. Let U = 10 major cities and T = cost of transportation that is more than $100. Describe verbally the fact that $U = T$.

Problems 45 and 46 refer to Table 2, which lists certain countries and the percentage of their population for two age groups as of 1986.

Let F be the three countries with the highest percentages aged 15–29 and let T be the three countries with the highest percentages aged 30–44.

45. Find (a) F (b) T.

46. Find (a) $F \cap T$ (b) $F \cup T$.

47. What is the difference between $\varnothing$ and $\{\varnothing\}$?

TABLE 2

Country	Brazil	Canada	China	Ireland	U.S.	USSR
Age 15–29	28.6	28.9	29.1	24.9	22.4	26.6
Age 30–44	16.4	20.0	17.5	17.2	19.1	19.1

Source: 1986 Britannica Book of the Year, pp. 826–831.

48. In music theory "the diatonic universe of seven pitch classes" consists of the notes A, B, C, D, E, F, and G, irrespective of inflection by sharps and flats.* (a) Use set notation to describe this universal set. (b) Use set notation to describe the set in which each note can also be sharp or flat (refer to a piano keyboard; that is, G, G-sharp, and G-flat are regarded as different note classes).

49. In 1988 the Office Network, a consortium of real estate firms, conducted a study of rental rates in the central business districts (CBDs) of several cities. Table 3 shows part of the results of the survey. (a) List the cities in the set A of those cities having the five highest CBD rates. (b) List the cities in the set B of those cities having the five highest rates outside CBDs. (c) What would it mean for A ∩ B = A? (d) For this study, is it true that A = B?

TABLE 3   Office Rental Costs
Typical annual rate, in dollars per square foot, for rent and operating expenses in space that is new or competitive with new space. Rates are for space inside central business district or outside CBD.

City	CBD	Outside
Baltimore	$23.38	$17.50
Boston	35.00	15.50
Chicago	28.95	19.56
Houston	12.00	9.50
Los Angeles	34.00	22.00
Miami	29.00	19.00
New York	40.00	35.00
Philadelphia	22.10	14.85
Princeton, N.J.	N.A.	18.50
San Francisco	23.65	22.25
Washington	32.00	22.50
Wilmington	18.00	16.10

N.A. = Not applicable.
Source: The Office Network

*John Clough, "Aspects of Diatonic Sets," *Journal of Music Theory,* Spring 1979, pp. 45–62.

6–2   VENN DIAGRAMS AND COUNTING

The word **count** usually means "one, two, three, four, ..." In mathematics, however, it has a wider usage. Perhaps a more appropriate word is "enumerate," meaning "to ascertain the number of." In mathematics the expression "count the number of . . ." does not mean to count the objects individually, but rather, to determine how many objects there are altogether. Sometimes enumerating a large set can be a complicated task. Indeed, the root of the word enumerate is the Latin *numerus,* meaning "nimble."

This section presents a basic counting principle that is explained and illustrated by Venn diagrams.

Number of Elements

Let $n(X)$ denote the number of distinct elements in a finite set X. The expression "count the number of elements in X" means to determine $n(X)$ by some method. Counting techniques are efficient methods for computing $n(X)$ directly.

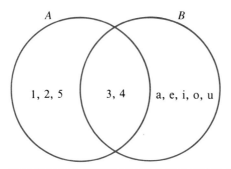

Figure 1

Let $A = \{1, 2, 3, 4, 5\}$ and $B = \{a, e, i, o, u, 3, 4\}$. Then $n(A) = 5$ and $n(B) = 7$. Since $A \cap B = \{3, 4\}$ it follows that $n(A \cap B) = 2$. The Venn diagram in Figure 1 shows the distribution of the elements in A and B. It also shows that $n(A \cup B) = 10$. Notice that the formula $n(A \cup B) = n(A) + n(B) - n(A \cap B)$ holds because $10 = 5 + 7 - 2$. This is a particular case of a basic counting technique called the Inclusion-Exclusion Principle. It holds in this case because when counting the elements in $A \cup B$ by first counting the elements in A and then counting the elements in B, the two elements in $A \cap B$ are counted twice, so they must be subtracted once in order to compensate.

Inclusion-Exclusion Principle

If X and Y are any sets, then

$$n(X \cup Y) = n(X) + n(Y) - n(X \cap Y)$$

Example 1 illustrates the reasoning behind this principle.

EXAMPLE 1

Suppose a survey of 100 people asks

Have you eaten at McDonald's this year?   Y   N
Have you eaten at Wendy's this year?   Y   N

The respondents circle Y or N for each question.

Problem

If 68 people circled Y for McDonald's, 42 circled Y for Wendy's, and 27 circled Y for both questions, how many respondents circled N for both questions?

SOLUTION Let M be the set of respondents who circled Y for McDonald's and let W be the set of respondents who circled Y for Wendy's. Then $n(M) = 68$ and $n(W) = 42$. Since 27 circled both Y, $n(M \cap W) = 27$. By the Inclusion-Exclusion Principle

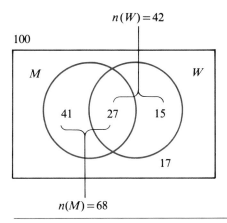

Figure 2

$$n(M \cup W) = n(M) + n(W) - n(M \cap W)$$
$$= 68 + 42 - 27$$
$$= 83$$

This means that 83 respondents circled Y for either McDonald's or Wendy's (or both). Since there were 100 respondents altogether, 17 respondents circled N for both questions. The Venn diagram is shown in Figure 2.

Recall that license plates in standard format S consist of three letters followed by three digits. License plates in *reverse format* R consist of three digits followed by three letters. Example 2 counts the number of license plates in either one of these formats, where the numbers of license plates in each format will be derived in the Case Study.

EXAMPLE 2 **Problem**

If $n(S) = 14,866,000$ and $n(R) = 16,208,000$, how many license plates are in either the standard or reverse format?

SOLUTION The problem reduces to finding $n(S \cup R)$ because $S \cup R$ represents all license plates in either the standard format or the reverse format. Each license plate in S begins with a letter and each license plate in R begins with a number, so there are no license plates in both formats. Thus $n(S \cap R) = 0$. By the Inclusion-Exclusion Principle

$$n(S \cup R) = n(S) + n(R) - n(S \cap R)$$
$$= 14,866,000 + 16,208,000 - 0$$
$$= 31,074,000$$

This means that over 31 million license plates can be made using either the standard format or the reverse format.

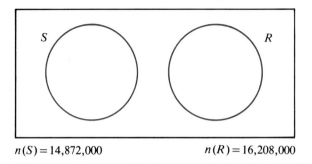

$n(S) = 14,872,000$ $n(R) = 16,208,000$

Figure 3

The sets S and R in Example 2 are disjoint. They are shown in Figure 3. The formula for $n(S \cup R)$ is a particular case of the Inclusion-Exclusion Principle.

> If X and Y are disjoint sets, then
> $$n(X \cup Y) = n(X) + n(Y)$$

Three Circles

One subset divides a universal set U into two regions. (See Figure 2 in Section 6–1.) Two subsets that are not disjoint divide U into four regions. (See Figure 2 in this section.) Three subsets in which no two subsets are disjoint divide U into eight regions, as Example 3 will show. The example refers to Table 1, which lists stock activity (sales, last price, and net change) during the week of April 11–15, 1988.

TABLE 1   **Most Active Stocks Week Ended April 15, 1988 (Consolidated)**

Company	Sales	Last	Net Chng
US West o	24,346,100	$51\frac{5}{8}$	$-1\frac{1}{4}$
NYSEG	13,857,000	$22\frac{1}{8}$	$-\frac{1}{8}$
IBM	11,068,100	$114\frac{1}{8}$	$+2\frac{1}{2}$
Varity	10,662,100	$3\frac{1}{2}$	$-\frac{1}{8}$
Gen El	10,515,100	40	$-2\frac{1}{2}$
Exxon	10,238,100	$43\frac{5}{8}$	-1
Texaco	10,083,400	$48\frac{5}{8}$	$-\frac{5}{8}$
AT&T	9,417,400	$26\frac{5}{8}$	$-1\frac{1}{2}$
Ford M	7,694,300	$45\frac{1}{4}$	$-1\frac{3}{8}$
Zayre	7,160,500	24	$+\frac{1}{8}$
Phil Pet	7,058,400	$17\frac{5}{8}$	$-\frac{7}{8}$
USX	7,019,100	32	$+\frac{5}{8}$
INCO	6,802,400	$26\frac{7}{8}$	$+1\frac{1}{4}$
Navistr	6,801,300	$6\frac{3}{8}$	$-\frac{3}{8}$
Am Exp	6,664,000	$24\frac{3}{4}$	-1

Source: The New York Times, April 17, 1988.

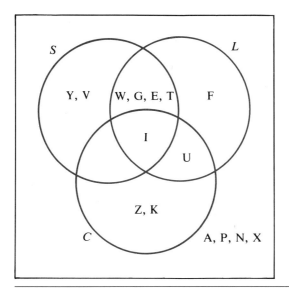

Figure 4

EXAMPLE 3 | Define these three sets of companies:

S = sales over $10,000,000
L = last price over $30
C = net change greater than $0

Problem

(a) Which companies are in all three sets? (b) Which companies are in C but in neither S nor L?

SOLUTION Refer to Figure 4, which partitions the companies into the three sets. From Table 1, using the first letter of each company, except X = AmExp, K = INCO, Y = NYSEG, and W = USWest o, we get

S = {W, Y, I, V, G, E, T}
L = {W, I, G, E, T, F, U}
C = {I, Z, U, K}

The set $S \cap L \cap C$ represents those companies in all three sets, so $S \cap L \cap C$ = { I }. This answers part (a).

To answer (b), note that $S \cap C$ = {I} and $L \cap C$ = {I, U}. The companies that are in C = {I, Z, U, K} but are in neither S nor L are obtained by removing from C those companies listed in either $S \cap C$ or $L \cap C$. This set is {I, U}, so the remaining set is {Z, K}, which answers part (b).

Example 4 is typical of problems involving the partition of a set by three of its subsets. It builds on the concepts introduced in Example 3.

EXAMPLE 4   A survey asked 200 people which sports they played regularly. It revealed that 30 played tennis, 25 played golf, 40 played softball, 10 played tennis and golf, 20 played tennis and softball, 15 played golf and softball, and 8 played all three sports.

Figure 5

(a)   (b)

(c)   (d)

(e)

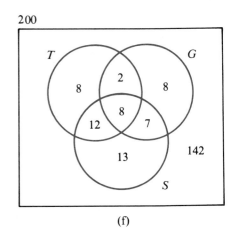

(f)

Problem

How many of those surveyed played (a) golf and softball but not tennis
(b) only golf (c) none of these sports?

SOLUTION A Venn diagram provides a convenient method for tabulating these
data. Let T = tennis, G = golf, and S = softball. Three circles are needed to
present these sets. [See Figure 5(a).] Since 8 people played all three sports, we
have $n(T \cap G \cap S) = 8$. This is shown in Figure 5(b).

Consider the fact that 15 people played golf and softball: $n(G \cap S) = 15$.
Restrict your attention to the G circle and the S circle in Figure 5(b). We have
already accounted for 8 of the 15 people in $G \cap S$. The remaining 7 lie in the
$G \cap S$ region but outside $T \cap G \cap S$. Figure 5(c) shows their proper alignment.
Notice that these 7 played golf and softball but not tennis, answering part (a).

The same kind of reasoning will place the numbers 2 and 12 in their
respective areas in Figure 5(d).

Consider only the G circle in Figure 5(d). Adding 2, 8, and 7 accounts for
17 people who played golf. Since $n(G) = 25$, 8 people in the G circle are
unaccounted for. These 8 are the people who played only golf, which answers
part (b).

Similar reasoning shows that 8 people played only tennis and 13 played
only softball. These numbers are shown in Figure 5(e). The sum of all of the
numbers in Figure 5(e) is 58. The remaining people did not play any of these
sports. Since 200 people were surveyed altogether, 142 people did not play these
sports. This answers part (c), which is shown in Figure 5(f).

EXERCISE SET 6–2

- In Problems 1 to 6 let $U = \{a, b, c, \ldots, z\}$, $V =$
  $\{a, e, i, o, u\}$, and $H = \{o, u, c, h\}$. Find the number of elements in the following sets:

  1. V 2. H 3. $V \cup H$

  4. $V \cup H'$ 5. $V \cap H'$ 6. $V \cap H$

In Problems 7 and 8 find $n(A \cup B)$.

7. $n(A) = 16$, $n(B) = 9$, and $n(A \cap B) = 5$

8. $n(A) = 6$, $n(B) = 12$, and $n(A \cap B) = 3$

In Problems 9 and 10 find $n(A \cap B)$.

9. $n(A) = 16$, $n(B) = 9$, and $n(A \cup B) = 21$

10. $n(A) = 6$, $n(B) = 12$, and $n(A \cup B) = 14$

In Problems 11 and 12 find $n(A \cup B \cup C)$.

11. $n(A) = 40$, $n(B) = 20$, $n(C) = 30$, $n(A \cap B)$
    $= 9$, $n(B \cap C) = 11$, $n(A \cap C) = 8$, and
    $n(A \cap B \cap C) = 3$

12. $n(A) = 60$, $n(B) = 50$, $n(C) = 70$, $n(A \cap B)$
    $= 19$, $n(B \cap C) = 20$, $n(A \cap C) = 13$, and
    $n(A \cap B \cap C) = 11$

13. The numbers of possible license plates in stan-
    dard format and in reverse format (without re-
    gard to any restrictions) are 17,576,000 each.
    How many license plates can be made in one
    or the other of these formats?

14. An English class consists of 5 freshmen, 8 soph-
    omores, and 22 upperclassmen (juniors and
    seniors). How many students are in the course
    altogether?

15. The combined membership of the Mathematical
    Association of America (MAA) and the Amer-
    ican Mathematical Society (AMS) is approxi-
    mately 45,000, of whom 27,000 belong to the
    MAA and 23,000 belong to the AMS. How
    many people belong to both organizations?

16. A history class consists of 35 students, of whom 20 are women. Two of the women and five of the men are veterans. How many of the men are not veterans?

In Problems 17 and 18 draw a Venn diagram to answer the questions.

17. A survey of 36 students revealed the following information:

    14 had studied French
    11 had studied Spanish
    14 had studied German
     4 had studied French and Spanish
     3 had studied Spanish and German
     5 had studied French and German
     1 had studied all three languages

    (a) How many had studied German and Spanish but not French?
    (b) How many had studied only French?
    (c) How many had not studied any of these languages?

18. A survey of 200 people revealed that 30 played tennis, 25 played golf, 40 played baseball, 10 played tennis and golf, 15 played golf and baseball, 20 played tennis and baseball, and 8 played all three sports.
    (a) How many played baseball and golf but not tennis?
    (b) How many played only tennis?
    (c) How many did not play any of these sports?

•• In Problems 19 and 20 A and B are disjoint sets. Find $n(A)$.

19. $n(A \cup B) = 10$ and $n(A) = n(B) + 2$

20. $n(A \cup B) = 8$ and $n(A) + 2 = n(B)$

In Problems 21 and 22 find $n(A \cap B)$.

21. $n(A \cup B) = 12$, $n(A) = 7$, and $n(B) = 8$

22. $n(A \cup B) = 4$, $n(A) = 3$, and $n(B) = 3$

In Problems 23 and 24 find $n(A \cup B \cup C)$ where A and B are disjoint sets.

23. $n(A) = 10$, $n(B) = 5$, $n(C) = 9$, $n(A \cap C) = 1$, and $n(B \cap C) = 3$

24. $n(A) = 8$, $n(B) = 9$, $n(C) = 12$, $n(A \cap C) = 7$, and $n(B \cap C) = 1$

In Problems 25 and 26 draw a Venn diagram of the universal set U and subsets A and B. Use the given information to fill in the number of elements for each region.

25. $n(U) = 20$, $n(A') = 11$, $n(B') = 15$, and $n(A \cap B) = 2$

26. $n(U) = 20$, $n(A) = 4$, $n(B') = 8$, and $n(A \cup B) = 13$

In Problems 27 and 28 find $n(B)$.

27. $n(A \cup B) = 22$, $n(A \cap B) = 6$, and $n(A \cap B') = 8$

28. $n(A \cup B) = 16$, $n(A \cap B) = 6$, and $n(A \cap B') = 8$

29. Is it possible for $n(A \cap B) = n(A \cup B)$?

30. Is it possible for $n(A \cap B) > n(A \cup B)$?

••• 31. If $n(X) = 3$, how many subsets does X have?

32. If $n(X) = k$, how many subsets does X have?

33. Is there a difference between $n(\varnothing)$ and $n(\{\varnothing\})$?

34. Is there a difference between $\varnothing$ and $\{\varnothing\}$?

Problems 35 to 38 refer to a study in sociology on the economic mobility of the poor.* The accompanying table presents the status of the one year poor, temporarily poor (1 to 2 years), and persistently poor (7 to 8 years) from that study. Find $n(X)$ where X is the set of

35. White people who are persistently poor.

36. Black people who are persistently poor.

	One Year	Temporarily Poor	Persistently Poor	Poor
Women				
White	494	112	327	933
Black	562	307	122	991
Men				
White	472	40	1102	1614
Black	247	198	225	670
Total	1775	657	1776	4208

*Greg J. Duncan, et al., *Years of Poverty, Years of Plenty*, Ann Arbor, Survey Research Center, Institute for Social Research, 1984, p. 49.

37. Men who are one year poor.

38. Women who are one year poor.

In 1985 New Zealand's prime minister, David Lange, proposed that his country withdraw from the Anzus pact instead of permitting U.S. nuclear submarines to land in its ports. A national opinion poll[†] revealed that 70 percent favored remaining in the alliance while 60 percent approved Mr. Lange's course. Problems 39 and 40 refer to the poll.

39. What percent agreed with both positions?

40. What percent favored Mr. Lange's proposal but did not want to remain in the alliance?

41. The three primary colors in the visible spectrum are red, blue, and green.[‡] Draw a Venn diagram with three overlapping circles, labeled R, B, and G, and color each circle with a primary color. Describe the color of each of these regions: (a) $R \cap G$ (b) $R \cap B$ (c) $B \cap G$ (d) $R \cap G \cap B$ (e) $(R \cup G \cup B)'$.

42. In the 1984–1985 National Hockey League season, there were 234 games in which at least one team scored exactly two goals.[§] The nonlosing team scored two goals in 38 games and the nonwinning team scored two goals in 217 games. How many games ended in a tie?

43. Draw a Venn diagram with six sets.[‖] How many regions are there?

[†]*The New York Times,* October 6, 1985, p. 24.

[‡]Raymond A. Serway, *College Physics,* Philadelphia, Saunders College Publishing, 1985, pp. 660–663.

[§]Benny Ercolani and Jeff Boyle, *NHL Data, 1984–85 Season,* Montreal, Canada, National Hockey League Communications Department, 1985.

[‖]The Venn diagram is drawn in J. Chris Fisher, E. Koh, and Branko Grünbaum, ''Diagrams Venn and How,'' *Mathematics Magazine,* February 1988, pp. 36–40.

6–3 TREE DIAGRAMS AND THE MULTIPLICATION PRINCIPLE

Hundreds of thousands of people plunk down a dollar or more daily in hopes of ''hitting the big one'' in the state lottery. Each day, as their ticket chunks out of the computer, visions of expensive cars, wall-to-wall stereo systems, and VCRs dance in their heads. The odds against winning are monumental, but on they play.

If you *knew* what the number was going to be, actually knew it by some foolproof method, would you only bet $1? Of course not! You would gather all available cash and purchase tickets in a flurry. This is unlikely, however, because state governments have taken great precautions to prevent even the slightest cheating.

Why then did a white Cadillac pull up in front of the Dew Drop Inn in South Philadelphia in 1980 and the driver purchase 904 lottery tickets? That was just the beginning. By the end of the day the driver purchased over 10,000 tickets, all on numbers that consisted of 4's and 6's. Throughout the state accomplices purchased another 10,000 tickets.

The Pennsylvania daily lottery is run by selecting a Ping-Pong ball from each of three urns with ten balls numbered 0 to 9. Violet Lowery, a senior citizen, drew the number 666, making many of those 20,000 tickets big winners. The gang held tickets worth over $1.75 million.

The next day many of the sellers of the tickets, most of whom are local shop owners who also pay out the winnings for a fee from the state, complained that their payoffs were unusually high. The number 666 was very suspicious. The Bible calls it "the number of the beast." But how suspicious was it?

Here are three questions raised by the possible rigging of the "daily number." What are the chances of the number 666 being drawn? Are they less likely, the same, or more likely to be drawn than any other three-digit number? What facts would the state need in order to launch a full-scale investigation to determine if the fix was on? All three questions will be answered in the epilogue to this section.

Tree Diagrams

A **tree diagram** is a convenient way to visualize and enumerate the number of outcomes in an event that occurs in two or more stages. It begins with the possible outcomes of the first stage joined by lines to one common point. These lines form the branches of the tree. Each of these outcomes is joined to the outcomes of the second stage by another set of branches. The tree, which is lying on its side, grows from left to right with each successive stage.

For instance, consider a dinner menu that offers these options:

Salad	Entrée	Dessert
House (h)	Turkey (t)	Pie (p)
Spinach (s)	Veal (v)	Cake (c)
	Fish (f)	Ice cream (i)
	Burger (b)	

If the restaurant wants its customers to order their meals by number, how many numbers are necessary? The tree diagram in Figure 1 portrays all of the possibilities. At the first stage there are two possibilities, the house salad and a spinach salad. This stage has two branches emanating from the starting point. Each of the salads can be ordered with one of the four entrées, resulting in eight selections for these two courses. There are three desserts that can be ordered for each of these eight selections, so altogether 24 complete meals can be ordered. They are numbered in Figure 1. Note that each meal is defined by three branches, so that meal 18 consists of a spinach salad, veal, and ice cream.

Slot machines, which are often called "one-armed bandits," consist of three sets of figures. The payoff, if any, is determined by which set of three figures the machine randomly selects. Example 1 shows how a tree diagram helps you to analyze this situation.

EXAMPLE 1

Suppose the first wheel in a slot machine has pictures of three fruits (cherries, pears, and oranges), the second has pictures of a man and a woman, and the third has pictures of three medals (gold, silver, and bronze).

Problem

How many groups of three figures, one from each wheel, can occur?

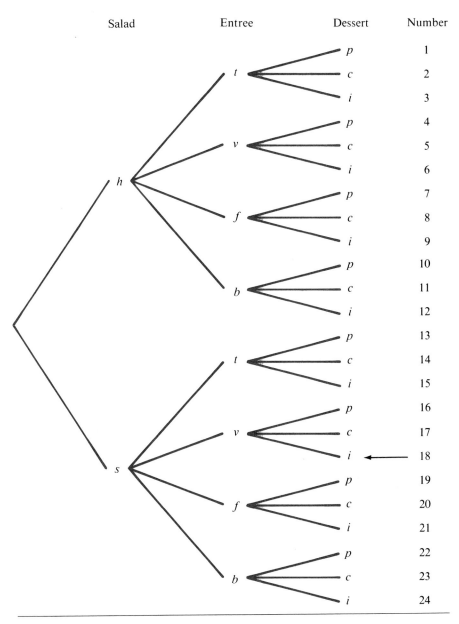

Figure 1

SOLUTION View each wheel in the slot machine as a stage in the tree diagram. The first stage will have three branches, one for each fruit. Each of these branches will have two branches corresponding to the second stage, resulting in six branches for the first two wheels. Each of these six branches will have three branches corresponding to the third stage. Thus there are 18 ways for the three wheels to occur. The tree diagram is shown in Figure 2.

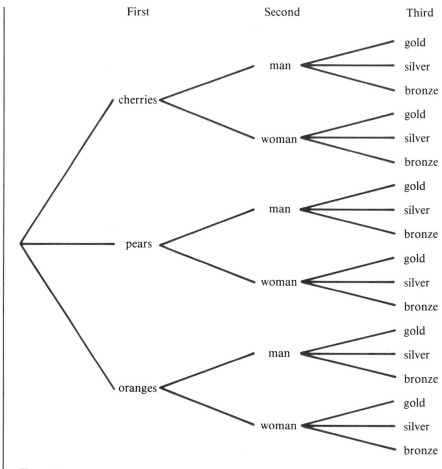

First   Second   Third

Figure 2

Trimmed Trees

Suppose a daily number is formed by drawing three balls from an urn containing ten balls numbered 0 to 9. The second and third balls are drawn from the urn without replacing those that have already been drawn. This is called **selection without replacement.** A tree diagram could present all the possibilities but it would be very unwieldy. We modify it by drawing only one line between the different stages and by writing above that line the number of branches that would have connected the two stages. The trimmed tree for the lottery is

start —— 10 —— 1st draw —— 9 —— 2nd draw —— 8 —— 3rd draw

The 10 represents the ten branches corresponding to the ten balls in the urn. Each of these branches is connected to nine branches leading to stage 2, and each of these is connected to eight branches leading to stage 3. Altogether there are $10 \cdot 9 \cdot 8 = 720$ ways of selecting the daily number without replacement.

Suppose the daily number is formed by selecting three balls from the urn in such a way that a ball is selected, its number is recorded, and then it is returned to the urn for another selection. This is called **selection with replacement.** The trimmed tree can be written with only the numbers and not the branches.

$$10 \quad \cdot \quad 10 \quad \cdot \quad 10 \; = \; 1000$$

first   second   third
stage   stage   stage

There are 1000 such daily numbers altogether.

The Multiplication Principle

The **Multiplication Principle** follows from the preceding examples, which were visualized in stages of two or more selections each.

Multiplication Principle

If a counting problem can be divided into two stages such that at stage 1 there are n_1 ways to make a selection and at stage 2 there are n_2 ways, then the total number of ways to make the two selections is

$$n_1 \cdot n_2$$

Similarly, if the problem requires a third stage in which there are n_3 ways to make the third selection, then the total number of ways to make the three selections is

$$n_1 \cdot n_2 \cdot n_3$$

In general if there are m stages and at the ith stage there are n_i ways to make a selection, then the total number of ways to make the m selections is

$$n_1 \cdot n_2 \cdot n_3 \cdots n_m$$

Example 2 applies the Multiplication Principle to license plates in standard format (three letters followed by three digits without regard to restrictions.)

EXAMPLE 2 | **Problem**

How many license plates are in standard format?

SOLUTION A license plate in standard format can be viewed as a tree with six stages: three letters (L) and three digits (D). There are 26 possible letters for each of the first three stages and 10 possible digits for each of the last three stages. The trimmed tree is

$$\underline{\quad 26 \quad}_{L} \underline{\quad 26 \quad}_{L} \underline{\quad 26 \quad}_{L} \underline{\quad 10 \quad}_{D} \underline{\quad 10 \quad}_{D} \underline{\quad 10 \quad}_{D}$$

Thus there are 17,576,000 license plates in standard format.

In 1982 California had more than 18 million licensed vehicles. Example 2 shows that they could not all have license plates in standard format. So California added the reverse format (three digits followed by three letters.) By the Multiplication Principle the reverse format also covers 17,576,000 license plates, allowing over 35 million license plates in these two formats. Why, then, did California adopt yet a third format? The answer lies in this chapter's Case Study.

The results of Example 3 will be needed repeatedly in Chapters 7 and 8.

EXAMPLE 3 | **Problem**

(a) If a coin is flipped two times, in how many ways can it land? (b) If a coin is flipped three times, in how many ways can it land? (c) If a coin is flipped ten times, in how many ways can it land?

SOLUTION (a) There are two choices for the first flip of the coin, heads (H) or tails (T). There are also two choices for the second flip. By the Multiplication Principle there are $2 \cdot 2 = 4$ ways that the coin can land. These ways are listed in the following tree diagram:

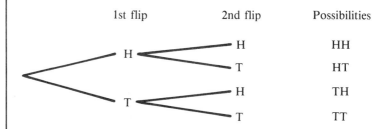

(b) By the Multiplication Principle there are $2 \cdot 2 \cdot 2 = 2^3$ ways the coin can land when it is flipped three times. Thus there are eight ways altogether.
(c) By the Multiplication Principle there are $2^{10} = 1024$ ways the coin can land when it is flipped ten times. Notice that the exponent of 2 is equal to the number of times the coin is flipped.

Epilogue: The Lottery Fix Revisited

We can now answer the questions posed in the introduction to this section. When the number 666 was drawn there should not have been any alarm that the lottery was fixed because 666 is simply one of the 1000 possibilities.

At first, state officials vehemently denied that foul play could have been associated with the drawing. They had made large payoffs before and there are certain numbers that people choose more often than others. The payoff for 666 was extremely large but it could be attributed to its unusual character.

Within a few days a number of investigative reporters were on the scent. They discovered that all of the numbers consisting of 4's and 6's were bet heavily under very similar wagering patterns. How many such numbers are there? By the

Multiplication Principle there are three stages with two selections (4 or 6) at each stage. So there are $2 \cdot 2 \cdot 2 = 8$ ways of arranging the two numbers in a three-digit number.

The gamblers were confident that one of these eight numbers would be drawn. They placed heavy bets on all eight numbers. This provided the authorities with good evidence of a conspiracy but it was still not known how they pulled it off.

Pennsylvania lost $1 million. Did state authorities recognize immediately that the fix was on? Are you kidding? Of course not! In fact State Revenue Secretary Howard Cohen, whose department was responsible for running the lottery, stated emphatically for months that it was "impossible" to rig the lottery. Six months later when the facts were uncovered, he was forced to resign.

Can you guess how the lottery was rigged? In each urn they weighted those balls numbered 4 and 6, ensuring that the tubes that pull one ball from each urn would pull a 4 or a 6. The nonweighted balls continued to whirl around, thus giving the appearance that the lottery was fair.

In 1982 two men were sentenced to prison terms of up to seven years for their roles in the lottery fix. Ironically, the day on which they went to prison Delaware's daily number came up 555. There was no unusually heavy betting on it, however, and no investigation was begun.

EXERCISE SET 6–3

• 1. A restaurant offers a breakfast menu with 2 juices (orange, apple), 4 eggs (scrambled, poached, soft-boiled, sunny), and 2 drinks (coffee, tea). Draw a tree diagram to list the possibilities. How many outcomes are there?

2. A study classifies human beings according to sex (male, female), height (tall, medium, short), and weight (heavy, normal, light). Draw a tree diagram to list the possibilities. How many outcomes are there?

3. Suppose that one coin is selected from each of two pockets, with each pocket containing a dime and a penny. Draw a tree diagram to list the possible ways of selecting the coins. How many selections are there?

4. Toss two dice. (This is equivalent to tossing one die two times.) Draw a tree diagram to list the possible outcomes. How many outcomes are there?

5. A competitor in the triathlon must choose from one of 3 swim suits (Jantzen, Laguna, Catalina),

2 bicycles (Schwinn, Raleigh), and 4 running shoes (New Balance, Nike, Adidas, Puma). Draw a tree diagram to list the possible choices. How many choices are there?

6. A coin is flipped and then a die is tossed. Draw a tree diagram to list the possible outcomes. How many outcomes are there?

7. Suppose that a family has five people in it. In how many ways is it possible for their birthdays to be in five different months?

8. The computer test bank for this chapter contains 12 problems on Section 6–1, 15 on Section 6–2, and 20 each on Sections 6–3 and 6–4. If a test has one problem from each section, how many different tests are possible?

9. A slot machine has the 12 zodiac signs on one wheel, 5 fruit on another wheel, and the 7 days of the week on the third. How many groups of three figures can occur?

10. A menu in a Chinese restaurant lists ten dinners in column A and seven dinners in column B.

How many selections can be made with one from column A and one from column B?

11. An urn contains 26 balls, each with a letter of the alphabet printed on it. How many three-letter "words" can be formed by selecting three balls without replacement from the urn?

12. A daily "word" is selected by drawing 1 ball from each of three urns. Each urn contains 26 balls with the letters of the alphabet printed on them. How many three-letter "words" can be formed?

13. License plates in Hawaii consist of two letters followed by three digits. How many such license plates are possible?

14. License plates in Louisiana consist of three digits, followed by one letter, followed by three digits. How many such license plates are possible?

15. A combination lock has a dial with ten digits. Three numbers are chosen to form the combination that opens the lock. (For example, the combination 3-5-3 means to turn the dial to the right to 3, then to the left to 5, and then to the right to 3.) How many combinations can be formed if the digits can be repeated?

16. In most locks whose combination has three digits the first and second digits cannot be the same but the second and third can. Given this restriction, how many combinations are possible if the dial has ten digits?

17. If the dial of a lock has 30 numbers and repetition of the digits is allowed, how many combinations with 3 numbers are possible?

18. If the dial of a lock has 30 numbers, how many combinations with 3 numbers are possible if the first and second numbers cannot be the same but the second and third can?

•• 19. A bag contains numerous $1 and $5 bills. You reach in and grab one bill at a time. You stop whenever you grab a $5 bill or whenever three bills have been grabbed. Draw a tree diagram to list the possible grabs. Write the amount of money grabbed beside each possibility.

20. The Miller's Lite teams (T = tastes great, L = less filling) decide to end their feud with a series in which the first team to win two games is the winner. Draw a tree diagram to list the possible series outcomes. How many possible outcomes are there?

21. If three dice are tossed, in how many ways can they land? (See Problem 4.)

22. If four dice are tossed, in how many ways can they land?

23. If ten dice are tossed, in how many ways can they land?

24. If n dice are tossed, in how many ways can they land?

In Problems 25 to 28 at each intersection the driver of the car can make either a left-hand turn or a right-hand turn or go straight ahead.

25. How many different routes are possible if two intersections are encountered?

26. How many different routes are possible if three intersections are encountered?

27. How many different routes are possible if ten intersections are encountered?

28. How many different routes are possible if n intersections are encountered?

29. License plates in several states consist of 6 characters, where a character is one of the 26 letters or one of the 10 digits. How many such license plates are possible?

30. License plates in Washington, D.C., consist of six digits. How many such license plates are possible if the first digit cannot be 0 and all six digits cannot be the same?

31. How many seven-digit telephone numbers can be formed if the first digit can be neither 0 nor 1?

32. How many seven-digit telephone numbers can be formed if the first three digits cannot be 800?

33. In how many ways can the letters of the name JACK be arranged?

34. In how many ways can the letters of the name VINCE be arranged?

35. Make a list of the different ways in which the letters of the name JILL can be arranged. How many ways are there?

36. Make a list of the different ways in which the letters of the name ANITA can be arranged. How many ways are there?

37. A contest has four cards with the letters A, B, C, D and four cards with the digits 6, 7, 8, 9. The object is to match one letter with one digit. How many different matchings are there altogether?

38. A contest has five cards with the letters A, B, C, D, E and five cards with the digits 5, 6, 7, 8, 9. The object is to match one letter with one digit. How many different matchings are there altogether?

39. How many different arrangements of four characters are there in which the first two characters are letters and the last two characters are digits?

40. How many different arrangements of four characters are there in which the first two characters are distinct letters and the last two characters are distinct digits?

41. Suppose that a hand consists of 2 cards drawn (without replacement) from an ordinary deck of 52 cards. How many hands can be drawn?

42. Suppose that a hand consists of 3 cards drawn (without replacement) from an ordinary deck of 52 cards. How many hands can be drawn?

●●● 43. How many different ways of choosing three ini-

tials are there that have two or three initials alike?

44. How many different ways of choosing three initials are there that have exactly two initials alike?

45. Over 1000 bombings take place in the United States each year. It is now possible for federal authorities to trace the explosives used in the bombs by identifying microscopic chips of layered paints called "taggants".* A typical identification taggant consists of 10 different colored layers. How many different taggants can be made from 10 different colors?

46. Maalox Plus is an antacid that is packaged with six tablets on a cardboard card.† Each tablet is white on one side and yellow on the other. Two possible arrangements on a card are

W	Y	W
W	Y	Y

and

Y	Y	W
W	Y	Y

How many different arrangements are possible on a card?

47. The IBM disk operating system for the PC-AT microcomputer makes use of *directories* to organize computer programs (called *files*) into groupings.‡ Draw a tree diagram which has one root directory that leads to four subdirectories (WP, DATA, SPREAD, and UTILITIES), where WP has four files (Electric, Volks, Star, and Perfect), DATA has three files (dBase, Reflex, and AskSam), SPREAD has three files (Lotus, VP, and VisiCalc), and UTILITIES has two files (Norton and SU).

*Richard D. Lyons, "Tagging Bombs, Trapping Bombers," *New York Times*, June 24, 1979.

†Bonnie Litwiller and David Duncan, "Maalox Lottery: A Novel Probability Problem," *The Mathematics Teacher*, September 1987, pp. 455–456.

‡IBM, *Reference Manual: Disk Operating System Version 3.20*, Boca Raton, FL, 1986.

6–4 PERMUTATIONS, COMBINATIONS, AND THE BINOMIAL THEOREM

Many useful counting techniques involve permutations and combinations. This section explains the meanings of these terms, develops rules for computing them, and applies them to the Binomial Theorem. Chapter 7 contains additional applications to probability problems.

Order

Suppose that a scholarship committee is going to award a total of $1200 to two students and that the committee has narrowed its list to five candidates. Suppose also that the committee is undecided whether to grant two $600 awards or to give one award of $800 and another award of $400. Examples 1 and 2 address these two cases.

EXAMPLE 1

Problem

In how many ways can the scholarship committee select two of the five students so that one student is awarded $800 and the other student $400?

SOLUTION Use the Multiplication Principle with two stages. The first stage has five choices for the $800 scholarship while the second stage has four choices for

Figure 1

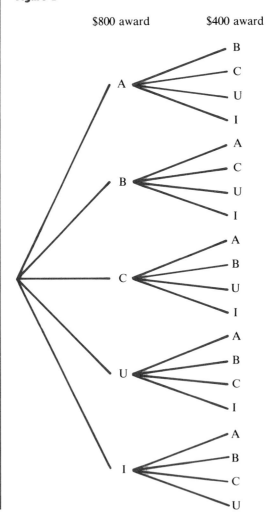

the $400 scholarship. No student can receive both scholarships. Figure 1 shows the $5 \cdot 4 = 20$ ways for students named A, B, C, U, and I. Note the difference between U getting $800 and I getting $400 versus I getting $800 and U getting $400.

EXAMPLE 2

Problem

In how many ways can the scholarship committee select two of the five students so that each one wins $600?

SOLUTION This problem cannot be solved by means of the Multiplication Principle alone. Certain pairs of students in Figure 1 reduce to one possibility here. For instance, the U-I and I-U pairs refer to separate awards, but here the order of the students is not relevant because each award is $600, so the two pairs reduce to the one set $\{U, I\}$. Originally Figure 1 displayed 20 pairs, with each pair of students listed twice. But when each pair is listed once, there are ten possibilities, as shown by the tree diagram in Figure 2.

The concept of order distinguishes Example 1 from Example 2. In Example 1 it mattered which name came first because one award was $800 and the other $400. In Example 2 the order of the names was immaterial because both awards were for $600. The following definition distinguishes these two concepts for objects that are selected without replacement:

_____ **Definitions** _____

A list of objects in a specific order is called a **permutation.** A list of objects in which the order is not considered is called a **combination.**

_____

Figure 2

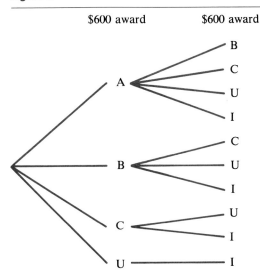

The 20 lists in Example 1 are permutations while the 10 lists in Example 2 are combinations.

Factorials

The expression $n!$ is read "n **factorial.**" It is defined by

$$n! = n(n - 1)(n - 2) \cdots 3 \cdot 2 \cdot 1$$

with $0! = 1$. Thus $5! = 120$ because

$$5! = 5 \cdot 4 \cdot 3 \cdot 2 \cdot 1$$

Many calculators have a key for computing factorials, so that pressing $\boxed{9}$ $\boxed{n!}$ yields 362,880. Factorials play a central role in permutations and combinations, so it is important to be able to evaluate them in various forms. For instance, $(7 - 2)! = 5! = 120$. Example 3 shows how to evaluate two other forms.

EXAMPLE 3 | **Problem**

Evaluate (a) $\dfrac{8!}{(8 - 5)!}$ (b) $\dfrac{8!}{(8 - 5)! \cdot 5!}$

SOLUTION Expand each factorial and cancel the common factors.

(a) $\dfrac{8!}{(8 - 5)!} = \dfrac{8!}{3!} = \dfrac{8 \cdot 7 \cdot 6 \cdot 5 \cdot 4 \cdot \cancel{3 \cdot 2 \cdot 1}}{\cancel{3 \cdot 2 \cdot 1}} = 8 \cdot 7 \cdot 6 \cdot 5 \cdot 4 = 6720$

(b) $\dfrac{8!}{(8 - 5)! \cdot 5!} = \dfrac{8!}{3! \cdot 5!} = \dfrac{8 \cdot 7 \cdot 6 \cdot \cancel{5 \cdot 4 \cdot 3 \cdot 2 \cdot 1}}{3 \cdot 2 \cdot 1 \cdot \cancel{5 \cdot 4 \cdot 3 \cdot 2 \cdot 1}} = \dfrac{8 \cdot 7 \cdot 6}{3 \cdot 2 \cdot 1} = 56$

Formulas for Permutations and Combinations

The notation $P(n, r)$ denotes the number of permutations of n objects taken r at a time. Thus $P(10, 4)$ is the number of permutations of ten objects taken four at a time. By the Multiplication Principle, $P(10, 4) = 10 \cdot 9 \cdot 8 \cdot 7$. This number can be written in terms of factorials.

$$P(10, 4) = 10 \cdot 9 \cdot 8 \cdot 7 = \dfrac{10!}{6!} = \dfrac{10!}{(10 - 4)!}$$

The final form is a specific case of the general formula for evaluating permutations. The derivation is similar.

$$P(n, r) = n \cdot (n - 1) \cdots (n - r + 1) = \dfrac{n!}{(n - r)!}$$

The number of ways of permuting n distinct objects taken r at a time is

$$P(n, r) = \dfrac{n!}{(n - r)!}$$

EXAMPLE 4 | **Problem**

How many permutations of three letters can be selected from the alphabet?

SOLUTION Since the alphabet has 26 letters, $n = 26$. Here $r = 3$. The number of permutations is

$$P(26, 3) = \frac{26!}{(26 - 3)!} = \frac{26!}{23!} = 26 \cdot 25 \cdot 24 = 15{,}600$$

A particular case of the formula for $P(n, r)$ occurs when $n = r$. Then

$$P(n, n) = \frac{n!}{(n - n)!} = \frac{n!}{0!} = \frac{n!}{1} = n!$$

Therefore the number of permutations of n objects is $n!$. Example 5 uses this fact. It also provides the reasoning for the way in which the formula for combinations is derived from the formula for permutations.

EXAMPLE 5 | **Problem**

(a) How many governing boards (president, vice-president, and secretary) can be formed from a law firm with eight partners?
(b) How many three-member investigative committees can be formed from a law firm with eight partners?

SOLUTION (a) The order in which partners are selected to the governing board is important because the three positions are distinguishable. Therefore the problem deals with permutations. There are eight partners so $n = 8$. There are three officers so $r = 3$. The number of governing boards is

$$P(8, 3) = \frac{8!}{(8 - 3)!} = 8 \cdot 7 \cdot 6 = 336$$

Thus 336 different governing boards can be formed.
 (b) In forming three-member committees the order of selection is not important because the positions are not distinguishable. The problem reduces to finding the number of combinations of eight partners taken three at a time. We derive this number from $P(8, 3)$.
 Consider a particular committee consisting of Adams (A), Buren (B), and Carter (C). The committee can be regarded as the set $\{A, B, C\}$. By the Multiplication Principle there are six different arrangements (permutations) of the partners in the set $\{A, B, C\}$.

$$A\text{-}B\text{-}C \qquad A\text{-}C\text{-}B \qquad B\text{-}A\text{-}C \qquad B\text{-}C\text{-}A \qquad C\text{-}A\text{-}B \qquad C\text{-}B\text{-}A$$

They are distinct arrangements; for instance, the arrangement $A\text{-}B\text{-}C$ (Adams is president, Buren vice-president, and Carter secretary) is different from $A\text{-}C\text{-}B$ (Adams is president, Carter vice-president, and Buren secretary). But as a three-member committee in which each member has equal standing, they reduce to the one set $\{A, B, C\}$ because, for instance, $\{A, B, C\} = \{A, C, B\}$. It follows that for every six permutations there is one combination, so the number of combinations is one-sixth the number of permutations.

$$\frac{P(8, 3)}{6} = \frac{336}{6} = 56$$

Altogether 56 committees of three members can be formed.

The notation $C(n, r)$ denotes the number of combinations of n objects taken r at a time. Another commonly used notation is $\binom{n}{r}$. Part (b) of Example 5 showed that $C(8, 3)$ was derived as follows:

$$C(8, 3) = \frac{P(8, 3)}{6} = \frac{P(8, 3)}{3!} = \frac{8!}{(8 - 3)! \cdot 3!}$$

To evaluate $C(n, r)$ for any n and r (where $r \leq n$) means to count the number of subsets containing r elements of a set with n elements. There are $P(n, r)$ arrangements that consist of r elements. However, because the order of the elements in a subset does not matter, each subset of r elements corresponds to $r!$ arrangements. Therefore the number of subsets is $P(n, r)$ divided by $r!$. This can be expressed in a more useful form as follows:

$$C(n, r) = \frac{P(n, r)}{r!} = \frac{P(n, r)}{1} \cdot \frac{1}{r!} = \frac{n!}{(n - r)!} \cdot \frac{1}{r!}$$

The number of combinations of n distinct objects taken r at a time is

$$C(n, r) = \frac{n!}{(n - r)! \cdot r!}$$

It is important to be able to distinguish between permutations and combinations because in many problems the essential ingredient is the proper identification of the concept being used. Keep in mind that a permutation is an "arrangement" that implies order while a combination is a "collection" without regard to order.

Example 6 applies the formulas for permutations and combinations to wagering. A "trifecta" refers to three objects that finish first, second, and third in a prescribed order. A "boxed trifecta" refers to three objects that finish first, second, and third in any order.

EXAMPLE 6 | **Problem**

How many trifecta and boxed trifecta tickets are possible in a horse race involving ten horses?

SOLUTION Since there are ten horses and three winners, $n = 10$ and $r = 3$. Trifecta tickets are permutations because the order must be exact. There are $P(10, 3) = 720$ trifecta tickets.

Boxed trifecta tickets are combinations because only the names of the three winners must be listed, not their order of finish. Thus there are $C(10, 3) = 120$ boxed trifecta tickets.

Example 7 combines the three counting techniques that were studied in the last two sections: the Multiplication Principle, permutations, and combinations.

EXAMPLE 7

A poker hand consists of 5 cards drawn from an ordinary deck of 52 cards.

Problem

How many poker hands consist of 3 aces and 2 kings?

SOLUTION The procedure will be to find the number of ways of selecting 3 aces, then to find the number of ways of selecting 2 kings. Each desired hand consists of 3 aces *and* 2 kings. By the Multiplication Principle the total number of such hands is obtained by multiplying these two numbers.

There are 4 aces, so there are $C(4, 3) = 4$ ways to select 3 aces from them. Similarly there are $C(4, 2) = 6$ ways to select 2 kings from the 4 kings. By the Multiplication Principle there are $6 \cdot 4 = 24$ hands consisting of 3 aces and 2 kings.

Binomial Theorem

Consider the expression $(x + y)^6$. It is called a "binomial form" because it contains two variables. The expression can be expanded by multiplying $(x + y)$ by itself six times.

$$(x + y)^6 = (x + y)(x + y)(x + y)(x + y)(x + y)(x + y)$$

This leads to an expression in the form

$$(x + y)^6 = x^6 + \_x^5y + \_x^4y^2 + \_x^3y^3 + \_x^2y^4 + \_xy^5 + y^6$$

The underlined spaces represent coefficients called **binomial coefficients,** which must be filled in with numbers.

Instead of multiplying the six factors one at a time to determine the binomial coefficients, it is instructive to compute one of them, say, $\_x^4y^2$. The term x^4y^2 arises every time the $(x + y)$ terms are multiplied to produce four x's and two y's. One such way is

$$(\underline{x} + y)(\underline{x} + y)(\underline{x} + y)(\underline{x} + y)(x + \underline{y})(x + \underline{y})$$

The appropriate variables have been underlined. Another way is

$$(\underline{x} + y)(\underline{x} + y)(\underline{x} + y)(x + \underline{y})(\underline{x} + y)(x + \underline{y})$$

The total number of ways to obtain four x's and two y's is equal to the number of ways of selecting four objects from a set of six objects. This is $C(6, 4)$. We use the conventional notation $\binom{6}{4}$ to denote this binomial coefficient. Therefore the appropriate term is $\binom{6}{4}x^4y^2$. Similarly, the next-to-last term in the expansion is $\binom{6}{1}x^1y^5$. The entire expansion can be written in the form

$$(x + y)^6 = \binom{6}{6}x^6 + \binom{6}{5}x^5y + \binom{6}{4}x^4y^2$$
$$+ \binom{6}{3}x^3y^3 + \binom{6}{2}x^2y^4 + \binom{6}{1}xy^5 + \binom{6}{0}y^6$$

Therefore

$$(x + y)^6 = x^6 + 6x^5y + 15x^4y^2 + 20x^3y^3 + 15x^2y^4 + 6xy^5 + y^6$$

The reasoning used to derive the binomial coefficients $\binom{n}{r}$ will be used again when we discuss binomial trials in the chapters on probability and statistics. For now we state the general result, called the **Binomial Theorem.**

Binomial Theorem

$$(x + y)^n = \binom{n}{n}x^n + \binom{n}{n-1}x^{n-1}y + \binom{n}{n-2}x^{n-2}y^2$$
$$+ \cdots + \binom{n}{2}x^2y^{n-2} + \binom{n}{1}xy^{n-1} + \binom{n}{0}y^n$$

EXAMPLE 8

Problem

What is the coefficient of the x^2y^6 term in the expansion of $(x + y)^8$?

SOLUTION The coefficient is $\binom{8}{2}$. The top number 8 comes from the exponent of $(x + y)^8$. The bottom number 2 comes from the exponent of x in x^2y^6. Since $\binom{8}{2} = \dfrac{8!}{2! \cdot 6!} = \dfrac{8 \cdot 7}{2} = 28$, the coefficient is 28.

The next example is central to the proof that a set with n elements has 2^n subsets. (See Exercise 45.)

EXAMPLE 9 | **Problem**

Evaluate

$$\binom{5}{5} + \binom{5}{4} + \binom{5}{3} + \binom{5}{2} + \binom{5}{1} + \binom{5}{0}$$

SOLUTION One approach is to evaluate each binomial coefficient and then add them. Alternately, by the binomial theorem

$$(x + y)^5 = \binom{5}{5}x^5 + \binom{5}{4}x^4y + \binom{5}{3}x^3y^2$$
$$+ \binom{5}{2}x^2y^3 + \binom{5}{1}xy^4 + \binom{5}{0}y^5$$

Let $x = 1$ and $y = 1$. Then the right-hand side of the equation is equal to the desired expression. The left-hand side equals $(1 + 1)^5 = 32$. Therefore

$$\binom{5}{5} + \binom{5}{4} + \binom{5}{3} + \binom{5}{2} + \binom{5}{1} + \binom{5}{0} = 32$$

EXERCISE SET 6–4

• In Problems 1 to 8 evaluate the given expression.

1. (a) $P(5, 2)$
   (b) $C(5, 2)$

2. (a) $P(5, 3)$
   (b) $C(5, 3)$

3. (a) $P(6, 2)$
   (b) $C(10, 2)$

4. (a) $P(10, 5)$
   (b) $C(10, 7)$

5. (a) $P(5, 0)$
   (b) $C(5, 0)$

6. (a) $P(5, 5)$
   (b) $C(5, 5)$

7. (a) $0!$
   (b) $(8 - 3)!$

8. (a) $1!$
   (b) $52!/(50!\cdot2!)$

In Problems 9 to 24 determine whether the solution involves the Multiplication Principle, permutations, or combinations, then use the appropriate formula to answer the question.

9. In how many ways can a club with ten members elect a president and a vice-president?

10. In how many ways can four people line up in a bakery?

11. If the call letters of a radio station must begin with the letter W, how many different stations could be designated by using four letters, such as WIND, without using repetitions?

12. The Coast Guard uses flags raised on a flagpole to signal ships concerning matters such as weather conditions. If three different flags are available, how many signals can be sent using all three flags if the order of the arrangement is important?

13. The number of signals available in Problem 12 can be increased by using signals of two or three flags. How many such signals are possible if the order of the flags is important?

14. In how many ways can ten volumes of a set of encyclopedias be arranged on a shelf?

15. In how many ways can the gold, silver, and bronze medals be awarded if ten women compete in a marathon?

16. In how many ways can four students be given assigned seats in a classroom with ten seats?

17. An agricultural expert has designed an experiment to test the effects of two types of insecticide and four types of fertilizer on six different types of crops. Each plot of land uses one insecticide, one fertilizer, and one crop. How many plots are needed?

18. A veterinarian wishes to test three different levels of dosage of a vaccine on four types of animals at two different age levels of the animals. How many classes of vaccinations are possible?

19. There are 20 candidates for an all-star basketball game and 5 will be selected for the first team. How many different first teams can be chosen?

20. Twelve people work as programmers in a computer center, with teams of three serving as consultants. In how many ways can teams be chosen?

21. Seven astronauts have reached the final training level for a particular space flight. (a) How many two-person crews can be chosen from the seven? (b) How many two-person crews will have one person designated captain and the other designated copilot?

22. How many 2-card hands can be dealt from an ordinary deck of 52 cards?

23. How many poker hands consist of three queens and two jacks?

24. How many poker hands consist of four queens and one jack?

25. What is the coefficient of the x^5y^3 term in the expansion of $(x + y)^8$?

26. What is the coefficient of the x^4y^4 term in the expansion of $(x + y)^8$?

27. Evaluate the expression

$$\binom{6}{6} + \binom{6}{5} + \binom{6}{4} + \binom{6}{3}$$
$$+ \binom{6}{2} + \binom{6}{1} + \binom{6}{0}$$

28. Evaluate the expression

$$\binom{4}{4} + \binom{4}{3} + \binom{4}{2} + \binom{4}{1} + \binom{4}{0}$$

•• 29. A basketball team has 5 starters selected from the 12 players. (a) How many starting teams are possible? (b) How many starting teams are possible if two of the players' contracts specify that they must start every game?

30. (a) How many poker hands can be drawn?
    (b) How many poker hands contain all four aces?

31. In how many ways can three initials be chosen so that two initials are the same but not all three?

32. If a penny, nickel, and dime are tossed in a fountain, in how many ways can they land so that two of them are heads?

33. How many poker hands consist of three red cards and two black cards?

34. How many poker hands consist of three clubs and two spades?

35. What is the coefficient of the x^7y^3 term in the expansion of $(x + y)^{10}$?

36. What is the coefficient of the x^3y^7 term in the expansion of $(x + y)^{10}$?

37. How many three-digit numbers can be chosen if repetition is allowed? (For example, 989 is permitted.)

38. How many three-digit numbers can be chosen if no repetitions are permitted?

39. How many boxed three-digit numbers can be chosen?

40. How many three-digit numbers have exactly two of the digits the same?

••• 41. In wagering, an "exacta" refers to two objects that finish first and second in a specified order. If a dog race contains nine greyhounds, how many exacta tickets can be purchased?

42. Many harness races consist of exactly nine horses. (a) How many different trifecta tickets can be purchased? (b) How many different boxed trifecta tickets can be purchased?

43. (a) Why is it true that $\binom{14}{5} = \binom{14}{9}$?

    (b) Why is it true that $\binom{n}{r} = \binom{n}{n-r}$?

44. Suppose a set S has 14 elements. Prove that the number of subsets of S with 5 elements is equal to the number of subsets of S with 9 elements.

45. (a) Use Example 9 and the Binomial Theorem to prove that a set with 5 elements has 2^5

subsets, including the empty set and the set itself.

(b) Use the Binomial Theorem to prove that a set with n elements has 2^n subsets.

46. An urn contains 26 red balls and 26 black balls. A "hand" consists of 5 balls selected from the urn. How many hands consist of 3 red balls and 2 black balls?

47. Use the Binomial Theorem to evaluate the expression

$$\left(\frac{1}{2}\right)^6 \left[\binom{6}{6} + \binom{6}{5} + \binom{6}{4} + \cdots + \binom{6}{0}\right]$$

48. Use the Binomial Theorem to evaluate the expression

$$\left(\frac{1}{3}\right)^4 + \binom{4}{3}\left(\frac{1}{3}\right)^3\left(\frac{2}{3}\right) + \binom{4}{2}\left(\frac{1}{3}\right)^2\left(\frac{2}{3}\right)^2$$
$$+ \binom{4}{1}\left(\frac{1}{3}\right)\left(\frac{2}{3}\right)^3 + \left(\frac{2}{3}\right)^4$$

49. (Pigs-get-fat-hogs-get-slaughtered department) It has been conjectured that the Pennsylvania lottery scam team could have milked the state for much more money by placing about one-half the number of bets and by repeating the procedure often. Why?

50. In music theory, chords and lines are basic terms.* A chord refers to a subset of the white keys on a piano, while a line is a sequence of distinct white keys. (a) Which concept refers to a permutation? (b) Which concept refers to a combination?

51. Campanology is the art of ringing bells.† Since modern bells often weigh more than a ton, it takes about two seconds to advance or retard each bell. Consequently a "ring" of bells is an orderly sequence of permutations on the bells instead of a melody. An *extent* is the maximum number of rings, so that the extent of 6 bells (called a Minor) is 6! = 720. What is the extent of a Major, which consists of 8 bells? (Plain Bob Major, which was played in 1963 and 1977, required over 18 hours to ring!)

Problems 52 and 53 require a knowledge of the programming language BASIC. They are based on an article that describes how computer programs can be used to introduce permutations and combinations.‡

52. What is the output of this program?

```
10 FOR I = 1 TO 4
20     FOR J = 1 TO 4
30         IF I = J THEN 50
40         PRINT I;J
50     NEXT J
60 NEXT I
```

53. What is the output of the program in Problem 52 if the equality sign in line 30 is replaced by $\geq$?

*John Clough and Gerald Myerson, "Musical Scales and the Generalized Circle of Fifths," *American Mathematical Monthly*, November 1986, pp. 695–701.

†Arthur T. White, "Ringing the Cosets," *American Mathematical Monthly*, October 1987, pp. 721–746.

‡Clark Kimberling, "Permutations and Combinations," *The Mathematics Teacher*, May 1987, pp. 403–407.

CASE STUDY   License Plates*

Figure 1 shows the license plates for passenger vehicles in each of the 50 states, the territories, and the Armed Forces of the United States.† Most of the states have adopted the *standard format,* which consists of three letters followed by

*Adapted from Edwin G. Landauer, "Counting Using License Plates and Phone Numbers: A Familiar Experience," *The Mathematics Teacher*, March 1984, pp. 183–187, 234.

†We are grateful to Carolyn Edwards and David R. McElhaney of the Office of Highway Information Management for supplying Figure 1 and other relevant information.

(a)

Figure 1

(b)

three digits. But some states, such as New Jersey, have adopted the *reverse format* as well. It consists of three digits followed by three letters.

There are about 20 million licensed vehicles in California, more than in any other state. The Golden State decided not only to use the standard format and the reverse format for its license plates, but also to use a third format. It consists of a single digit followed by three letters and three digits; we will call it the *7-Cal format*.

How many usable license plates do these three formats allow? We will answer this question by defining the term "usable license plate" and by applying the counting techniques we have developed in this chapter.

Usable license plates   A convenient way to view the standard format is *LLLDDD*, where *L* stands for the 26 letters and *D* for the 10 digits. By the Multiplication Principle there are 17,576,000 possible license plates in standard format.

$$\underline{\quad 26 \quad}_{L} \underline{\quad 26 \quad}_{L} \underline{\quad 26 \quad}_{L} \underline{\quad 10 \quad}_{D} \underline{\quad 10 \quad}_{D} \underline{\quad 10 \quad}_{D}$$

This is the *theoretical number* of license plates because not all license plates are usable. For instance, JKI123 is not usable because the third letter can easily be confused with the first digit. This is an example of a license plate being restricted for reasons of "clarity." Other license plates are reserved for disabled persons and disabled veterans. Two examples are DPW123 and VET789.

The set of *usable license plates* refers to those license plates that are available to the general public. This set excludes license plates that are restricted for clarity or reserved for disabled persons and disabled veterans. Example 1 counts the number of usable license plates in standard format.

EXAMPLE 1   The number of theoretical license plates in standard format *LLLDDD* is 17,576,000. The number of usable license plates is $17,576,000 - n(C \cup D)$ where

   C = license plates restricted for clarity
   D = license plates reserved for the disabled

The clarity rule is as follows:

1. The letters I, O, Q, and Z are not allowed in the third-letter position.

   Since four letters cannot be used in the third position

   $n(C) = 26 \cdot 26 \cdot 4 \cdot 10 \cdot 10 \cdot 10 = 2,704,000$

   There are two rules for reserved license plates.
2. The combinations DPW, DPX, DPY, and DPZ are designated for vehicles of disabled persons.
3. The combinations VET, VTN, and VTR are designated for vehicles of disabled veterans.

By the Multiplication Principle there are 1000 license plates of the form DPW_ _ _ because each blank can have ten possible digits. Similarly, each combination DPX, DPY, and DPZ eliminates 1000 license plates, making 4000 license plates that are reserved for disabled persons by rule 2. Three such combinations are also reserved for disabled veterans by rule 3, so $n(D) = 7000$.

It follows from rules 1 and 2 that a license plate lies in both C and D only if the letters are DPZ. There are 1000 license plates with the letters DPZ, so

$$n(C \cap D) = 1000$$

By the Inclusion-Exclusion Principle

$$n(C \cup D) = 2{,}704{,}000 + 7000 - 1000 = 2{,}710{,}000$$

Thus the number of usable license plates is

$$17{,}576{,}000 - 2{,}710{,}000 = 14{,}866{,}000$$

The theoretical number of license plates in reverse format is the same as the theoretical number in standard format. But the number of restrictions is different. Exercises 1 to 3 count the number of usable license plates in reverse format.

The 7-Cal format   The 7-Cal format consists of a digit from 1 to 9 followed by the standard format. It can be written *ELLLDDD*, where *E* represents one of the digits from 1 to 9. The theoretical number of license plates in 7-Cal format is

$$9{\cdot}26{\cdot}26{\cdot}26{\cdot}10{\cdot}10{\cdot}10 = 158{,}184{,}000$$

This number is roughly three-fourths of the population of the entire United States! It stands to reason that the 7-Cal format can accommodate all of the licensed passenger vehicles in the state of California.

However, the 7-Cal format has four restrictions.

1. The letters I and O are not allowed in the first-letter position.
2. The letters I, O, Q, and Z are not allowed in the third-letter position.
3. The three-character combination 1DP is reserved for vehicles of disabled persons.
4. The three-character combination 1VT and the four-character combination 1VET are reserved for vehicles of disabled veterans.

Example 2 counts the number of usable license plates in the 7-Cal format by an approach that is quite different from the approach in Example 1.

EXAMPLE 2   By the Multiplication Principle the number of usable license plates in the 7-Cal format *ELLLDDD* is

$$9{\cdot}24{\cdot}26{\cdot}22{\cdot}10{\cdot}10{\cdot}10 = 123{,}552{,}000$$

The number 24 comes from rule 1 and the number 22 from rule 2. This total includes the restrictions for clarity but not the reservations for disabled persons.

By rule 3 the characters 1DP are followed by one letter and three digits. The letter cannot be Z because of rule 2, so the number of license plates reserved for disabled persons is

TABLE 1

TABLE 1

Format	Number of Usable License Plates
Standard	14,866,000
Reverse	16,208,000
7-Cal	123,501,000
Total	154,575,000

$$25 \cdot 10 \cdot 10 \cdot 10 = 25,000$$

Next consider rule 4. There are 25,000 license plates beginning with 1VT. Also 1000 license plates begin with 1VET. Thus 26,000 license plates are reserved for disabled veterans. The sets of license plates that are reserved for disabled persons and reserved for disabled veterans are disjoint. Hence the number of reserved license plates is $25,000 + 26,000 = 51,000$.

Altogether there are $123,552,000 - 51,000 = 123,501,000$ usable license plates in the 7-Cal format.

Table 1 summarizes the number of usable license plates in California.

Future needs   More than 31 million usable license plates are available in either the standard or the reverse format. Yet in 1980 California, with less than 18 million licensed vehicles, began issuing license plates in the 7-Cal format. Why did the Golden State adopt a third format?

For two reasons: First, California does not reissue regular license plates, except for vanity plates. This reason alone creates the need for a new format, but it is indigenous to California. Second, all states have lists of unacceptable combinations of letters and numbers. Most of them are obscene, so we will not list them here (for reasons of taste as well as litigation).

Will California ever have the need for a fourth format? We doubt it. After all the population of the state is approximately 25 million, so the 150 million usable license plates allow six vehicles per resident. However, if its population should happen to increase dramatically, or its residents take up the hobby of collecting cars, California can always adopt a new format of a single letter followed by the reverse format. Exercise 4 discusses this possibility.

What format does your state use? How many usable license plates are available? The exercises examine Louisiana, Montana, and California.

Case Study Exercises

1. For reasons of clarity the letters I and O are not allowed in the first-letter position of license plates in the reverse format *DDDLLL*. How many license plates does this restriction eliminate?

2. Two sets of three-letter combinations are reserved for license plates in the reverse format *DDDLLL*. Those designated for vehicles of disabled persons are DPW, DPX, DPY, DPZ, RDP,

SDP, TDP, UDP, VDP, WDP, XDP, YDP, and ZDP. Those designated for vehicles of disabled veterans are VET, VTN, and VTR. How many license plates do these reservations eliminate?

3. How many usable license plates are in the reverse format *DDDLLL?* (You must make use of the answers to Problems 1 and 2.)

4. How many usable license plates consist of a letter followed by three digits and three letters, subject to the restrictions stated in Problems 1 and 2?

5. License plates in Louisiana consist of three digits, a letter, and three digits. (See Figure 1.) The letter designates the state police troop area that issued the plate (A, B, C, D, E, F, G, H, I, K, and L). The letter X means that the plate was issued by mail. How many usable license plates does this format allow?

6. Can you think of any restrictions in Louisiana's format that are due to reasons of clarity?

7. Louisiana currently has about 2,700,000 licensed vehicles. Will a new format be needed soon?

8. Montana's license plate format consists of one or two digits, followed by a dot, and then up to five more digits. (See Figure 1.) The first group of digits runs from 1 to 56; each one designates the county in which the owner resides. The second group runs from 1 to the number of cars licensed in that county. How many usable license plates does this format allow?

CHAPTER REVIEW

Key Terms

6–1 Sets

Set	Complement
Standard Format	Universal Set
Element	Intersection
Natural Number	Union
Integer	Irrational Number
Rational Number	Real Number
Subset	Empty (Null) Set
Venn Diagram	Disjoint Sets
Equal Sets	

6–2 Venn Diagrams and Counting
"Count"

6–3 Tree Diagrams and the Multiplication Principle

Tree Diagram	Selection With Replacement
Selection Without Replacement	The Multiplication Principle

6–4   Permutations, Combinations, and the Binomial Theorem

Permutations   Binomial Coefficients
Combinations   Binomial Theorem
Factorial

Summary of Important Concepts

Notation

$\{ \quad \}$ $\not\subset$
$\{ \quad : \quad \}$ $\varnothing$
$\in$ $n(X)$
$\notin$ $n!$
$\subset$

Formulas

Inclusion-Exclusion Formula

$$n(A \cup B) = n(A) + n(B) - n(A \cap B)$$

Multiplication Principle

$$n_1 \cdot n_2 \cdots n_r$$

Permutations

$$P(n, r) = \frac{n!}{(n - r)!}$$

Combinations

$$C(n, r) = \frac{n!}{(n - r)!r!}$$

REVIEW PROBLEMS

1. Which of these collections is well defined?
   (a) nice numbers.
   (b) U.S. cities whose population in the 1980 census exceeded 100,000.

In Problems 2 and 3 let $U = \{a, b, c, \ldots, z\}$, $V = \{a, e, i, o, u\}$, and $A = \{a, b, c, d, e\}$.

2. Find (a) $V \cup A$ (b) $V \cap A$.

3. Find (a) A' (b) $V' \cap A'$ (c) $(V \cup A)'$.

4. Determine whether A or B is a subset of the set of rational numbers Q where $A = \{x \in Z : x^2 = -1\}$ and $B = \{x \in Z : x^2 > 9\}$.

5. Find $n(A \cup B)$ where $n(A) = 9$, $n(B) = 5$, and $n(A \cap B) = 4$.

6. Find $n(A \cup B)$ where $n(A \cap B') = 9$, $n(A' \cap B) = 5$, and $n(A \cap B) = 4$.

7. What is X' if X is the set of rational numbers and the universal set is the set of real numbers?

8. The 30,000 members of the MAA can subscribe to three journals: *Monthly* (M), *Math Magazine* (G), and *College Math Journal* (C). In 1988 the numbers of subscribers to the journals (in thousands) were $n(M \cap G \cap C) = 1$,

$n(M \cap G) = 5$, $n(M \cap C) = 7$, $n(G \cap C) = 1$, $n(M) = 18$, $n(G) = 7$, and $n(C) = 11$. How many members subscribed to
(a) Both *Math Magazine* and the *College Math Journal?*
(b) The *College Math Journal* only?
(c) None of the three journals?

9. Refer to the figure to determine the number of elements in each set.
(a) $A \cap B$ (b) $A \cup B$ (c) A' (d) U

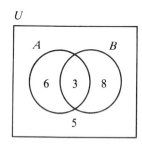

10. In how many different ways can four people be seated in four seats in a row?

11. In the game of "Clue" a suggestion consists of a suspect, a weapon, and a room. There are 6 suspects, 6 weapons, and 9 rooms. How many suggestions are possible?

12. How many arrangements of four letters begin with one consonant, end with a different consonant, and have two distinct vowels in between?

13. Four people enter the elevator in a building on the ground floor. The elevator can stop at floors 1 through 12.
(a) In how many different ways can the four people exit?
(b) In how many different ways can the four people exit if no two exit on the same floor?

14. Each ticket in the game of "Lotto" has six different numbers punched from the numbers 1 to 40. How many different tickets are possible?

15. In how many different ways can the letters of the word MAJORITY be arranged?

In Problems 16 and 17 a tournament with four prizes starts with 64 teams. Each losing team is eliminated and a winning team continues to play until it loses a game.

16. How many orderings of the first through the fourth place teams are possible?

17. How many combinations of teams can make the final four?

18. The final four teams in the 1988 NCAA basketball tournament were Duke, Oklahoma, Arizona, and Kansas. In how many different ways could these four teams have finished first through fourth?

19. Twenty dancers audition for roles in *A Chorus Line*.
(a) How many different selections of 3 of them can be made for the first, second, and third leading roles?
(b) How many different selections of 3 of the remaining 17 dancers can be made for 3 indistinguishable understudy roles?

20. Two dice (one red and the other green) are rolled and their sum is recorded. In how many different ways can the dice land so that the sum is equal to 10?

21. An urn contains 26 balls, each with a letter of the alphabet printed on it. How many three-letter "words" can be formed by drawing 3 balls from the urn without replacement?

22. What is the coefficient of the $x^2 y^5$ term in the expansion of $(x + y)^7$?

7 Probability

CHAPTER OVERVIEW In many aspects of mathematics a formula or process is used to predict the future. Sometimes the prediction is exact, but often events take place randomly and it is not possible to predict exactly what will happen. A model must be constructed that takes this random nature of outcomes into account. For example, a meteorologist cannot tell precisely when, or even if, it will rain, but instead, assigns probabilities to the possible types of weather to expect. Sections 7–1 and 7–2 show how to construct probability models and also illustrate their fundamental properties. The later sections cover more specific topics that govern when we are able to add, subtract, multiply, and divide probabilities. We also calculate probabilities using the counting techniques of Chapter 6.

Throughout the chapter games of chance are used as examples because they are easy to explain and understand. We also mention several significant applications of probability theory in the real world. In fact the first example is from a study of automobile accidents on a major Boston thoroughfare that used probability theory in its analysis. The last example, the case study, gives a more thorough description of an application of probability.

CASE STUDY PREVIEW

When Jane Blalock, star professional golfer, was banned from participating in tournaments by the Ladies' Professional Golf Association, she sued the organization under the federal antitrust laws. The court reinstated her but the decision came too late for her to participate in the tournament. She then sued for damages. But how much money would she have earned? She might have won the tournament; however, she just as well might have finished out of the money. Her court battle was based on probability theory. The case study gives the surprising result of her suit.

7–1 SAMPLE SPACES

Many events in the real world seem to happen by chance. By studying the behavior of some of these seemingly unpredictable events, we can often determine patterns. The theory of probability is the branch of mathematics that helps find these patterns. In order to describe the basic ideas of probability, let us look at a concrete example of such a study.

Introduction: An Example

In the early 1970s the Massachusetts Department of Public Works conducted a study of speed limits in an effort to improve motor vehicle safety. Highway accident data were analyzed on a suburban stretch of Route 2, a major road entering Boston, Massachusetts, from the west. In an article about the study William B. Fairley raised the critical question, What season of the year had the most accidents? The year was divided into four quarters, the first being January to March, then April to June, and so forth. Data were gathered for each quarter from 1970 to 1972. Table 1 gives the quarterly count, called the *frequency,* and the average for each quarter. For example, during the three years of the study there was a total of 302 accidents in the first quarter: 138 in 1970, 74 in 1971, and 90 in 1972. The averages are computed by dividing each quarterly count by the total number of accidents in all quarters, 1269. Thus the quarterly average for quarter 1 is $302/1269 = 0.24$ rounded to two decimal places.

By looking at the averages, we can easily determine that not only were there more accidents in the fourth quarter, but there were almost twice as many in quarter 4 as in quarter 2. The percentages in the "Quarterly Average" column are called **relative frequencies.** They arise from the fraction, m/N, where m is the number of occurrences of accidents in a quarter and N is the total number of accidents. If

TABLE 1   Accidents on Route 2, 1970–1972

Quarter	Quarterly Count	Quarterly Average
1	302	0.24
2	244	0.19
3	283	0.22
4	440	0.35
Total	1269	

Source: William B. Fairley, "Accidents on Route 2: Two-Way Structures for Data," *Statistics and Public Policy,* by Fairley and Mosteller, eds., Reading, MA, Addison-Wesley, 1977, pp. 23–50.

we select an accident at random from the total of 1269, we say that the **probability** that it occurred in quarter 4 is .35.

To say that the probability of this event is .35 means that since it has occurred 35% of the time in the past, it is reasonable to predict that it will happen 35% of the time in the future. In this way probability is used to predict future trends. That is, under the same conditions 35% of the accidents on Route 2 will occur in quarter 4.

The phrase ''under the same conditions'' spells the difference between studies of real-world phenomena and games of chance, such as tossing coins or rolling dice, or playing cards or roulette. Games are repeated exactly, but it is much more difficult to repeat real-world experiments under the same conditions. Therefore games are used to illustrate the concepts of probability because they are familiar and easy to demonstrate, and on each play they are repeated identically. In this chapter we refer to specific experiments using simple descriptive models. In the next chapter we will study experiments in statistics that involve data analysis.

Sample Spaces

The basic idea in probability is to study outcomes of experiments. By an **experiment** we mean an action whose possible outcomes can be determined and recorded. For example, four simple experiments are

Experiment *A*: Toss a coin and observe heads or tails.
Experiment *B*: Roll a die and record the number on the top face.
Experiment *C*: The tie-breaking number in a lottery is chosen by selecting one ball from an urn containing ten balls numbered 0, 1, . . . , 9.
Experiment *D*: In the Fairley study of accidents on Route 2 a researcher chooses an accident at random and records the quarter in which it occurs.

Each time an experiment is repeated an outcome is observed. The set of possible outcomes is called the **sample space** for the experiment. For instance, Experiment *A* has two outcomes: heads and tails. If we let ''*H*'' represent heads and ''*T*'' represent tails, then $S = \{H, T\}$ is the sample space for Experiment *A*.

EXAMPLE 1

Problem

Find the sample space for Experiments *B*, *C*, and *D*.

SOLUTION Experiment *B* has six possible outcomes: 1, 2, 3, 4, 5, 6. Hence the sample space for Experiment *B* is

$$S = \{1, 2, 3, 4, 5, 6\}$$

Experiment *C* has ten outcomes. Its sample space is

$$S = \{0, 1, 2, 3, 4, 5, 6, 7, 8, 9\}$$

Experiment *D* has four outcomes. Its sample space is

$$S = \{\text{quarter } 1, \text{ quarter } 2, \text{ quarter } 3, \text{ quarter } 4\}$$

_____ **Definition** _____

The **sample space** of an experiment is the set of all possible outcomes of the experiment. Each time the experiment is conducted exactly one outcome is recorded.

_____

Thus an experiment is described by its sample space. The next two examples illustrate how to determine the sample space. An experiment can be viewed as consisting of two steps, an action followed by recording something about the action. The possible quantities that can be recorded are the outcomes. The set of outcomes is the sample space. In Experiment D the action is to select an accident at random, and then the quarter in which it occurred is recorded. Example 2 shows that even if the same action is performed, a new experiment is defined if the quantity recorded is changed.

EXAMPLE 2

Problem

An experiment consists of selecting at random an accident from the Fairley study on Route 2 and recording the month in which it occurred. What is the sample space of the experiment?

SOLUTION The outcomes of the experiment are the 12 months of the year. Thus the sample space S is

S = {Jan., Feb., Mar., Apr., May, Jun., Jul., Aug., Sep., Oct., Nov., Dec.}

The next example shows that the same action can generate two different experiments by recording different outcomes, thus producing two separate sample spaces.

EXAMPLE 3

Toss two coins. Record (a) the sequence of H's and T's, and (b) the number of H's.

Problem

Find the sample space for each experiment.

SOLUTION (a) There are four outcomes for the experiment: HH, HT, TH, and TT. Thus the sample space is

S = {HH, HT, TH, TT}

(b) There are three outcomes: 0, 1, and 2. Hence the sample space is

S = {0, 1, 2}

The significance of the distinction between the two experiments in Example 3 will be explained in Section 7–2.

Events

Often we are interested in looking at the possibility of more than one outcome being observed. For example, in Experiment D suppose we wanted to determine how many accidents occurred in the first six months of the year. The corresponding outcomes would be quarter 1 and quarter 2. In mathematical language this is the set {quarter 1, quarter 2}, which is a subset of the sample space. In general a subset of the sample space is called an **event.** The next two examples describe events associated with Experiments B and C.

EXAMPLE 4

Problem

Describe in terms of sets the two events E and F associated with Experiment B, where

E: An odd number occurs.
F: A number less than 5 occurs.

SOLUTION Event E consists of the three outcomes, 1, 3, and 5. Thus

$$E = \{1, 3, 5\}$$

Event F consists of the outcomes 1, 2, 3, and 4. Thus

$$F = \{1, 2, 3, 4\}$$

EXAMPLE 5

Problem

Describe in terms of sets the three events E, F, and G that are associated with Experiment C whose sample space is {0, 1, 2, 3, 4, 5, 6, 7, 8, 9}.

E: An odd number occurs.
F: A number less than 5 occurs.
G: Either an odd number or a number less than 5 occurs.

SOLUTION Event E consists of the five outcomes: 1, 3, 5, 7, and 9. Thus

$$E = \{1, 3, 5, 7, 9\}$$

Event F consists of the outcomes 0, 1, 2, 3, and 4. Thus

$$F = \{0, 1, 2, 3, 4\}$$

Event G consists of the odd numbers 1, 3, 5, 7, and 9, together with the numbers less than 5, which are 0, 1, 2, 3, and 4. Notice that 1 and 3 are mentioned twice, but it makes sense to write them only once in the set G. Thus

$$G = \{0, 1, 2, 3, 4, 5, 7, 9\}$$

_____ **Definition** _____

An **event** of an experiment is a subset of the sample space. The event E *occurs* if the outcome of the experiment is an element of the subset E.

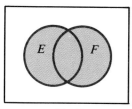

$E \cup F$

FIGURE 1

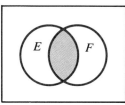

$E \cap F$

FIGURE 2

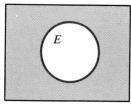

E'

FIGURE 3

Since events are defined in terms of sets, we can use set theory to define new events from other events. For instance, in Example 5, event G is the union of E and F, because $E \cup F$ occurs when E or F (or both) occurs. Thus $G = E \cup F$. We can also define the event $E \cap F$, which is the event that occurs when both E and F occur at the same time. In Example 5, $E \cap F = \{1, 3\}$.

Often in an experiment an event is defined by a description rather than a list of its elements. In this case the word "or" generally represents the union of two events while "and" refers to the intersection. For example, consider the experiment to roll a die and observe the number on the top face. Let A be "observe an odd number" and B be "observe a prime number." Then the event "observe a number that is odd and a prime" is the intersection of the two events, $A = \{1, 3, 5\}$ and $B = \{2, 3, 5\}$. It is $A \cap B = \{3, 5\}$. The event "observe a number that is odd or a prime" is the union of A and B. It is $A \cup B = \{1, 2, 3, 5\}$.

Figures 1 and 2 give the Venn diagrams for $E \cup F$ and $E \cap F$.

In a probability experiment the universal set is the entire sample space. In Experiment C, whose sample space is given in Example 5, consider the event "a number not less than 5 occurs." This set is the **complement** of F.

$$F' = \{5, 6, 7, 8, 9\}$$

Figure 3 gives the Venn diagram for the complement, E', of an event E. Thus the three set operations of union, intersection, and complement allow us to build new events from given events.

_____ **Definition** _____

If E and F are two events of an experiment, then

1. $E \cup F$ occurs when E or F, or both, occur.
2. $E \cap F$ occurs when both E and F occur.
3. E' occurs when E does not occur.

_____

Two events that cannot both occur at the same time are called **mutually exclusive events.** For example, when a die is tossed, the events "an even number" and "an odd number" cannot both occur, so they are mutually exclusive. In terms of sets, two events E and F are mutually exclusive if $E \cap F = \varnothing$. Figure 4 gives the Venn diagram for mutually exclusive events. Let us look at two examples to illustrate these points.

FIGURE 4

_____

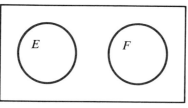

$E \cap F = \varnothing$

EXAMPLE 6

Let the sample space for an experiment be $S = \{a, b, c, d, e\}$. Define the events

$$E = \{a, b\} \quad F = \{b, c, d\} \quad G = \{a, b, c, e\}$$

Problem

Find (a) $E \cup F$, (b) $E \cap F$, (c) G', (d) $E \cap G'$.

SOLUTION
(a) $E \cup F = \{a, b, c, d\}$
(b) $E \cap F = \{b\}$
(c) $G' = \{d\}$
(d) $E \cap G' = \varnothing$, so E and G' are mutually exclusive.

In the next example we use the notation $n(A)$ to represent the number of elements in the set A. This notation was introduced in Chapter 6.

EXAMPLE 7

The second part of the Fairley study of accidents on Route 2 centered on the incidence of accidents per month to see if there was a finer trend to be discovered. The data are as follows:

Month	Jan.	Feb.	Mar.	Apr.	May	Jun.
Number of accidents	120	85	97	81	88	75

Month	Jul.	Aug.	Sep.	Oct.	Nov.	Dec.
Number of accidents	95	84	104	149	143	148

The experiment is to select an accident at random and to record its month. Define the events

   J: The accident occurred in January.
   F: The accident occurred in February.
   A: The accident occurred in quarter 2 (Apr., May, Jun.).
   S: The accident occurred in Jun., Jul., or Aug.

Problem

Determine the sample space and find how many accidents are associated with each of the following events: (a) $J \cap F$, (b) $J \cup F$, (c) $A \cap S$, (d) J'.

SOLUTION The sample space consists of the 12 months, so it is

   $S = \{\text{Jan., Feb., Mar., Apr., May, Jun., Jul., Aug., Sep., Oct., Nov., Dec.}\}$

(a) J and F are mutually exclusive since $J \cap F = \varnothing$ so $n(J \cap F) = 0$.
(b) There were 120 accidents in January and 85 in February, so $n(J) = 120$ and $n(F) = 85$. Since $J \cap F = \varnothing$, $n(J \cup F) = 120 + 85 = 205$.
(c) $A \cap S = \{\text{Jun.}\}$, so $n(A \cap S) = n(\text{Jun.}) = 75$.
(d) There were 1269 accidents altogether and 120 occurred in January, so $n(J') = 1269 - 120 = 1149$.

EXERCISE SET 7–1

• In Problems 1 to 14 find the sample space of the experiment.

1. Roll a die and record whether the number on the top face is even or odd.

2. Roll a die and record whether the number on the top face is less than 3, equal to 3, or greater than 3.

3. Select a letter from the word "probability."

4. Select a letter from the word "dzo."

5. Select a card from an ordinary deck and record its color.

6. An ordinary deck of cards is shuffled and a card is selected at random and its suit is recorded.

7. A survey of voters is taken to determine their affiliation. They are classified as Democratic, Republican, or independent. A voter is selected at random and the voter's affiliation is recorded.

8. A person is selected and his or her birth month is recorded.

9. At a weekly meeting of weightwatchers, members record whether they gained weight, lost weight, or stayed the same. Two members are selected at random and their weight changes are recorded.

10. A manufacturer chooses a sample of four items from a shipment and records the number of defective items in the sample.

11. A financial analyst keeps a record of whether the closing price of a particular stock increases, decreases, or is unchanged from the previous day's close. The record is kept for four days.

12. The analyst in Problem 11 records the number of days of the four days that the closing price of the stock increases.

13. A farmer records over a three-year period whether corn, potatoes, or alfalfa is planted in a particular field in a given year. Each crop is planted once in the field in the three-year period.

14. The farmer in Problem 13 records whether the same three crops were planted in a field. In this field no crop was planted in successive years.

In Problems 15 to 24 find $E \cup F$, $E \cap F$, and E'.

15. A die is rolled and the number on the top face is recorded.

$$E = \{1, 2, 3\}, \qquad F = \{2, 3, 4\}$$

16. A die is rolled and the number on the top face is recorded.

$$E = \{1, 2, 3, 4\}, \qquad F = \{2, 4, 6\}$$

17. A letter is selected from the word "probability."

E: a vowel, F: a consonant

18. A letter is selected at random from the word "dzo."

$$E = \{d, z\}, \qquad F = \{z, o\}$$

19. An ordinary deck of cards is shuffled and a card is selected and its suit is recorded.

E: heart, F: red card

20. The 12 picture cards of an ordinary deck of cards are shuffled and one of them is selected at random.

E: heart, F: king

21. The sample space is $\{2, 3, 4, 5, 6, 7, 8, 9, 10, 11, 12\}$.

$$E = \{2, 3, 4\}, \qquad F = \{3, 4, 5, 6, 7\}$$

22. The sample space is $\{-3, -2, -1, 0, 1, 2, 3, 4\}$.

$$E = \{-3, -2, 2, 3\}, \qquad F = \{2, 3, 4\}$$

23. A coin is tossed two times and the number of heads is recorded.

E: at most one H,
F: at least one H

24. A coin is tossed two times and the number of heads is recorded.

E: at most one H,
F: exactly one H

In Problems 25 to 28 let S be the sample space of an experiment. Determine whether E and F are mutually exclusive events.

25. $S = \{a, b, c, d\}$, $E = \{a, c\}$, $F = \{b, d\}$

26. $S = \{a, b, c, d, e, f\}$, $E = \{a, b, c\}$, $F = \{b, d, f\}$

27. $S = \{2, 3, 4, 5, 6, 7, 8, 9, 10, 11, 12\}$, $E = \{2, 3, 4\}$, $F = \{4, 5, 6, 7\}$

28. $S = \{-3, -2, -1, 0, 1, 2, 3, 4\}$, $E = \{-3, -2, 0, 1\}$, $F = \{2, 3, 4\}$

•• 29. An experiment consists of choosing a committee of two individuals from a list of six people whose last initials are A, B, C, R, S, T. Determine the sample space of the experiment and list the elements of the events E, F, and G.

E: A, B, or C is on the committee.
F: Neither A nor B is on the committee.
G: Either C or S is on the committee.

30. The experiment consists of selecting a marble from an urn containing red, white, and blue marbles and recording its color. Define the events

E: It is not red.
F: It is red or blue.

Determine the sample space S of the experiment. Find (a) $E \cup F$, (b) $E \cap F$, (c) E'.

31. A survey asks people whether they (1) smoke at least a pack of cigarettes a day, (2) take vitamin C regularly, and (3) have more than two colds in a year. The experiment consists of selecting a person and recording their answers of "yes" or "no". Define the events

E: At least two questions were answered "Yes."
F: At most two questions were answered "Yes."

Determine the sample space S of the experiment. Find (a) $E \cup F$, (b) $E \cap F$, (c) E'.

32. Two urns contain red, white, and blue marbles. The experiment consists of first selecting an urn and then a marble from the urn and recording its color. Define the events

E: It is not red.
F: It is red or blue.

Determine the sample space S of the experiment. Find (a) $E \cup F$, (b) $E \cap F$, (c) E'.

33. Let E and F be events. The following events are described in words; express them in terms of $\cup$, $\cap$, and $(\ )'$.
(a) Both E and F occur.
(b) E occurs but F does not.
(c) E does not occur.
(d) E or F occurs, perhaps both.

34. Let E, F, and G be events. The following events are described in words; express them in terms of $\cup$, $\cap$, and $(\ )'$.
(a) All three occur.
(b) E occurs but F and G do not.
(c) Neither E nor F occurs.
(d) At least one of the three occurs.

35. An experiment consists of tossing a coin three times and recording the number of heads. Find the sample space and determine the elements in the two events

E: More than one heads
F: An even number of heads

Find (a) $E \cup F$, (b) $E \cap F$, (c) E'.

36. An experiment consists of tossing a coin five times and recording the number of heads. Find the sample space and determine the elements in the two events

E: More than one heads
F: An even number of heads

Find (a) $E \cup F$, (b) $E \cap F$, (c) E'.

••• 37. For any event E, is it true that E and E' are mutually exclusive? Why?

38. Let E and F be mutually exclusive events.
(a) Must E' and F' be mutually exclusive?

(b) Must $E \cap F$ and $(E \cup F)'$ be mutually exclusive? Why?

39. Determine all possible events that are associated with the sample space S where (a) $S = \{1, 2, 3\}$, (b) $S = \{1, 2, \ldots, n\}$.

40. In a study* designed to explore the possibility of a link between air pollution and ill-health the investigators discussed the effects of their research on mortality rates characterized by age, race, and sex. They defined the following events:

I: Infant (under 1 year)
C: Child (from 1–14 years old)
A: Adult (over 14 years old)
N: Nonwhite W: White
F: Female M: Male

Define the following events in words.
(a) $I \cup A$, (b) $A \cup F$, (c) $C \cap M$, (d) $N \cap F$, (e) F', (f) $A' \cap W'$, (g) $I \cap N \cap M$

41. In a medical experiment† designed to study the effect of Reye's syndrome on the inner ear, patients were asked whether they had contracted an influenza infection, had taken aspirin during their illness, and had an arsenine-deficient diet. Define the events I, A, and D as follows:

I: The patient had influenza.
A: The patient took aspirin.

D: The patient had an arsenine-deficient diet.

Define the following events in words:
(a) $I \cup A$, (b) $A \cup D$, (c) $I \cap D$, (d) $A \cap D$, (e) I', (f) $I' \cap A'$, (g) $I \cap A \cap D$.

42. Farmers classify damage to crops in three categories: damage due to insects, disease, and weather. In an article illustrating methods farmers use to estimate damage probabilities, Pingali and Carlson gathered data from 47 apple orchards in Henderson County, North Carolina.‡ Farmers were asked to estimate the percentage of the total damage done to the crop by each type of damage. (a) Find the sample space of the experiment, that is, determine the outcomes the farmers were asked to consider. (b) Suppose the farmers estimate that the percentages of each type of damage will be equal. Find the fraction of the total damage assigned to each individual type of damage. (c) Suppose the farmers estimate that the damage due to insects and disease will be equal but there will be twice as much damage due to weather as to insects. Find the fraction of the total damage assigned to each type of damage. (d) Suppose the farmers estimate that the damage due to insects will be twice as high as the damage due to disease, and the damage due to disease will be twice as much as the damage due to weather. Find the fraction assigned to each type of damage.

*John W. Pratt (ed.), *Statistical and Mathematical Aspects of Pollution Problems*, New York, Marcel Dekker, Inc., 1974, pp. 223–247.

†K. E. Rarey, et al., "Effects of Influenza Infection, Aspirin, and an Arsenine-Deficient Diet on the Inner Ear in Reye's Syndrome," *Annals of Otology, Rhinology, and Laryngology*, Vol. 93, November–December 1984, pp. 551–557.

‡Prabhu L. Pingali and Gerald A. Carlson, "Human Capital, Adjustments in Subjective Probabilities, and the Demand for Pest Controls," *American Journal of Agricultural Economics*, 67:853–61, 1985.

7–2 FUNDAMENTAL PROBABILITY PRINCIPLES

In this section we assign probabilities to the outcomes of a sample space. For example, when a coin is tossed we assume that heads and tails are equally likely to occur so each will happen half of the time. If H represents heads and T represents tails, then we say that the *probability of the event H* is $\frac{1}{2}$, which we express as $Pr(H) = \frac{1}{2}$. Also, $Pr(T) = \frac{1}{2}$. This means that if we conducted the experiment many times, we would expect H to occur half of the time and T to occur half of the time.

If the outcomes of the sample space are **equally likely,** the probability of an event can be determined by the following fundamental principle.

_____ **Definition** _____

Suppose a sample space of an experiment has N equally likely outcomes. Then the probability of each outcome is $1/N$. If the event E consists of m distinct outcomes, then the **probability of E**, written $Pr(E)$, is

$$Pr(E) = \frac{m}{N}$$

For example, when a die is rolled and the number on the top face is recorded, the six outcomes are equally likely, so each has the probability $\frac{1}{6}$. For instance, we write $Pr(2) = \frac{1}{6}$. The next example shows how to compute the probability of events that consist of more than one outcome for this experiment.

EXAMPLE 1   The experiment is to toss a die and record the number on the top face. Define the events E, F, and G as follows:

$E = \{3, 4\}$
$F = \{1, 3, 4, 5, 6\}$
G: An even number occurs

Problem

Find (a) $Pr(E)$, (b) $Pr(F)$, (c) $Pr(G)$.

SOLUTION (a) In the formula $m = 2$ and $N = 6$, so $Pr(E) = \frac{2}{6} = \frac{1}{3}$.
(b) Here $m = 5$ and $N = 6$, so $Pr(F) = \frac{5}{6}$.
(c) There are three outcomes that are even, so $G = \{2, 4, 6\}$. Thus $m = 3$ and so $Pr(G) = \frac{3}{6} = \frac{1}{2}$.

Probability Distributions

It is sometimes helpful to list the outcomes together with their associated probabilities in a table, called a **probability distribution.** For example, the probability distributions for the experiment of tossing a coin and rolling a die are as follows:

Toss a Coin		Roll a Die	
Outcome	Probability	Outcome	Probability
H	$\frac{1}{2}$	1	$\frac{1}{6}$
T	$\frac{1}{2}$	2	$\frac{1}{6}$
		3	$\frac{1}{6}$
		4	$\frac{1}{6}$
		5	$\frac{1}{6}$
		6	$\frac{1}{6}$

Sometimes the sample space does not have equally likely outcomes. In this case we usually have to picture the experiment in a different setting, one in which the outcomes are equally likely. For example, the experiment of tossing two coins and recording the number of heads has three outcomes; 0, 1, 2. They are not equally likely, as seen in the next example.

EXAMPLE 2

Two coins are tossed. There are two natural sample spaces: we can record (a) the sequence of heads and tails or (b) the number of heads that occur.

Problem

Assign probabilities to the elements of each sample space.

SOLUTION (a) To see the difference between the sample spaces, we need to distinguish the coins. Suppose the first coin is a nickel and the second is a quarter. The sample space is {*HH, HT, TH, TT*}, where *HH* represents *H* on both coins, *HT* represents *H* on the nickel and *T* on the quarter, and so on. The following tree diagram gives a pictorial view of the sample space:

Nickel	Quarter	Outcome
H	H	HH
	T	HT
T	H	TH
	T	TT

From experience we know that these outcomes are equally likely, so the fundamental principle says that each has a probability $\frac{1}{4}$. The probability distribution is given as follows:

Toss Two Coins

Record the *H*'s and *T*'s

Outcome	Probability
HH	$\frac{1}{4}$
HT	$\frac{1}{4}$
TH	$\frac{1}{4}$
TT	$\frac{1}{4}$

(b) When the number of heads is recorded the sample space is {0, 1, 2}. The event "0," meaning "0 heads," corresponds to the event "*TT*" in the previous sample space, so the probability of "0" is the same as the probability of "*TT*." Therefore $Pr(0) = \frac{1}{4}$. Similarly, the event "2" corresponds to "*HH*," so $Pr(2) = \frac{1}{4}$. For the event "1" there are two ways to have exactly one heads occur; *H* on the nickel and *T* on the quarter, or *T* on the nickel and *H* on the quarter. Thus the event "1" corresponds to "*TH* or *HT*." Thus $Pr(1) = \frac{1}{2}$. The probability distribution for this sample space is given here, alongside the one in part (a).

Toss Two Coins		Toss Two Coins	
Record the *H*'s and *T*'s		Record the Number of Heads	
Outcome	Probability	Outcome	Probability
TT	$\frac{1}{4}$	$\rightarrow 0$	$\frac{1}{4}$
HT	$\frac{1}{4}$		
		$\rightarrow 1$	$\frac{1}{2}$
TH	$\frac{1}{4}$		
HH	$\frac{1}{4}$	$\rightarrow 2$	$\frac{1}{4}$

Let us look at another example of how to assign probabilities when the outcomes are not equally likely. This example has a little twist to the familiar experiment of tossing a die.

EXAMPLE 3

A fair die has three sides colored red, two sides colored white, and one side colored blue. The experiment consists of tossing the die and recording the color of the top face.

Problem

Find the sample space and assign probabilities to the outcomes.

SOLUTION There are three possible outcomes—red, white, and blue. So the sample space is {red, white, blue}. Of the six equally likely faces, only one is colored blue so $Pr(\text{blue}) = \frac{1}{6}$. Two of the faces are white so $Pr(\text{white}) = \frac{2}{6} = \frac{1}{3}$. Three of the faces are red so $Pr(\text{red}) = \frac{3}{6} = \frac{1}{2}$. The assignment of probabilities can be easily read from the following table:

Outcome	Probability
Red	$\frac{3}{6} = \frac{1}{2}$
White	$\frac{2}{6} = \frac{1}{3}$
Blue	$\frac{1}{6}$

Basic Properties of Distributions

Probability distributions have several important basic properties. When we assign a probability to an outcome, it must be a number between 0 and 1, inclusive. An event that is certain to happen, called a "**certain event**," has probability 1, while an "**impossible event**" has probability 0. For example, if a die is rolled, a certain event would be "observe a number less than 7" and an impossible event would be "observe a number greater than 6."

The sample space must exhaust all possible outcomes. This means that the sum of the probabilities of the outcomes in the sample space must add up to 1. Finally, the outcomes in the sample space must be mutually exclusive.

7 Probability

_____ **Basic Properties of Probability Distributions** _____

Let S be the sample space for an experiment and let the probabilities assigned to the outcomes in S be $p_1, p_2, \ldots, p_n$. Then

1. For all $i = 1, \ldots, n,$ $0 \le p_i \le 1$.
2. $p_1 + p_2 + \cdots + p_n = 1$.
3. If E is a certain event, then $Pr(E) = 1$.
4. If E is an impossible event, then $Pr(E) = 0$.

Examples 2 and 3 demonstrate that it is sometimes necessary to examine an experiment from a different perspective to assign probabilities. As another example that also illustrates the properties of sample spaces, consider again the study of accident rates on Route 2 in the next example.

EXAMPLE 4

In Section 7–1 the idea of a probability distribution was introduced by looking at the Fairley study of accidents on Route 2, a major road into Boston. The number of accidents in a three-year span was gathered per quarter. Recall that the quarterly average is m/N, where m is the number of accidents that occurred in the quarter and N is the total number of accidents, 1269. For instance, in quarter 1, $m = 302$, and so $m/N = 302/1269 = .24$. Table 1 contains the probability distribution.

TABLE 1   Accidents on Route 2, 1970–1972

Quarter	Quarterly Count	Quarterly Average
1	302	0.24
2	244	0.19
3	283	0.22
4	440	0.35
Total	1269	

Source: William B. Fairley, "Accidents on Route 2: Two-Way Structures for Data," *Statistics and Public Policy,* by Fairley and Mosteller, eds., Reading, MA, Addison-Wesley, 1977, pp. 23–50.

This study can be expressed in terms of probabilities by regarding it as an experiment whose sample space is {quarter 1, quarter 2, quarter 3, quarter 4}. The quarterly averages are probabilities. Thus the probabilities of the four outcomes are, respectively, .24, .19, .22, and .35. That is, if we select an accident at random, then the probability that it occurred in quarter 1 is Pr(quarter 1) $= .24$, while the probability that it occurred in quarter 2 is Pr(quarter 2) $= .19$. The four outcomes are not equally likely.

To picture the experiment with equally likely events, think of it in this way. Let the sample space consist of the 1269 individual accidents. If we select one at random, each is as likely to be chosen as another, so each has probability 1/1269. With this sample space the event "quarter 1" has 302 outcomes, each with probability 1/1269. To get the probability of quarter 1, we add them to get

302/1269. We are allowed to add the probabilities because they are mutually exclusive.

Let us compare the two sample spaces by taking a closer look at the event quarter 1 in each sample space. The first sample space has 4 outcomes that are not equally likely. The second sample space has 1269 outcomes that are equally likely. In the first sample space quarter 1 is a single outcome. In the second sample space quarter 1 consists of 302 distinct outcomes. In each sample space its probability is the same, 302/1269.

Example 4 suggests that if two or more events are mutually exclusive, the probability of their union is the sum of their probabilities. This is expressed in the next principle.

_____ **Addition Principle for Mutually Exclusive Events** _____

If two events E and F are mutually exclusive; that is, $E \cap F = \emptyset$, then

$$Pr(E \cup F) = Pr(E) + Pr(F)$$

The next two examples illustrate this principle.

EXAMPLE 5

Two coins are tossed and the number of heads is recorded.

Problem

Find the probability of observing at most one heads.

SOLUTION Let E represent "observing at most one heads." If we let F and G represent the two events "exactly 0 heads" and "exactly 1 heads," respectively, then $E = F \cup G$ and $F \cap G = \emptyset$. Hence by the Addition Principle, $Pr(E) = Pr(F) + Pr(G) = \frac{1}{4} + \frac{1}{2} = \frac{3}{4}$.

In Example 6 remember that an ace is not considered a picture card, which is also referred to as a face card. The picture cards are the kings, queens, and jacks.

EXAMPLE 6

The experiment is to draw a card from an ordinary deck. The sample space consists of the 52 cards, each assigned the probability $\frac{1}{52}$. Define the events

E: A ten
F: A red ace
G: A picture card

Problem

Find $Pr(E \cup F)$, $Pr(E \cup G)$, and $Pr(F \cup G)$.

SOLUTION The events are mutually exclusive since no two can occur at the same time. Since there are 4 tens, 2 red aces, and 12 picture cards, we have $Pr(E) = \frac{4}{52}$, $Pr(F) = \frac{2}{52}$ and $Pr(G) = \frac{12}{52}$. Then

$$Pr(E \cup F) = Pr(E) + Pr(F) = \frac{4}{52} + \frac{2}{52} = \frac{6}{52} = \frac{3}{26}$$
$$Pr(E \cup G) = Pr(E) + Pr(G) = \frac{4}{52} + \frac{12}{52} = \frac{16}{52} = \frac{4}{13}$$
$$Pr(F \cup G) = Pr(F) + Pr(G) = \frac{2}{52} + \frac{12}{52} = \frac{14}{52} = \frac{7}{26}$$

Complements

Suppose an urn contains 52 marbles, 4 of which are red and the rest are various other colors. If a marble is drawn at random from the urn, what is the probability that it is not red? There are 4 red marbles, so there are $52 - 4 = 48$ marbles that are not red, implying that the probability is $48/52 = 12/13$. This idea can be expressed in terms of complements. Let E represent "the marble is red." Then the desired event, "the marble is not red," is the complement, E'. In terms of probabilities, the last computation can be expressed as

$$Pr(E') = \frac{52 - 4}{52} = \frac{48}{52}$$

Still another way to write it is

$$Pr(E') = \frac{52}{52} - \frac{4}{52} = 1 - \frac{4}{52} = 1 - Pr(E)$$

This generalizes into the **rule for complements.**

_____ **Rule for Complements** _____

If E is any event and E' is its complement, then

$$Pr(E') = 1 - Pr(E)$$

Since E is the complement of E', this rule can also be written as $Pr(E) = 1 - Pr(E')$.

Verification of the Rule for Complements

Since $E \cap E' = \varnothing$, the addition rule for mutually exclusive events yields

$$Pr(E \cup E') = Pr(E) + Pr(E')$$

By definition of E', $E \cup E' = S$, and since S is the certain event, we have

$$Pr(E \cup E') = Pr(S) = 1$$

Hence, putting these two equations together, we get

$$Pr(E) + Pr(E') = 1$$

Therefore $Pr(E') = 1 - Pr(E)$.

The next example shows how to apply the complement rule.

EXAMPLE 7 | **Problem**

In the experiment of Example 4, find the probability that a particular accident occurred in neither quarter 1 nor quarter 2.

SOLUTION Let E be the desired event.

E: The accident occurred in neither quarter 1 nor quarter 2.

Then E' is "the accident occurred in quarter 1 or quarter 2" and so $Pr(E') = .24 + .19 = .43$. By the rule for complements

$$Pr(E) = 1 - Pr(E') = 1 - .43 = .57$$

EXERCISE SET 7–2

• In Problems 1 to 6 a die is rolled and the number on the top face is recorded. Find the probability of the event.

1. 6.

2. 5 or 6.

3. An odd number.

4. A number greater than 4.

5. A number between 2 and 5 inclusive.

6. A number less than 5.

In Problems 7 to 10 two coins are tossed and the number of heads is recorded. Find the probability of the events.

7. One heads.

8. Two heads.

9. Less than two heads.

10. An odd number of heads.

In Problems 11 to 14 the experiment is the Fairley study of accidents on Route 2. Find the probability of the events.

11. The accident occurred in quarter 1 or quarter 2.

12. The accident occurred in quarter 2, quarter 3, or quarter 4.

13. The accident occurred in the first three quarters.

14. The accident occurred in the last two quarters.

In Problems 15 to 18 a card is drawn from an ordinary deck. Define the events

E: A red card	A: The ace of spades
F: A spade	B: A red 2
G: A club	C: A picture card

Find the probability of the events.

15. $E \cup F$.

16. $E \cup G$.

17. $A \cup B$.

18. $B \cup C$.

•• In Problems 19 and 20 a card is drawn from an ordinary deck. Determine $Pr(E)$ by first finding $Pr(E')$.

19. E: The card is not a black ace.

20. E: The card is not the king of hearts.

In Problems 21 to 24 an experiment has a sample space $S = \{a, b, c, d, e, f\}$. Is the table a possible probability distribution for the experiment? If not, state why.

21.
outcomes	a	b	c	d	e	f
probability	.1	.1	.2	.3	.2	.1

22.
outcomes	a	b	c	d	e	f
probability	.2	.3	.1	.1	.2	.1

23.
outcomes	a	b	c	d	e	f
probability	.4	.1	.2	.1	.2	.1

24.
outcomes	a	b	c	d	e	f
probability	.11	.1	.38	.1	.21	.1

25. Three coins are tossed and the number of heads is recorded. Find the probability of the events (a) less than two heads, (b) an odd number of heads.

26. Four coins are tossed and the number of heads is recorded. Find the probability of the events (a) at least three heads, (b) no more than one head.

27. American roulette is a game where a ball is rolled on a spinning wheel and it lands in one of 38 slots, 18 of which are red, 18 are black, and 2 are green. What is the probability that the ball lands on a red number?

28. Select a ball from an urn containing ten balls numbered 0, 1, . . . , 9. What is the probability that the number selected is (a) even, (b) greater than 7, (c) less than 3 or greater than 7?

29. A marble is chosen from an urn containing 3 red and 7 white marbles. What is the probability that it is red?

30. A marble is chosen from an urn containing 3 red, 2 blue, and 7 white marbles. What is the probability that it is (a) red, (b) blue, (c) red or blue?

31. A card is drawn from an ordinary deck. Find the probability that the card is (a) a king or a jack, (b) a red ace or a black king, (c) a club or a diamond.

In Problems 32 to 34 an experiment has the sample space $S = \{a, b, c, d\}$ with the following probability distribution:

Outcome	Probability
a	.32
b	.51
c	.07
d	p

Define the events $E = \{a, b\}$ and $F = \{c, d\}$.

32. Find p.

33. Find $Pr(E)$ and $Pr(F)$.

34. Find $Pr(E \cup F)$.

In Problems 35 to 38 an experiment has the sample space $S = \{a, b, c, d, e, f\}$ with the following probability distribution:

Outcome	Probability
a	.25
b	.20
c	.15
d	.05
e	.10
f	.25

Define the events $E = \{a, b\}$, $F = \{c, d\}$, and $G = \{e, f\}$.

35. Find $Pr(E)$, $Pr(F)$, and $Pr(G)$.

36. Find $Pr(E \cup F)$ and $Pr(E \cup G)$.

37. Find $Pr(E')$ and $Pr(F')$.

38. Find $Pr[(E \cup F)']$.

In Problems 39 and 40 let E and F be events such that $Pr(E) = .3$, $Pr(F) = .4$, and $Pr(E \cap F) = 0$.

39. Find $Pr(E \cup F)$.

40. Find $Pr[(E \cup F)']$.

41. In an introductory business course students were asked whether or not they were accounting majors. The survey also recorded the sex of the student. The results are given in the table.

	Accounting Major	Not Accounting Major
Men	10	18
Women	6	16

Define the events

E: The applicant was a man and was an accounting major.

F: The applicant was a woman and was not an accounting major.

Find (a) $Pr(E)$, (b) $Pr(F)$, (c) $Pr(E \cup F)$.

42. In an introductory political science course students were asked whether they were registered Democrats or Republicans and whether or not they were in favor of a business tax increase. The results are given in the table.

	Business Tax Increase	
	In Favor	Not in Favor
Democrat	13	7
Republican	8	12

Define the events

E: The applicant was a Democrat and in favor of the increase.

F: The applicant was a Republican and in favor of the increase.

Find (a) $Pr(E)$, (b) $Pr(F)$, (c) $Pr(E \cup F)$, (d) What percentage of Democrats were in favor of the increase?

●●● 43. Suppose an experiment has sample space $S = \{a, b, c\}$ such that $Pr(a) = 2Pr(b) = 3Pr(c)$. Find the probability distribution.

44. Suppose an experiment has sample space $S = \{a, b, c\}$ such that $Pr(a) = .1 + Pr(b)$. Define the event $E = \{a, b\}$. If $Pr(E') = .7$, find the probability distribution.

45. An important issue in our society is determining whether discrimination because of sex or ethnic identity is being practiced. In an article by Peter Bickel et al., the following data were gathered while studying the graduate admissions at the University of California, Berkeley, for the fall quarter, 1973.

	Outcome	
Applicants	Admit	Deny
Men	3738	4704
Women	1494	2827

Source: From Peter Bickel et al., "Sex Bias in Graduate Admissions: Data from Berkeley," *Science*, Vol. 187, February 7, 1975, pp. 398–404.

The question studied was whether the decision to admit or deny was influenced by the sex of the applicant. Define the events

E: The applicant was a man and was admitted.

F: The applicant was a man and was denied admittance.

G: The applicant was a woman and was admitted.

H: The applicant was a woman and was denied admittance.

Select an applicant at random and find (a) $Pr(E \cup F)$, (b) $Pr(G \cup H)$, (c) $Pr(E \cup G)$, (d) $Pr(F \cup H)$.

46. In an article on the possible underestimating of catastrophic accident probabilities, William B. Fairley reviewed data supplied to the Federal Power Commission by importers of liquefied natural gas. A large spill in a port could cause a catastrophic fire, resulting in the loss of lives and property similar to a major accident at a nuclear power plant. Fairley defined the following probability distribution based on natural gas shipments to Staten Island and Providence in a ten-year period.

Outcome	Probability
Accident occurred based on officially reported data	8/10,000
Accident occurred based on unreported accounts	32/10,000
No accident occurred	

Source: From William Fairley, "Evaluating the 'Small' Probability of a Catastrophic Accident from the Marine Transportation of Liquefied Natural Gas," *Statistics and Public Policy* by Fairley and Mosteller, eds., Reading, MA, Addison-Wesley, 1977, pp. 331–353.

Determine (a) $Pr($No accident occurred.$)$, (b) $Pr($An accident occurred.$)$. (c) What is the factor by which Fairley multiplied the historical probability of an accident in order to obtain the probability of an unreported accident; that is, how much larger is the latter than the former?

47. Probability theory is often used to assess the potential outcomes when certain risks are taken. Farming is a profession where risks are taken as a matter of course. In an article investigating the practical selection of which types of risks

to take, Atwood studied Philippine rice yields. The table gives the yield in pounds per acre and the corresponding probability of attaining that yield. (a) Find the probability of attaining at most 40 pounds per acre. (b) Find the probability of attaining more than 90 pounds per acre.

Yield	0–40	40–50	50–60	60–70	70–80	80–90	>90
Probability	.03	.05	.09	.13	.18	.23	

Source: From Joseph Atwood, ''Demonstration of the Use of Lower Partial Moments to Improve Safety-first Probability Limits,'' *American Journal of Agricultural Economics*, 67:787–95, 1985.

7–3 THE ADDITION PRINCIPLE

In the previous section some elementary rules allowed us to compute probabilities of various events. In all cases the events were mutually exclusive. In this section we consider events that have a nonempty intersection.

For example, select a card from a deck and define the two events:

     E: A spade   *F*: A king

To find $Pr(E \cup F)$, we count the number of outcomes favorable to the event $E \cup F$. There are 13 spades and 4 kings; but the king of spades was counted twice, once as a spade and once as a king. Thus in $E \cup F$ there are 13 spades and only 3 additional kings, the king of spades having already been counted. Therefore

$$Pr(E \cup F) = \frac{16}{52} = \frac{4}{13}$$

A Venn diagram helps to explain where the king of spades causes the difficulty—in $E \cap F$. Hence to find the probability of the union of two events that are not mutually exclusive, the intersection of the events is critical. This discussion leads to the general result called the **Addition Principle.** The Addition Principle for Mutually Exclusive Events presented in the previous section is a special case of this result.

FIGURE 1

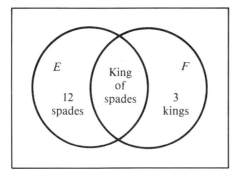

_____ **The Addition Principle** _____

If E and F are any two events in a sample space, then

$$Pr(E \cup F) = Pr(E) + Pr(F) - Pr(E \cap F)$$

_____

The Addition Principle for Mutually Exclusive Events is a special case of the Addition Principle because it is used when $E \cap F$ is the empty set. The Addition Principle can be used in all cases.

To illustrate how to use the Addition Principle, we consider the above discussion and find $Pr(E \cup F)$. We have $Pr(E) = 13/52$, $Pr(F) = 4/52$ and $Pr(E \cap F) = 1/52$. Thus

$$Pr(E \cup F) = \frac{13}{52} + \frac{4}{52} - \frac{1}{52} = \frac{16}{52}$$

It is necessary to subtract $Pr(E \cap F)$ because it was accounted for twice, once in $Pr(E)$ and again in $Pr(F)$.

The fraction 16/52 can be reduced to 4/13 as was done earlier. However, sometimes it is advantageous to leave the fraction in its unreduced form. In this case the fraction 16/52 illustrates that 16 of the 52 cards are in $E \cup F$. In general we reduce a fraction unless it is more instructive not to. For instance, in the next two examples that further illustrate the Addition Principle, some fractions will be left in their unreduced form because they will be added to other fractions and their least common denominator is easier to find in this form.

EXAMPLE 1   The experiment is to draw a card from a deck. Define

E: A heart F: A picture card G: A red card

Problem

Find (a) $Pr(E \cup F)$, (b) $Pr(F \cup G)$.

SOLUTION There are 13 hearts, 12 picture cards, and 26 red cards, so $Pr(E) = 13/52$, $Pr(F) = 12/52$, and $Pr(G) = 26/52$. (a) There are 3 hearts that are picture cards, so $Pr(E \cap F) = 3/52$. Therefore

$$Pr(E \cup F) = \frac{13}{52} + \frac{12}{52} - \frac{3}{52} = \frac{22}{52} = \frac{11}{26}$$

(b) There are 6 red cards that are picture cards, so $Pr(F \cap G) = \frac{6}{52}$. Therefore

$$Pr(F \cup G) = \frac{12}{52} + \frac{26}{52} - \frac{6}{52} = \frac{32}{52} = \frac{8}{13}$$

The next example deals with the experiment of tossing two dice and recording the numbers on the top two faces. What is the sample space, S? It might seem as though there are 11 elements in S, the numbers 2, 3, 4, 5, 6, 7, 8, 9, 10, 11, 12.

These numbers, however, are the sums of the top faces. The experiment is to record the numbers on each face. It is helpful to distinguish between the two dice. Let one be white and the other black (see Figure 2a), or let one be the first die and the other be the second (see Figure 2b). In any case, the sample space consists of 36 elements. Thus "5 on the white die and 6 on the black" is represented by the ordered pair $(5, 6)$. Note that $(5, 6) \neq (6, 5)$. The 36 outcomes are equally likely so the probability of each is $1/36$.

In these examples it is assumed that the dice are "fair," meaning that they are well balanced and each face is equally likely to occur. If a die is "loaded," meaning that one or more sides have extra weight so that one number is more likely to occur, the probabilities would change. Casinos take great precaution to emphasize to their customers that their dice are fair. After a set amount of plays the dice are replaced and the old ones are destroyed.

If the experiment is to toss two dice and record the sum of the top faces, the sample space consists of the 11 sums, 2 through 12. These outcomes are not equally likely. A 2 comes up much less frequently than a 7. To find the probability of each

FIGURE 2

(a)

Second die

(1, 1)	(1, 2)	(1, 3)	(1, 4)	(1, 5)	(1, 6)
(2, 1)	(2, 2)	(2, 3)	(2, 4)	(2, 5)	(2, 6)
(3, 1)	(3, 2)	(3, 3)	(3, 4)	(3, 5)	(3, 6)
(4, 1)	(4, 2)	(4, 3)	(4, 4)	(4, 5)	(4, 6)
(5, 1)	(5, 2)	(5, 3)	(5, 4)	(5, 5)	(5, 6)
(6, 1)	(6, 2)	(6, 3)	(6, 4)	(6, 5)	(6, 6)

First die

(b)

outcome, we determine how many ordered pairs are associated with each sum. For instance, a sum of 2 can occur in only one way, that is, observing (1, 1). So $Pr(2) = 1/36$. But a 5 can occur in four ways; (1, 4), (2, 3), (3, 2), or (4, 1), so $Pr(5) = 4/36 = 1/9$. The probability distribution is as follows. Notice the symmetry of the distribution; 2 and 12 have the same probability, as do 3 and 11, and so forth.

outcome	2	3	4	5	6	7
probability (reduced)	$\frac{1}{36}$	$\frac{2}{36}$ $\frac{1}{18}$	$\frac{3}{36}$ $\frac{1}{12}$	$\frac{4}{36}$ $\frac{1}{9}$	$\frac{5}{36}$	$\frac{6}{36}$ $\frac{1}{6}$

outcome	8	9	10	11	12
probability (reduced)	$\frac{5}{36}$	$\frac{4}{36}$ $\frac{1}{9}$	$\frac{3}{36}$ $\frac{1}{12}$	$\frac{2}{36}$ $\frac{1}{18}$	$\frac{1}{36}$

EXAMPLE 2

Toss two dice and record the sum of the top faces. Define the events

E: A sum of 3, 4, or 5 occurs.
F: A sum less then 7 occurs.
G: An odd sum occurs.

Problem

Find (a) $Pr(E)$, $Pr(F)$, and $Pr(G)$; (b) $Pr(E \cup G)$

SOLUTION (a) The events 3, 4, and 5 are mutually exclusive so $Pr(E)$ is the sum of their respective probabilities. Hence $Pr(E) = Pr(3) + Pr(4) + Pr(5) = \frac{2}{36} + \frac{3}{36} + \frac{4}{36} = \frac{9}{36} = \frac{1}{4}$. Similarly, $Pr(F) = Pr(2, 3, 4, 5, 6) = Pr(2) + Pr(3) + Pr(4) + Pr(5) + Pr(6) = \frac{1}{36} + \frac{2}{36} + \frac{3}{36} + \frac{4}{36} + \frac{5}{36} = \frac{15}{36} = \frac{5}{12}$. Also $Pr(G) = Pr(3, 5, 7, 9, 11) = \frac{2}{36} + \frac{4}{36} + \frac{6}{36} + \frac{4}{36} + \frac{2}{36} = \frac{18}{36} = \frac{1}{2}$.
(b) There are two outcomes, 3 and 5, in the intersection of E and G, so $Pr(E \cap G) = Pr(3, 5) = Pr(3) + Pr(5) = \frac{2}{36} + \frac{4}{36} = \frac{6}{36} = \frac{1}{6}$. In Figure 3

FIGURE 3

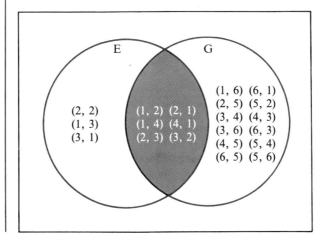

the ordered pairs corresponding to E and G are indicated with $E \cap G$ shaded. From the Addition Principle

$$Pr(E \cup G) = \frac{9}{36} + \frac{18}{36} - \frac{6}{36} = \frac{21}{36} = \frac{7}{12}$$

Verification of the Addition Principle

Recall that if two events A and B are mutually exclusive, then $Pr(A \cup B) = Pr(A) + Pr(B)$. Consider the Venn diagram in Figure 4. The set $E \cup F$ is the union of three disjoint sets.

$$E \cup F = (E \cap F') \cup (E \cap F) \cup (E' \cap F)$$

Because these sets are disjoint, we write

$$Pr(E \cup F) = Pr(E \cap F') + Pr(E \cap F) + Pr(E' \cap F)$$

Also from Figure 4 notice that

$$E = (E \cap F') \cup (E \cap F) \qquad \text{and} \qquad F = (F \cap E') \cup (F \cap E)$$

In each equality the sets are disjoint so

$$Pr(E) = Pr(E \cap F') + Pr(E \cap F)$$
$$Pr(F) = Pr(F \cap E') + Pr(F \cap E)$$

Solving for $Pr(E \cap F')$ and $Pr(F \cap E')$, respectively, yields

$$Pr(E \cap F') = Pr(E) - Pr(E \cap F)$$
$$Pr(F \cap E') = Pr(F) - Pr(F \cap E)$$

Substituting these equations into the original statement for $Pr(E \cup F)$ gives us

$$Pr(E \cup F) = [Pr(E) - Pr(E \cap F)] + Pr(E \cap F)$$
$$+ [Pr(F) - Pr(E \cap F)]$$

FIGURE 4

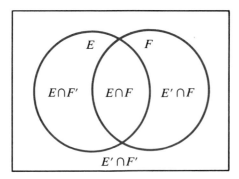

Two of the three terms with $Pr(E \cap F)$ cancel and we get

$$Pr(E \cup F) = Pr(E) + Pr(F) - Pr(E \cap F)$$

An Application to Social Science

The following application from political science provides another example of the Addition Rule. In an article* that studied many uses of probability in the law, Zeisel and Kalvern investigated the perennial debate over the merits of the jury system. The question they raised is, Would there be a difference between verdicts reached by a jury and verdicts that the judge would have reached alone if there were no jury? The argument is, If the judge and jury hardly ever differed, the jury would be a wasteful institution; if they differed too often, serious questions would be raised about the rationality of the system, including juries, judges, and the very meaning of justice itself. They studied 3576 jury trials in which the judge reported how he or she would have ruled if there were no jury. The data are as follows: in 17% of the cases the judge would have acquitted the defendant, in 33% of the cases the jury acquitted the defendant, and in 14% of the cases they both reached the verdict of acquittal.

EXAMPLE 3

Problem

A case from the Zeisel and Kalvern data is chosen at random. What is the probability that
(a) Either the jury acquitted or the judge would have acquitted the defendant?
(b) The jury convicted the defendant and the judge would have acquitted the defendant?

SOLUTION Define the events

E: The judge would have acquitted.
F: The jury acquitted.

Then $Pr(E) = .17$, $Pr(F) = .33$, and $Pr(E \cap F) = .14$. See the Venn diagram in Figure 5.
(a) The desired event is $E \cup F$. By the Addition Principle

$$Pr(E \cup F) = .17 + .33 - .14 = .36$$

(b) The desired event is $E \cap F'$. From the Venn diagram in Figure 5 we see that $E = [E \cap F'] \cup [E \cap F]$. Since these sets are disjoint, we have

$$Pr(E) = Pr(E \cap F') + Pr(E \cap F)$$

*Hans Zeisel and Harry Kalvern, "Parking Tickets and Missing Women: Statistics and the Law," *Statistics: A Guide to the Unknown,* by J. A. Tanur, et al., eds., San Francisco, Holden-Day, Inc., 1972, pp. 102–111. The authors also investigated the 1968 trial of the pediatrician-author Dr. Benjamin Spock for conspiracy to violate the Selective Service Act by encouraging resistance to the war in Vietnam. They asked the question, "What is the probability that 'this total absence of women jurors was an accident'?" They computed the probability that it was an accident to be 1/1,000,000,000,000,000,000,000.

Solving for $Pr(E \cap F')$ yields

$$Pr(E \cap F') = Pr(E) - Pr(E \cap F)$$
$$= .17 - .14 = .03$$

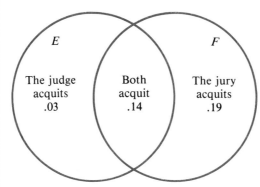

FIGURE 5

A qualified conclusion can be reached from Example 3. If all other factors are insignificant (which is a significant assumption, of course) a defendant has a better chance for acquittal with a jury trial. In only 3% of the cases would a judge have acquitted the defendant when the jury found the defendant guilty. However, in 19% of the cases the jury reached acquittal when the judge would have found the defendant guilty.

Odds

When probabilities are assigned in empirical studies, they are usually based on past experience. Theoretically, if the probability of $\frac{1}{4}$ is assigned to the event E, it means that if the exact experiment were performed many times, E would occur about one-fourth of the time. But often the experiment will be conducted only once, as when a weather forecaster predicts the likelihood that a hurricane will hit land, or an economist predicts the likelihood that the Dow-Jones average will increase by at least 20%. Frequently this type of probability is expressed in terms of "odds." Consider the statements

The odds against a hurricane reaching land are 3 to 1.
The odds in favor of a 20% or more increase in the Dow-Jones average are 5 to 3.

The first statement means that, under exactly the same weather conditions, three times the hurricane will not reach land and once it will. Let E be the event that the hurricane will not reach land. Then E will occur three out of four times, so $Pr(E) = 3/4$. Notice that the first statement mentions the "odds against" the event occurring, while the second statement mentions the "odds in favor." One is the complement of the other, so when computing odds we must be careful which odds are desired. If the odds against an event are 3 to 1, then the odds in favor of the event are 1 to 3. We generalize these remarks as follows:

Definition

If the **odds in favor of an event E** are m to n, then $Pr(E) = \dfrac{m}{m + n}$.

If $Pr(E) = m/N$, then the odds in favor of E are m to $N - m$.
If the odds in favor of E are m to n, then the odds against E are n to m.

EXAMPLE 4

Problem

The odds in favor of a general strike are 6 to 5.
(a) What is the probability that a strike will take place? (b) What are the odds against a strike?

SOLUTION (a) Let E be the event that a strike will take place. Then $m = 6$ and $n = 5$ in the formula, so $Pr(E) = 6/(6 + 5) = 6/11$. (b) The probability that a strike will not occur is $1 - 6/11 = 5/11$ so the odds against a strike are 5 to 6.

An alternate formula for odds in favor of event E is

$$\text{odds in favor of } E = \frac{Pr(E)}{Pr(E')} = \frac{Pr(E)}{1 - Pr(E)}$$

This definition is used if $Pr(E)$ is given and the odds are to be computed. The original definition is used when the odds are given and $Pr(E)$ is to be computed. The next example illustrates that the two definitions are equivalent.

EXAMPLE 5

Problem

The odds in favor of an increase tomorrow in the Dow-Jones average are 5 to 3. Let E represent the event that an increase will take place tomorrow.
(a) Use the original formula to compute $Pr(E)$.
(b) Use $Pr(E)$ computed in (a) in the alternate formula to compute the odds in favor of E.

SOLUTION (a) From the original formula $m = 5$ and $n = 3$, so $Pr(E) = 5/(5 + 3) = 5/8$.
(b) Since the odds are given as 5 to 3, the answer should be 5/3 if the definitions are equivalent. Substituting $Pr(E) = 5/8$ and $Pr(E') = 1 - Pr(E)) = 1 - 5/8 = 3/8$ into the alternate formula yields

$$\text{odds in favor of } E = \frac{5/8}{3/8} = \frac{5}{3}$$

Theoretical Versus Empirical Probability

There are several ways to assign probabilities to events. Everyone uses probabilistic reasoning in making daily decisions. If you are in a hurry to cross a busy intersection, you weigh the chance of being injured with the consequence of being late. If time is running short when studying for an exam, you make a decision on what to concentrate based on the likelihood that the material will be on the exam. These are called **subjective probabilities** because there is no concrete formula for assigning a probability to the event. Decisions are made subjectively based on experience.

Another way to assign probabilities is to rely on concrete data. This is called empirical probability. Most of the experiments in this text are theoretical rather than empirical. Let us explain the difference between these two types of probabilities.

In Section 7–1 we mentioned the difference between games of chance and studies of real-world phenomena. In assigning probabilities to the outcomes of games, we have **theoretical probabilities,** because we compute them by theoretical reasoning rather than by gathering data. For example, when we say the probability of observing a 2 when tossing a die is $\frac{1}{6}$, we recognize that there are six equally likely outcomes. Experience at tossing dice bears this out, but the number $\frac{1}{6}$ came from theoretical reasoning rather than actual data. These types of experiments can be repeated exactly any number of times.

However, most studies of real-world phenomena cannot be repeated under exactly the same conditions. In this case the only way to assign probabilities is to look at past experience. The probabilities are then relative frequencies, often called **empirical probabilities.** Each view of probability has advantages in particular situations.

One of the difficulties with empirical probability is that two studies might yield significantly different results. The science of statistics establishes safeguards so that separate studies will produce as similar results as possible.

EXERCISE SET 7–3

- In Problems 1 to 4 two dice are tossed and the sum of the top faces is recorded. Find $Pr(E \cup F)$ for the given events E and F.

1. E: A sum of at most 4 F: A sum of 4 or 5

2. E: A sum of at most 7 F: A sum of 6, 7, or 8

3. E: An even sum F: A sum divisible by 3

4. E: A sum less than 7 F: An odd sum

In Problems 5 to 8 a card is chosen from an ordinary deck. Find $Pr(E \cup F)$ for the given events E and F.

5. E: A picture card F: A black card

6. E: A red picture card F: A king

7. E: An ace or a queen F: A picture card

8. E: A jack or a red ace

     F: A card that is not a king

In Problems 9 and 10 a marble is selected at random from an urn with 3 red, 4 white, 5 blue, and 6 green marbles. Find $Pr(E \cup F)$ for the given events E and F.

9. *E*: A red or white marble
   F: A blue or red marble

10. *E*: A red or blue marble
    F: A green or blue marble

In Problems 11 to 14 an experiment is conducted with the following probability distribution:

Outcome	Probability
a	.25
b	.30
c	.10
d	.35

Define the events

$$E = \{a, b, c\} \quad F = \{b, c, d\} \quad G = \{b, d\}$$

Find the probability of the event.

11. $E \cup F$.   12. $E \cup G$.

13. $E' \cup F$.   14. $F' \cup G'$.

In Problems 15 to 18 suppose $Pr(E) = .16$, $Pr(F) = .38$, and $Pr(E \cap F) = .05$. Use a Venn diagram to find the probability of the event.

15. $E' \cap F$.   16. $E' \cup F$.

17. $E \cap F'$.   18. $E \cup F'$.

In Problems 19 to 22 find $Pr(E)$.

19. The odds in favor of *E* are 4 to 7.

20. The odds in favor of *E* are 8 to 3.

21. The odds against *E* are 5 to 2.

22. The odds against *E* are 3 to 7.

In Problems 23 to 26 find the odds in favor of *E* and the odds against *F*.

23. $Pr(E) = \dfrac{2}{9}, Pr(F) = \dfrac{3}{7}$

24. $Pr(E) = \dfrac{3}{4}, Pr(F) = \dfrac{1}{8}$

25. $Pr(E) = .26, Pr(F) = .38$

26. $Pr(E) = .42, Pr(F) = .05$

•• 27. Five coins are tossed and the number of heads is recorded. Find the probability of observing at least three heads or at most four heads.

28. Suppose $Pr(E) = .45$, $Pr(F) = .31$, and

$Pr(E \cap F) = .22$. Find (a) $Pr(E' \cap F')$ (b) $Pr(E' \cup F')$.

29. Suppose $Pr(E \cap F) = .15$, $Pr(E \cup F) = .77$, and $Pr(E \cap F') = .30$. Find (a) $Pr(E' \cap F)$ (b) $Pr(E \cap F')$.

30. In a lottery a three-digit number is selected with replacement. What is the probability that it is (a) less than 100, (b) greater than 50, (c) greater than 50 or less than 100?

31. An American roulette wheel has 38 slots. Two slots are green and labeled 0 and 00. The other slots are numbers 1 to 36 and are evenly divided between red and black. If the ball lands in a slot at random, what is the probability that the slot is (a) red, (b) black, (c) red or black, (d) red or green?

32. In American roulette, what is the probability that the slot in which the ball lands is (a) red or a number greater than 30 (b) black or a number less than 5.

33. Suppose $Pr(E) = .35$ and $Pr(F) = .65$. Find (a) the odds in favor of *E*, (b) the odds against *F*.

34. The odds in favor of the first proposal passing are 7 to 6, the odds of the second proposal passing are 9 to 4, and the odds that both will pass are 5 to 8. What are the odds that at least one will pass?

The following table gives the results of a survey that measured how many people from three levels of income regularly invest in the stock market:

Regularly Invest in the Stock Market

Income Level	Yes	No
High	18	6
Medium	21	16
Low	15	24

A person in the survey is chosen at random. Define the events

H: The person has a high income.
L: The person has a low income.

Y: The person regularly invests in the stock market.
N: The person does not regularly invest in the stock market.

In Problems 35 to 40 find the probability of the event.

35. $H \cup L$.

36. $H' \cup L$.

37. $L \cap Y$.

38. $L' \cap Y$.

39. $(H \cup L) \cap N$.

40. $(H' \cup L) \cap Y$.

●●● 41. Suppose *E*, *F*, and *G* are events. Show that

$$Pr(E \cup F \cup G) = Pr(E) + Pr(F) + Pr(G) - Pr(E \cap F)$$
$$- Pr(E \cap G) - Pr(F \cap G)$$
$$+ Pr(E \cap F \cap G)$$

42. Show that the odds in favor of *E* can be expressed as $Pr(E)/Pr(E')$, that is, the ratio of "$Pr(E)$ to $Pr(E')$."

43. An article by Colton and Buxbaum* on state-mandated vehicle inspection studied the effectiveness (as measured by loss of life, personal injury, and property damage) of those states that had mandatory vehicle inspection versus those that did not. One measure they considered was the percentage of mortality rates (rate per 100,000 population) by sex. They found that 58% of the mortalities were in noninspection states, 24% were women, and 14% were women in noninspection states. Define the events

E: The accident occurred in a noninspection state.
F: The victim was a woman.

Find $Pr(E \cup F)$ and $Pr(E \cap F')$.

44. The admission process at many universities is a very complex operation, especially in the face of considerable uncertainty of student performance. In an article by Holtzman and Johnson† designed to use probabilistic models to help colleges choose the right admissions policy for a particular situation, computer simulation was used to compare student performance of those students who were accepted at a major university versus the predicted performance of those students who were rejected. Define the events

E: The applicant was accepted.
F: The applicant's GPA (grade-point average) was (or was predicted to be) greater than or equal to 2.0.

The event *F* includes the first year performance of those students who were accepted as well as the predicted GPA of those who were rejected. The data showed that $Pr(E) = .64$, $Pr(F) = .60$, and $Pr(E \cap F) = .50$. Express these three events in words and then find (a) $Pr(E \cup F)$, (b) $Pr(E' \cap F)$, (c) $Pr(E \cap F')$.

45. Probability theory assigns numbers from 0 to 1 to events in order to measure the likelihood of the event occurring. English words, such as "seldom" and "toss-up," also are used to describe the chance of an event occurring. In an article studying whether individuals agree on the meaning of such words, Budescu and Wallsten asked subjects to assign numbers from 0 to 1 to 19 words and phrases that connote a probabilistic meaning.‡ Thirty-two subjects were divided into two groups, and each person was asked to give a rank ordering to the probability phrases. For one phrase, 94% of the individuals in group 1 agreed with its ordering while 81% of group 2 agreed with its ordering. Find the probability that an individual selected at random from either group would agree with its ordering.

*Theodore Colton and Robert Buxbaum, "Motor Vehicle Inspection and Motor Vehicle Accident Mortality," *American Journal of Public Health*, Vol. 58, No. 6, pp. 1090–1099.

†Albert G. Holzman and Donald B. Johnson, "A Simulation Model of the College Admission Process," *Interfaces*, Vol. 5, No. 3, May 1975, pp. 34–43.

‡David Budescu and Thomas Wallsten, "Consistency in Interpretation of Probabilistic Phrases," *Organizational Behavior and Human Decision Processes*, 36:391–405, 1985.

7–4   COUNTING TECHNIQUES

In this section we use the counting techniques introduced in Chapter 6 to solve more complex problems in probability. Their solutions depend on the equally likely outcomes principle, first covered in Section 7–2. Let us recall it here.

_____ **Definition** _____

If a sample space S contains N equally likely outcomes and an event E contains m outcomes, then

$$Pr(E) = \frac{m}{N}$$

In other words, if $n(E)$ is the number of outcomes in E and $n(S)$ is the number of outcomes in S, then

$$Pr(E) = \frac{n(E)}{n(S)}$$

We make use of the Multiplication Principle and the formulas for permutations and combinations, $P(n, r)$ and $C(n, r)$, respectively. The first example uses two of these formulas.

EXAMPLE 1   Five coins are tossed and the number of heads is recorded.

Problem

What is the probability of observing (a) exactly two heads, (b) at most two heads?

SOLUTION The sample space is $S = \{0, 1, 2, 3, 4, 5\}$, but these outcomes are not equally likely. To get equally likely outcomes, distinguish the coins and consider the sequence of H's and T's.

We first find the denominator, $n(S)$, in each problem. It is the total number of ways that five coins can land. There are two choices, H and T, for each of the five coins. By the Multiplication Principle presented in Chapter 6, there are $2^5 = 32$ outcomes or sequences of H's and T's. Thus the denominator for each problem is $n(S) = 32$.

(a) From Chapter 6 we see the number of ways of getting exactly two heads with five coins is $C(5, 2)$. Thus

$$Pr(2H) = \frac{C(5, 2)}{2^5} = \frac{10}{32} = \frac{5}{16}$$

(b) The event "at most two heads" is the union of the three mutually exclusive events "exactly 0 heads," "exactly one heads" and "exactly two heads." The denominator for the probability of each event is 32 and their numerators are $C(5, 0)$, $C(5, 1)$, and $C(5, 2)$, respectively. Thus

$$Pr(\text{at most } 2\ H\text{'s}) = Pr(0\ H) + Pr(1\ H) + Pr(2\ H)$$

$$= \frac{C(5,\ 0)}{32} + \frac{C(5,\ 1)}{32} + \frac{C(5,\ 2)}{32}$$

$$= \frac{1}{32} + \frac{5}{32} + \frac{10}{32} = \frac{16}{32} = \frac{1}{2}$$

Many problems in probability can be reduced to urn problems since an experiment usually consists of selecting m objects from a total of N objects. For instance, the coin tossing problem in Example 1 can be thought of as an urn problem in two ways. Think of the experiment as consisting of 5 urns, each with 2 marbles labeled H and T (or red and white). The experiment is to select a marble from each urn. Equivalently, consider the experiment as consisting of 1 urn with 2 marbles labeled H and T, and then select 1 marble five times with replacement. The sample space in each of these urn problems is identical with the sample space in Example 1. The Multiplication Principle is used for the denominator because the selection is done with replacement. The numerators use combinations because we are selecting m of the urns to have the outcome H, so that the remaining $5 - m$ urns will have T.

In previous sections we selected only one object from an urn. But now counting techniques will enable us to tackle more complex situations.

EXAMPLE 2

An urn contains 3 red marbles and 9 white marbles. Three marbles are selected at random without replacement.

Problem

Find the probability of observing (a) 3 red marbles, (b) 3 white marbles, (c) 2 red and 1 white marbles (in any order).

SOLUTION There are 12 marbles in all, so the denominator for each probability is $C(12, 3)$. To determine the numerators, think of the marbles as distinguished, say, each has a different number. The order in which they are selected is not important.
(a) There is one way to select all 3 red marbles. By formula $C(3, 3) = 1$. Hence

$$Pr(3 \text{ red}) = \frac{C(3,\ 3)}{C(12,\ 3)} = \frac{1}{220}$$

(b) The number of ways of selecting 3 white marbles is $C(9, 3)$. Hence

$$Pr(3 \text{ white}) = \frac{C(9,\ 3)}{C(12,\ 3)} = \frac{84}{220} = \frac{21}{55}$$

(c) There are $C(3, 2)$ ways to select 2 of the 3 red marbles and $C(9, 1)$ ways to select 1 of the 9 white marbles. By the Multiplication Principle there are

$C(3, 2) \cdot C(9, 1)$ ways to select 2 red and 1 white marble. Hence

$$Pr(2 \text{ red and } 1 \text{ white}) = \frac{C(3, 2) \cdot C(9, 1)}{C(12, 3)}$$

$$= \frac{3 \cdot 9}{220} = \frac{27}{220}$$

Poker Hands

When the big loser inevitably screams ''Wait! What beats what?'' he is demanding to know which poker hand is better than another. One hand beats another when the probability of its occurring is smaller. In fact the rules of poker are based on probability. To illustrate we consider three hands: a pair (2 matching cards with 3 remaining unmatched cards), a full house (3 of a kind with a pair), and a flush (all the cards the same suit). The order in which the cards appear is immaterial.

EXAMPLE 3

Problem

Compute the probability of being dealt the following 5-card poker hands: (a) a flush, (b) a full house, (c) a pair.

SOLUTION The number of possible hands with 5 cards is the denominator. Since we need 5 of the 52 cards the denominator will be $C(52, 5)$.
(a) For a flush we need to draw 5 of the 13 cards in one of the 4 suits. Thus for each suit there are $C(13, 5)$ ways to get a flush.* Since there are four distinct suits, use the Multiplication Principle to show that there are $4 \cdot C(13, 5)$ ways to draw a flush. Therefore

$$Pr(\text{a flush}) = \frac{4 \cdot C(13, 5)}{C(52, 5)} \approx .002$$

(b) To get 3 of a kind, we must draw 3 of the 4 cards of a particular type, and there are 13 different types of cards. Thus there are $13 \cdot C(4, 3)$ ways of getting 3 of a kind. But we must also fill out the hand with another pair. There are 12 remaining types of cards. Each type has 4 cards and we must draw 2 of them. Thus there are $12 \cdot C(4, 2)$ ways to fill out the hand. Therefore

$$Pr(\text{full house}) = \frac{13 \cdot C(4, 3) \cdot 12 \cdot C(4, 2)}{C(52, 5)} \approx .0014$$

(c) There are 13 different types of cards, each in a separate suit. To get a pair, we must draw 2 of the 4, which can be done in $C(4, 2)$ ways. Since this can be done in 13 ways, there are $13 \cdot C(4, 2)$ ways to get a pair. The other 3 cards must not match. In each suit there are $C(12, 3)$ ways to select 3 different types from

---

*To be precise we should subtract the 40 ways of getting a straight flush. That is, in each suit there are 10 ways the cards can also be a straight, like 2 through 6, and a straight flush beats every other hand.

the remaining 12 types of cards. The 3 cards can be in any suit, so there are 4^3 ways of selecting each of the $C(12, 3)$ possibilities. Hence there are $C(12, 3) \cdot 4^3$ ways to fill out the hand. Therefore

$$Pr(\text{a pair}) = \frac{13 \cdot C(4, 2) \cdot C(12, 3) \cdot 4^3}{C(52, 5)} \approx .42$$

Thus a full house beats a flush, which in turn beats a pair.

One of the most useful applications of probability theory is sampling. When an entire set of data cannot be inspected, usually because of time or cost considerations, a sample of the data is selected and studied. Then an inference about the whole set is made from the sample; for example, when a firm is shipped goods often a sample is chosen at random and tested for defectives. If more than a prescribed number of defectives is found, the entire shipment is rejected. The next example illustrates how probability and sampling are used in accounting. (See Problem 42 for another application with more realistic data.)

EXAMPLE 4 An auditor generally does not study all the accounts of a firm, but rather, selects a representative sample. In a certain category the auditor chooses a sample of 4 from a total of 20 accounts to see how many are delinquent. Suppose that, unknown to the auditor, 5 of the 20 are delinquent. If the auditor finds that 1 out of the 4 in the sample is delinquent, the sample will have exactly the same percentage of delinquencies as the entire set—25%. If the sample is off by 1, so that 0 or 2 delinquencies are found, the sample will not be too far off. But if 3 or 4 are discovered, then the sample will be a very poor representation of the data and the auditor's conclusions will be grossly in error.

Problem

Given that 5 out of 20 accounts are delinquent, if a sample of 4 is selected, what is the probability that 3 or 4 of the accounts are delinquent?

SOLUTION There are $C(20, 4)$ ways to select 4 accounts at random, which is the denominator. There are $C(5, 4)$ ways to select 4 of the 5 delinquent accounts. So $Pr(4 \text{ delinquencies}) = C(5, 4)/C(20, 4)$. There are $C(5, 3)$ ways to select 3 of the 5 delinquencies, and for each of these there are $C(15, 1)$ ways to select 1 of the other 15 accounts. Hence $Pr(3 \text{ delinquencies}) = C(5, 3) \cdot C(15, 1)/C(20, 4)$. Therefore

$$Pr(3 \text{ or } 4 \text{ delinquencies}) = Pr(4) + Pr(3)$$

$$= \frac{C(5, 4)}{C(20, 4)} + \frac{C(5, 3) \cdot C(15, 1)}{C(20, 4)}$$

$$= \frac{5}{4845} + \frac{30}{969} = \frac{31}{969} \approx .03$$

Thus even with a small sample of 4 out of 20, there is only a 3% chance that the sample will yield a large discrepancy from the whole set of data.

The next example demonstrates how to use the complement principle to simplify some computations. Recall that if E' is the complement of E, then $Pr(E) = 1 - Pr(E')$.

EXAMPLE 5

An urn contains 10 marbles numbered 1 to 10. Select 3 marbles from the urn with replacement and record the numbers.

Problem

What is the probability that at least 2 will have the same number?

SOLUTION Let E represent the event that at least 2 will have the same number. Then E' represents the event that the 3 marbles do not have the same number. Think of the experiment in three stages; at each stage, select a marble, record its number, and then replace it. For E' to occur the three numbers must be different—at the first stage there are ten possible numbers, at the second stage there are nine numbers, and eight at the third stage. So the number of ways of selecting 3 marbles with different numbers is $P(10, 3) = 10 \cdot 9 \cdot 8 = 720$. Permutations are used because the order of the digits is important; a different order can produce a different number, such as 123 versus 321. There are 10^3 total ways to select three numbers. Hence

$$Pr(E') = Pr(\text{three different numbers})$$

$$= \frac{P(10, 3)}{10^3}$$

$$= \frac{720}{1000} = \frac{18}{25}$$

Therefore the probability of getting at least two numbers the same is

$$Pr(E) = Pr(\text{at least two numbers the same})$$

$$= 1 - Pr(E') = 1 - Pr(\text{three different numbers})$$

$$= 1 - \frac{18}{25} = \frac{7}{25}$$

Most people find this result surprising. It seems as if the probability would be smaller. Here is another result that is even more perplexing. How many people would you have to select so that the probability of two of them having the same birthday is greater than 50%? After the next example we answer this question.

EXAMPLE 6

Problem

Four people are selected at random. What is the probability that at least two of them have the same birthday?

SOLUTION This problem is similar to Example 5. Here we have four stages instead of three and at each stage there are 365 possibilities instead of 10. Let E

be the event that at least two of them have the same birthday. Then E' is the event that all four have different birthdays. Think of the possibilities at each of the four stages for E' to occur.

stage	1	2	3	4
possibilities	365	364	363	362

Hence there are $P(365, 4)$ ways that the four can have different birthdays. There are 365^4 total ways of selecting four birthdays. Thus

$$Pr(E') = Pr(\text{four different birthdays})$$

$$= \frac{P(365, 4)}{365^4}$$

$$\approx .984$$

Therefore the probability of at least two people having the same birthday is

$$Pr(E) = Pr(\text{at least two birthdays the same})$$
$$= 1 - Pr(E') = 1 - Pr(\text{four different birthdays})$$
$$= 1 - .984 = .016$$

Example 6 says that if 4 people are chosen at random, the probability that at least 2 will have the same birthday is just 1.6%. As expected, this is a small number. The same process can be used to compute the probability that at least 2 people have the same birthday if 10, 20, or any number of people were selected. What is the smallest number of people that we can choose so that the probability is greater than 50%? Suppose n people are selected. Let E be the event that at least 2 of the n people have the same birthday. Then the formula for $Pr(E)$ is

$$Pr(E) = 1 - \frac{P(365, n)}{365^n}$$

The following table gives the values for various choices of n:

n	5	10	15	20	22	23	25	30	40	50
$Pr(E)$	.027	.117	.253	.411	.476	.507	.569	.706	.891	.970

The probability for 22 people is less than .5, but for 23 people the probability is .507, greater than 50%. Many people guess that the number needed for the probability to exceed .5 is closer to 50 or more, but notice that if 50 people are selected, it is almost certain that at least 2 will have the same birthday. It is not difficult to conduct this experiment, and a large group of people is not needed. Simply write the numbers from 1 to 365 on slips of paper and select them from a hat with replacement until the same number appears. The number of draws will most likely be close to 23.

EXERCISE SET 7–4

• In Problems 1 to 6 a coin is tossed n times and the number of heads is recorded. Find the probability of observing the given event E.

1. $n = 6$, E is "exactly one heads."

2. $n = 6$, E is "exactly two heads."

3. $n = 6$, E is "at most two heads."

4. $n = 6$, E is "at least two heads."

5. $n = 7$, E is "at least two heads."

6. $n = 7$, E is "at most two heads."

In Problems 7 to 10 an urn contains 7 red and 5 white marbles. Four marbles are selected without replacement. Find the probability of the event.

7. Exactly 4 are red.

8. Three are red and 1 is white.

9. At least 3 are red.

10. At most 1 is white.

11. A manufacturer ships a package of $n = 12$ automobile batteries. The customer selects $m = 4$ of them and tests for defectives. Suppose that the package has two defectives. Find the probability that the sample will contain at most one defective.

In Problems 12 to 14 solve Problem 11 with the different choices of n and m.

12. $n = 20$, $m = 4$.

13. $n = 20$, $m = 6$.

14. $n = 30$, $m = 5$.

15. An auditor selects a sample of 5 accounts from a total of $n = 20$ accounts and examines them for delinquency. Suppose there are $m = 5$ delinquent accounts. Find the probability that the sample will have 4 or 5 delinquent accounts.

In Problems 16 to 18 solve Problem 15 with the different choices of n and m.

16. $n = 20$, $m = 6$.

17. $n = 50$, $m = 5$.

18. $n = 100$, $m = 6$.

In Problems 19 and 20 a 3-card poker hand is dealt. Find the probability of observing the described hand.

19. All 3 cards are the same color.

20. Two of a kind.

In Problems 21 and 22 a 3-card poker hand is dealt. Find the probability of observing the described hand.

21. All 3 cards are the same suit.

22. Three of a kind.

In Problems 23 to 26 an urn contains n marbles numbered 1 to n. Select m marbles with replacement and record the number. Find the probability that at least 2 of them are the same number.

23. $n = 10$, $m = 4$.

24. $n = 15$, $m = 4$.

25. $n = 20$, $m = 5$.

26. $n = 50$, $m = 4$.

•• 27. Ten coins are tossed and the number of heads is recorded. Find the probability of observing (a) exactly three heads, (b) at most three heads, (c) at least three heads.

28. A true-false test consists of ten questions. If the student guesses at each answer, find the probability that the number of correct answers is at least seven.

29. An urn contains 10 red, 6 white, and 4 blue marbles. Four marbles are selected without replacement. Find the probability that (a) all 4 are red, (b) at least 3 are red.

30. In the game of Lotto the player selects six numbers from 1 to 40 inclusive, without repetition. The order in which the numbers are selected is not important. What is the probability of winning?

31. In poker, what is the probability of drawing (a) 4 of a kind, (b) a straight (the 5 cards are in sequence, but not necessarily dealt in sequence, so ace-2-3-4-5 in any order is a straight).

32. For a group of 10 people compute the probability that at least 2 of them have the same birthday.

33. There are 50 members in a union and 10 agree with the president on whether to take a pay cut. The president selects 4 at random and asks their opinion. What is the probability that at least 1 will agree with the president?

34. Seven letters are selected from the alphabet with replacement to form a code. Find the probability that the code has at least one letter repeated.

35. There are 15 banks in a city and 8 of them offer students a discount on checking accounts. A student calls 6 banks at random and asks if they offer a discount. Find the probability that at least 3 of the banks questioned offer a discount.

36. Thirty people are called for jury duty, 18 of whom are women. Find the probability that at most 1 woman is selected for the 12-person jury.

37. A shipment of 100 engines contains five defectives. The quality control inspector selects 3 engines at random to be tested. Find the probability that none of the engines tested is defective.

38. A batch of 50 computers contains five defectives. The quality control inspector selects 4 computers at random to be tested. Find the probability that at least 1 of the computers tested is defective.

●●● 39. A true-false test consists of 20 questions. If the student guesses at each answer, find the probability that the number of correct answers is at least 14.

40. An urn contains 20 red, 12 white, and 8 blue marbles. Five marbles are selected without replacement. Find the probability that (a) all 5 are red, (b) at least 3 are red, (c) 2 are red and at least 2 are white.

41. A study of how a bank should best improve waiting time in its drive-in facility found that bank officials define service as poor if a customer must wait more than 4 minutes.* This is because people base their impatience on the movement of the minute hand between two marks, which represents 5 minutes on a clock. The average service time was just less than 2 minutes, and out of every 14 customers, 3 would require more than $2\frac{1}{2}$ minutes. Poor service would result if 2 of these people entered the same line. Assume that a bank has 7 teller stations and 14 people are in line, two at each teller station. Suppose 3 of the 14 require more than $2\frac{1}{2}$ minutes. Find the probability that 2 of these 3 are in the same line.

42. In an experiment, reported in an article by Neter, to test the accuracy of their sampling method of accounting, executives of the Chesapeake & Ohio Railroad compared the actual amount of revenue due them on interline freight shipments versus the amount computed by sampling.† Sampling techniques have many checks to ensure that the sample contains percentages similar to the actual data. This is where probability is used. In the six-month period of the study there were 23,000 total bills, so they divided them into categories according to the size of the bill. In category 3, $11 to $20, the sample consisted of 20% of the bills. There were 8000 total bills in category 3, so 1600 bills were in the sample. If too many bills in the sample were greater than $18, the sample would give a much larger gross figure than the actual amount. There were 800 bills in category 3 that were greater than $18. Set up the formula for the probability that the sample had at most 200 bills that were greater than $18.

43. How much better are experts than nonexperts at making subjective probabilities? This is the subject of an article studying ways to measure

*B. L. Foote, "A Queuing Case Study of Drive-In Banking," *Interfaces*, Vol. 6, No. 4, August 1976.

†John Neter, "How Accountants Save Money by Sampling," *Statistics: A Guide to the Unknown*, by J. A. Tanur et al., (eds.), San Francisco, Holden-Day, Inc., 1976.

how well experts make probability assessments about events in their field of expertise.‡ The researchers questioned 22 casino dealers and 24 others on various situations in the card game blackjack. They posed the same 19 questions to each group. An example of a question is: "Out of 1000 cases in which the dealer's first card is a 4, how often will the dealer finally bust (mean- ing get a score of more than 21)?" Of the 46 individuals, five answered this question cor- rectly, where "correctly" means their answer came within 100 cases of the correct answer. Given that each person has the same probability of answering correctly, find the probability that all 5 who answered the question correctly were dealers.

‡W. A. Wagenaar and G. B. Keren, "Calibration of Probability Assessments by Professional Blackjack Dealers, Statistical Experts and Lay People," *Organizational Behavior and Human Decision Processes*, 36:406–416, 1985.

7–5 CONDITIONAL PROBABILITY

Probability theory is most often used to make predictions about the future, using past performance as a guide. Sometimes the likelihood of an event occurring is altered by new information. We often get a different perspective of a problem by looking at it from another angle. In this section we see how probability handles the situation when a new or different piece of information is included in the problem.

To illustrate, let us consider two problems. The experiment is to draw a card from a deck.

Problem *A*: Find the probability of observing the king of clubs.
Problem *B*: Suppose the card is known to be a picture card, find the probability that it is the king of clubs.

To solve Problem *A* note that there are 52 cards, they are equally likely to occur, and there is one king of clubs, so the probability is 1/52. To solve Problem *B* note that there is a restricted sample space. Since we know the card is a picture card, we are restricted to those 12 cards. One of them is the king of clubs so the probability is 1/12.

The solution to Problem *B* is called a *conditional probability*. If E and F are two events, then we write $E \mid F$ to represent the event "E given that F has occurred" or simply "E given F" or "E after F." For example, in Problem *B* if we let

E: The king of clubs   and F: A picture card

then $Pr(E \mid F) = 1/12$. This expression is called a conditional probability and is read "the probability of E given F is 1/12."

A conditional probability can be expressed in terms of the intersection of the two events. To demonstrate, consider Problem *B* again and note that $Pr(E \cap F) = 1/52$ and

$$Pr(E \mid F) = \frac{Pr(E \cap F)}{Pr(F)} = \frac{1/52}{12/52} = \frac{1}{12}$$

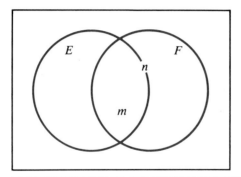

FIGURE 1

This is true for any two events E and F. Let us generalize this result for events in a sample space S with N equally likely outcomes. Refer to the Venn diagram in Figure 1. Suppose F has n outcomes and $E \cap F$ has m outcomes. Thus $Pr(F) = n/N$ and $Pr(E \cap F) = m/N$. One way to find $Pr(E \mid F)$ is to consider the restricted sample space of F. That is, since we know that F has occurred, one of the n outcomes in F must occur; there are m outcomes that are also in E, so $Pr(E \mid F) = m/n$. The preceding expression gives the same answer.

$$Pr(E \mid F) = \frac{Pr(E \cap F)}{Pr(F)} = \frac{m/N}{n/N} = \frac{m}{n}$$

This leads us to the definition of $Pr(E \mid F)$.

_____ **Definition** _____

If E and F are two events with $Pr(F) \neq 0$, then the **conditional probability** *of E given F* is

$$Pr(E \mid F) = \frac{Pr(E \cap F)}{Pr(F)}$$

The discussion preceding the definition implies that there is an alternative way to view conditional probabilities in a sample space with equally likely outcomes. In this case compute $Pr(E \mid F)$ by dividing the number of outcomes in $E \cap F$ by the number of outcomes in F. Thus *for equally likely outcomes*

$$Pr(E \mid F) = \frac{n(E \cap F)}{n(F)}$$

In the first example we use the definition because the sample space does not have equally likely outcomes. The second example demonstrates the use of the alternate formula.

EXAMPLE 1

Roll two dice and record the sum of the top faces. Define the two events.

E: Observe an even sum.
F: Observe a sum of 5, 6, or 7.

Problem

Find (a) $Pr(E \mid F)$, (b) $Pr(F \mid E)$.

SOLUTION We need to compute $Pr(E)$, $Pr(F)$, and $Pr(E \cap F)$. We have

$$Pr(E) = Pr(2, 4, 6, 8, 10, 12) = \frac{1 + 3 + 5 + 5 + 3 + 1}{36}$$

$$= \frac{18}{36} = \frac{1}{2}$$

$$Pr(F) = Pr(5, 6, 7) = \frac{4 + 5 + 6}{36} = \frac{15}{36} = \frac{5}{12}$$

$$Pr(E \cap F) = Pr(6) = \frac{5}{36}$$

(a) By the definition

$$Pr(E \mid F) = \frac{Pr(E \cap F)}{Pr(F)} = \frac{5/36}{15/36} = \frac{5}{15} = \frac{1}{3}$$

(b) Since $E \cap F = F \cap E$, by the definition we get

$$Pr(F \mid E) = \frac{Pr(E \cap F)}{Pr(E)} = \frac{5/36}{18/36} = \frac{5}{18}$$

EXAMPLE 2

A composition course has 25 students of whom 14 are male and 9 are sophomores. In addition, 5 of the sophomores are men.

Problem

A student is chosen at random. Find the probability that the student is
(a) a sophomore given that the student is a man.
(b) a woman given that the student is a sophomore.

SOLUTION Define the events

M: The student is a man.
S: The student is a sophomore.

Then we must compute (a) $Pr(S \mid M)$ and (b) $Pr(M' \mid S)$. Refer to the Venn diagram in Figure 2.

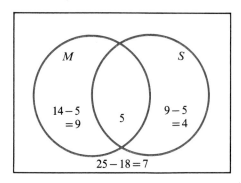

FIGURE 2

(a) We have $n(S \cap M) = 5$ and $n(M) = 14$. Hence

$$Pr(S \mid M) = \frac{n(S \cap M)}{n(M)} = \frac{5}{14}$$

(b) We have $n(M' \cap S) = 4$ and $n(S) = 9$. Hence

$$Pr(M \mid S) = \frac{n(M' \cap S)}{n(S)} = \frac{4}{9}$$

Product Rule

In applications sometimes a conditional probability is easy to compute but the probability of the intersection is not. In this case the formula in the definition of conditional probability can be altered to give the probability of the intersection. If we multiply each side of the equation in the definition by $Pr(F)$ we obtain the **Product Rule.**

_____ **Product Rule** _____
If E and F are any two events, then

$$Pr(E \cap F) = Pr(F) \cdot Pr(E \mid F)$$

The Product Rule is especially effective when solving problems with tree diagrams. The next example shows how tree diagrams can help solve probability problems that occur in stages.

EXAMPLE 3 A utility company submits a rate increase proposal to the state utility commission. The proposal will go to one of two commission members, either Ms. Allen or Mr. Barnes. In the past Ms. Allen has accepted 60% of the company's proposals, while Mr. Barnes has accepted 75%. There is a 70% chance that Ms. Allen will receive the proposal.

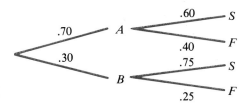

FIGURE 3

Problem

Determine the probability that the proposal will be accepted.

SOLUTION Consider the tree diagram in Figure 3, where A represents the result that Ms. Allen receives the proposal and B represents Mr. Barnes receiving it. Let S represent success and F failure of the proposal. On the branch of the tree to A we put $Pr(A) = .70$, the probability that A gets the proposal. On the branch to B we put $Pr(B) = .30$. On the branch from A to S we put $Pr(S \mid A) = .60$, so on the branch from A to F we put $Pr(F \mid A) = .40$. On the branches from B to S and from B to F we put $Pr(S \mid B) = .75$ and $Pr(F \mid B) = .25$, respectively. There are two paths that correspond to success; (1) from the start to A and then to S and (2) from the start to B and then to S. They are the events $A \cap S$ and $B \cap S$. The Product Rule is used to compute their probabilities by multiplying the numbers on the corresponding branches. Hence

$$Pr(A \cap S) = Pr(S \mid A) \cdot Pr(A) = (.60)(.70) = .42$$
$$Pr(B \cap S) = Pr(S \mid B) \cdot Pr(B) = (.75)(.30) = .225$$

The event S can occur in one of the two mutually exclusive ways, $S \cap A$ and $S \cap B$. Therefore

$$Pr(S) = Pr(S \cap A) + Pr(S \cap B) = .42 + .225 = .645$$

Hence there is a 64.5% chance of the proposal being accepted.

The next example shows how conditional probability is used in medicine. The article by Gilbert et al. concerns one of the first attempts by the medical profession to distinguish well-designed experiments from poor ones. By well-designed is meant a **controlled experiment,** that is, one in which the patients who get the treatment are compared with some who do not. In particular, does a controlled experiment give more reliable results?

The investigators reviewed 53 studies of a particular type of operation, the portacaval shunt, which treats bleeding in the esophagus. They measured the degree of control of the experiment versus the enthusiasm of the physician. They measured whether the study had a good control group, a poor control group, or none at all. They compared that measure with whether the physician was very enthusiastic, had moderate enthusiasm, or showed no enthusiasm for the operation.

EXAMPLE 4 | The following table contains the data of the study of the portacaval shunt operation.

Degree of Control	Marked	Moderate	None	Totals
Well-controlled	0	3	3	6
Poorly controlled	10	3	2	15
Uncontrolled	24	7	1	32
Totals	34	13	6	53

Degree of Enthusiasm spans Marked, Moderate, None columns.

Source: John P. Gilbert, et al., "Assessing Social Innovations: An Empirical Base for Policy," *Statistics and Public Policy,* by Fairley and Mosteller, eds., Reading, MA, Addison-Wesley, 1977, pp. 185–241.

Define the events

W: The experiment was well controlled.
P: The experiment was poorly controlled or uncontrolled.
M: The physician showed marked enthusiasm.
N: The physician showed no enthusiasm.

Problem

Find (a) $Pr(M \mid W)$, (b) $Pr(N \mid W)$, (c) $Pr(W \mid N)$, (d) $Pr(P \mid M)$

SOLUTION For the probabilities of W, M, and N we look at the totals and compute $Pr(W) = 6/53$, $Pr(M) = 34/53$, $Pr(N) = 6/53$. To calculate the probabilities of the intersections we look at the intersection of the column and row described. We get $Pr(W \cap M) = 0/53 = 0$, $Pr(W \cap N) = 3/53$, $Pr(P \cap M) = (10 + 24)/53 = 34/53$. Now we compute the conditional probabilities.

(a) $Pr(M \mid W) = Pr(W \cap M)/Pr(W) = 0/(6/53) = 0$
(b) $Pr(N \mid W) = Pr(W \cap N)/Pr(W) = (3/53)/(6/53) = 3/6 = 1/2$
(c) $Pr(W \mid N) = Pr(W \cap N)/Pr(N) = (3/53)/(6/53) = 3/6 = 1/2$
(d) $Pr(P \mid M) = Pr(P \cap M)/Pr(M) = (34/53)/(34/53) = 34/34 = 1$

The significance of this study lies in the fact that the investigators found that after four years the survival rates of the patients in the control groups were the same as those who elected to have the operation. Because $Pr(P \mid M) = 1$, the study shows that uncontrolled and poorly controlled experiments led to far greater enthusiasm than well-controlled experiments. Later investigation showed, however, that this enthusiasm was unwarranted. On the other hand, since $Pr(M \mid W) = 0$, the experimenters that used well-controlled studies had a more accurate opinion of effectiveness of the operation. This shows that performing poorly controlled experiments wastes time and can lead to false conclusions.

Independent Events

Each spin of a roulette wheel is a new experiment where the probabilities remain the same. But try to convince gamblers of that! If red has come up ten straight times, the amount of money placed on black is enormous. Yet the probability that black occurs is still $18/38 = 9/19$. Each spin of the wheel is said to be independent of the others; that is, what happens on one spin does not affect the probability of any other spin. Two events E and F are said to be independent if the likelihood of one occurring does not affect the chance of the other.

Definition

Two events E and F are **independent** if

$$Pr(E \mid F) = Pr(E)$$

EXAMPLE 5

Two dice are rolled and the sum of the top faces is recorded. Define the three events

E: A sum of 2, 7, or 11
F: An odd number
G: A sum of 4, 7, or 10

Problem

Determine whether (a) E and F are independent, (b) F and G are independent.

SOLUTION We compute $Pr(E) = 9/36 = 1/4$, $Pr(F) = 18/36 = 1/2$, $Pr(G) = 12/36 = 1/3$, $Pr(E \cap F) = Pr(7 \text{ or } 11) = 8/36 = 2/9$, and $Pr(F \cap G) = Pr(7) = 6/36 = 1/6$.
(a) We must compare $Pr(E)$ and $Pr(E \mid F)$. By definition

$$Pr(E \mid F) = \frac{(8/36)}{(18/36)} = \frac{8}{18} = \frac{4}{9}$$

Since $Pr(E) = 1/4 \neq 4/9$, E and F are not independent.
(b) We must compare $Pr(F)$ and $Pr(F \mid G)$. By definition

$$Pr(F \mid G) = \frac{(6/36)}{(12/36)} = \frac{6}{12} = \frac{1}{2}$$

Since $Pr(F) = 1/2 = Pr(F \mid G)$, F and G are independent.

In part (b) of Example 5 we could just as well have shown that $Pr(G) = Pr(G \mid F)$. That is, in the definition of independent events, one equality implies the other. (See Problem 35.) This shows that the expression "independent events" is not a one-sided relationship. That is, E and F are independent if $Pr(E \mid F) = Pr(E)$, which also implies that $Pr(F \mid E) = Pr(F)$.

The definition of independence can be coupled with the Product Rule to form another useful rule. It governs when we can multiply the probabilities of independent events. Suppose E and F are independent. Then $Pr(E \mid F) = Pr(E)$, and so the Product Rule yields

$$Pr(E \cap F) = Pr(E \mid F) \cdot Pr(F) = Pr(E) \cdot Pr(F)$$

The same steps in reverse show that if

$$Pr(E) \cdot Pr(F) = Pr(E \cap F)$$

then E and F are independent. This is an important rule so we single it out.

_____ **The Product Rule for Independent Events** _____

Two events E and F are independent if and only if

$$Pr(E) \cdot Pr(F) = Pr(E \cap F)$$

For instance, in Example 5 we have

$$Pr(F) \cdot Pr(G) = \left(\frac{1}{2}\right) \cdot \left(\frac{1}{3}\right) = \frac{1}{6} = Pr(F \cap G)$$

Thus the Product Rule confirms that F and G are independent.

The rule can be used in the opposite direction also. That is, when two events E and F are known to be independent, $Pr(E \cap F)$ can be computed by multiplying $Pr(E)$ and $Pr(F)$. An important application of this idea is in reliability. The **reliability** of a machine or process is defined to be the probability that it will not fail during an operation.*

For example, suppose the probability of a computer failing is .05. Then its reliability is the probability that it will not fail, which is $1 - .05 = .95$. Suppose this computer operates independently with another computer; that is, they both perform the same operation but independently. If the probability of the second failing is .02, then the probability of both failing is the intersection of the two independent events that each fails individually. Since we assume that they are independent, to find the probability that both fail we multiply the probabilities and get $(.05)(.02) = .001$.

This demonstrates the need for what is often called a *backup* system. One computer backs up another in the sense that it can handle the load if one fails. A concrete example of such a system is the computer facility aboard the space shuttle. Each shuttle is equipped with four computers that operate independently. Each has its own tasks but they also serve as backups for the others, especially for essential computations. If two do not agree with a calculation, the third and fourth are consulted. The historic first flight of the space shuttle *Columbia* was delayed for several days when the countdown was halted in the last few minutes because the reliability of the computer system was too low.

*Gerald J. Lieberman, ''Striving for Reliability,'' *Statistics: A Guide to the Unknown*, by J. A. Tanur et al., (eds.), San Francisco, Holden-Day, Inc., 1972, pp. 400–406.

EXERCISE SET 7–5

• In Problems 1 to 4 a card is chosen from a deck. Find the conditional probability for the following events:

E: An ace or a king
F: A picture card
G: A club or a red card

1. $Pr(E \mid F)$.

2. $Pr(F \mid E)$.

3. $Pr(E \mid G)$.

4. $Pr(F \mid G)$.

In Problems 5 to 8 a club has 45 members, 20 of whom are women, 25 are less than 20 years old, and 7 women are less than 20 years old. A member is chosen at random. Find the conditional probability.

5. Pr(the person is a woman given that the person is less than 20).

6. Pr(the person is less than 20 given that the person is a woman).

7. Pr(the person is a man given that the person is 20 or more).

8. Pr(the person is less than 20 given that the person is a man).

In Problems 9 and 10 a company sends a package by an overnight courier. The package will be picked up by one of three delivery vans, depending on which is nearest. The chances are equal for any of the vans to make the pickup. The problem gives the past record for each van's success in delivering packages on time. Find the probability that the package will be delivered on time.

9. Pr(success for van 1) $= .95$.
   Pr(success for van 2) $= .96$.
   Pr(success for van 3) $= .90$.

10. Pr(success for van 1) $= .90$.
    Pr(success for van 2) $= .93$.
    Pr(success for van 3) $= .99$.

In Problems 11 and 12 urn 1 and urn 2 have the indicated number of white and red marbles. The experiment consists of first selecting 1 of 2 urns and then selecting a marble from the urn and recording its color. For the given number of marbles in each urn, find the probability of observing a red marble.

11. Urn 1 has 5 white and 7 red marbles; urn 2 has 3 white and 9 red marbles.

12. Urn 1 has 9 white and 8 red marbles; urn 2 has 2 white and 8 red marbles.

Problems 13 to 16 refer to the table whose entries are the data from the Zeisel and Kalvern study of criminal jury trials, first mentioned in Section 7–3. After the jury gave its verdict the judge was asked what verdict he or she would have given if there had been no jury. The percentages of agreement and disagreement are in the table. One way to analyze the data is by conditional probabilities.

| | **Jury Verdict** | |
Judge Verdict	**Acquitted**	**Convicted**
Acquitted	14	3
Convicted	19	64

Source: Data from Hans Zeisel and Harry Kalvern, "Parking Tickets and Missing Women: Statistics and the Law," *Statistics: A Guide to the Unknown*, by J. A. Tanur et al., eds., San Francisco, Holden-Day, Inc., 1972, pp. 102–111.

In Problems 13 to 16 find the indicated conditional probability.

13. Pr(judge would have convicted $\mid$ jury convicted).

14. Pr(judge would have acquitted $\mid$ jury convicted).

15. Pr(jury acquitted $\mid$ judge would have convicted).

16. Pr(jury convicted $\mid$ judge would have acquitted).

In Problems 17 to 20 two dice are rolled and the sum of the top faces is recorded. For the given events E and F find $Pr(E \mid F)$ and determine if E and F are independent.

17. E: A sum of 2, 3, 4, or 5
    F: A sum of 4, 5, 6, or 7.

18. E: A sum of 2, 6, 8, or 12
    F: A sum of 2, 3, 11, or 12.

19. E: An odd sum
    F: A sum of 2, 4, or 5.

20. E: A sum greater than 5
    F: A sum less than 6.

In Problems 21 and 22 two machines operate independently and have reliability .90 and .95, respectively.

21. Find the probability that they both fail.

22. Find the probability that neither fails.

•• 23. Let E and F be events with $Pr(E) = .35$, $Pr(F) = .67$, and $Pr(E \cap F) = .24$. Find $Pr(E \mid F')$ and $Pr(E' \mid F')$.

24. An economics class has 40 students, of whom 10 are marketing majors, 23 are seniors, 12 are juniors, and 5 of the juniors and 3 of the seniors are marketing majors. A student is selected at random. Find the probability that the student is (a) a marketing major given that the student is a senior, (b) not a senior given that the student is a marketing major.

25. There are 3 urns: urn 1 has 1 red and 2 white marbles, urn 2 has 1 red and 3 white marbles, urn 3 has 3 red and 2 white marbles. A die is rolled: if 1 occurs, then a marble is selected from urn 1; If 2 or 3 occurs, then a marble is selected from urn 2; if 4, 5, or 6 occurs, then a marble is selected from urn 3. Find the probability that the marble is red.

26. For the experiment in Problem 25, if a red marble is selected, then it is replaced and a second marble is selected from urn 1, and if the first marble is white, a second marble is selected from urn 2. Find the probability that the second marble is red.

27. An urn contains 6 red and 9 white marbles. Two marbles are selected without replacement. Find the probability that (a) both are red, (b) both are white, (c) the second is red given that the first is white.

28. A factory has three assembly lines; line 1, line 2, and line 3. The reliability of line 1 is .90, meaning that 90% of its products are acceptable, 10% are defective. The reliability of line 2 is .85 and the reliability of line 3 is .95. If line 1 makes 60%, line 2 30%, and line 3 10% of the products, find the probability that a product selected at random is defective.

29. A survey shows that 20% of a population are left-handed, 60% of the left-handed people are men while 55% of the right-handed people are women. What percentage of the people are (a) left-handed women, (b) right-handed men?

Problems 30 to 33 refer to the table whose entries are the data from a large metropolitan university. Each entry is the percentage of students enrolled in the university that is described by the category heading of its row and column. The categories are full-time and part-time students, and students in the College of Business and students in the College of Arts and Sciences.

College	Full-Time	Part-Time
Business	25	18
Arts and Sciences	45	12

A student is selected at random. In Problems 30 to 33 find the indicated conditional probability.

30. Pr(student is full-time | student is a business major).

31. Pr(student is part-time | student is a business major).

32. Pr(student is full-time | student is an arts and science major).

33. Pr(student is part-time | student is an arts and science major).

••• 34. Suppose E and F are mutually exclusive events and $Pr(F)$ is nonzero. Find $Pr(E \mid F)$.

35. By definition, if E and F are independent, then $Pr(E \mid F) = Pr(E)$. Show that it is also true that $Pr(F \mid E) = Pr(F)$.

36. For any two events E and F show that $Pr(F' \mid E) = 1 - Pr(F \mid E)$.

37. Let E and F be events with nonzero probabilities. Show that, in general, only one of the following statements is true.
(a) $Pr(E \mid F) + Pr(E' \mid F') = 1$.
(b) $Pr(E \mid F) + Pr(E' \mid F) = 1$.
(c) $Pr(E \mid F) + Pr(E \mid F') = 1$.

38. Suppose E and F are independent. Determine whether E' and F' are independent.

39. Suppose E and F are independent. Show that $Pr(E \cup F) = 1 - Pr(E') \cdot Pr(F')$.

40. A family has two children. Suppose that each child is equally likely to be a boy or a girl.

Assume that the conditional probability that both children are boys given that the older child is a boy is $\frac{1}{2}$. What is the conditional probability that both are boys given that at least one of the children is a boy?

41. Brightman and Haslanger studied whether past performance of portfolio managers for mutual funds is a good indication of their ability to perform well in the future. They collected data on 52 portfolio managers over a six-year period. They ranked each manager's performance according to which quartile they were in (i.e., a ranking in the first quartile means the manager was ranked in the top quarter, or top 13). They first measured the performance in the 1971–1973 period and then compared that with the performance in the 1974–1976 period. The data are given in the following table:

1971–1973 Performance Quartile

		1	2	3	4	Total
1974–1976	1	3	1	1	8	13
Performance	2	2	3	5	3	13
Quartile	3	4	3	5	1	13
	4	4	6	2	1	13
Total		13	13	13	13	52

Source: Brightman and Haslanger, "Past Investment Performance: Seductive but Deceptive," *Journal of Portfolio Management,* Vol. 6, 1980, pp. 43–45.

Find the probability that a manager was ranked
(a) in the first quartile in the 1974–1976 period given that the manager was ranked in the first quartile in the 1971–1973 period.
(b) in the first quartile in the 1974–1976 period given that the manager was ranked in the first or second quartile in the 1971–1973 period.

(c) in the third or fourth quartile in the 1974–1976 period given that the manager was ranked in the first or second quartile in the 1971–1973 period.

42. In one of the first studies of the side effects of oral contraceptive users, Jick et al. compared the incidence of thromboembolic disease with the user's blood type. A control group of healthy women was studied for comparison. The data are in the following table:

Blood Group	Women with Thromboembolism	Healthy Women
A	32	51
B	8	19
AB	6	5
O	9	70

Source: Jick et al., "Venous Thromboembolic Disease and ABO Blood Types," *Lancet,* Vol. 1, No. 1, 1969, p. 539.

If a woman is selected at random, find the probability that she
(a) had thromboembolism given that she had blood group A.
(b) had thromboembolism given that she had blood group O.
(c) had blood group A given that she had thromboembolism.

43. One of the most perplexing and agonizing problems in our society is whether capital punishment deters homicide. In an article studying data from Vital Statistics, Stephen K. Layson analyzed arrest, conviction and execution frequencies based on conditional probability.* Let C represent the event "conviction" and E represent "execution." Then $Pr(C) = .0046$ and $Pr(E \mid C) = .00016$. Find $Pr(C \cap E)$ from the formula for conditional probability. Use the fact that $C \cap E = E$ to determine whether C and E are independent.

*Stephen K. Layson, "Homicide and Deterrence: A Reexamination of the United States Time-series Evidence," *Southern Economics Journal,* 52:68–77, 1985.

7-6 BAYES' THEOREM

The previous section dealt with conditional probability, the probability of an event E occurring given that event F has occurred. In this section the process is reversed. We use conditional probability to determine the probability that an earlier event F occurred, given that a later one E has happened. In other words, if we know $Pr(E \mid F)$, can we find $Pr(F \mid E)$?

For example, suppose a medical experiment separates patients into two groups, the treatment group and the control group. Then the treatment is administered to the former group and a placebo is given to the control group. A typical question answered in the last section would be, "What is the probability that a patient was cured given that the patient was in the control group?" In this section a typical question would be, "What is the probability that a patient was in the control group given that the patient was cured?"

The experiments in this section consist of two stages. Tree diagrams will be useful. Let us look at a simple two-stage problem to illustrate the type of problem and its solution.

EXAMPLE 1

The experiment is to choose a marble from either an urn or a hat. The urn contains 1 red and 2 white marbles and the hat contains 3 red and 2 white marbles. (See Figure 1.)

Problem

Find the probability that the marble came from the urn given that it is a red marble.

SOLUTION The tree diagram in Figure 1 shows that the experiment consists of two stages, where the first stage is to choose either the urn or the hat and the second is to select a marble. Let U, H, R, and W represent selecting the urn, the hat, a red marble, and white marble, respectively. Then we assign the following

FIGURE 1

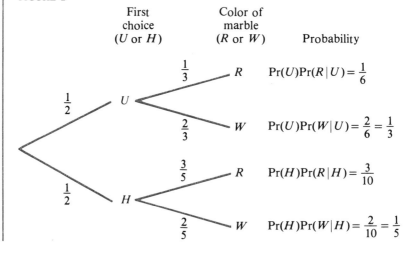

| First choice (U or H) | Color of marble (R or W) | Probability |

$$Pr(U)Pr(R \mid U) = \frac{1}{6}$$

$$Pr(U)Pr(W \mid U) = \frac{2}{6} = \frac{1}{3}$$

$$Pr(H)Pr(R \mid H) = \frac{3}{10}$$

$$Pr(H)Pr(W \mid H) = \frac{2}{10} = \frac{1}{5}$$

probabilities: $Pr(U) = Pr(H) = \frac{1}{2}$, $Pr(R \mid U) = \frac{1}{3}$ and $Pr(R \mid H) = \frac{3}{5}$, $Pr(W \mid U) = \frac{2}{3}$, and $Pr(W \mid H) = \frac{2}{5}$. The problem is to find $Pr(U \mid R)$.

Notice that R can occur in two ways, from U or from H. The corresponding branches are U-R and H-R. In terms of sets this can be expressed as $R = (U \cap R) \cup (H \cap R)$. Because these two sets are mutually exclusive, we have

$$Pr(R) = Pr(U \cap R) + Pr(H \cap R) \qquad (1)$$

From the definition of conditional probability we have

$$Pr(U \mid R) = \frac{Pr(U \cap R)}{Pr(R)}$$

Substitute equation 1 into this equation to get

$$Pr(U \mid R) = \frac{Pr(U \cap R)}{Pr(U \cap R) + Pr(H \cap R)} \qquad (2)$$

These probabilities can be computed from the tree diagram by using the Product Rule. Thus

$$Pr(U \cap R) = Pr(U) \cdot Pr(R \mid U) = \frac{1}{2} \cdot \frac{1}{3} = \frac{1}{6}$$

$$Pr(H \cap R) = Pr(H) \cdot Pr(R \mid H) = \frac{1}{2} \cdot \frac{3}{5} = \frac{3}{10}$$

Substituting these numbers into equation 2 yields

$$Pr(U \mid R) = \frac{1/6}{1/6 + 3/10} = \frac{1/6}{28/60} = \frac{5}{14}$$

Therefore the probability that the marble came from the urn given that it is red is 5/14.

The key step in Example 1 is equation 2. It expresses the conditional probability of an earlier event given a later event in terms of known probabilities. A Venn diagram helps to get a handle on the meaning of equation 2. In Figure 2 event R intersects U and H, and $U \cap R$ and $H \cap R$ are mutually exclusive sets. This shows that equation 1 holds.

FIGURE 2

FIGURE 3

$$Pr(R) = Pr(U \cap R) + Pr(H \cap R) \tag{1}$$

Now suppose that R intersects three mutually exclusive sets, say, U_1, U_2, and U_3. The counterpart to equation 1 for three sets is

$$Pr(R) = Pr(U_1 \cap R) + Pr(U_2 \cap R) + Pr(U_3 \cap R) \tag{3}$$

(See Figure 3.) Just as we substituted equation 1 into the Product Rule to get equation 2, so too we can substitute equation 3 into the Product Rule to find $Pr(U_1 \mid R)$.

$$Pr(U_1 \mid R) = \frac{Pr(U_1 \cap R)}{Pr(R)}$$

$$= \frac{Pr(U_1 \cap R)}{Pr(U_1 \cap R) + Pr(U_2 \cap R) + Pr(U_3 \cap R)}$$

In the same way we can extend this result to any finite number of mutually exclusive sets. The theorem is called **Bayes' Formula,** after the Presbyterian minister and part-time mathematician, Reverend Thomas Bayes (1702–1736).*

_____ **Bayes' Formula** _____

Suppose U_1, U_2, . . . , U_n are mutually exclusive events whose union is the entire sample space. Let E be an event with nonzero probability. Then

$$Pr(U_i \mid E) = \frac{Pr(U_i \cap E)}{Pr(E)}$$

$$= \frac{Pr(U_i \cap E)}{Pr(U_1 \cap E) + Pr(U_2 \cap E) + \cdots + Pr(U_n \cap E)}$$

$$= \frac{Pr(U_i)Pr(E \mid U_i)}{Pr(U_1)Pr(E \mid U_1) + \cdots + Pr(U_n)Pr(E \mid U_n)}$$

The next example demonstrates the use of Bayes' Formula in an experiment with three possibilities at the first stage.

*Recent evidence indicates that Bayes' Formula was known before Bayes was born. See Stephen H. Stigler, ''Who Discovered Bayes' Theorem?'' *The American Statistician,* November 1983, pp. 290–296.

EXAMPLE 2

A department store classifies credit card customers according to the maximum amount of credit allowed per month. Preferred customers are allowed over $500 credit per month, regular customers are allowed a maximum of $500 per month, and basic customers are allowed a maximum of $50 per month. A minimum monthly charge is assessed each account. If the minimum is not paid, the account is frozen and no new charges are allowed. Past experience shows that 5% of the preferred accounts are frozen, 10% of the regular accounts are frozen, and 20% of the basic accounts are frozen each month. Suppose 20% of the accounts are preferred, 50% are regular accounts, and 30% are basic accounts.

Problem

Find the probability that a particular frozen account is a preferred account.

SOLUTION The first step is to construct a tree diagram. (See Figure 4.) At the first stage let P, R, and B represent preferred, regular, and basic accounts, respectively. At the second stage let F and NF represent frozen and not frozen, respectively. Label each branch with its corresponding probability. The probability that a frozen account is a preferred account is $Pr(P \mid F)$. Using Bayes' Formula yields

$$Pr(P \mid F) = \frac{Pr(F \cap P)}{Pr(F \cap P) + Pr(F \cap R) + Pr(F \cap B)}$$

The probabilities on the tree diagram can be used to get

$$Pr(P \mid F) = \frac{(.20)(.05)}{(.20)(.05) + (.5)(.1) + (.3)(.2)}$$

$$= \frac{.01}{.01 + .05 + .06} = \frac{.01}{.12}$$

$$= \frac{1}{12} \approx .0833$$

FIGURE 4

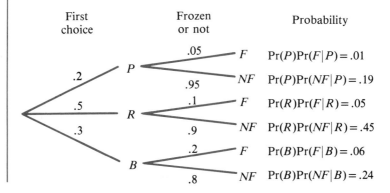

First choice	Frozen or not	Probability
	.05 F	$Pr(P)Pr(F \mid P) = .01$
.2 P	.95 NF	$Pr(P)Pr(NF \mid P) = .19$
.5 R	.1 F	$Pr(R)Pr(F \mid R) = .05$
.3	.9 NF	$Pr(R)Pr(NF \mid R) = .45$
B	.2 F	$Pr(B)Pr(F \mid B) = .06$
	.8 NF	$Pr(B)Pr(NF \mid B) = .24$

Therefore the probability that an account selected at random is preferred given that it is frozen is $\frac{1}{12}$, or about 8.3% of the frozen accounts are preferred.

The next example illustrates how to use Bayes' Formula when there are more than three possibilities at the first stage.

EXAMPLE 3

A television manufacturer purchases cabinets from five different suppliers. The following table gives the percentage of the manufacturer's business that each supplier has as well as the percent of defective cabinets from each supplier.

Supplier	Percent of Business	Percent Defectives	Product
1	.35	.02	.007
2	.25	.04	.010
3	.20	.06	.012
4	.10	.07	.007
5	.10	.09	.009
Total			.045

Problem

Find the probability that a defective cabinet was manufactured by the first supplier.

SOLUTION Let S_i be the event that the cabinet was manufactured by supplier i. Let D be the event that the cabinet is defective. We want to compute $Pr(S_1 \mid D)$. In Bayes' Formula let $n = 5$ and replace U_i by S_i and E by D. The denominator is the sum of the products of the two columns of percentages. Hence

$$Pr(S_1 \mid D) = \frac{Pr(S_1 \cap D)}{Pr(D)}$$

$$= \frac{Pr(S_1)Pr(D \mid S_1)}{Pr(S_1)Pr(D \mid S_1) + \cdots + Pr(S_5)Pr(D \mid S_5)}$$

$$= \frac{(.35)(.02)}{(.35)(.02) + (.25)(.04) + (.2)(.06) + (.1)(.07) + (.1)(.09)}$$

$$= \frac{.007}{.007 + .01 + .012 + .007 + .009} = \frac{.007}{.045}$$

$$\simeq .1555$$

Thus about 15.6% of the defectives are made by supplier 1.

The problem in Example 3 can also be solved by a tree diagram. (See Figure 5.) The first stage corresponds to the percent of business of each supplier. The second stage is the percent of defectives for each supplier. The product of the numbers on the branches of each path is then formed and the sum of the products is the solution.

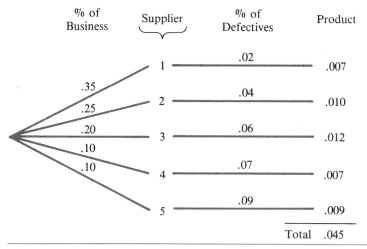

FIGURE 5

EXERCISE SET 7–6

- In Problems 1 to 4 two mutually exclusive events U_1 and U_2 and their probabilities are given together with the conditional probabilities $Pr(E \mid U_1)$ and $Pr(E \mid U_2)$. Find $Pr(U_1 \mid E)$.

  1. $Pr(U_1) = .6$, $Pr(U_2) = .4$
     $Pr(E \mid U_1) = .3$, $Pr(E \mid U_2) = .2$

  2. $Pr(U_1) = .35$, $Pr(U_2) = .65$
     $Pr(E \mid U_1) = .1$, $Pr(E \mid U_2) = .2$

  3. $Pr(U_1) = .4$, $Pr(U_2) = .6$
     $Pr(E \mid U_1) = .1$, $Pr(E \mid U_2) = .7$

  4. $Pr(U_1) = .51$, $Pr(U_2) = .46$
     $Pr(E \mid U_1) = .3$, $Pr(E \mid U_2) = .1$

In Problems 5 and 6 three mutually exclusive events U_1, U_2, and U_3 and their probabilities are given together with the conditional probabilities $Pr(E \mid U_1)$, $Pr(E \mid U_2)$, and $Pr(E \mid U_3)$. Find $Pr(U_1 \mid E)$.

  5. $Pr(U_1) = .2$, $Pr(U_2) = .5$, $Pr(U_3) = .3$
     $Pr(E \mid U_1) = .1$, $Pr(E \mid U_2) = .2$,
     $Pr(E \mid U_3) = .05$

  6. $Pr(U_1) = .35$, $Pr(U_2) = .45$, $Pr(U_3) = .20$
     $Pr(E \mid U_1) = .2$, $Pr(E \mid U_2) = .1$,
     $Pr(E \mid U_3) = .25$

In Problems 7 to 10 find the indicated probabilities by referring to the accompanying tree diagram:

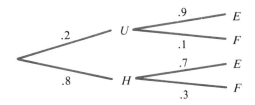

  7. $Pr(U \cap E)$, $Pr(H \cap E)$
  8. $Pr(U \cap F)$, $Pr(H \cap F)$
  9. $Pr(U \mid E)$, $Pr(H \mid E)$
  10. $Pr(U \mid F)$, $Pr(H \mid F)$

In Problems 11 to 18 find the indicated probabilities by referring to the accompanying tree diagram:

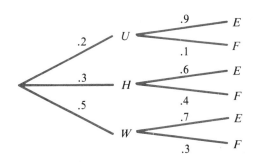

  11. $Pr(U \cap E)$, $Pr(H \cap E)$, $Pr(W \cap E)$

12. $Pr(U \cap F)$, $Pr(H \cap F)$, $Pr(W \cap F)$

13. $Pr(U \mid E)$ 15. $Pr(W \mid E)$ 17. $Pr(H \mid F)$

14. $Pr(H \mid E)$ 16. $Pr(U \mid F)$ 18. $Pr(W \mid F)$

In Problems 19 to 22 refer to the Venn diagram. The number of elements in each set is given. For example, $n(U) = 15$ and $n(U \cap E) = 5$. Find the indicated probabilities.

U	10	H	4	W	1
	5	E 6		2	

19. $Pr(U \mid E)$

20. $Pr(H \mid E)$

21. $Pr(W \mid E)$

22. $Pr(U \mid E')$

In Problems 23 and 24 the experiment is first to choose at random either an urn or a hat, each containing some red and white marbles, and then to select a marble. The urn has 4 red and 6 white marbles and the hat has 7 red and 5 white marbles. Find the probability of each event.

23. The urn was chosen given that the marble is red.

24. The hat was chosen given that the marble is white.

In Problems 25 to 27 the experiment is first to choose at random one of 3 urns, each containing some red

and white marbles, and then to select a marble. Urn 1 has 4 red and 8 white marbles, urn 2 has 3 red and 7 white marbles, and urn 3 has 7 red and 5 white marbles. Find the probability of each event.

25. Urn 1 was chosen given that the marble is red.

26. Urn 2 was chosen given that the marble is white.

27. Urn 3 was chosen given that the marble is white.

28. A firm that specializes in filling out income tax forms has two types of accounts—large (gross income greater than \$50,000) and small (gross income less than \$50,000). They find that 5% of the large accounts and 2% of the small accounts are audited by the IRS. If 40% of their accounts are large accounts, and an account is audited, what is the probability that it is large?

29. A medical research company conducted a survey of a new ointment that claims to reduce swelling if applied immediately after an injury. They choose 100 people to be in the treatment group and 50 in the control group. Those in the control group were given a similar looking ointment that is known to be ineffective in reducing swelling. Of those in the treatment group, 80 said that the swelling was reduced significantly. Of those in the control group, 30 reported that the swelling was reduced significantly. Find the probability that a person chosen at random was in the control group given that the person reported that the swelling was reduced significantly.

•• Problems 30 and 31 refer to the accompanying Venn diagram, where U, V, W, A, and B are mutually exclusive sets whose union is the sample space:

U	10	V	5	W	3	A	1	B	4
	5	E 6		2		1		1	

30. Compute $Pr(U \mid E)$.

31. Compute $Pr(V \mid E)$.

32. The experiment consists of choosing at random 1 of 5 urns, each containing some red and white marbles, and then selecting a marble. Urn i has i red marbles and $1 + i$ white marbles. (Thus urn 1 has 1 red and $1 + 1 = 2$ white marbles; urn 2 has 2 red and $1 + 2 = 3$ white marbles, and so on.) Find the probability that urn 1 was chosen given that the marble is red.

33. The following table gives the distribution of family income, the proportion of the families with that income, and the proportion of families with two wage earners. If a family with two wage earners is selected at random, what is the probability that the family income is at least $30,001?

Family Income (per year)	Proportion of Families	Proportion with Two Wage Earners
<10,000	.08	.1
10,000–20,000	.20	.2
20,001–30,000	.40	.3
30,001–40,000	.20	.2
>40,000	.12	.2

34. In a survey of skiing accidents at five hospitals, the incidence of severe fractures (a fracture that requires surgery) was as follows: hospital 1, 22% of the fractures were severe; hospital 2, 12%; hospital 3, 5%; hospital 4, 8%; hospital 5, 3%. Assuming that there were the same number of accidents at each hospital, find the probability that a severe fracture occurred at hospital 1 or hospital 2.

35. In order to measure the degree of visibility of an ad campaign, interviewers often ask consumers if they can remember seeing certain ads. An advertising firm questioned 1000 TV viewers and asked if they recalled seeing a particular commercial. They were then asked if they had bought the product recently. Of the 110 people who definitely remembered the commercial, 60 had bought the product recently. Of the 330 who vaguely remembered the commercial, 66 had bought the product recently. Of the 560 who did not remember seeing the commercial, 40 had bought the product recently. Find the prob-

ability that a person definitely remembered seeing the ad given that the person bought the product recently.

36. Red-green color-blind people see these two colors the same. The probability that this type of color blindness occurs is .049. The proportion of men who are red-green color-blind is .085. Suppose a person is selected at random. Assume the probability that the person is a man is .5. Find the probability that the person is a man given that the person is red-green color-blind.

●●● 37. Suppose $A, B, C,$ and D are mutually exclusive events in a sample space S where $S = A \cup B \cup C \cup D$. Let E be any event. Show that

$$Pr(E) = Pr(E \cap A) + Pr(E \cap B) \\ + Pr(E \cap C) + Pr(E \cap D)$$

38. Suppose A, B, and C are mutually exclusive events in a sample space S where $S = A \cup B \cup C$. Let E be any event. Show that

$$Pr(E) = Pr(E \mid A)Pr(A) + Pr(E \mid B)Pr(B) \\ + Pr(E \mid C)Pr(C)$$

39. In a study of 1405 students, participation in religious activities was compared with the answer to the question, "How happy has your home life been?" C. L. Stone found the following:

Religious Participation	Answer to Question	
	Happy	Not Happy
Not at all	183	25
Very little	406	30
Somewhat	521	24
Very much	203	13

Adapted from C. L. Stone, "Church Participation and Social Adjustment of High School and College Youth," *Rural Sociology Series on Youth*, No. 12, Bulletin 550, Agricultural Research Center, Washington State University, 1954.

Find the probability that a person who did not have a happy home life participated very little or not at all in religious activities.

40. In an article that studied the effects of seeding hurricanes, Howard et al. used Bayes' Formula

extensively.* The possibility of mitigating the destructive force of hurricans by seeding them with silver oxide was suggested as early as 1961, but strong evidence was not obtained until 1969 when "Hurricane Debbie" was seeded. The researchers defined the events: H_1—seeding reduces wind speed; H_2—seeding has no effect; H_3—seeding increases wind speed. They determined the theoretical probabilities $Pr(H_1) = .49 = Pr(H_2)$, $Pr(H_3) = .02$. They defined the event "Debbie" to be that wind reduction of at least 31% was observed. They computed $Pr(\text{Debbie} \mid H_1) = .6$, $Pr(\text{Debbie} \mid H_2) = .25$, and $Pr(\text{Debbie} \mid H_3) = .4$. Find the probability that H_1 occurred given that Debbie occurred.

41. The Equal Employment Opportunity Act of 1972 extended Title VII of the Civil Rights Act protection against sex discrimination in employ-ment to academic personnel. In a paper studying the extent to which the salary structure of a large public university has changed since 1972, the researchers applied their findings to the University of Arizona.† In one part of the study the percentage of female faculty in the College of Liberal Arts was compared with the percentage of female faculty in other colleges in 1982. These data are given in the table below. Use Bayes' theorem to find the probability that a faculty member selected at random is in the college of liberal arts given that the person is female.

| | College | |
	Liberal Arts	Other
Female	51	106
Male	408	572

*Ronald A. Howard et al., "The Decision to Seed Hurricanes," *Science*, Vol. 176, June, 1972, pp. 1191–1202.

†Sharon B. Megdal and Michael R. Ransom, "Longitudinal Changes in Salary at a Large Public University: What Response to Equal Pay Legislation?" *American Economics Review*, 75:271–274, 1985.

7–7 BERNOULLI TRIALS AND EXPECTED VALUE

A common type of problem in probability is one in which an experiment is repeated several times. To illustrate, consider the following problems.

1. Toss a coin five times and record H or T each time.
2. Roll a die four times and record whether the number on the top face of each roll is 1 or not.
3. Select ten patients and determine if each is cured.

Each repetition is called a **trial.** In Problem 1 the trial is tossing a coin and is repeated five times. In Problem 3 the trial is to select a patient and is repeated ten times. Each problem has specific characteristics: a physical experiment or trial is repeated several times; the trials are identical; each trial has the same two outcomes. The outcomes are usually called success and failure, or simply S and F. In most experiments the connotation of "success" is arbitrary. Thus S can be assigned to either outcome and then F is its complementary event. Therefore if $Pr(S) = p$, then $Pr(F) = 1 - p$.

Bernoulli Trials and Binomial Experiments

Experiments of this kind are called *binomial experiments* and the trials themselves are called *Bernoulli trials*. Thus a binomial experiment consists of the repetition of several Bernoulli trials. In Problem 2 the Bernoulli trial is to roll a die, and the binomial experiment is the repetition of the trial four times. Let us summarize.

_____ **Definition** _____

A **binomial experiment** is a probability problem consisting of one experiment repeated several times. Each repetition is called a **Bernoulli trial.** The experiment has the following properties:

1. The experiment in each trial remains the same when it is repeated.
2. There are two outcomes, called success and failure, or simply S and F.
3. The trials are independent.
4. The probability of the outcomes remain the same for each trial.

Sometimes a binomial experiment is called a **Bernoulli process.**
    We considered experiments like this in earlier sections, but we had to solve each problem from scratch. Example 1 illustrates the derivation of a formula that will help solve this common type of problem.

EXAMPLE 1 | **Problem**

Roll a die four times. Find the probability of observing (a) exactly four 1's (b) exactly three 1's.

SOLUTION Let us assign S to the event of observing 1. For each trial $Pr(S) = 1/6$ and $Pr(F) = 5/6$. Hence
(a) Pr(exactly four 1's) $= Pr(SSSS) = (1/6)(1/6)(1/6)(1/6) = 1/1296$.
(b) There are $C(4, 3) = 4$ ways to observe exactly three 1's: $SSSF$, $SSFS$, $SFSS$, and $FSSS$. The probability of each is the product of three $(1/6)$'s and one $5/6$. Hence

$$Pr(\text{exactly three 1's}) = Pr(\text{exactly three } S\text{'s})$$

$$= C(4, 3) \left(\frac{1}{6}\right)^3 \left(\frac{5}{6}\right)$$

$$= 4\left(\frac{1}{216}\right)\left(\frac{5}{6}\right) = \frac{20}{1296} = \frac{5}{324}$$

    The second part of Example 1 provides a model for the general case. Suppose a binomial experiment consists of n repetitions of a Bernoulli trial with $Pr(S) = p$. The problem is to find the probability of observing exactly x successes. One way to observe x successes is for the S's to come first followed by the F's. There would be x S's and $n - x$ F's.

$$\underbrace{SS \cdots SS}_{x \text{ times}}\underbrace{FF \cdots FF}_{n - x \text{ times}}$$

Since $Pr(S) = p$, we have $Pr(F) = q = 1 - p$. Hence

$$Pr(\underbrace{SS \cdots SS}_{x \text{ times}}\underbrace{FF \cdots FF}_{n-x \text{ times}}) = \underbrace{Pr(S)Pr(S) \cdots Pr(S)}_{x \text{ times}}\underbrace{Pr(F) \cdots Pr(F)}_{n-x \text{ times}}$$

$$= \underbrace{pp \cdots pp}_{x \text{ times}}\underbrace{(1-p) \cdots (1-p)}_{n-x \text{ times}}$$

$$= p^x (1-p)^{n-x}$$

The event "exactly x S's" in n trials occurs with any arrangement of the x S's and $n - x$ F's, and there are $C(n, x)$ such arrangements. This leads to the following general result, where we substitute q for $1 - p$:

_____ **Probabilities of Bernoulli Trials** _____

In a binomial experiment consisting of n repetitions of a Bernoulli trial with $Pr(S) = p$ and $Pr(F) = 1 - p = q$, the probability of x successes (implying there are $n - x$ failures) is

$$Pr(\text{exactly } x \text{ } S\text{'s}) = C(n, x)p^x q^{n-x}$$

_____

The next two examples illustrate the use of the formula.

EXAMPLE 2   An urn contains 3 red and 7 white marbles. Five marbles are drawn with replacement.

Problem

Find the probability of observing (a) exactly 3 red marbles, (b) at least 3 red marbles.

SOLUTION This is a binomial experiment with five repetitions of the Bernoulli trial of drawing a marble. Let S represent drawing a red marble. Then $p = Pr(S) = \frac{3}{10} = .3$ and $q = Pr(F) = 1 - \frac{3}{10} = \frac{7}{10} = .7$.
(a) We need to compute "exactly three successes." From the formula with $n = 5$ and $x = 3$ we have

$$Pr(x = 3) = C(5, 3)(.3)^3(.7)^{5-3}$$
$$= 10(.027)(.49) = .1323$$

(b) The event "at least 3 red marbles" consists of the three mutually exclusive events "exactly 3," "exactly 4," and "exactly 5" marbles. As in part (a), we find the latter two probabilities from the formula with $n = 5$ and $x = 4$ and 5, respectively.

$$Pr(x = 4) = C(5, 4)(.3)^4(.7)^1 = .02835$$
$$Pr(x = 5) = C(5, 5)(.3)^5(.7)^0 = .00243$$

Therefore if we let "at least three S's" be represented by "$x \geq 3$"

$$Pr(x \geq 3) = Pr(3) + Pr(4) + Pr(5)$$
$$= .1323 + .02835 + .00243 = .16308$$

EXAMPLE 3

Food companies use what is known as a "triangle taste test" to judge the significance of the taste of one product over another. For example, when the Coca-Cola Co. was preparing to introduce a new brand of Coke they used this test to evaluate the difference in taste between the old brand, now called "Coke Classic," against the new version. In the triangle taste test the taster is given three samples, two of which are alike, and asked to choose the odd food by tasting. Suppose the taster does not have a well-developed sense of taste for this product and chooses at random.

Problem

If the taster makes six identical taste tests of the product, find the probability that the taster will make (a) no correct choices, (b) exactly one correct choice, (c) at least three correct choices.

SOLUTION This is a binomial experiment with six repetitions and with $Pr(S) = 1/3$ where S represents choosing the odd food. In the formula let $n = 6$ and $x = 0$, $x = 1$, and $x \geq 3$, respectively. We have
(a) $Pr(x = 0) = C(6, 0)(1/3)^0(2/3)^6 = 64/729 \approx .09$
(b) $Pr(x = 1) = C(6, 1)(1/3)^1(2/3)^5 = 192/729 \approx .26$
(c) We need to compute

$$Pr(x \geq 3) = 1 - [Pr(0) + Pr(1) + Pr(2)]$$

Hence we need to calculate

$$Pr(x = 2) = C(6, 2) \left(\frac{1}{3}\right)^2 \left(\frac{2}{3}\right)^4 = \frac{240}{729} \approx .33$$

Therefore

$$Pr(x \geq 3) = 1 - [Pr(0) + Pr(1) + Pr(2)]$$
$$\approx 1 - [.09 + .26 + .33] = 1 - .68 = .32$$

Thus there is only about a 32% chance that the taster will get at least three of six choices correct.

Expected Value

An important topic in many applications of probability is that of the "expected value" of an experiment. The expected value is very similar to the concept of "average value." Intuitively, the average is used to measure the center of a set of numbers, while the expected value measures the center of a probability distribution whose outcomes are numbers. In the next chapter we will show the close connection between average and expected value.

For example, consider the experiment of tossing a coin 100 times. How many times would we expect heads to appear? While we might not get exactly 50 heads each time the experiment is run, we would certainly expect the number of heads to be close to 50, since $Pr(H) = \frac{1}{2}$ and $(\frac{1}{2}) \cdot 100 = 50$. If a die were rolled 600 times, about how many 2's would appear? Since $Pr(2) = \frac{1}{6}$, we would expect about

$(\frac{1}{6})\cdot 600 = 100$ 2's. Therefore for the binomial experiment of rolling a die 600 times and recording the number of 2's the expected value is 100.

These examples suggest the following result concerning the expected value of a binomial experiment.

If a binomial experiment has n repetitions with $Pr(S) = p$, then the expected value E is

$$E = np$$

The expected value is a useful property for any probability distribution. To see how to compute it, we look at a simple experiment whose outcomes are not equally likely. An urn contains 10 marbles. One marble has the number 1 marked on it, 3 marbles have the number 2, and 6 have the number 3. A marble is selected from the urn and its number is recorded. What is the expected value? If the experiment were done many times, the number 1 would be expected to appear one-tenth of the time, 2 would appear three-tenths of the time, and 3 would appear six-tenths of the time. The probability distribution is

x	1	2	3
Pr	$\frac{1}{10}$	$\frac{3}{10}$	$\frac{6}{10}$

The **expected value** is then the sum of the products of the outcomes and their respective probabilities. Hence the expected value is

$$1(\tfrac{1}{10}) + 2(\tfrac{3}{10}) + 3(\tfrac{6}{10}) = \tfrac{25}{10} = 2.5$$

This means that if the experiment were done, say, 100 times, we would expect the average of the 100 outcomes, that is, their sum divided by 100, to be 2.5. Thus we would expect the sum of the 100 outcomes to be 250.

This expression demonstrates that the expected value is the sum of the products of the outcome and their corresponding probabilities. We can now formally state the definition.

--- **Definition** ---

If an experiment has the probability distribution, where the x_i are numbers,

x_i	x_1 x_2 ... x_n
p_i	p_1 p_2 ... p_n

then the expected value, E, of the experiment is

$$E = x_1 p_1 + x_2 p_2 + \cdots + x_n p_n$$

The next two examples illustrate the use of this definition. Example 4 discusses the concept of a "fair game," meaning that each person has an equal chance to win. That is, if the game was played many times, each player would expect to win about the same amount of money. This means that a game is fair if the expected value is 0.

EXAMPLE 4

Two players, A and B, play the following game. A draws a card from a shuffled deck. If a picture card is drawn, A wins $3 and if another kind of card is drawn, B wins $1.

Problem

Is the game fair?

SOLUTION We compute the expected value and then interpret its meaning. There are two outcomes: A either wins $3 or loses $1, so the two outcomes are 3 and -1. The outcome 3 occurs if one of the picture cards is drawn so the corresponding probability is $\frac{3}{13}$. Hence $Pr(3) = \frac{3}{13}$. Similarly, $Pr(-1) = \frac{10}{13}$. Using the notation in the definition, we have $x_1 = 3$, $p_1 = \frac{3}{13}$, $x_2 = -1$, and $p_2 = \frac{10}{13}$. Thus the probability distribution is

x_i	3	-1
p_i	$\frac{3}{13}$	$\frac{10}{13}$

Hence the expected value E for player A is

$$E = x_1 p_1 + x_2 p_2$$
$$= 3(\tfrac{3}{13}) + (-1)\tfrac{10}{13}$$
$$= \tfrac{9}{13} - \tfrac{10}{13} = -\tfrac{1}{13}$$

Therefore the game is not fair because A will likely lose $1 every 13 plays in the long run.

The next example shows how the expected value is often used in business applications. When someone starts a new business a good knowledge of the prospects for success is necessary. Often the expected value can give a good indication of the likelihood for success.

EXAMPLE 5

A person has applied for a donut shop franchise. There are two locations that look profitable. Each has a good traffic pattern indicating that a heavy volume of customers is possible. Site 1 has most of its traffic in the morning and afternoon: 50% of the customers will come in the morning, 40% in the afternoon, and 10% in the evening. Site 2 has a more homogeneous clientele: 30% will come in the morning, 30% in the afternoon, and 40% in the evening. Each site has the same projected volume—about 1000 customers per day. The parent company has determined that the average customer spends $5 in the morning, $4 in the afternoon, and $3 in the evening.

Problem

Which site has the larger expected profit?

SOLUTION The problem consists of finding the expected value for two experiments: site 1 and site 2. Here are the probability distributions for each experiment.

Period of Day	Revenue per Customer (x_i)	Percentage of Volume	
		Site 1 p_i	Site 2 p_i
Morning	$5	.5	.3
Afternoon	$4	.4	.3
Evening	$3	.1	.4

The expected value for site 1 is

$$E = 5(.5) + 4(.4) + 3(.1) = 2.5 + 1.6 + .3 = 4.4$$

The expected value for site 2 is

$$E = 5(.3) + 4(.3) + 3(.4) = 1.5 + 1.2 + 1.2 = 3.9$$

Hence site 1 has a better expected value of $4.40 per customer versus $3.90 per customer for site 2. With a volume of 1000 customers per day this means that the owner can expect a revenue of about $4400 per day from site 1 and about $3900 per day from site 2.

EXERCISE SET 7–7

In Problems 1 to 4 the experiment is to toss a coin six times and record the sequence of heads and tails. Find the probability of the event.

1. Exactly four heads.

2. Exactly five heads.

3. At least four heads.

4. At least five heads.

In Problems 5 to 8 an urn contains 5 red and 15 white marbles. Select n marbles with replacement and find the probability of the event.

5. $n = 5$, exactly 4 red marbles.

6. $n = 5$, at least 4 red marbles.

7. $n = 6$, at least 5 red marbles.

8. $n = 7$, at most 2 red marbles.

In Problems 9 and 10 a card is drawn from a deck five times with replacement and it is recorded whether or not it is a club. Find the probability of the event.

9. At least 4 clubs.

10. At most 2 clubs.

In Problems 11 to 14 find the expected value of the binomial experiment.

11. Toss a coin 50 times and record the number of heads.

12. Toss two coins 200 times and record the number of times two heads occurs.

13. Roll a die 12 times and record the number of times 6 occurs on the top face.

14. Roll two dice ten times and record the number of times the sum of 2 occurs.

In Problems 15 to 18 find the expected value for the games. Is the game fair? (A game is fair if the expected value is 0.)

15. Player A chooses a card from a deck of 52 cards and wins $4 if a picture card appears and loses $1 if it is not a picture card.

16. Player A chooses a card from a deck of 52 cards and wins $13 if a picture card appears and loses $3 if it is not a picture card.

17. Player A rolls two dice and records the sum of the top faces. A wins $9 if 7 or 11 appears and loses $2 otherwise.

18. Player A rolls two dice and records the sum of the top faces. A wins $3 if 2, 7, or 11 appears and loses $1 otherwise.

In Problems 19 and 20 determine which site is predicted to produce the larger revenue for a proposed franchise.

		Percentage	
Expenditure		Site 1	Site 2
19. Morning	$5	.5	.6
Afternoon	$3	.1	.1
Evening	$6	.2	.2
Late night	$2	.2	.1

		Percentage	
Expenditure		Site 1	Site 2
20. Morning	$4	.4	.3
Afternoon	$3	.3	.4
Evening	$5	.1	.2
Late night	$2	.2	.1

•• 21. It is known that 80% of the population in a ward favor an upcoming bond issue. Eight people are selected. Find the probability that (a) exactly six of them will favor the bond issue, (b) at least six of them will favor the bond issue.

22. A student takes a ten-question true-false test and guesses at the answers. What is the probability of getting (a) exactly seven correct, (b) at least seven correct?

23. The past performance of a machine shows that it produces 5% defectives. What is the probability that in a run of ten items (a) exactly eight will be defective, (b) at least eight will be defective?

24. A die is rolled eight times. What is the probability of observing (a) exactly five 6's, (b) at most five 6's?

25. A part of a final exam consists of five questions that the student has not studied. If the questions are multiple choice, each with four answers, and the student guesses, find the probability that the student gets (a) exactly three correct, (b) at least three correct.

26. Find the expected value for the probability distribution.

x	2	3	4	5	6
Pr	.05	.25	.35	.15	.20

27. An urn contains 3 red and 7 white marbles. Ten marbles are selected with replacement and the number of red marbles is recorded. Find the expected value of the experiment.

28. A raffle offers a first prize of $5000, two second prizes of $1000, four third prizes of $500, and ten prizes of $100. One thousand tickets are sold at $20 each. Find the expected value for the person who buys one ticket.

29. A lottery consists of a player choosing a three-digit number. Repetitions are allowed. Each ticket costs $1 and it pays $500 to win. Compute the expected earnings of the player who purchases one ticket.

30. Compute the expected earnings of the lottery player in Problem 29 if repetitions are not allowed.

31. In a lottery a player purchases a ticket with a three-digit number for $2. Repetitions are allowed. If the number is drawn, the player wins $400. If the number drawn is not the player's but the digits are the same except for the order, the player wins $200. Find the expected earnings.

32. The annual premium for a $6000 insurance policy against the loss of jewelry is $100. If the probability that the jewelry will be lost is .005, what is the expected gain or loss on the insurance policy?

33. A common way that salespeople assess their potential is to assign probabilities to their success at selling certain volumes of business. For

the following month a salesperson determines from past records that the probability of selling (a) less than $10,000 worth of business next month is .10, (b) between $10,000 and $20,000 is .45, (c) between $20,000 and $30,000 is .40, and (d) greater than $30,000 is .05. Find the salesperson's expected earnings.

34. A professional tennis player is asked what he expects to win each week. Of course, it is impossible to determine how he will perform in any given week, but it is possible to calculate the player's expected winnings from past performances. The following table gives the likelihood that the tennis player finishes the next tournament in one of the first five places or in a lower position. It also gives the average prize for that place. Find his expected winnings for the next tournament.

place	1	2	3	4	5	>5
probability	.02	.04	.08	.12	.20	.54
prize ($1000)	50	25	12	5	2.5	1

In Problems 35 to 38 we consider games that are similar to those encountered in Chapter 10, Game Theory. The outcomes in the games are given along with their probabilities. Find the expected value.

35.

x_i	-1	0	2	3
p_i	.3	.1	.5	.1

36.

x_i	-1	-5	1	10
p_i	.2	.3	.4	.1

37.

x_i	-10	2	3	3
p_i	.15	.75	.08	.02

38.

x_i	-10	1	2	3
p_i	.15	.7	.1	.05

●●● 39. There is a difference between the game of roulette played in America versus Europe. The American roulette wheel has 38 slots, numbered 1 to 36 with one labeled 0 and another 00. Of the 36 number slots, 18 are red and 18 are black. The other 2 are green. The European wheel does not have the 00 slot and the betting differs. The differences can be compared via the expected value to see which version gives the player the better advantage. Consider first the version played in America. If you bet $1 on red, you have two possibilities: you may win $1 (in addition to your initial bet of $1) with probability $\frac{18}{38} = \frac{9}{19}$, or you may lose your bet of $1 with probability $\frac{20}{38} = \frac{10}{19}$. Here is the probability distribution.

American Roulette

Outcomes x_i	Probability p_i	$x_i p_i$
1	$\frac{9}{19}$	$\frac{9}{19}$
-1	$\frac{10}{19}$	$-\frac{10}{19}$

In European roulette there is a third outcome besides the same two for the American version—if the ball lands on green, you do not necessarily lose your bet. Your bet is carried to the next spin where there are again three possibilities: if the ball lands on red, you get your $1 bet back, so you break even; if it lands on black, you lose $1; and if it lands on green again, you get back $.50, which means you lose $.50 of your initial bet. There are four outcomes: 1 (or win $1), 0 (or break even), $-.5$ (or lose $.50), and -1 (or lose $1). Construct the probability distribution for European roulette and find the expected value for each game.

40. In an attempt to determine if performance on ESP tests is affected by the person's belief in ESP, Ryzl* scored people according to their relative ability on an ESP test. Forty percent of the people performed well on the ESP test. Suppose 10 are selected at random. Find the probability that at least 9 of them performed well.

41. Land allocation for farmers is always done under risk. They are guided by return on their crops, but various conditions force decisions to be made whose outcomes are unknown: What

---

*M. Ryzl, "Precognition Scoring and Attitude Toward ESP," *Journal of Parapsychology,* Vol. 32, pp. 1–8, 1968. Ryzl found that 96% of the people who scored well on the test also believed in ESP. Do you think there is a connection between scoring well on the ESP test and believing in ESP?

will the weather conditions be during the season? How much fertilizer will produce the best harvest? In an article weighing various options for farmers, R. N. Collender and D. Zilberman use the expected value to predict the return for several different strategies. They center their attention on farmland in the lower Mississippi drainage area. The land can be used to grow corn or cotton. Two types of plantings of each crop are to be made. Type 1 corn and cotton do well under good weather conditions while type 2 corn and cotton do well under good weather conditions. Certain percentages of the available land are planted with each type of crop. There are constraints dictated by the marketplace, for instance, no more than 60% of the land can be used for cotton, and each crop must be planted on at least 10% of the land. The table gives the return per acre expected from each type of crop and two strategies for the percent of land used for each type of crop. Which strategy will give the highest return?

Type of Crop	Return ($/acre)	Percent Land Planted	
		Strategy A	Strategy B
Type 1 corn	53	10	10
Type 2 corn	77	30	30
Type 1 cotton	244	20	30
Type 2 cotton	351	40	30

Source: R. N. Collender and D. Zilberman, "Land Allocation under Uncertainty for Alternative Specifications of Return Distributions," *American Journal Agricultural Economics,* 67; 1985.

42. Probabilistic inferences and forecasts are an integral part of many decision-making situations.

In an article evaluating such forecasts, Keith Levi studies the decision-making process of a patient who has found a lump under the skin and wants two specialists to determine whether the tumor is malignant or benign. To evaluate the accuracy of the two physicians, data from past diagnoses are studied. There are four possibilities for each case: the diagnosis was that the tumor was benign and it was actually either benign or malignant, or the diagnosis was that it was malignant and it was actually benign or malignant. The best case is when it is benign and it is diagnosed as such. This is referred to as event A. The next best case, event B, is when the tumor is malignant and it is diagnosed as such. An incorrect diagnosis is made otherwise. The worst error, event C, would be that the tumor was malignant and was diagnosed as benign. The remaining possibility is event D, when the tumor is benign but diagnosed as malignant. The next step in the analysis is to assign a weight, called a *utility,* to each possibility. The worst case, event C, is arbitrarily assigned 0, and 1 is assigned to the best case, event A. Then the relative merits, or demerits, of the remaining cases are determined. The patient in the study decided to assign .7 to event B and .2 to event D. The percentage of cases handled in each category by each physician was then determined, using 180 cases for each physician. The data are given in the table. Compute the expected value for each physician to evaluate the relative performance of each.

Diagnosis of Tumors: Physician Performance

Diagnosis: Actual:	Benign Benign	Malignant Malignant	Benign Malignant	Malignant Benign
Event	A	B	C	D
Utility	1	.7	0	.2
Past performance				
Physician 1	.54	.09	.01	.36
Physician 2	.29	.35	.27	.09

Source: Keith Levi, "A Signal Detection Framework for the Evaluation of Probabilistic Forecasts," *Organizational Behavior and Human Decision Processes,* 36, 1985, pp. 143–166.

In a celebrated court case in 1974 Ms. Jane Blalock sued a group of her peers for damages, claiming they had deprived her of her means of income. Ms. Blalock joined the Ladies Professional Golf Association (LPGA) in 1969 and soon became a star performer. By May 1972 she had become the year's leading money winner on the tour. Then during the final round of the Bluegrass Invitational Tournament she was disqualified from the tournament for an alleged infraction of the rules. A committee of her competitors, selected from the LPGA, also suspended her from the next tournament, the Ladies Carling Open. Blalock sought relief in court under the Sherman Antitrust Act, which holds that individuals cannot be suspended from their profession by their peers because it would tend to lessen competition. This legal precedent appeared justified because her peers would clearly benefit by her absence from the tournament. Blalock received an injunction enabling her to play in the Carling Open, but it was issued too late for her to join the tournament.

Besides being denied the chance to earn prize money in the Carling Open, Blalock had incurred a large legal fee, so she again sued the LPGA for damages under the Sherman Act. The only legal point the court could rule on was her loss of potential earnings from being denied the right to participate in the tournament. This is a very sticky issue in most antitrust cases because the court must decide what the earnings of the claimant, if any, would have been had the competitors not interfered, in this case by suspending the claimant. Thus her problem was to convince the court that, had she been allowed to play, she would most likely have scored low enough to win prize money—and how much (remember that in golf the lower the score the better). Therefore she had to estimate her score in a tournament in which she never played—and most important, her estimate and her means of obtaining it, had to convince a jury that it was valid.

One of the key issues for Blalock was to use a measure that would not only give a clear indication of her most probable score, but would also be easy for a jury to understand. She used the expected value.

A wide range of statistics was gathered and her legal team decided to use the nine tournaments before the disqualification to compare Blalock's performance with that of her competitors. Some fairly complicated statistical measures were used to measure the probability of Blalock's attaining a particular score. These probabilities were obtained by comparing Blalock's scores in the previous nine tournaments with those of her main competitors. We summarize the results in Table 1. The possible scores, or outcomes, are the integers from 209 (that being the lowest since it was the winning score in the tournament) to 232 (that being the highest because anyone with a higher score did not win any money). The data become more manageable if we limit the 24 outcomes to 8, so we let the outcome 210 represent the three possible scores, 209, 210, or 211. The corresponding probability, .07, is therefore the probability that she would score either 209, 210, or 211. Thus the probability that she would score between

TABLE 1

Possible Outcome x_i	Probability p_i	$x_i p_i$
210	.07	14.70
213	.16	34.08
216	.23	49.68
219	.24	52.56
222	.17	37.74
225	.09	20.25
228	.03	6.84
231	.01	2.31
Total		218.19

Source: Adapted from Ferdinand K. Levy, "Anti-trust and the Links—Estimating a Golfer's Tournament Score," *Interfaces,* 6:6–17, 1976.

212 and 214, represented by the outcome 213, is .16. In Table 1 the possible outcomes are the x_i's and the probabilities are represented by p_i's. In the third column we compute $x_i p_i$. Then the sum of the $x_i p_i$'s is the expected value. Thus the expected value is 218.19, or a score of about 218. Blalock reasoned that this was the best estimate of her score for the tournament in which she was disqualified, and thus she should be awarded a judgment equal to the prize money given to the person(s) who scored 218 in the tournament. Three women shot 218 and tied for fifth place, each earning $1427.50. If the jury ruled in her favor, these damages would be tripled and her legal fees included under the provisions of the Sherman Act.

This argument was so overwhelmingly convincing to the jury that they decided that not only were Blalock's chances of winning the prescribed amount very good, but the likelihood of her winning the tournament and garnering the first place prize was great enough that they awarded her the first place money— $4500. This was then tripled and her legal expenses were included.

Case Study Exercises

In Problems 1 to 4 refer to Table 1 and find the probability of Jane Blalock's attaining the described score.

1. A score less than or equal to 220.

2. A score less than or equal to 226.

3. A score between 218 and 226 inclusively.

4. A score between 221 and 229 inclusively.

In Problems 5 and 6 assume that Jane Blalock's probability of attaining a particular outcome in a tournament is given in the table. As in Table 1, assume that the outcomes represent the probability of her getting that score, or the one lower or the one higher. Thus the outcome of 222 represents the probability of her getting 221, 222, or 223. Find the expected value.

5. **Outcome**	**Probability**		6. **Outcome**	**Probability**
210	.05		210	.10
213	.20		213	.15
216	.25		216	.20
219	.22		219	.25
222	.15		222	.22
225	.08		225	.04
228	.03		228	.03
231	.02		231	.01

CHAPTER REVIEW

In Section 6–1 we said that set theory forms the basis for probability. In the list of key terms for this chapter we put the corresponding terminology from Chapter 6 in parentheses.

Key Terms	Notation	Formulas
7–1 Sample Spaces		
Frequency		
Relative Frequency		
Probability	$Pr(\ )$	
Experiment		
Sample Space (universal set)		
Outcome (element)		
Event (subset)		
One or the Other Occurs (union)	$\cup$	
Both Occur (intersection)	$\cap$	
Mutually Exclusive Events (disjoint subsets)		
Does Not Occur (complement)	E'	
Rule for Complements		
7–2 Fundamental Probability Principles		
Equally Likely Events		
Probability Distribution		
Addition Principle for Mutually Exclusive Events		$Pr(E) + Pr(F)$
Certain Event		$Pr(E) = 1$
Impossible Event		$Pr(E) = 0$
7–3 The Addition Principle		
Addition Principle		$Pr(E) + Pr(F) - Pr(E \cap F)$

Odds in Favor of an Event E	m to n	$\dfrac{Pr(E)}{Pr(E')}$
		$Pr(E) = \dfrac{m}{m + n}$

Subjective Probability
Theoretical Probability
Empirical Probability

7–5 Conditional Probability

Conditional Probability	$Pr(E \mid F)$	$\dfrac{Pr(E \cap F)}{Pr(F)}$
Product Rule		$Pr(E \cap F) =$ $Pr(E \mid F) \cdot Pr(F)$
Controlled Experiment		
Independent Events		$Pr(E \mid F) = Pr(E)$
Reliability		

7–6 Bayes' Formula

Bayes' Formula

$$Pr(U_i \mid E) = \frac{Pr(U_i \cap E)}{Pr(E)}$$

$$\frac{Pr(U_i)Pr(E \mid U_i)}{Pr(U_1)Pr(E \mid U_1) + \cdots + Pr(U_n)Pr(E \mid U_n)}$$

7–7 Bernoulli Trials and Expected Value

Trial
Binomial Experiment

Bernoulli Trial

$Pr(x\ S\text{'s in } n \text{ trials})$

Bernoulli Process

$= C(n, x)p^x q^{n-x}$

Expected Value

$E = p_1 x_1 + \cdots + p_n x_n$

REVIEW PROBLEMS

In Problems 1 and 2 find the sample space of the experiment.

1. Select a letter in the word "mathematics."

2. Three coins are tossed and the number of heads is recorded.

3. Three urns contain red and blue marbles. The experiment consists of first selecting an urn and then a marble from the urn and recording its color. Define the events

   E: It is not red.

   F: It is red or blue.

   Determine the sample space S of the experiment. Find (a) $E \cup F$, (b) $E \cap F$, (c) E'.

4. A marble is selected from an urn containing 4 red, 5 white, and 6 blue marbles, and its color is recorded. Find (a) $Pr(\text{red})$, (b) $Pr(\text{red or blue})$.

5. A card is drawn from a deck. Define the events $E = \{\text{king}\}$, $F = \{\text{black ace}\}$. Find (a) $Pr(E)$, (b) $Pr(F)$, (c) $Pr(E \cup F)$.

6. A die is rolled and the number on the top face

is recorded. Define the two events $E = \{2, 3\}$ and $F = \{3, 4, 5\}$. Find $Pr(E \cup F)$.

In Problems 7 and 8 a sample space has the following probability distribution.

outcome	a	b	c	d
probability	.6	.1	.2	p

7. (a) Find $Pr(d)$, (b) Define the two events, $E = \{a, c\}$, and $D = \{c, d\}$. Find $Pr(E \cup F)$.

8. Define the events

$$E = \{a, b, c\} \qquad F = \{b, c, d\}$$

Find $Pr(E \cup F')$.

9. If $Pr(E) = .27$, $Pr(F) = .65$, and $Pr(E \cup F) = .78$, find $Pr(E \cap F)$ and $Pr(E' \cup F')$.

10. Suppose $Pr(E) = .38$, $Pr(F) = .68$, and $Pr(E \cup F) = .84$. Find $Pr(E' \cup F')$.

11. If the odds in favor of E are 6 to 7, find the probability of E occurring.

12. If a 3-card hand is dealt, find the probability that all of the cards are picture cards.

13. If 12 coins are tossed, find the probability that at least 10 are heads.

14. An auditor selects a sample of 5 accounts from a total of 20 accounts and examines them for delinquency. Suppose there are 4 delinquent accounts. Find the probability that the sample will have 3 or 4 delinquent accounts.

15. A club has 40 members, 25 of whom are seniors and 18 of whom are women. If 12 of the women are seniors, find the probability that a person selected is a man given that the person is a senior.

16. Urn 1 has 4 red marbles and 8 white marbles and urn 2 has 6 red marbles and 4 white marbles. The experiment consists of selecting one of the two urns and then drawing a marble from the

urn and recording its color. Find the probability of selecting a red marble.

17. Two dice are rolled and the sum of the top faces is recorded. Define the two events $E = \{odd\ sum\}$ and $F = \{a\ sum\ less\ than\ 7\}$. Are E and F independent?

18. Refer to the accompanying Venn diagram. The number of elements in each set is given. For example, $n(A) = 12$ and $n(A \cap E) = 3$.

Find $Pr(A \cap E)$, $Pr(H \cap E)$, $Pr(W \cap E)$, $Pr(A \mid E)$.

19. Choose 1 of 3 urns, each containing some red and white marbles, and then select a marble. Urn 1 has 1 red and 9 white marbles, urn 2 has 4 red and 6 white marbles, and urn 3 has 2 red and 8 white marbles. Find the probability that urn 1 was chosen given that the marble is red.

20. The three mutually exclusive events A, B, and C have probabilities $Pr(A) = .4$, $Pr(B) = .1$, and $Pr(C) = .5$, and $Pr(E \mid A) = .1$, $Pr(E \mid B) = .2$, and $Pr(E \mid C) = .1$. Find $Pr(A \mid E)$.

21. Find the expected value of the experiment of tossing two coins 300 times and recording the number of times two heads occurs.

22. A raffle offers a first prize of $15,000, 2 prizes of $10,000, 4 prizes of $5000, and 20 prizes of $500. Two thousand tickets are sold at $50 each. Find the expected value for the person who buys one ticket.

8 **Statistics**

CHAPTER OVERVIEW Statistics is concerned with presenting, describing, and interpreting sets of data. This chapter begins by placing probability distributions in a statistical setting. It shows how histograms present various kinds of distributions in a graphic manner. Next, five units of measure are introduced. The mean, median, and expected value describe the "average" of a set of numbers, while the variance and standard deviation describe the spread from the average. The normal curve is introduced to interpret data from the binomial experiments discussed in Section 7–7.

Statistics is one of the most useful areas of mathematics. This chapter considers applications to diverse fields such as marketing, law, medicine, sociology, ecology, agriculture, industrial accidents, gambling, drug studies, and the overbooking of reservations by airlines.

CASE STUDY PREVIEW

The 13 colonies in America joined forces from 1776 to 1783 to fight for their independence from Great Britain. But then a family squabble broke out as the independent colonies wrestled with their own self-governance. The writing of a constitution was finally completed in September of 1787. The next step was its ratification by the colonies, and one of the most instrumental documents supporting it was The Federalist papers.

As we all know, the Constitution was ratified unanimously. However, for almost 200 years the identity of the authors of some of The Federalist papers remained unknown. Conclusive evidence was first published in 1964. The statistical methods described in this chapter will enable you to understand the results of that historic study.

8–1 PROBABILITY DISTRIBUTIONS

This section bridges the gap between probability and statistics. First we present some topics from Chapter 7 in a statistical setting. Then it is shown how certain graphs called histograms represent data geometrically. Finally we demonstrate how histograms are related to probability.

Histograms can be seen in most daily newspapers, especially in the business section. In addition, when major manufacturers of personal computers market a new model they frequently picture it with a full-color display of a histogram.

Triskaidekaphobia

Many superstitions are associated with Friday the 13th. Hollywood has exploited a fear of this date by filming several horror movies with the same title, yet the thirteenth of a month is more likely to fall on a Friday than any other day. Let us explain.

The calendar repeats itself every 400 years. There are 4800 months in this time period since every year has 12 months. Table 1 lists the days of the week that the 13th falls on for all 4800 months. It shows that the 13th occurs on a Friday more frequently than on any other day.*

Table 1 is called a **frequency distribution.** The sample space consists of the seven days of the week. Since numerical outcomes are more convenient to handle, we assign integers 1 through 7 to the days of the week. They are denoted by the letter X. The number of months that the 13th falls on each day is called the **frequency.** The frequencies are listed in the column under f.

*For an interesting history of the role mathematics played in the calendar, see Jacques Dutka, ''On the Gregorian Revision of the Julian Calendar,'' *The Mathematical Intelligencer*, Vol. 7, No. 1, 1985, pp. 56–64.

TABLE 1

Day	X	f
Sunday	1	687
Monday	2	685
Tuesday	3	685
Wednesday	4	687
Thursday	5	684
Friday	6	688
Saturday	7	684
Total		4800

Source: From V. Frederick Rickey, ''Mathematics of the Gregorian Calendar,'' *The Mathematical Intelligencer*, Vol. 7, No. 1, 1985, pp. 53–56.

TABLE 2

X	f	p
1	687	.1431
2	685	.1427
3	685	.1427
4	687	.1431
5	684	.1425
6	688	.1433
7	684	.1425
Total	4800	.9999

A **random variable** is a rule that assigns a real number to each outcome in the sample space. The term "random variable" refers to the unpredictable nature of the outcomes. In the example above, the variable X is a random variable.

In Table 1 the frequency of $X = 6$ is 688. This means that the 13th of a month falls on a Friday 688 of the 4800 months. The proportion of these occurrences is called the **relative frequency.** The relative frequency for $X = 6$, rounded to four decimal places, is $688/4800 = 0.1433$. Since the relative frequency can be interpreted as a probability, we write

$$Pr(X = 6) = .1433$$

In general $Pr(X = x)$ stands for the probability that the random variable X is equal to x. Table 2 is formed from Table 1 by omitting the names of the days and by adding a column of relative frequencies under the letter p (for probability). It is called a **probability distribution.**

Example 1 shows how to find the probability that a value of X falls between two numbers a and b. The notation for this is $Pr(a \leq X \leq b)$.

EXAMPLE 1

Problem

What is the probability that the 13th of a month falls on a weekday?

SOLUTION Weekdays are assigned integers between 2 and 6 inclusive. Thus the problem is to find $Pr(2 \leq X \leq 6)$. It is obtained by summing the relative frequencies.

$$
\begin{aligned}
Pr(2 \leq X \leq 6) &= Pr(X = 2) + Pr(X = 3) + Pr(X = 4) \\
&\quad + Pr(X = 5) + Pr(X = 6) \\
&= .1427 + .1427 + .1431 + .1425 + .1433 \\
&= .7143
\end{aligned}
$$

The probability of the 13th of a month falling on a weekday is .7143.

Example 2 applies these concepts to coin-tossing experiments that will occur repeatedly in this chapter. The notation $Pr(X \geq c)$ refers to the probability that X is greater than or equal to the number c.

EXAMPLE 2 | Toss three coins and let X represent the number of heads that occur.

Problem

Construct the probability distribution and find $Pr(X \geq 1)$.

SOLUTION The random variable X assumes the values 0, 1, 2, and 3 for the number of heads that can occur. From Chapter 7, the probability distribution is

X	f	p
0	1	$\frac{1}{8}$
1	3	$\frac{3}{8}$
2	3	$\frac{3}{8}$
3	1	$\frac{1}{8}$

The notation $Pr(X \geq 1)$ stands for the probability of tossing at least one head. It can be obtained directly.

$$Pr(X \geq 1) = Pr(X = 1) + Pr(X = 2) + Pr(X = 3)$$
$$= \tfrac{3}{8} + \tfrac{3}{8} + \tfrac{1}{8} = \tfrac{7}{8}.$$

Alternately, it can be obtained from the rule for complementary probabilities:
$$Pr(X \geq 1) = 1 - Pr(X < 1) = 1 - Pr(X = 0) = 1 - \tfrac{1}{8} = \tfrac{7}{8}.$$

Histograms

Statistics makes vital use of various graphing techniques to analyze data. A **histogram** represents data by means of a bar graph. Figure 1 shows the histogram for the days on which the 13th of a month occurs. The horizontal axis represents the

FIGURE 1

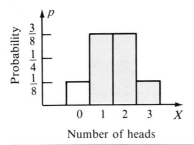

FIGURE 2

days of the week and the vertical axis represents the frequencies. The jagged line indicates that an interval of data has been omitted, resulting in a graph whose true height is distorted.

A set of data can have a histogram for its frequency distribution and a histogram for its probability distribution. In both cases the units along the horizontal axis are the same. However, the units along the vertical axis are either the frequencies or the relative frequencies. The shapes of the two histograms are identical, but the histogram with relative frequencies has a direct connection with probability. When the widths of the bars are equal to one unit and the heights of the bars are the relative frequencies, the probability of each outcome is equal to the area of that bar. In this way the probability of an event E is equal to the sum of the areas of the bars of all outcomes in E. Example 3 illustrates this property.

EXAMPLE 3 | **Problem**

Construct a histogram for the probability distribution in Example 2. Which bars represent the probability of tossing at least one head?

SOLUTION The histogram for the probability distribution is shown in Figure 2. The shaded portion represents the fact that $Pr(X \geq 1) = \frac{7}{8}$ because the total area of the shaded rectangles equals $(\frac{3}{8})(1) + (\frac{3}{8})(1) + (\frac{1}{8})(1) = \frac{7}{8}$.

Histograms are often used to compare different sets of frequencies of the same sample space. The Contraceptive Drug Study, which examined the possible side effects of oral contraceptives on blood pressure, made use of such a comparison. Table 3 presents the results of the study for women aged 17 to 24 who were neither

TABLE 3

Nonusers		Users	
X	f	X	f
100	14	100	9
110	26	110	21
120	33	120	27
130	18	130	25
140	7	140	14
150	2	150	4

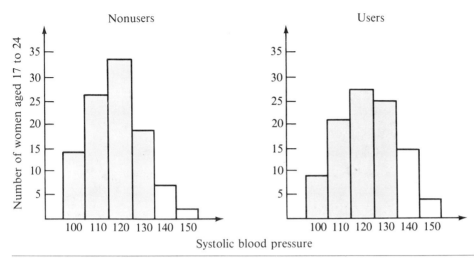

Nonusers   Users

FIGURE 3

pregnant nor taking hormonal medication. It gives one frequency distribution for women who used the Pill and another frequency distribution for women who did not. The random variable X represents systolic blood pressure (in millimeters) and the relative frequency represents the percentage of women with the given blood pressure.

EXAMPLE 4 | **Problem**

Draw the histograms for each frequency distribution in Table 3. Shade the portion that represents blood pressures that are 130 or lower.

SOLUTION The histograms are drawn in Figure 3. They show graphically that women aged 17 to 24 who used the Pill had a much higher probability of having a blood pressure over 130 than women who did not use the Pill.

Intervals

Sometimes intervals of numbers are tabulated instead of the numbers themselves. For statistical purposes the entries become the midpoints of the intervals. Table 4 gives the weights of the 1988 Super Bowl champion team, the Washington Redskins. Example 5 shows how to draw the histogram for these data.

EXAMPLE 5 | **Problem**

Construct the frequency distribution and draw the histogram for the weights of the Washington Redskins.

SOLUTION The random variable X assumes values that are midpoints of the intervals. The interval 170–179 includes all players who weigh between 170 and 180, so the midpoint is 175. The remaining entries are 185, 195, and so on.

TABLE 4

Weight	Frequency
170–179	4
180–189	5
190–199	3
200–209	2
210–219	3
220–229	5
230–239	5
240–249	3
250–259	1
260–269	2
270–279	6
280–289	1
290–299	3

This leads to the accompanying frequency distribution. The histogram is drawn in Figure 4.

X	f
175	4
185	5
195	3
205	2
215	3
225	5
235	5
245	3
255	1
265	2
275	6
285	1
295	3

FIGURE 4

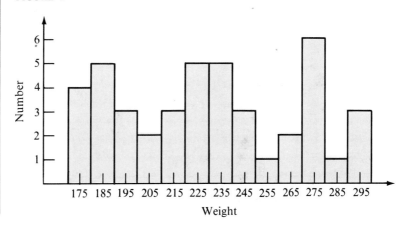

The histogram in Figure 4 is nearly level. It is close to what is called a **uniform distribution.** Contrast it with the histogram in Figure 2, which is close to what is called a **normal distribution** (or a **bell-shaped curve**). For the general population the distribution of weight is a normal distribution. It is somewhat surprising that the distribution of weights on an NFL team is not a normal distribution. We certainly expect the weights to be higher than the general population, but we do not expect them to be so uniform.

The distribution of income is an example of a skewed distribution. It will be illustrated in the exercises.

EXERCISE SET 8–1

• Problems 1 to 4 refer to Example 1. Find

1. Pr(the 13th of a month occurs on a weekend).

2. Pr(the 13th of a month does not occur on a Friday).

3. $Pr(X \geq 5)$.   4. $Pr(X \leq 6)$.

In Problems 5 to 8 toss three coins and let X be the number of heads that appear. Find

5. $Pr(X = 1)$.   6. $Pr(X = 3)$.

7. $Pr(X \geq 2)$.   8. $Pr(X \leq 2)$.

In Problems 9 to 12 draw the histogram for the experiment of tossing three coins and shade the stated portion.

9. exactly one head.   10. exactly three heads.

11. at least two heads.   12. at most two heads.

In Problems 13 and 14 toss two coins and let X be the number of heads that appear.

13. Construct the frequency distribution and draw its histogram.

14. Construct the probability distribution and draw its histogram.

Problems 15 to 18 refer to the Contraceptive Drug Study tabulated in Table 3.

15. Draw the two histograms and shade the portions for blood pressures that are 120 or less.

16. Draw the two histograms and shade the portions for blood pressures that are 110, 120, or 130.

17. What conclusions can be drawn about blood pressures that are 120 or less? (Use Problem 15.)

18. What conclusions can be drawn about blood pressures that are 110, 120, or 130? (Use Problem 16.)

19. Draw the histogram for the ages of students in a marketing course from this table.

age	17–19	20–22	23–25	26–28	29–31
frequency	10	14	7	2	1

20. Draw the histogram for the scores on an exam as tabulated in this table.

score	50–59	60–69	70–79	80–89	90–99
frequency	4	5	6	5	5

•• In Problems 21 to 26 construct the probability distribution, draw its histogram, and find the stated probability.

21. Toss four coins and let X be the number of heads that occur. Find $Pr(X \geq 1)$.

22. A jar contains four vanilla cookies and three chocolate cookies. Two cookies are selected at random without replacement. Let X be the number of vanilla cookies. Find $Pr(X \geq 1)$.

23. Toss two dice and let X be the sum of the top faces. Find $Pr(6 \leq X \leq 8)$.

24. Toss two dice and let X be the sum of the top faces. Find $Pr(7 \leq X \leq 10)$.

25. Select 3 cards from an ordinary deck without

replacement and let X be the number of hearts. Find $Pr(X \leq 2)$.

26. Select 2 cards from an ordinary deck without replacement and let X be the number of kings. Find $Pr(X \geq 1)$.

Problems 27 to 32 refer to this table of scores on two quizzes for a class of 24 students.

score	10	9	8	7	6	5
quiz 1	7	7	5	1	1	3
quiz 2	2	3	8	6	1	4

27. Find $Pr(X \geq 8)$ for quiz 1.

28. Find $Pr(X \geq 8)$ for quiz 2.

29. Find $Pr(6 \leq X \leq 8)$ for quiz 1.

30. Find $Pr(6 \leq X \leq 8)$ for quiz 2.

31. Draw the histogram for each quiz.

32. Which quiz scores are higher? Do the histograms drawn in Problem 31 convey this visually?

Problems 33 to 36 refer to the histogram shown.

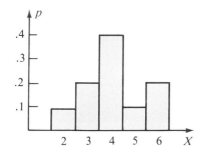

33. Find $Pr(X \geq 4)$.

34. Find $Pr(2 \leq X \leq 4)$.

35. Find $Pr(3 \leq X \leq 5)$.

36. Find $Pr(X \geq 5)$.

●●● 37. This table lists the total attendance (in millions) and average ticket price (in dollars) for the Broadway theater over a 12-year span. Draw the histograms for total attendance and ticket price. What conclusion can be drawn from comparing the histograms?

year	74–75	75–76	76–77	77–78	78–79	79–80
attendance	6.6	7.3	8.8	9.6	9.6	9.6
$	8.63	9.72	10.56	11.87	13.95	15.20

year	80–81	81–82	82–83	83–84	84–85	85–86
attendance	11.0	10.1	8.4	7.9	7.4	6.5
$	17.90	22.07	24.88	28.73	28.78	28.96

Source: From The League of American Theatres and Producers.

38. Draw the histogram for Table 5, which lists the distribution of income of families in the United States. What kind of distribution is it?

TABLE 5 Distribution of Income of Families in the United States, 1979

Income Level ($)	Percent
0– 5,000	4
5,000–10,000	13
10,000–15,000	17
15,000–20,000	20
20,000–25,000	16
25,000–30,000	12
30,000–35,000	7
35,000–40,000	4
40,000–45,000	3
45,000–50,000	2
50,000–55,000	1

Source: From The Current Population Survey.

39. Table 6 lists the height of men and women in the United States. Draw the histogram for each distribution. What kind of distributions are they?

40. Table 7 traces the size of Hershey's standard milk chocolate bar from its introduction in 1950 to its discontinuation in 1974. Draw its histogram with the horizontal axis for the years from 1950 to 1974.

41. What does the word "triskaidekaphobia" mean?

Problems 42 and 43 refer to Table 8, which lists weights, in intervals, of the Denver Broncos, the losing team in the 1988 Super Bowl.

TABLE 6 Height of U.S. Men and Women, 1979

Men		Women	
Height (in.)	Number (in thousands)	Height (in.)	Number (in thousands)
60	1	53	0.5
61	5	54	0.5
62	9	55	0.2
63	17	56	0.2
64	37	57	10
65	35	58	13
66	70	59	38
67	63	60	45
68	94	61	83
69	54	62	104
70	63	63	73
71	32	64	90
72	28	65	47
73	11	66	44
74	6	67	14
75	1	68	12
		69	1
		70	1

Source: From The Current Population Survey.

TABLE 8

Weight	Frequency
170–179	4
180–189	5
190–199	9
200–209	2
210–219	2
220–229	3
230–239	5
240–249	3
250–259	2
260–269	8
270–279	2

TABLE 7 Prices of Hershey's Standard Milk Chocolate Bar

Common Retail Price: 10¢	Ounces
1950	2
July 1951	$1\frac{7}{8}$
February 1954	$1\frac{5}{8}$
August 1954	$1\frac{1}{2}$
July 1955	$1\frac{3}{4}$
February 1958	$1\frac{1}{2}$
June 1960	2
August 1963	$1\frac{3}{4}$
August 1965	2
June 1966	$1\frac{1}{4}$
February 1968	$1\frac{1}{2}$
March 1970	$1\frac{3}{8}$
January 1973	1.26
Discontinued 1-1-74	

Source: From H. J. Maidenberg, "Misadventures of Cocoa Trade," *New York Times,* February 25, 1979, and Hershey Foods Corp.

42. (a) Construct a frequency distribution with the midpoints of the intervals as entries.
    (b) Draw the histogram.

43. Does the histogram resemble the uniform curve drawn in Figure 4?

44. This problem describes the way in which probability distributions are used in financial management to assess a firm's operating leverage. Figure 5 shows two typical EBIT (earnings before interest and taxes) distributions for Industrial Technology, Inc.* The vertical axis represents the probability of achieving profits under Plans A and B.
    (a) What is the EBIT at Plan A's highest probability?

*B. J. Campsey and Eugene F. Brigham, *Introduction to Financial Management,* Hinsdale, Ill., The Dryden Press, 1985, pp. 226–230.

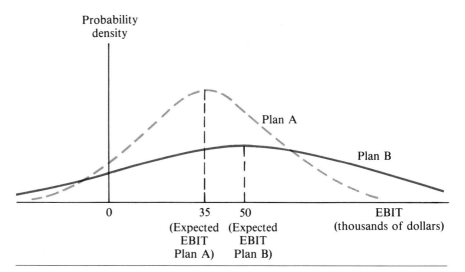

FIGURE 5 EBIT probability distribution.

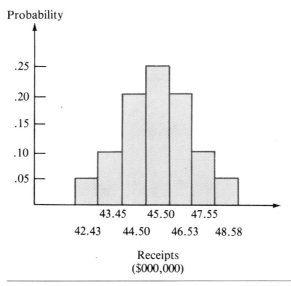

FIGURE 6

(b) What is the EBIT at Plan B's highest probability?

(c) Is it possible to conclude from Figure 5 which Plan is more volatile, A or B? If so, which one?

45. The histogram in Figure 6 is taken from a study that contrasted several forecasting models for estimating the expected cash receipts of a company.* Does it represent a probability distribution? Why?

*James L. Pappas and George P. Huber, "Probabilistic Short-term Financial Planning," *Financial Management*, August 1973, pp. 36–44.

8–2 MEASURES OF CENTRAL TENDENCY

Section 8–1 showed that a histogram provides a quick overall impression of a set of data, giving it a form and structure that is sometimes missing from the mere listing of the individual entries. A histogram reveals other characteristics of the data as well, like the average. This section examines three meanings of the term "average": mean, median, and expected value.

Average

The League of American Theatres and Producers reported that the average price for a ticket to a Broadway show during the 1985–1986 season was $28.96. By the term "average" the League meant the mean, which was obtained by dividing the total ticket revenue by the total attendance. However, another type of average is the median ticket price that is directly in the middle of all ticket prices, so that half the tickets cost more and half the tickets cost less than the median price.

The mean and median are numerical values that represent an entire collection of data. Since averages measure the center of a set of data, they are called "measures of central tendency." The discussion of Broadway theater ticket prices illustrates these two measures of central tendency, the mean and the median. It is important to state precisely which one is being used. The exercises examine a federal court ruling in a civil rights discrimination suit that revolved around the distinction between these two terms.

The *mean* of a set of data, denoted by $\bar{x}$, is defined in the display.

Definition

The **mean** of a set of n numbers $x_1, x_2, \ldots, x_n$, is

$$\bar{x} = \frac{x_1 + x_2 + \cdots + x_n}{n}$$

Calculators with a $\bar{x}$ key can find the mean of any set of data directly. The stated average ticket price of $28.96 to a Broadway show is, more precisely, the mean price. As another example, the mean of five test grades 82, 84, 78, 76, 95 is

$$\frac{82 + 84 + 78 + 76 + 95}{5} = 83$$

The **median** of a set of data is the middle number when the entries are arranged in order. Curiously there is no special notation for this concept. To find the median, first determine if there is an odd or even number of elements in the set. If it is odd, the median is the middle number. If it is even, the median is the mean of the middle two numbers. Thus the median grade on tests 82, 84, 78, 76, 95 is 82 because when these grades are listed in order, 82 is the middle grade.

76, 78, 82, 84, 95
↑
median

Notice that for this set of grades the mean is 83 and the median is 82. The median grade on the first four tests is equal to the mean of 78 and 82, which is 80.

76, 78, 82, 84
median = 80

For this set of four grades the mean and median are equal.

Example 1 shows how to compute the mean and median for sets with repeated entries.

EXAMPLE 1 A branch of a bank has 8 tellers who earn $15,000 per year, 2 managers who earn $35,000, a vice-president who earns $70,000, and a president who earns $100,000.

Problem

What are the mean and median salaries?

SOLUTION The mean salary is equal to the total salaries divided by the number of employees.

$$\text{total salary} = 15 + 15 + 15 + 15 + 15 + 15 + 15 + 15$$
$$\text{(in 1000's)} \quad + 35 + 35 + 70 + 100$$
$$= 8(15) + 2(35) + 1(70) + 1(100) = 360$$
$$\text{number of employees} = 8 + 2 + 1 + 1 = 12$$

Thus

$$\bar{x} = \frac{8(15) + 2(35) + 1(70) + 1(100)}{8 + 2 + 1 + 1} = \frac{360}{12} = 30$$

The mean salary is $30,000.

Since there are 12 employees, the median salary is the mean of the sixth and seventh numbers.

15, 15, 15, 15, 15, 15, 15, 15, 35, 35, 70, 100
median

Thus the median salary is $15,000.

Example 1 suggests the derivation of an alternate formula for the mean in terms of frequencies.

> **Alternate Formula for the Mean**
>
> If a set has f_1 entries equal to x_1, f_2 entries equal to x_2, . . . , and f_n entries equal to x_n, then
>
> $$\bar{x} = \frac{f_1 x_1 + f_2 x_2 + \cdots + f_n x_n}{f_1 + f_2 + \cdots + f_n}$$

A convenient way to evaluate $\bar{x}$ from this expression is to make a frequency table with a third column for the products of each entry x and its frequency f.

x	f	xf
15	8	120
35	2	70
70	1	70
100	1	100
Totals	12	360

$$\bar{x} = \frac{360}{12} = 30$$

EXAMPLE 2

A psychology professor enters the 100 scores on a ten-question test into a grade book. To find the mean, the grades are grouped as follows:

number correct	10	9	8	7	6	5	4	3	2	1	0
number of students	11	16	20	15	13	10	10	4	1	0	0

Problem

What is the mean?

SOLUTION Write the entries x and their frequencies f in a table, with a third column of the products xf.

x	f	xf
10	11	110
9	16	144
8	20	160
7	15	105
6	13	78
5	10	50
4	10	40
3	4	12
2	1	2
Total	100	701

$$\bar{x} = \frac{701}{100} = 7.01$$

The mean number of correct answers is slightly more than 7.

Expected Value

The alternate formula for the mean leads directly to its analogue for probability distributions. In the alternate formula for $\bar{x}$ set $N = f_1 + f_2 + \cdots + f_n$. Then

$$\bar{x} = \frac{f_1 x_1 + f_2 x_2 + \cdots + f_n x_n}{f_1 + f_2 + \cdots + f_n}$$

$$= \frac{f_1 x_1 + f_2 x_2 + \cdots + f_n x_n}{N}$$

$$= \left(\frac{f_1}{N}\right) x_1 + \left(\frac{f_2}{N}\right) x_2 + \cdots + \left(\frac{f_n}{N}\right) x_n$$

Each fractional part f_i/N is a relative frequency, so it can be interpreted as a probability. The mean of a probability distribution is called the *expected value*, denoted by $E(X)$. By setting $p_i = f_i/N$, the formula for the expected value becomes

$$E(X) = p_1 x_1 + p_2 x_2 + \cdots + p_n x_n$$

We encountered expected value in Chapter 7, where it referred to the average number of outcomes that occur when a probability experiment is repeated many times. To compute it, extend the probability distribution of a random variable X to a third column containing the products Xp.

X	p	Xp
x_1	p_1	$x_1 p_1$
x_2	p_2	$x_2 p_2$
$\vdots$	$\vdots$	$\vdots$
x_n	p_n	$x_n p_n$
	Total $=$	$E(X)$

The formula for $E(X)$ shows that it is equal to the sum of the numbers in the Xp column.

The **expected value** of the random variable X is

$$E(X) = x_1 p_1 + x_2 p_2 + \cdots + x_n p_n$$

Both the mean and the expected value describe the average of a set of numbers, but the mean is computed from a frequency distribution, whereas the expected value is computed from a probability distribution. Example 3 finds the expected value for the sum of two dice.

EXAMPLE 3 Toss two dice and let X be the sum of the faces.

Problem

Find $E(X)$.

SOLUTION The probability distribution given here is taken from the table before Example 2 in Section 7–3.

X	p	Xp
2	1/36	2/36
3	2/36	6/36
4	3/36	12/36
5	4/36	20/36
6	5/36	30/36
7	6/36	42/36
8	5/36	40/36
9	4/36	36/36
10	3/36	30/36
11	2/36	22/36
12	1/36	12/36
	Total =	252/36

The expected value is $E(X) = 252/36 = 7$.

Example 3 means that if you toss two dice many times, and form the sum each time, the mean of all the sums will be very close to 7. Example 4 computes an expected value directly from the probability distribution.

EXAMPLE 4 Suppose that the New Jersey Turnpike Commission reports the number of accidents during the Friday evening rush hour to be 0, 1, 2, or 3 with probabilities .92, .03, .02, and .03, respectively.

Problem

What is the expected number of accidents during the Friday evening rush hour?

SOLUTION The term "expected number" refers to the expected value $E(X)$, where the random variable X represents the number of accidents. Form the expanded probability distribution and total the Xp column.

X	p	Xp
0	.92	.00
1	.03	.03
2	.02	.04
3	.03	.09
	Total =	.16

Thus $E(X) = .16$. Since the expected number of accidents is approximately $\frac{1}{6}$ per Friday evening rush hour, one accident will occur about every six weeks.

EXERCISE SET 8–2

• In Problems 1 to 8 find the mean and median.

1. 2, 5, 7, 9, 2, 1, 13, 12, 50, 60

2. 2, 5, 7, 9, 2, 1, 13, 12, −50, −60

3. .06, .07, .02, .10, .05, .02

4. 5.2, 6.1, 10.8, 11.2, 6.3, .5, 8.7, 7.2

5.
value	0	1	2	3	
frequency	2	5	3	4	

6.
value	1	2	3	4	
frequency	8	2	1	5	

7.
value	3	5	7	9	
frequency	2	1	0	4	

8.
value	11	12	13	14	15
frequency	3	5	2	0	1

In Problems 9 to 12 find the expected value from the probability distribution.

9.
x	p
2	$\frac{1}{8}$
3	$\frac{1}{2}$
5	$\frac{1}{4}$
7	$\frac{1}{8}$

10.
x	p
2	$\frac{1}{2}$
4	$\frac{1}{3}$
6	$\frac{1}{6}$

11.
x	p
2	.2
3	.3
4	.5

12.
x	p
1	.2
3	.4
5	.1
7	.3

In Problems 13 and 14 find the mean from the given histogram.

13.

14.

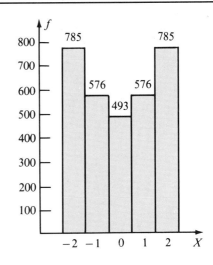

In Problems 15 and 16 find the expected value from the given histogram.

15.

16.

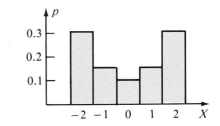

•• In Problems 17 to 22 find the mean and median.

17. −3, 2, −3, 1, −3, 1, −1, 0, 0, 1

18. 3, −3, 2, 2, 2, −3, −1, 0, 0, −2

19.
value	4.4	5.1	6.4	7.3	8.2
frequency	3	4	1	5	7

20.
value	0.5	0.6	0.7	0.9	
frequency	9	5	2	4	

21.

value	−3	−2	−1	0	2	3
frequency	2	1	1	2	3	1

22.

value	−3	−1	0	1	2
frequency	3	1	2	3	1

In Problems 23 to 26 find the expected value from the probability distribution.

23.

x	p
−3	.2
−1	.2
0	.1
1	.4
3	.1

24.

x	p
−2	.2
−1	.1
0	.4
1	.2
2	.1

25.

x	p
−10	.2
−5	.4
0	.1
5	.2
10	.1

26.

x	p
−5	.1
−3	.3
0	.1
1	.2
4	.3

27. Toss three coins and let X be the number of heads that occur. Find $E(X)$.

28. Toss four coins and let X be the number of heads that occur. Find $E(X)$.

29. Find the mean and median weights of the Washington Redskins from the table in Example 5 of Section 8–1.

30. What is the median grade on the psychology test described in Example 2?

In Problems 31 and 32 the data are tabulated in intervals. Find the mean using the midpoints of the intervals as the data.

31.

interval	0–2	2–4	4–6	6–8	8–10
frequency	4	5	3	0	1

32.

interval	0–5	5–10	10–15	15–20	20–25	25–30
frequency	16	14	10	8	2	1

33. A set of data has five entries. Each entry is either 0, 1, or 2.
(a) What is the set if the mean is 0?
(b) What is the set if the mean is 2?
(c) Can the mean be 3?

34. A set of data has five entries. Each entry is either 10, 11, or 12. What is the set if the mean is 10.8 and the median is 10?

35. A psychologist studies whether reward or punishment is a more effective method of inducing learning by running 11 mice through two sets of mazes. The differences (in seconds) between the rewarded trials and the punished trials are −1.1, 2.0, 1.3, −0.9, −1.9, 0.8, 1.2, −1.2, 4.0, 0.2, −0.1.
(a) Find the mean of these rates.
(b) Which method is more effective?

36. An entomologist studying the morphological variation in a species of mosquito selects ten samples and measures their body lengths (in centimeters): 1.2, 1.4, 1.3, 1.6, 1.0, 1.5, 1.7, 1.1, 1.2, 1.3. What are the mean and median body lengths?

●●● 37. The 1964 Civil Rights Act prohibits discrimination by race, sex, or national origin in hiring and promotions. In 1981, 20 people filed suit that they were discriminated against when they were not hired in 1975 by the Federal Reserve Bank. The court agreed. The attorneys for the two sides disagreed on the amount of money the bank should pay each plaintiff. A sample of 319 employee records revealed this pertinent information about the job turnover rate, where M denotes the number of months worked before termination and E denotes the number of employees. (See Table 1.)
(a) Find the mean number of months worked.
(b) Find the median number of months worked.

TABLE 1

M	0.5	1	2	3	4	5	8	10	11	14	18	20	24	30	35	40	60
E	15	21	27	33	42	9	17	32	18	12	22	26	11	9	6	8	11

Source: Data from EEOC vs. *Federal Reserve Bank* 30 FEP Cases 1138–1169.

TABLE 2

B	−6	−5	−4	−3	−2	−1	0	1	2	3	4	5
D	90	100	87	96	101	86	119	118	121	114	113	106

Source: Data from J. M. Tanur et al. (eds.), *Statistics: A Guide to the Unknown,* San Francisco, Holden-Day Inc., 1978, pp. 71–85.

(c) Which side argued for which average?

(d) Each of the 20 plaintiffs was to be compensated at a rate of seven (years) times $9500 (the yearly salary), divided by the average number of months worked. What is the difference in the amount of money the bank would have to pay according to whether the average is interpreted as the mean or the median?

38. J. E. Kerrich described a coin-tossing experiment while interned during World War II.* He tossed a coin 1000 times and recorded the number of heads. He conducted the experiment ten times, obtaining these results.

502, 511, 497, 529, 504, 476, 507, 528, 504, 529

What are the mean and median number of heads?

39. Sociologist David Phillips designed a study to determine whether some individuals have a will that is strong enough to prolong life. He recorded the number of months between the birthday and day of death of 1251 famous people, reasoning that famous people anticipate being honored on their birthday and might ward off death for a period of time. Table 2 lists the results, where B is the birth month and D is the number of deaths. What is the expected death month?

40. What is the mean income of families in the United States according to Table 5 in Section 8–1?† What is the median income?

41. Table 6 in Section 8–1 lists the height of women in the United States.† What are the mean and median heights?

42. Toss n coins and let X be the number of heads that occur. Find $E(X)$.

43. The probability distribution for the number of games played in each World Series from 1922 to 1987 is given below. What is the expected number of games played?

x	4	5	6	7
$P(X = x)$	0.167	0.197	0.197	0.439

Source: Craig Carter (ed.), *The Complete Baseball Record Book,* St. Louis, The Sporting News Publishing Co., 1988, p. 323.

44. The probability distribution for the number x of children of the presidents of the United States from George Washington to Ronald Reagan is given in Table 3. What is the expected number of children of a president of the United States based on this information?

*J. E. Kerrich, *An Experimental Introduction to the Theory of Probability,* Copenhagen, Belgisk Import Co.

†Source, The Current Population Survey, 1982.

TABLE 3

x	0	1	2	3	4	5	6	7	8	10	15
$P(X = x)$	.154	.026	.179	.179	.179	.077	.102	.026	.026	.026	.026

Source: Joseph N. Kane, *Facts About the Presidents,* (4th ed.), New York, H. W. Wilson Co., 1981, pp. 348–349.

8-3 STANDARD DEVIATION

The mean and expected value are numerical values that describe the central tendency. The Greek letter μ (mu) is the conventional symbol for this measure, so $\mu = \bar{x}$ for a frequency distribution and $\mu = E(X)$ for a probability distribution. We refer to it as "the mean μ."

In this section we discuss two values, the variance and standard deviation, which measure the spread of the data from the mean μ. A knowledge of a set's mean and standard deviation enables many conclusions about a study to be drawn.

Measures of Dispersion

Consider these three sets of data.

> set A: 6, 5, 5, 4, 5, 3, 5, 7, 5, 5
> set B: 6, 9, 7, 6, 5, 5, 2, 1, 4, 5
> set C: 6, 2, 10, 9, 4, 1, 0, 1, 7, 10

Their histograms are shown in Figure 1. The mean of each set is 5, but their spreads are quite different. The entries in set C are more dispersed than the entries in set B, and the entries in set B are more dispersed than the entries in set A. The variance and standard deviation are measures that describe these dispersions quantitatively.

In Table 1 we list the entries of set B and find $\mu = 5$. To calculate the deviation of each entry x from the mean, we compute $x - \mu$ and list it in the second column. Notice that the sum of this column is 0; the definition of μ ensures that it will be 0 for every set of data. To compensate for the negative numbers canceling out the positive numbers, we compute the square of each deviation, $(x - \mu)^2$, and list it in the third column. The *variance*, denoted by $V(X)$, is the mean of the numbers in this column. The name "variance" refers to the variability of the data. For set B, $V(X) = \frac{48}{10} = 4.8$. The *standard deviation*, which is denoted by the Greek letter σ (sigma), is equal to the square root of the variance. Here $\sigma = \sqrt{4.8} \approx 2.2$.

FIGURE 1

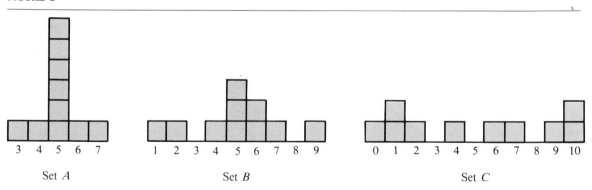

Set A   Set B   Set C

TABLE 1   Set B

x	$x - \mu$	$(x - \mu)^2$
6	1	1
9	4	16
7	2	4
6	1	1
5	0	0
5	0	0
2	−3	9
1	−4	16
4	−1	1
5	0	0

Total   50   0   48

$$\mu = \tfrac{50}{10} = 5 \qquad V(X) = \tfrac{48}{10} = 4.8 \qquad \sigma = \sqrt{4.8} \approx 2.2$$

_____ **Definition** _____

For a set of data with n entries $x_1, x_2, \ldots, x_n$ and with mean μ, the **variance** is

$$V(X) = \frac{(x_1 - \mu)^2 + (x_2 - \mu)^2 + \cdots + (x_n - \mu)^2}{n}$$

The **standard deviation** is $\sigma = \sqrt{V(X)}$

Some calculators have a σ key for computing the standard deviation. However, there are two types of standard deviation, the population standard deviation, which we are using, and the sample standard deviation. If you are using a calculator, you should check your manual on the set B to see which standard deviation is displayed.

Formulas

The formulas for the variance and standard deviation are appropriate for describing their meanings as measures of the spread of data from the mean. However, another approach is more convenient for computations. It can be obtained by simplifying the formula for $V(X)$.

Computational Formula for Variance

$$V(X) = \frac{x_1^2 + x_2^2 + \cdots + x_n^2}{n} - \mu^2$$

Example 1 contrasts the two methods of computing $V(X)$. In both cases $\sigma = \sqrt{V(X)}$.

EXAMPLE 1 | **Problem**

Find the standard deviation of set C.

6, 2, 10, 9, 4, 1, 0, 1, 7, 10

SOLUTION The tables list the data for the two formulas. The table on the left contains the calculation by definition while the table on the right contains the calculation by formula.

x	$x - \mu$	$(x - \mu)^2$		x	x^2
6	1	1		6	36
2	-3	9		2	4
10	5	25		10	100
9	4	16		9	81
4	-1	1		4	16
1	-4	16		1	1
0	-5	25		0	0
1	-4	16		1	1
7	2	4		7	49
10	5	25		10	100
Totals 50	0	138		Totals 50	388

$$\mu = \frac{50}{10} = 5 \qquad\qquad \mu = \frac{50}{10} = 5$$

$$V(X) = \frac{138}{10} = 13.8 \qquad\qquad V(X) = \frac{388}{10} - (5)^2 = 13.8$$

Both formulas yield $V(X) = 13.8$, so $\sigma = \sqrt{13.8} \approx 3.7$.

The frequency formula for the variance is used when data are given by a frequency distribution.

Frequency Formula for the Variance

If the entries $x_1, x_2, \ldots, x_n$ occur with frequencies $f_1, f_2, \ldots, f_n$, respectively, then

$$V(X) = \frac{f_1 x_1^2 + f_2 x_2^2 + \cdots + f_n x_n^2}{f_1 + f_2 + \cdots + f_n} - \mu^2$$

EXAMPLE 2 | An ecologist investigates the reproductive behavior of certain water fowl to determine whether DDT is causing infertile eggs. The species could become endangered if the eggs are infertile. A frequency distribution gives the number of nests that hold between zero and five eggs.

eggs in nest	0	1	2	3	4	5
frequency	42	44	32	19	11	2

Problem

Compute the mean and standard deviation.

SOLUTION Table 2 contains the essential ingredients.

TABLE 2

x	f	fx	x^2	fx^2
0	42	0	0	· 0
1	44	44	1	44
2	32	64	4	128
3	19	57	9	171
4	11	44	16	176
5	2	10	25	50
Totals	150	219		569

$$\mu = \frac{219}{150}$$

$$V(X) = \frac{569}{150} - \left(\frac{219}{150}\right)^2 \approx 1.66$$

$$\sigma = \sqrt{1.66} \approx 1.3$$

Thus the mean number of bird eggs in a nest is 1.46 with a standard deviation of 1.3.

Probability Distributions

The Frequency Formula for $V(X)$ leads to an interpretation of the variance for probability distributions. The procedure is analogous to the derivation of the formula for $E(X)$ from the formula for $\bar{x}$ that was carried out in Section 8–2. The Frequency Formula is

$$V(X) = \frac{f_1x_1^2 + f_2x_2^2 + \cdots + f_nx_n^2}{N} - \mu^2$$

where $N = f_1 + f_2 + \cdots + f_n$. Then

$$V(X) = \left(\frac{f_1}{N}\right)x_1^2 + \left(\frac{f_2}{N}\right)x_2^2 + \cdots + \left(\frac{f_n}{N}\right)x_n^2 - \mu^2$$

Therefore

$$V(X) = p_1x_1^2 + p_2x_2^2 + \cdots + p_nx_n^2 - \mu^2$$

This is the desired computational form of the variance for probability distributions.

—————————— **Variance for Probability Distributions** ——————————

If the probabilities of the outcomes $x_1, x_2, \ldots, x_n$ are $p_1, p_2, \ldots, p_n$, respectively, and μ is the mean, then

$$V(X) = p_1 x_1^2 + p_2 x_2^2 + \cdots + p_n x_n^2 - \mu^2$$

EXAMPLE 3

Problem

Find the standard deviation for the probability distribution given in this table.

x	p
4	.1
5	.3
6	.2
7	.4

SOLUTION The procedure is to compute $V(X)$ and to set $\sigma = \sqrt{V(X)}$. Expand the table to include columns for xp, x^2, and $x^2 p$. The total of the xp column gives the mean $\mu = 5.9$. The total of the $x^2 p$ column is 35.9. By the formula for probability distributions, $V(X) = 35.9 - (5.9)^2 = 1.09$. Then $\sigma = \sqrt{1.09} \approx 1.04$.

x	p	xp	x^2	$x^2 p$
4	.1	0.4	16	1.6
5	.3	1.5	25	7.5
6	.2	1.2	36	7.2
7	.4	2.8	49	19.6
Totals		5.9		35.9

Example 4 uses the expected value and standard deviation to compare two sets of data.

EXAMPLE 4

A farmer must determine which of two hybrid grains, A or B, has the higher yield and smaller variation. The following distributions are known, where the units for x are in thousands of bushels.

Type A

x	p
1	.2
2	.1
3	.4
4	.2
5	.1

Type B

x	p
2	.4
3	.4
4	.1
5	.1

Problem

Determine the expected value and standard deviation for each type of hybrid grain. Which type should the farmer plant?

SOLUTION Expand the tables for both types of hybrid grain.

Type A

x	p	xp	x^2	x^2p
1	.2	.2	1	.2
2	.1	.2	4	.4
3	.4	1.2	9	3.6
4	.2	.8	16	3.2
5	.1	.5	25	2.5
Totals		2.9		9.9

Type B

x	p	xp	x^2	x^2p
2	.4	.8	4	1.6
3	.4	1.2	9	3.6
4	.1	.4	16	1.6
5	.1	.5	25	2.5
Totals		2.9		9.3

The two grains have the same expected value $\mu = 2.9$, so they produce the same yield. For type A, $V(X) = 9.9 - (2.9)^2 = 1.49$ and $\sigma = \sqrt{1.49} \approx 1.22$. For type B, $V(X) = 9.3 - (2.9)^2 = 0.89$ and $\sigma = \sqrt{0.89} \approx 0.94$. Since the variation in type B's yield is smaller than the variation in type A's yield, the farmer should plant type B.

Chebyshev's Theorem

The expression "within k standard deviations of the mean" refers to those entries whose distance to the mean μ is less than or equal to $k\sigma$. We write "within $k\sigma$" for short. The entries can lie to the left or right of μ, as Figure 2 shows.

Recall set B, which was given at the beginning of this section.

set B: 6, 9, 7, 6, 5, 5, 2, 1, 4, 5

We showed that $\mu = 5$ and $\sigma \approx 2.2$. An entry lies within 2σ of the mean if its value is between $\mu - 2\sigma \approx 5 - 4.4 = 0.6$ and $\mu + 2\sigma \approx 5 + 4.4 = 9.4$. All ten entries of B lie within this interval.

Since the standard deviation measures the spread of data, it seems reasonable that as k increases, the number of entries within $k\sigma$ of the mean will also increase.

FIGURE 2

FIGURE 3

This fact, which was proved by the Russian mathematician Pafnuti Chebyshev (1821–1894), is called **Chebyshev's Theorem.***

—————————————————— **Chebyshev's Theorem** ——————————————————

In any experiment the probability that an entry lies within k standard deviations of the mean is at least $1 - (1/k)^2$.

For $k = 2$ Chebyshev's Theorem asserts that the proportion of entries within 2σ of the mean is at least $1 - (\frac{1}{2})^2 = \frac{3}{4}$, or 75%. Indeed, for set B, 100% of the entries lie in this interval. The importance of Chebyshev's result is that it refers to *any* distribution. Later we will improve the approximation for data with a normal distribution.

EXAMPLE 5

When a coin is tossed 100 times, the expected number of heads is $\mu = 50$. It will be shown in Section 8–5 that $\sigma = 5$.

Problem

What is the lower bound for the probability of tossing between 35 and 65 heads?

SOLUTION We must first determine the number of standard deviations from the mean for $x = 35$ and $x = 65$. Set $x = 65$. Then $x - \mu = 65 - 50 = 15$. Since $\sigma = 5$ and $k\sigma = 15$, we get $k = \frac{15}{5} = 3$. This means that $x = 65$ lies 3σ to the right of the mean. (See Figure 3.) Similarly, $x = 35$ lies 3σ to the left of the mean. By Chebyshev's Theorem the probability of an entry lying within 3σ of the mean is at least $1 - (\frac{1}{3})^2 = 1 - (\frac{1}{9}) = \frac{8}{9}$.

*Stephen M. Stigler has discovered that the French mathematician Bienaymé "stated and proved the Chebyshev inequality a decade before and in greater generality than Chebyshev's first work on the topic." See S. M. Stigler, "Stigler's Law of Eponymy," *Transactions of the New York Academy of Sciences*, April 1980, p. 148.

EXERCISE SET 8–3

• In Problems 1 to 10 calculate the variance and standard deviation.

1. 0, 1, 3, 1, 2, 0, 6, 2, 7, 3

2. −1, 0, −3, 1, 4, 0, −2, 5, 1, −3

3. 10, 15, 18, 11, 12, 15, 20, 10, 12, 18

4. 25, 20, 18, 15, 24, 22, 16, 20, 17, 23

5. 22, 18, 32, 40, 20, 28, 38, 42, 15, 45

6. 40, 48, 60, 55, 58, 65, 42, 45, 52, 35

7.

x	1	3	6	9	11
f	3	7	5	10	5

8.

x	2	6	10	14	18
f	15	18	20	28	20

9.

x	11	13	15	17	19	21
f	1	8	10	5	15	3

10.

x	1	2	3	4	5	6
f	10	8	7	8	6	3

In Problems 11 to 14 find the variance and standard deviation of the probability distribution. The expected values were calculated in Problems 9 to 12 in Section 8–2.

11.

x	p
2	$\frac{1}{8}$
3	$\frac{1}{2}$
5	$\frac{1}{4}$
7	$\frac{1}{8}$

12.

x	p
2	$\frac{1}{2}$
4	$\frac{1}{3}$
6	$\frac{1}{6}$

13.

x	p
2	.2
3	.3
4	.5

14.

x	p
1	.2
3	.4
5	.1
7	.3

15. An automobile dealer has four types of cars selling for $21,000, $17,000, $14,000, and $9000. The probability that a shopper will buy each type of car is, respectively, .1, .2, .4, and .3. What is the expected sales price? What is the standard deviation?

16. A student taking an economics exam has a probability of .75 of scoring 80%, .15 of scoring 85%, .05 of scoring 90%, and .05 of scoring 100%. What is the expected score? What is the standard deviation?

In Problems 17 and 18 a coin is tossed 100 times.

17. What is the lower bound for the probability that it will land heads between 40 and 60 times?

18. What is the lower bound for the probability that it will land heads between 20 and 80 times?

•• In Problems 19 to 24 calculate the variance and standard deviation.

19. 2.5, 2.0, 1.8, 1.5, 2.4, 2.2, 1.6, 2.0, 1.7, 2.3

20. 1.0, 1.5, 1.8, 1.1, 1.2, 1.5, 2.0, 1.0, 1.2, 1.8

21. 4.0, 4.8, 6.0, 5.5, 5.8, 6.5, 4.2, 4.5, 5.2, 3.5

22. 2.2, 1.8, 3.2, 4.0, 2.0, 2.8, 3.8, 4.2, 1.5, 4.5

23.

x	20	60	100	140	180
f	15	18	20	28	20

24.

x	110	130	150	170	190	210
f	1	8	10	5	15	3

In Problems 25 to 28 find the variance and standard deviation of the probability distribution. The expected values were calculated in Problems 23 to 26 in Section 8–2.

25.

x	p
−3	.2
−1	.2
0	.1
1	.4
3	.1

26.

x	p
−2	.2
−1	.1
0	.4
1	.2
2	.1

27.

x	p
−10	.2
−5	.4
0	.1
5	.2
10	.1

28.

x	p
−5	.1
−3	.3
0	.1
1	.2
4	.3

In Problems 29 to 32 find the standard deviation from the given histogram. The mean or expected value was computed in Problems 13 to 16 of Section 8–2.

29.

30.

31.

32.

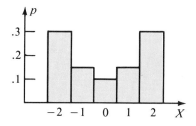

33. Toss a coin three times and let X be the number of heads. In Problem 27 of Section 8–2 it was found that $E(X) = \frac{3}{2}$. Find the standard deviation.

34. Two cookies are selected without replacement from a jar with four vanilla cookies and three chocolate cookies. Find the expected number of vanilla cookies and the standard deviation.

35. A biologist investigating airborne bacteria places a petri plate containing glucose under a venti-

lation duct. The number of bacteria colonies that appear per square millimeter is tabulated, as follows:

colonies per cell	0	1	2	3	4	5
cells	5	7	10	6	2	1

(a) What is the average number of colonies per cell?

(b) What is the standard deviation?

36. An economist determines trends in the stock market by comparing the average price per share of ten selected stocks from month to month. The prices for the first month are $15.75, $17, $6.25, $7.25, $47, $32, $21.50, $19, $40, and $12.88. What is the average price per share, and how much variation is there?

••• 37. Prove that the formula for variance by definition and the computational formula are equivalent for $n = 2$. (*Hint:* Prove this equality.)

$$\frac{(x_1 - \mu)^2 + (x_2 - \mu)^2}{2} = \frac{x_1^2 + x_2^2}{2} - \mu^2$$

38. State the equality for the fact that the variance by definition is equal to the computational formula for the variance for the case $n = 3$.

39. The expected birth month for the Phillips study on strong wills prolonging life was found in Problem 39 of Section 8–2. Calculate the standard deviation.

40. Problem 38 in Section 8–2 computed the mean number of heads in J. E. Kerrich's coin-tossing experiment from these data: 502, 511, 497, 529, 504, 476, 507, 528, 504, 529. Find the standard deviation.

41. If a coin is tossed n times, the expected number of heads is $n/2$. What is the standard deviation?

42. Many studies use statistical analyses, based on the mean and standard deviation, to draw conclusions. One such study conducted by Paula Cook asked whether teaching geometry to general students in high school using computers (the logo group) would improve student achievement over the standard pencil-and-paper approach (the

TABLE 3

Group	Mean number of correct answers
Logo	29.28
Traditional	27.47

traditional group).* This problem asks you to derive conclusions from Dr. Cook's findings.

(a) An achievement test was administered to determine if the two groups of students were equivalent at the beginning of the study. The relevant results are tabulated in Table 3. Do you think that the groups were equivalent?

(b) The students took both a pretest and a posttest on achievement in geometry. The relevant data are tabulated in Table 4. Do you think that the use of the computer improved the scores over the standard approach?

(c) The students took both a pretest and a posttest on their attitude toward mathematics.

TABLE 4

Group	Pretest $\bar{x}$	σ	Posttest $\bar{x}$	σ
Logo	10.47	2.51	17.03	4.40
Traditional	9.57	3.33	16.27	3.38

TABLE 5

Group	Pretest $\bar{x}$	σ	Posttest $\bar{x}$	σ
Logo	55.00	20.07	60.31	17.93
Traditional	67.37	13.24	61.87	19.11

The relevant data are tabulated in Table 5. Do you think that the use of a computer changed their attitudes significantly toward mathematics?

43. Table 6 presents three methods that were used to forecast the consumption of gasoline during

*Paula J. Cook, "The Effects of an Instructional Unit Utilizing Logo and the Computer on Achievement in Geometry and Attitude Toward Mathematics of Selected High School General Mathematics Students," doctoral dissertation, Temple University, Philadelphia, PA, 1988.

TABLE 6

	Econometric Forecast	Time Series Forecast	Composite Forecast	Actual Gasoline Consumption
		(million barrels per month)		
		1982		
4	195.93	194.38	200.30	206.7
5	210.36	204.03	202.92	206.15
6	207.12	206.96	202.06	204.36
7	217.03	206.36	205.01	210.77
8	220.03	209.09	206.14	206.31
9	202.20	199.11	196.79	195.21
10	211.54	207.64	201.19	198.12
11	200.70	194.63	194.54	196.77
12	213.96	208.96	203.06	202.99
		1983		
1	193.56	191.02	192.46	185.41
2	175.85	180.26	182.02	174.64
3	200.60	201.18	196.56	212.13

Source: From Anthony E. Bopp, "On Combining Forecasts: Some Extensions and Results," *Management Science*, December 1985, pp. 1492–1498.

TABLE 8   Projected Rates of Return for the Standard and Designer Phone Lines

State of the Economy	Probability	Designer Line Rate of Return (%)	Standard Line Rate of Return (%)
Boom	0.3	100	20
Normal	0.4	15	15
Recession	0.3	−70	10

Source: From B. J. Campsey and Eugene F. Brigham, *Introduction to Financial Management,* Hinsdale, Ill. The Dryden Press, 1985, pp. 426–432.

1982–1983. The actual gasoline consumption is listed in the last column. Compute the mean and standard deviation for each forecast and for the actual consumption. Which method of forecasting is most accurate?

44. The national speed limit was reduced to 55 miles per hour in 1974. Table 7 would seem to confirm the reasonableness of this policy, yet some critics object to it. Arnold Barnett, of the Sloan School of Management at MIT, used elementary statistics to rebut the critics.

(a) Compute the mean number of road fatalities and the standard deviation for the years 1970–1973.

(b) Within how many standard deviations of the mean does the number of deaths in 1974 lie?

TABLE 7

Year	Deaths
1970	54,800
1971	54,700
1972	56,600
1973	55,700
1974	46,600

Source: From Arnold Barnett, "Misapplications Reviews: Speed Kills?," *Interfaces,* March–April 1986, pp. 63–68.

45. For assets which are not held in a diversified portfolio, it is commonplace in business to identify the standard deviation σ with the measure of risk. Eastern Communications, Inc., a manufacturer and retailer of business and consumer telephones, projected its rate of return on two new lines, Standard and Designer, over a wide range of states of the economy. Table 8 summarizes the projections. (a) Compute σ (as a percent) for the Standard Phone line. (b) Compute σ (as a percent) for the Designer Phone line. (c) Which line had the smaller chance of significant loss from investing in that product line?

46. Table 9 lists the expected EPS (earnings-per-share) and the standard deviation of the EPS for various D/A (debts/assets) ratios for Santa Clara Industries.

(a) For which D/A ratios is there a negative EPS within one standard deviation of the expected EPS?

(b) For which D/A ratios is there a negative EPS within 1.5 standard deviations of the expected EPS?

TABLE 9

D/A Ratio (%)	Expected EPS	Standard Deviation of EPS
0	$2.40	$1.52
10	$2.56	$1.69
20	$2.75	$1.90
30	$2.97	$2.17
40	$3.20	$2.53
50	$3.36	$3.04
60	$3.30	$3.79

Source: From Campsey and Brigham, idem, p. 539.

47. The well-known writer Stephen Jay Gould made use of statistics to support his claim that oldtime baseball players, several of whom hit .400, were no better than the best players of today.* Gould wrote, "As (standard) variation shrinks around a constant mean batting average, .400 hitting disappears. It is, I think, as simple as that." Are you convinced by this argument?

*Stephen Jay Gould, "Entropic Homogeneity Isn't Why No One Hits .400 Any More," *Discover,* August 1986, pp. 60–66.

8–4 THE NORMAL CURVE

Suppose an expert archer and a complete novice shoot at a target on an archery range. Although the expert's shots will be more accurate, the novice's shots will land in a surprisingly similar manner. This does not mean that the outcome of each single shot can be predicted. Rather, if each archer shoots 1000 arrows and the distance between the hits and the target is measured, then each histogram of the distances will assume a definite shape that does not depend on their relative abilities.

The shape of each histogram will approximate an ideal distribution known as the normal curve. The more shots that are made, the closer the histogram will be to this curve. Because the normal curve resembles a bell, it is sometimes called a "bell-shaped curve."

The normal curve describes many mental and physical characteristics of human beings. This is fortunate because many important facts are known about the normal curve, including its mean and standard deviation. This section examines these facts and shows how to use tables of values to answer questions involving heights and IQs (intelligence quotient).

Fundamental Properties

Consider a probability experiment of tossing six coins and recording the number of heads. Its histogram is shown in Figure 1. We have superimposed upon it a curve that passes through the midpoints at the top of the rectangles. It resembles a curve known as the **normal curve.**

The normal curve has an unwieldy looking equation.

$$y = \frac{1}{\sqrt{2\pi}} e^{-z^2/2}$$

where $\pi \approx 3.1416$ and $e \approx 2.7183$. The graph of the normal curve is given in Figure 2. The values on the z-axis are standard deviations. The units on the y-axis correspond to the relative frequencies on a histogram, and therefore a normal curve can be used to approximate the histogram.

There are four basic properties of the normal curve.

1. The mean corresponds to the value $z = 0$.
2. The curve is symmetric about the mean.
3. The curve always lies above the z-axis.
4. The area under the curve and above the z-axis is exactly 1.

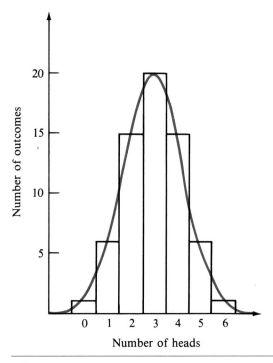

Number of heads

FIGURE 1

We make use of these properties to find areas that lie under the curve and above the z-axis. The reason for the emphasis on areas is that the area under a normal curve corresponds to the probability of an event. There are three fundamental areas.

The area between $z = -1$ and $z = 1$ is about .68.
The area between $z = -2$ and $z = 2$ is about .95.
The area between $z = -3$ and $z = 3$ is about .99.

FIGURE 2

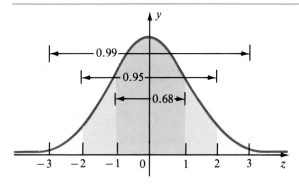

These areas refer to the proportion of entries that lie within 1σ, 2σ, and 3σ of the mean. Thus for a normal curve about two-thirds of the entries will lie within 1σ of the mean, about 95% will lie within 2σ, and virtually all will lie within 3σ. These properties are shown in Figure 2.

According to Chebyshev's Theorem for $k = 2$, at least three-fourths of the entries in any distribution lie within 2σ of the mean. For data with a normal distribution this approximation can be improved to 95%. For 3σ Chebyshev's Theorem gives a figure of at least $\frac{8}{9} = .89$, while the corresponding figure for the normal curve is known to be .99.

Examples 1 and 2 combine the four basic properties of the normal curve with the three designated areas to derive other areas under the normal curve.

EXAMPLE 1

Problem

What is the area under the normal curve between $z = 0$ and $z = 1$?

SOLUTION The area under the normal curve between $z = -1$ and $z = 1$ is about .68. Since the normal curve is symmetric about the mean, the mean divides this area into two equal parts. (See Figure 3.) The area between $z = 0$ and $z = 1$ is one-half of the area between $z = -1$ and $z = 1$, so its area is about $(.5)(.68) = .34$.

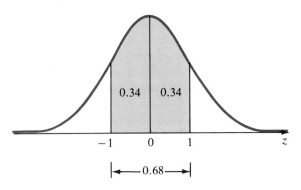

FIGURE 3

EXAMPLE 2

Problem

What is the area under the normal curve to the right of $z = 2$?

SOLUTION The area under the normal curve to the right of $z = 2$ is derived from the fact that the area between $z = -2$ and $z = 2$ is about .95. There is about .05 remaining. There are two regions under the curve that comprise this area, the region to the right of $z = 2$ and the region to the left of $z = -2$. These regions have equal areas because of the symmetry of the normal curve. Hence the area under the normal curve to the right of $z = 2$ is one-half of .05, or .025. (See Figure 4.)

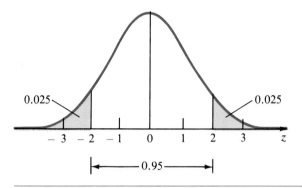

FIGURE 4

The area under the normal curve to the left of $z = 2$ can be obtained from Example 2 by using property 4. Since the total area is 1 and the area to the right is about .025, the area to the left is approximately $1 - .025 = .975$. An alternate way to obtain this area is to look it up in a table.

Using a Table

TABLE 1

z	$A(z)$
2.00	.9773
1.50	.9332
1.00	.8413
0.50	.6915
0.00	.5000
−0.50	.3085
−1.00	.1587
−1.50	.0668
−2.00	.0227

The values we have supplied for areas under the normal curve are very limited; they are rough approximations that apply only to integral multiples of σ. Table 1 contains several more values that are approximated to four decimal places. A more complete table is printed in the Appendix. For each value of z, called the **z-score,** $A(z)$ gives the area under the normal curve to the left of z. The area $A(z)$ is shaded in Figure 5.

There are three kinds of areas under the normal curve.

L: The area to the left of z
R: The area to the right of z
B: The area between two values of z

Table 1 accounts for L-areas. For instance, the area to the left of $z = 2$ is $A(2) = .9773$. This is a better approximation than the value .975 obtained after Example 2. Example 3 shows how to compute R-areas.

FIGURE 5

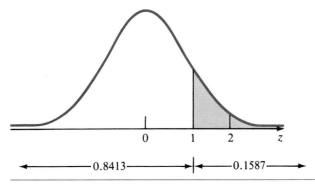

FIGURE 6

EXAMPLE 3 | **Problem**

Find the area under the normal curve to the right of $z = 1$.

SOLUTION From Table 1 the area to the left of $z = 1$ is $A(1) = .8413$. Since the total area under the normal curve is 1, the area to the right of $z = 1$ is $1 - .8413 = .1587$. See Figure 6.

The following rule emerges from Example 3.

The area under the normal curve to the right of z is $1 - A(z)$.

EXAMPLE 4 | **Problem**

Find the area under the normal curve between $z = -0.5$ and $z = 0.5$.

SOLUTION First set $z_1 = 0.5$, then $A(0.5) = .6915$ is the area to the left of $z_1 = 0.5$. Similarly, $A(-0.5) = .3085$ is the area to the left of $z_2 = -0.5$. The desired area is the difference between these two areas. See Figure 7 for a geometric view of why this is true. Thus the area between $z_1 = -0.5$ and $z_2 = 0.5$ is $A(0.5) - A(-0.5) = .6915 - .3085 = .3830$.

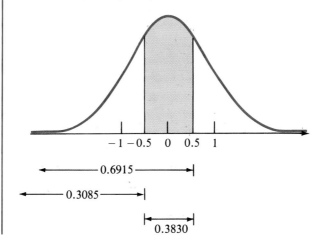

FIGURE 7

The following rule emerges from Example 4.

> The area under the normal curve between z_1 and z_2, where $z_1 < z_2$, is $A(z_2) - A(z_1)$.

The normal curve is a mathematical abstraction that is not derived from any specific real-world phenomena. However, it is very useful for describing any set whose histogram resembles the normal curve. The main difference between the normal curve and such a set is that the normal curve is *continuous*, while data sets are necessarily *discrete*. In spite of this difference the normal curve's utility is vast. Let us turn to its applications.

Applications

So far we have referred to *the* normal curve. Actually, it is called the **standard normal curve.** Any list of data that is assumed to be normally distributed will have a histogram, and hence a normal curve, that resembles the standard normal curve. The **normal approximation** is a procedure for finding the area under any normal curve by adjusting the units of the horizontal and vertical axes, so that the resulting curve resembles the standard normal curve. Let us explain this procedure.

The Health Examination Survey, conducted by the Department of Health and Human Services, found that the average height of women in the United States forms a normal curve, with $\mu = 64''$ and $\sigma = 2.5''$. *What percentage of women are less than $67\frac{3}{4}''$ tall?* To answer this question, draw an x-axis and mark the heights at μ, $\mu \pm 1\sigma$ (66.5″ and 61.5″), and $\mu \pm 2\sigma$ (69″ and 59″). Draw a z-axis below it, marking the corresponding number of standard deviations. (See Figure 8.) The value $x = 67.75''$ on the upper line lies between 66.5″ and 69″, so the corresponding z-score on the lower line is between 1 and 2. Since $\mu = 64''$ and $\sigma = 2.5$, the precise z-score is obtained from this formula.

$$z = \frac{\text{distance between } x \text{ and } \mu}{\text{standard deviation}} = \frac{67.75 - 64}{2.5} = \frac{3.75}{2.5} = 1.5$$

This means that $x = 67.5''$ is $z = 1.5$ standard deviations from the mean $\mu = 64''$. The formula for the z-score is described in the display.

> If a set of data is normally distributed with mean μ and standard deviation σ, the z-score corresponding to a value of x is
>
> $$z = \frac{x - \mu}{\sigma}$$

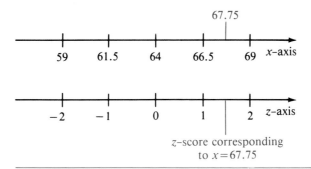

FIGURE 8

The question about the percentage of women whose height is less than $x = 67.75''$ translates into z-scores less than $z = 1.5$. This is an L-area problem. From Table 1 the area to the left of $z = 1.5$ is $A(1.5) = .9332$. Thus about 93.32% of American women are less than $67.75''$ tall.

In summary, the normal approximation is a three-part procedure. First the x-scores are converted to z-scores. Then the problem is translated into one of the fundamental areas: L, R, or B. Finally, a table of z-scores for the standard normal curve provides the area, which is interpreted as either a probability or a percentage. Example 5 illustrates this procedure for a B-area.

EXAMPLE 5

Assume that the heights of women in the United States are normally distributed with $\mu = 64''$ and $\sigma = 2.5''$.

Problem

What percentage of heights are between $5'$ and $5'2''$?

SOLUTION Since $5' = 60''$, set $x_1 = 60$. The corresponding z-score is $z_1 = (60 - 64)/2.5 = -1.6$. From the Appendix, $A(-1.6) = .0548$. Next, set $x_2 = 62$. Then $z_2 = (62 - 64)/2.5 = -0.8$ and $A(-0.8) = .2119$.

The percentage of heights between $x_1 = 60''$ and $x_2 = 62''$ translates to the area between $z_1 = -1.6$ and $z_2 = -0.8$. This area is $A(z_2) - A(z_1) = A(-0.8) - A(-1.6) = .2119 - .0548 = .1571$. Consequently, 15.71% of American women are between $5'$ and $5'2''$. This percentage is roughly 1 in 6.

Example 6 applies the normal approximation procedure to IQ (intelligence quotient) tests under the usual assumption that IQ scores are normally distributed. It illustrates the method for R-areas.

EXAMPLE 6 | Many IQ tests are designed so that the mean score is 100 and the standard deviation is 20.

Problem

If a person is selected at random, what is the probability that the IQ score is greater than 130?

SOLUTION The z-score corresponding to $x = 130$ is $z = (130 - 100)/20 = 1.5$. From Table 1, $A(1.5) = .9332$. Scores above 130 translate to the area to the right of $z = 1.5$. This is an R-area, which is equal to $1 - A(1.5) = 1 - .9332 = .0668$. Thus the probability is .0668.

In many public schools the cutoff point for a student to be classified as "gifted" is an IQ score of 130. Example 6 shows that about 7% of the population is gifted under the assumption that $\mu = 100$ and $\sigma = 20$.

EXERCISE SET 8–4

• In Problems 1 to 8 find the area under the normal curve within the given limits, using the four basic properties and three fundamental areas of the normal curve.

1. $z = 0$ and $z = 2$.

2. $z = 0$ and $z = 3$.

3. $z = -1$ and $z = 0$.

4. $z = -2$ and $z = 0$.

5. to the right of $z = 1$.

6. to the right of $z = 3$.

7. to the left of $z = -2$.

8. to the left of $z = -1$.

In Problems 9 to 12 use Table 1 to find the area under the normal curve within the given limits.

9. to the right of $z = 1.5$.

10. to the right of $z = -0.5$.

11. between $z = -1$ and $z = 1$.

12. between $z = -0.5$ and $z = 1.5$.

In Problems 13 to 16 assume that the heights of women in the United States are normally distributed with $\mu = 64''$ and $\sigma = 2.5''$. Determine the percentage of heights that are in the range

13. between 5′3″ and 5′6″.

14. between 5′1″ and 5′5″.

15. taller than 5′7″.

16. shorter than 5′.

In Problems 17 to 20 the mean IQ is 100 and the standard deviation is 20. Determine the percentage of IQs that are in the given range.

17. between 110 and 130.

18. between 90 and 115.

19. greater than 95.

20. less than 110.

•• In Problems 21 to 26 find the area under the normal curve within the given limits, using the four basic properties and three basic areas of the normal curve.

21. $z = 1$ and $z = 2$.

22. $z = 2$ and $z = 3$.

23. $z = -1$ and $z = 2$.

24. $z = -2$ and $z = 1$.

25. to the right of $z = -1$.

26. to the left of $z = 3$.

In Problems 27 to 30 use the table in the Appendix to find the area under the normal curve within the given limits.

27. to the left of $z = 0.38$.

28. to the left of $z = -1.05$.

29. between $z = -1.74$ and $z = 0.51$.

30. between $z = -0.56$ and $z = 2.15$.

••• The mean and median scores on the SAT exams vary slightly from year to year.* In Problems 31 to 34 assume that they form a normal distribution and that, for the mathematics section, $\mu = 470$ and $\sigma = 100$.

31. What percentage of students scored above 500?

32. What percentage of students scored between 400 and 600?

33. If a student is selected at random, what is the probability that the student scored between 430 and 530?

34. If a student is selected at random, what is the probability that the student scored above 440?

Problems 35 to 38 refer to the Stanford study on the effects of long-distance running on the joints. Table 2, which is taken from that study, tabulates the effects on the knee joint using three measures.

35. For nonrunners the mean sclerosis measure is $\mu = 5.2$ with $\sigma = 0.3$. What percentage of the nonrunners had a sclerosis measure greater than 5.9, which is the mean sclerosis measure for runners?

36. For nonrunners the mean spur measure is $\mu = 5.7$ with $\sigma = 0.7$. What percentage of the nonrunners had a spur measure greater than 7.3, which is the mean spur measure for runners?

37. For runners the sclerosis measure is $\mu = 5.9$ and $\sigma = 0.3$.
    (a) What percentage of runners had a sclerosis measure less than 5.5?
    (b) What percentage of nonrunners had a sclerosis measure less than 5.5?

*William H. Angoff, "How We Calibrate College Board Scores," *The College Board Review*, No. 68, Summer 1968.

TABLE 2   Knee Joint Roentgenograms (Mean Values)

| | All Subjects | | | | Matched Pairs | | | | | |
| | | | | | All Pairs | | Female Pairs | | Male Pairs | |
	Runners ($n = 65$)	Nonrunners ($n = 33$)	50+ Runners ($n = 41$)	Controls ($n = 57$)	50+ Runners ($n = 41$)	Controls ($n = 41$)	50+ Runners ($n = 18$)	Controls ($n = 18$)	50+ Runners ($n = 23$)	Controls ($n = 23$)
Sclerosis										
Mean	5.9	5.2	5.9	5.5	6.0	5.3	6.7	5.1	5.5	5.5
SE	0.3	0.3	0.3	0.3	0.3	0.3	0.5	0.3	0.4	0.5
					(0.44)		(0.63)		(0.56)	
Spurs										
Mean	7.3	5.7	7.4	6.3	7.5	6.0	8.4	5.1	6.7	6.8
SE	0.5	0.7	0.6	0.5	0.6	0.6	0.9	0.7	0.8	0.9
					(0.81)		(0.96)		(1.16)	
Narrowing (mm)										
Mean	7.0	6.9	7.0	6.9	7.0	6.7	6.6	6.3	7.4	7.1
SE	0.2	0.3	0.3	0.2	0.3	0.3	0.3	0.4	0.4	0.4
					(0.32)		(0.57)		(0.36)	

Source: From Nancy E. Lane et al., "Long-Distance Running, Bone Density, and Osteoarthritis," *Journal of the American Medical Association*, March 7, 1986, pp. 1148–1151.

TABLE 3   Sample Data

	Proportion of Wins	Mean Win (chips)	Standard Deviation (chips)	Mean Player Edge (%)	Largest Player Deficit (chips)
Las Vegas	0.87	2084	1786	1.35	4961
Atlantic City	0.79	1108	1471	0.91	3621

Source: From Martin H. Millman, "A Statistical Analysis of Casino Blackjack," *American Mathematical Monthly,* August–September 1983, pp. 431–436.

38. Refer to Table 2 for the mean and standard deviation of the joint space narrowing.
    (a) What percentage of runners had a joint space narrowing between 6.8 and 7.2 mm (millimeters)?
    (b) What percentage of nonrunners had a joint space narrowing between 6.8 and 7.2 mm?

Problems 39 and 40 refer to Table 3, which lists the data from a computer simulation of 20,000 games of blackjack in Atlantic City and Las Vegas. (The actual playing of 20,000 hands would consume some 300 to 400 human hours.)

39. What is the probability of a player winning at least 300 chips in (a) Las Vegas and (b) Atlantic City?

40. What is the probability of a player winning less than 100 chips in (a) Las Vegas and (b) Atlantic City?

8–5   THE NORMAL APPROXIMATION TO BINOMIAL EXPERIMENTS

When you apply to a college the admissions office either accepts or rejects your application. Each decision from a college represents a **Bernoulli trial** because it consists of two possible outcomes. Some other examples of Bernoulli trials are the effectiveness of a medication (the patient is or is not cured), the movement of the Dow-Jones average (up or down), and the firing of a missile (hit or miss).

A **binomial experiment** consists of n Bernoulli trials. In Section 7–7 we saw that if the probability of success is p and the probability of failure is $q = 1 - p$, then the probability of obtaining exactly x successes in n trials is

$$Pr(X = x) = C(n, x)p^x q^{n-x}$$

This formula is useful when the value of n is relatively small, but the computations become tedious when n is large. Fortunately the normal curve provides a very close approximation to a binomial experiment in such a case. This section shows how to carry out and apply the normal approximation procedure.

Coin Tossing

Coin tossing serves as a convenient model for binomial experiments. As an example, toss a coin six times and let the random variable X represent the number of heads

TABLE 1

x	p	xp	x^2p
0	1/64	0/64	0/64
1	6/64	6/64	6/64
2	15/64	30/64	60/64
3	20/64	60/64	180/64
4	15/64	60/64	240/64
5	6/64	30/64	150/64
6	1/64	6/64	36/64
Totals	1	192/64	672/64

that appear. Then $n = 6$ and $p = \frac{1}{2}$. The probability of tossing exactly four heads is

$$Pr(X = 4) = C(6, 4)\left(\frac{1}{2}\right)^4\left(\frac{1}{2}\right)^2 = \frac{15}{64}$$

The entire probability distribution, including the particular case $X = 4$, is given in Table 1. Notice the values in the row of totals. The column of probabilities p adds up to 1, which serves as a check on the individual probabilities. The sum of the xp column is the expected value $\mu = 192/64 = 3$. Common sense confirms that six tosses of a coin should result in three heads on the average. Recall from Section 7–7 that another way of obtaining this value is from the formula $\mu = np$. Here $\mu = (6)(1/2) = 3$.

Recall that the variance $V(X)$ is equal to the total of the x^2p column of Table 1 minus the square of μ.

$$V(X) = \frac{672}{64} - 3^2 = \frac{21}{2} - 9 = \frac{3}{2}$$

For a binomial experiment it is possible to compute this value from the formula $V(X) = npq$. In this example $V(X) = 6(1/2)(1/2) = 3/2$, which confirms the value derived from Table 1. The appropriate formulas are given in the display.

If a binomial experiment consists of n Bernoulli trials, each with probability of success p and probability of failure q, then

The mean is $\mu = np$.
The variance is $V(X) = npq$.
The standard deviation is $\sigma = \sqrt{npq}$.

EXAMPLE 1

Toss a coin 100 times and record the number of heads that appear.

Problem

What are the mean and standard deviation?

SOLUTION We have $n = 100$ and $p = \frac{1}{2}$ so $\mu = np = 50$. Since $p = \frac{1}{2}$, it follows that $q = 1 - p = \frac{1}{2}$, so $\sigma = \sqrt{npq} = \sqrt{25} = 5$.

Figure 1 shows the histogram for the probability distribution of tossing six coins and recording the number of heads. The histogram can be approximated by the normal curve that is superimposed upon it. The normal curve has $\mu = 3$ and $\sigma = \sqrt{6/4} \approx 1.225$. Let us see how to use this normal curve to approximate probabilities.

Consider the value of $X = 4$ heads. Its probability is equal to the area of the shaded bar above 4 in the histogram. This area can be approximated by the area under the normal curve between $x = 3.5$ and $x = 4.5$. The procedure for finding this area, as explained in Section 8–4, uses the formula $z = (x - \mu)/\sigma$.

$$\text{for } x_1 = 3.5, \qquad z_1 = \frac{(3.5 - 3)}{1.225} \approx 0.41$$

$$\text{for } x_2 = 4.5, \qquad z_2 = \frac{(4.5 - 3)}{1.225} \approx 1.22$$

From the table in the Appendix, $A(z_1) = .6591$ and $A(z_2) = .8888$. The area under the normal curve between 3.5 and 4.5 is approximately equal to $A(z_2) - A(z_1) = .8888 - .6591 = .2297$. This value compares favorably with the theoretical probability of $15/64 \approx .2344$ listed in Table 1.

The probability of tossing at least four heads in six tosses is given by $Pr(4) + Pr(5) + Pr(6) = 22/64 \approx .3438$. This value can be approximated by the area under the normal curve to the right of $x = 3.5$ since the area of the histogram in Figure 1 corresponds to this probability. For $x = 3.5$ we found that $z = 0.41$

FIGURE 1

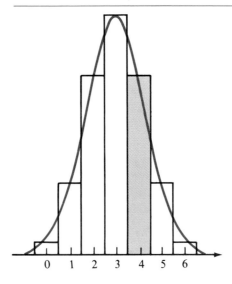

and $A(0.41) = .6591$. The desired area, which lies to the right of $z = 0.41$, is $1 - A(z) = 1 - .6591 = .3409$. This value compares favorably with the theoretical probability of .3438.

Example 2 applies a similar approximation to the question of whether a coin is fair. Do you think a coin that lands heads 60 times in 100 tosses is fair?

EXAMPLE 2 | **Problem**

What is the probability of a coin landing heads 60 or more times in 100 tosses?

SOLUTION From Example 1, $\mu = 50$ and $\sigma = 5$. The problem is to find $Pr(X \geq 60)$. The normal curve is shown in Figure 2, where the desired area is shaded. The value $x = 59.5$ is used for $X = 60$ because the bar corresponding to 60 runs from 59.5 to 60.5. Then $z = (59.5 - 50)/5 = 9.5/5 = 1.9$. From the table in the Appendix, the area to the left of $z = 1.9$ is $A(1.9) = .9713$. The area to the right is $1 - .9713 = .0287$. This is the desired probability.

FIGURE 2

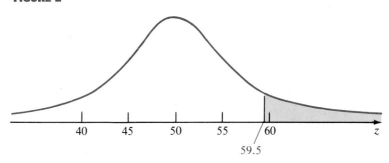

It follows from Example 2 that if a fair coin is tossed 100 times, it will land heads 60 or more times about 3% of the time. It is reasonable to conclude that if a coin produces 60 or more heads in 100 tries, it is probably not fair.

The display lists the steps for applying the normal approximation procedure to a binomial experiment. To appreciate its power, notice that the alternative to Example 2 is to carry out these computations.

$$Pr(X \geq 60) = Pr(X = 60) + Pr(X = 61) + \cdots + Pr(X = 99) + Pr(X = 100)$$
$$= C(100, 60)(\tfrac{1}{2})^{60}(\tfrac{1}{2})^{40} + C(100, 61)(\tfrac{1}{2})^{61}(\tfrac{1}{2})^{39} + \cdots + C(100, 99)(\tfrac{1}{2})^{99}(\tfrac{1}{2})^{1} + C(100, 100)(\tfrac{1}{2})^{100}(\tfrac{1}{2})^{0}$$

Procedure for Approximating a Binomial Experiment by the Normal Curve

1. Compute $\mu = np$ and $\sigma = \sqrt{npq}$.
2. Sketch the normal curve and shade the desired area.
3. Convert X to x, then x to z using $z = (x - \mu)/\sigma$.
4. Use a table to determine the area.

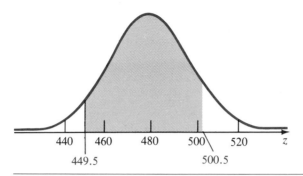

440 | 460 | 480 | 500 | 520 | z

449.5   500.5

FIGURE 3

EXAMPLE 3 | **Problem**

If a pair of dice is tossed 2880 times, what is the probability of obtaining between 450 and 500 doubles inclusive?

SOLUTION We have $n = 2880$. The probaoility of tossing doubles is 6/36, so $p = 1/6$ and $q = 5/6$. Then $\mu = 480$ and $\sigma = \sqrt{npq} = 20$. The problem is to find $Pr(450 \leq X \leq 500)$. The desired area is sketched in Figure 3. Set $x_1 = 449.5$ and $x_2 = 500.5$. Then $z_1 = (449.5 - 480)/20 = -1.525$ and $z_2 = (500.5 - 480)/20 = 1.025$. The area between z_1 and z_2 is $A(z_2) - A(z_1) = .8461 - .0643 = .7818$. Thus the probability of obtaining between 450 and 500 doubles is .7818.

Accidents and Drugs

Many situations involve either one outcome or another. We label each outcome a "success" or "failure" even though the meanings of these terms need not apply. Consider the cause of accidents. Fatigue causes a certain percent p of accidents depending on the particular occupation. An accident is either due or not due to fatigue, so it represents a Bernoulli trial. If an accident is due to fatigue, we label it a success and assign it the probability p, even though in the human arena an accident caused by fatigue could hardly be called a success.

Example 4 considers industrial accidents.

EXAMPLE 4 | Fatigue is the cause of 20% of all industrial accidents. Last year there were 2500 accidents in the steel industry.

Problem

What is the probability that at least 525 of the accidents were caused by fatigue?

SOLUTION Since $p = 0.2$ and $n = 2500$, $\mu = (2500)(0.2) = 500$ and $\sigma = \sqrt{(2500)(0.2)(0.8)} = \sqrt{400} = 20$. The problem is to find $Pr(X \geq 525)$. The desired area is sketched in Figure 4. Set $x = 524.5$. Then $z =$

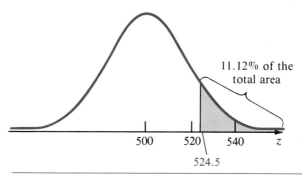

11.12% of the total area

500   520   540 z

524.5

FIGURE 4

$(524.5 - 500)/20 = 1.225$. From the table in the Appendix, $A(1.22) = .8888$. The desired area lies to the right of $z = 1.225$, so

$$Pr(X \geq 525) = 1 - .8888 = .1112$$

Thus the probability of at least 525 of the accidents being caused by fatigue is about 11%.

Example 5 shows how statistical studies allow some professions to prepare for certain events beforehand.

EXAMPLE 5

Next year 62,500 people will use a drug that has caused severe side effects (loss of weight, rash, depression) in 10% of all people who have already taken it.

Problem

What is the probability that at least 6200 people will suffer side effects?

SOLUTION We have $p = 0.1$ and $n = 62,500$. Then $\mu = 6250$ and $\sigma = \sqrt{npq} = \sqrt{5625} = 75$. The problem is to compute $Pr(X \geq 6200)$. The desired area is sketched in Figure 5. Set $x = 6199.5$. Then

FIGURE 5

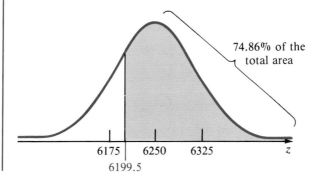

74.86% of the total area

6175   6250   6325 z

6199.5

$$z = \frac{6199.5 - 6250}{75} = \frac{-50.5}{75} \approx -.67$$

The table in the Appendix gives $A(-0.67) = .2514$. Therefore the probability is approximately equal to $1 - .2514 = .7486$, which is about 75%.

EXERCISE SET 8–5

• In Problems 1 to 4 toss n coins and record the number of heads. Find μ and σ.

1. $n = 144$ 2. $n = 36$

3. $n = 81$ 4. $n = 121$

In Problems 5 to 8 a coin is tossed 100 times. Find the probability of each event.

5. at least 55 heads.

6. at least 45 heads.

7. between 45 and 50 heads.

8. at most 60 heads.

In Problems 9 to 12 a coin is tossed 36 times. Find the probability of each event.

9. less than 15 heads.

10. at least 20 heads.

11. between 15 and 20 heads.

12. more than 20 heads.

In Problems 13 to 16 a pair of dice is tossed 2880 times. Find the probability of each event.

13. at least 450 doubles.

14. at least 510 doubles.

15. between 500 and 540 doubles.

16. at most 490 doubles.

In Problems 17 to 20 a pair of dice is tossed 180 times. Find the probability of each event.

17. at most 25 doubles.

18. at most 35 doubles.

19. between 25 and 35 doubles.

20. at least 20 doubles.

Problems 21 and 22 refer to an industry where 20% of the accidents are caused by fatigue. Suppose that 900 accidents occur. Compute the probability of each event.

21. at least 150 accidents were caused by fatigue.

22. between 162 and 192 accidents were caused by fatigue.

Problems 23 and 24 refer to a drug that causes severe side effects in 10% of the people who take it. Suppose that 100 people take the drug. Compute the probability of each event.

23. Between 5 and 15 people suffer severe side effects.

24. At least 12 people suffer severe side effects.

•• In Problems 25 to 28 a coin is tossed 400 times. Find the probability of each event.

25. at most 205 heads.

26. at most 195 heads.

27. between 195 and 220 heads.

28. at least 210 heads.

In Problems 29 to 34 a survey of 1200 families with exactly two children is conducted. Find the probability of each event.

29. At most 315 of the families have two boys.

30. At most 290 of the families have two girls.

31. At least 310 of the families have two girls.

32. At least 285 of the families have two boys.

33. Between 270 and 330 of the families have two girls.

34. Between 300 and 345 of the families have two girls.

35. About 2% of a dentist's patients cancel their

appointments. What is the probability that out of 10,000 patients, at least 180 appointments will be canceled?

36. A clairvoyant draws a card from an ordinary deck of 52 cards and announces the suit. This is performed 48 times. What is the probability that the clairvoyant will name the correct suit at least 18 times?

37. A jar contains 3 red jelly beans and 9 black jelly beans. Select 75 jelly beans with replacement. What is the probability that between 15 and 20 are red?

38. A fluoride treatment is rated 98% effective. If it is applied in 100 cases, what is the probability that between 90 and 98 are effective?

39. A student known as a "partier" must guess at every answer on a 50-question multiple-choice test, which has five possible answers to every question. What is the probability that the student will obtain a grade of 40% or better?

40. A manufacturer of disks has found that only 1% of the disks is defective. What is the probability that a shipment of 1100 disks will contain less than 12 defective disks?

41. The Dow-Jones average will rise on a given day about 55% of the time. What is the probability that it will rise on at least 210 days during a regular year?

42. Of 1000 two-children families, what is the probability that 600 or more of them will have one boy and one girl?

●●● 43. USAir knows from experience that 15% of the reservations will not be honored.*
   (a) The airline sells 160 tickets for a flight that has 148 seats. What is the probability that more than 148 people will show up?
   (b) What is the maximum number of tickets the

airline should sell so that there is a 95% chance that everyone who shows up will get a seat?

44. In Dade County, Florida, the number of malpractice lawsuits against doctors doubled from the year 1975 to the year 1985. At present, 25% of all physicians in the state of Florida are sued annually.† If a hospital has 48 physicians on staff, what is the probability that at least 15 of them will be sued during a year?

45. When Hewlett-Packard Company (H-P) entered into a joint venture with Yokogawa Corporation in 1983 to produce circuit boards for computers, it found that boards produced at the Tokyo plant contained eight defects per million connections compared with the rate of six defects per thousand connections at the H-P plant in Cupertino, California.‡
   (a) If a batch of 50 million connections is produced at the Tokyo plant, what is the probability that less than 460 connections will be defective?
   (b) If a batch of 50 million connections is produced at the Cupertino plant, what is the probability that less than 298,360 connections will be defective?

46. In American jury trials, prospective jurors can be "peremptorily" challenged by the counsel for either side. Although the state of Florida ordinarily allows six peremptory challenges for each side, the judge in the Theodore Bundy trial permitted 20 because of the national publicity surrounding Bundy's brutal murder of two sorority sisters. A pertrial survey conducted by the Public Defender's office revealed that 40 percent of the eligible jurors felt that Bundy was guilty.§ If 30 people were questioned, what is the probability that at least 12 of them did not already feel that Bundy was guilty?

*We are grateful to Dennis Dombroski of USAir for supplying us with these figures.

†William F. Buckley, Jr., "Malpractice Brings a Truce, of Sorts," Nationally Syndicated Column, April 4, 1988.

‡John Schneidawind, "Computer Company Capitalizes on Japanese Methods," Knight-Ridder News Service, May 13, 1988.

§LeRoy A. Franklin, "Bayes' Theorem, Binomial Probabilities, and Fair Numbers of Peremptory Challenges in Jury Trials," *College Mathematics Journal,* September 1987, pp. 291–299.

Let us recall some events from the American Revolution. In the 1760s the British government adopted the policy of taxing the American colonies. The most famous and most important of these measures was the Stamp Act, which was later repealed and replaced by the Townshend Acts, both of which provided taxes to maintain the British army in America. This led to the Boston Tea Party, and the decision by King George III that the colonies should be subdued. Revolt followed. Hostilities broke out in April 1775, and on July 4 of the following year the delegates to the second Continental Congress approved the declaration that ''These United Colonies are, and of right ought to be, free and independent states.'' The war came to an official end in 1783 with the Treaty of Paris, in which Britain recognized the independence of the United States.

That sequence of events answered one of the two great questions from the revolutionary era: How should power be distributed between Britain and America? But another question remained unresolved: How should power be divided among the Americans? It is instructive to think not of *the* revolution, but of two revolutions, one external and the other internal. We just outlined the external revolution; now let us describe the internal revolution.

The Continental Congress functioned as a national government until 1781, when the Articles of Confederation, the first national constitution, were adopted. However, the Articles contained one major defect: they did not provide for a strong central government. This is understandable; American heads of state had no intention of replacing one form of despotism by another. Consequently, the central government's lack of power to regulate commerce caused grave commercial problems, and its lack of power to tax led to a financial disaster. The situation was so serious that historians have come to call the period from 1783 to 1787 not *a* critical period but *the* critical period in American history.

A constitutional convention was convened in hot, steamy Philadelphia from May to September of 1787. There the delegates hammered out a document that would prove to be the cornerstone of our government—the Constitution. The 13 colonies were much less united than the 50 states are today, yet politicians from all ideologies recognized the importance of the passage of the Constitution. The Federalist papers were published by Alexander Hamilton, John Jay, and James Madison to persuade the citizens of New York State to vote for ratification. They have earned an important niche in political philosophy.

A problem arose afterward because The Federalist papers had been published anonymously over the signature ''Publius.'' Of the 85 papers there was general agreement that Hamilton wrote 51, Madison wrote 14, and Jay wrote 5. Three were of joint authorship. The remaining 12 were written by either Madison or Hamilton, but it was not clear which of these distinguished statesmen wrote them.

Madison or Hamilton—who was the author of the disputed Federalist papers? Even though they held opposing political views, neither would claim the honor in deference to the greater good of convincing the New York electorate to vote for ratification of the new Constitution. The usual means for determining authorship by political content and writing style failed to solve the problem. Besides, Madison and Hamilton consciously adopted and perfected a common legal writing style. Thus the usual linguistic methods of studying sentence length and word usage shed no further light on the problem. The papers became known as the "disputed papers."

The first conclusive evidence was provided nearly two centuries later in a historic study by Mosteller and Wallace.* It was based on certain words, called *marker words,* which are exceptionally common to one author's style and not the other's. However, the use of marker words presents two big difficulties: finding them and proving that the difference in their usage is significant. The nagging uncertainty never disappears, as perfect markers can never be found, but sometimes a nearly conclusive case can be made.

The question of the authorship of The Federalist papers forged an important link between statistics and history. The standard deviation was the key element. We present the data gathered by Mosteller and Wallace for the marker word "upon" from 50 essays written by Madison, 48 essays written by Hamilton, and the 12 disputed papers. Tables 1 and 2 present the distribution for the essays by Hamilton and Madison, respectively, while Table 3 gives the distribution for the 12 disputed papers.

The means and standard deviations are summarized in Table 4. They show that the usage of the marker word "upon" in the disputed papers fits the usage in the Madison essays much more closely than it fits the Hamilton essays. The same conclusion was obtained for many other marker words. The inference from the Mosteller and Wallace study is clear: Madison is the author of the disputed Federalist papers.

*Frederick Mosteller and David L. Wallace, *Inference and Disputed Authorship: The Federalist Papers,* Reading, MA, Addison-Wesley, 1964. This study was updated in a second edition of the book under the title *Applied Bayesian and Classical Inference: The Case of the Federalist Papers,* New York, Springer-Verlag, 1984.

TABLE 1 Frequency Distribution of Rate per 1000 Words in 48 Hamilton Papers for the Word "Upon"

Rate per 1000 Words		Number of			
Interval	Midpoint x_i	Papers f_i	$f_i x_i$	x_i^2	$f_i x_i^2$
0.5–1.5	1	5	5	1	5
1.5–2.5	2	10	20	4	40
2.5–3.5	3	12	36	9	108
3.5–4.5	4	10	40	16	160
4.5–5.5	5	8	40	25	200
5.5–6.5	6	2	12	36	72
6.5–7.5	7	1	7	49	49
		48	160	140	634

$$\bar{x} = \frac{f_1 x_1 + \cdots + f_7 x_7}{f_1 + \cdots + f_7} = \frac{160}{48} \approx 3.33$$

$$\sigma = \sqrt{\frac{f_1 x_1^2 + \cdots + f_7 x_7^2}{f_1 + \cdots + f_7} - \bar{x}^2} = \sqrt{\frac{634}{48} - \left(\frac{160}{48}\right)^2} \approx 1.45$$

TABLE 2 Frequency Distribution of Rate per 1000 Words in 50 Madison Papers for the Word "Upon"

Rate per 1000 Words		Number of Papers f_i	$f_i x_i$	x_i^2	$f_i x_i^2$
Interval	Midpoint x_i				
0.0–0.4	0.2	41	8.2	0.04	1.64
0.4–0.8	0.6	2	1.2	0.36	0.72
0.8–1.2	1.0	4	4.0	1.00	4.00
1.2–1.6	1.4	1	1.4	1.96	1.96
1.6–2.0	1.8	2	3.6	3.24	6.48
		50	18.4	6.60	14.80

$$\bar{x} = \frac{18.4}{50} \approx 0.37$$

$$\sigma = \sqrt{\frac{14.8}{50} - \left(\frac{18.4}{50}\right)^2} \approx 0.40$$

TABLE 3 Frequency Distribution of Rate per 1000 Words in 12 Disputed Federalist Papers for the Word "Upon"

Rate per 1000 Words		Number of Papers f_i	$f_i x_i$	x_i^2	$f_i x_i^2$
Interval	Midpoint x_i				
0.0–0.4	0.2	11	2.2	0.04	0.44
0.4–0.8	0.6	0	0.0	0.36	0.00
0.8–1.2	1.0	0	0.0	1.00	0.00
1.2–1.6	1.4	1	1.4	1.96	1.96
		12	3.6	3.36	2.40

$$\bar{x} = \frac{3.6}{12} = 0.3$$

$$\sigma = \sqrt{\frac{2.4}{12} - \left(\frac{3.6}{12}\right)^2} \approx 0.33$$

TABLE 4 Mean and Standard Deviation for 48 Hamilton, 50 Madison, and 12 Disputed Papers for the Word "Upon"

	Hamilton	Madison	Disputed
Mean	3.33	0.37	0.30
σ	1.45	0.40	0.33

Case Study Exercises

Problems 1 to 10 provide the data for ten marker words in The Federalist papers. The problems are adapted from Table 4.8–3 of the 1984 edition of the Mosteller and Wallace book, pp. 167–172. H stands for Hamilton, M for Madison, and D for disputed. The intervals refer to the rate per 1000 words, and the numbers in the other columns give the number of papers having the stated rate. (There were 48 papers of Hamilton, 50 of Madison, and 12 disputed.) For each marker word, determine whether the results are strong enough to assign the authorship of the disputed papers.

1. **Marker Word: "also"**

Interval	H	M	D
0 –0.4	28	18	0
0.4–0.8	13	11	8
0.8–1.2	5	16	3
1.2–1.6	2	1	0
1.6–2.0	0	3	1
3.2–3.6	0	1	0

2. **Marker Word: "an"**

Interval	H	M	D
0– 2	1	1	2
2– 4	7	20	5
4– 6	20	19	4
6– 8	15	8	1
8–10	4	2	0
10–12	0	0	0
12–14	1	0	0

3. **Marker Word: "by"**

Interval	H	M	D
0– 2	1	0	0
2– 4	3	0	8
4– 6	14	4	3
6– 8	12	2	0
8–10	12	11	1
10–12	6	11	0
12–14	0	12	0
14–16	0	4	0
16–18	0	4	0
18–20	0	2	0

4. **Marker Word: "of "**

Interval	H	M	D
32–36	0	1	0
44–48	0	1	0
48–52	2	4	0
52–56	2	10	3
56–60	13	16	7
60–64	10	8	0
64–68	8	7	1
68–72	3	3	1
72–76	6	0	0
76–80	4	0	0

5. **Marker Word: "on"**

Interval	H	M	D
0– 2	9	0	0
2– 4	23	2	1
4– 6	13	14	1
6– 8	2	14	2
8–10	1	9	4
10–12	0	8	4
12–14	0	2	0
14–16	0	1	0

6. **Marker Word: "there"**

Interval	H	M	D
0– 1	2	24	6
1– 2	8	16	3
2– 3	12	5	3
3– 4	10	4	0
4– 5	10	1	0
5– 6	4	0	0
6– 7	1	0	0
9–10	1	0	0

7. **Marker Word: "this"**

Interval	H	M	D
0– 2	1	2	0
2– 4	1	8	1
4– 6	8	13	3
6– 8	14	18	5
8–10	19	9	2
10–12	5	0	1

8. **Marker Word: "to"**

Interval	H	M	D
23–26	0	3	3
26–29	2	2	2
29–32	2	11	1
32–35	4	11	3
35–38	6	7	2
38–41	10	7	1
41–44	8	6	0
44–47	10	1	0
47–50	3	2	0
50–53	1	0	0
53–56	1	0	0
56–59	1	0	0

9. **Marker Word: "both"**

Interval	H	M	D
0 −0.4	23	10	3
0.4–0.8	10	12	4
0.8–1.2	11	10	2
1.2–1.6	3	6	3
1.6–2.0	1	6	0
2.0–2.4	0	1	0
2.4–2.8	0	2	0
2.8–3.2	0	1	0
3.2–3.6	0	2	0

10. **Marker Word: "though"**

Interval	H	M	D
0 −0.4	9	21	8
0.4–0.8	13	22	3
0.8–1.2	12	5	1
1.2–1.6	6	1	0
1.6–2.0	5	0	0
2.0–2.4	1	0	0
2.4–2.8	2	0	0
2.8–3.2	0	0	0
3.2–3.6	0	1	0

CHAPTER REVIEW

Key Terms

8–1 Probability Distributions

Frequency Distribution   Probability Distribution
Frequency   Histogram
Random Variable   Uniform Distribution
Relative Frequency   Normal Distribution

8–2 Measures of Central Tendency

Mean   Expected Value
Median

8–3 Standard Deviation

Variance   Chebyshev's Theorem
Standard Variation

8–4 The Normal Curve

Normal Curve   Standard Normal Curve
z-Score   Normal Approximation

8–5 The Normal Approximation to Binomial Experiments

Bernoulli Trial   Binomial Experiment

Summary of Important Concepts

Formulas

Mean: $\bar{x} = \dfrac{x_1 + x_2 + \cdots + x_n}{n}$

Mean: $\bar{x} = \dfrac{f_1 x_1 + f_2 x_2 + \cdots + f_n x_n}{f_1 + f_2 + \cdots + f_n}$

Expected value: $E(X) = x_1 p_1 + x_2 p_2 + \cdots + x_n p_n$

$$\text{Variance: } V(X) = \frac{x_1{}^2 + x_2{}^2 + \cdots + x_n{}^2}{n} - \mu^2$$

$$V(X) = \frac{f_1 x_1{}^2 + f_2 x_2{}^2 + \cdots + f_n x_n{}^2}{f_1 + f_2 + \cdots + f_n} - \mu^2$$

$$V(X) = p_1 x_1{}^2 + p_2 x_2{}^2 + \cdots + p_n x_n{}^2 - \mu^2$$

Standard deviation: $\sigma = \sqrt{V(X)}$

Area within 1σ of mean is approximately .68.

Area within 2σ of mean is approximately .95.

Area within 3σ of mean is approximately .99.

Normal approximation: $z = \dfrac{x - \mu}{\sigma}$

Binomial experiments

Mean $\mu = np$

Variance $V(X) = npq$

Standard deviation $\sigma = \sqrt{npq}$

REVIEW PROBLEMS

1. Draw the histogram for the experiment of tossing four coins and recording the number of heads. Shade the portion corresponding to less than three heads.

Problems 2 to 4 refer to the probability experiment of tossing two dice and letting X be the sum of the top faces.

2. Construct the probability distribution.

3. Draw the histogram.

4. Find $Pr(3 \le X \le 7)$.

Problems 5 to 7 refer to the probability experiment of choosing 3 cards from an ordinary deck without replacement and letting X be the number of clubs selected.

5. Construct the probability distribution.

6. Draw the histogram.

7. Find $Pr(0 \le X \le 2)$.

8. Find the (a) mean and (b) median using the midpoints of the intervals as the data given in the table.

interval	0–2	2–4	4–6	6–8	8–10
frequency	3	1	3	2	1

Problems 9 and 10 refer to the frequency distribution in the table.

value	1	2	3	4	5
frequency	6	7	2	3	2

9. Find the (a) mean and (b) median.

10. Find the (a) variance and (b) standard deviation.

Problems 11 and 12 refer to the probability distribution in the table.

x	p
-2	$\frac{1}{8}$
-1	$\frac{1}{2}$
1	$\frac{1}{4}$
2	$\frac{1}{8}$

11. Find the expected value.

12. Find (a) the variance and (b) the standard deviation.

13. Ten employees in a company have a mean salary of $15,000.
  (a) If another person is hired at a salary of $15,000, what is the mean salary of the 11 employees?
  (b) If a twelfth person is hired at a salary of $50,000, what is the mean salary of the 12 employees?

14. A leap year has 4 months with 30 days, 7 months with 31 days, and February with 29 days.
  (a) Find the mean number of days in a month.
  (b) Find the standard deviation.

15. A college wishes to estimate the average cost of books per student during the year. Compute the average cost and the standard deviation from the table.

class	freshmen	sophomore	junior	senior
percent	35%	25%	21%	19%
cost of books	$100	$160	$180	$190

In Problems 16 and 17 find the area under the normal curve within the given limits.

16. (a) between $z = -2$ and $z = 2$.
  (b) to the right of $z = 2$.

17. (a) between $z = -0.73$ and $z = 1.44$.
  (b) to the right of $z = -1.05$.

In Problems 18 and 19 suppose that the SAT-Math scores form a normal distribution with $\mu = 470$ and $\sigma = 20$.

18. What percentage of scores are between 430 and 480?

19. If a student is selected at random, what is the probability that the SAT-Math score is at least 500?

20. If a pair of dice is tossed 180 times, what is the probability of obtaining a sum of 7 at least 25 times?

Problems 21 to 24 refer to a binomial experiment consisting of 18 trials, each with a probability of success equal to $\frac{2}{3}$.

21. Compute the probability of obtaining exactly 14 successes.

22. What is the mean number of successes?

23. What is the standard deviation?

24. Compute the probability of obtaining at least 14 successes.

Problems 25 to 27 refer to the probability experiment of drawing 1 card from each of 4800 decks of ordinary playing cards and recording the number of hearts drawn.

25. What is the expected value?

26. What is the standard deviation?

27. What is the probability that more than 1150 hearts will be drawn?

9 Markov Processes

CHAPTER OVERVIEW

Many real-world situations are processes that can be viewed as a sequence of experiments. Some examples are the administering of a vaccine to a large population, the voting patterns of a particular populace over a series of elections, and the managing of a portfolio of stocks by a stockbroker over a short time period. The outcomes of these experiments depend on probabilities so that probability theory can be used effectively to predict the future.

In this chapter we are interested in certain types of processes called "Markov processes" or "Markov chains." A Markov process is a sequence of identical experiments in which the outcomes of the current experiment depend only on the outcomes of the previous experiment. The first section is an introduction to the terminology and elementary skills. Section 9–2 deals with the most common type of Markov chain, regular chains. The last section studies the long-term trend of Markov chains by looking at absorbing chains.

One of the primary purposes of Markov processes is to predict the future. Political analysts study the voting trends of certain types of people to help them predict the outcome of future elections; investment counselors study the trends of the value of stocks to improve the total value of their customers' portfolios; and advertisers monitor the effectiveness of their ads in order to achieve the best impact. Therefore, when we view these real-world phenomena as Markov processes we see that long-term trends of the probabilities are important. This is our ultimate goal in this chapter.

CASE STUDY PREVIEW

What happens when someone, for whatever reason, fails to pay you for something you did? You try various means to collect the money due. In business an unpaid bill is called a bad debt. Some bad debts are never collected. Large retail firms must account for a percentage of money due that they never will collect. The case study shows how accountants use Markov chains to give firms a structured means of predicting the amount of bad debts to account for in the future, based on past records.

9–1 BASIC PROPERTIES

Political scientists have learned that people with a strong religious affiliation tend to support one political party and rarely split their vote. Thus if they voted Democratic in one election, it is very likely that they will vote Democratic in the next election. In a study* of voting trends in Elmira, New York, researchers found that the Christian community's vote, as a whole, depended primarily on the previous election's results. The sequence of elections forms a Markov process. A **Markov process** is a probability experiment consisting of a sequence of identical trials or stages. The outcome of the current stage of the process depends only on the outcome of the previous stage. The outcomes at each stage of the preceding process are "Democratic" and "Republican." The probabilities of each outcome, which we study in Example 1, are the percentages of people who voted Democratic and Republican, respectively.

Definitions

Each trial of a Markov process is called a **stage** of the process; the present stage is called the **current stage** and the following one is the **next stage.** The outcomes are called **states.** The first time the experiment is performed it is called the **initial stage.** As the process goes from the current stage to the next stage, the state either changes or it stays the same. This is called the **transition** from state to state. We will see how it can be described by tree diagrams and matrices.

EXAMPLE 1

In the Elmira example researchers found that of the people who voted Democratic in a given election, 90% voted Democratic in the next election and 10% voted Republican. Of those who voted Republican in the current election, 70% voted Republican in the next election and 30% voted Democratic. The tree diagrams in Figure 1 graphically depict these numbers. There are two states for

FIGURE 1

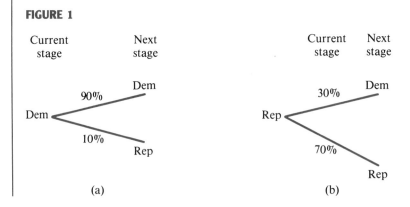

(a)   (b)

*P. Lazarsfeld, B. Berelson and W. McPhee, *Voting: A Study of Opinion Formation in a Presidential Election*, Chicago, IL, University of Chicago Press, 1954.

this experiment: the first state is "voted Democratic," represented by "Dem.," and the second is "voted Republican," represented by "Rep." These numbers form a matrix called the **transition matrix.** We label the matrix T and the rows and columns of T with the two states, "Dem." and "Rep."

$$
\begin{array}{cc}
 & \overbrace{\begin{matrix} \text{state 1} & \text{state 2} \\ \text{Dem.} & \text{Rep.} \end{matrix}}^{\text{Current Stage}} \\
\begin{matrix} \text{Next} \\ \text{Stage} \end{matrix} \begin{cases} \text{state 1} \mid \text{Dem.} \\ \text{state 2} \mid \text{Rep.} \end{cases} & \begin{bmatrix} .9 & .3 \\ .1 & .7 \end{bmatrix} = T
\end{array}
$$

To see how to interpret the transition matrix, view the tree diagram from a different perspective.

The branches of the tree diagram become the columns of the transition matrix. To construct the matrix T, delete the lines and move the column headings to the right.

$$
T = \begin{bmatrix} .9 & .3 \\ .1 & .7 \end{bmatrix}
$$

In the following sections we drop the column and row headings and simply refer to the matrix as T.

Each entry in the transition matrix is a conditional probability. For example, the probability that a person will vote Democratic in the next election given that the person voted Democratic in the current election is .9, and the probability that a person will vote Republican in the next election given that the person voted Democratic in the current election is .1. Thus the entries are conditional probabilities. Let the event "will vote Democratic in the next election" be represented by "D-next" and the event "voted Democratic in the current election" be represented by "D-current." Define "R-next" and "R-current" similarly. With this notation the event "will vote Democratic in the next election given that the person voted Democratic in the current election" is "D-next | D-current." Thus $Pr(\text{D-next} \mid \text{D-current}) = .9$. Each of the entries in T can be represented as

$$
T = \begin{bmatrix} .9 & .3 \\ .1 & .7 \end{bmatrix} = \begin{bmatrix} Pr(\text{D-next} \mid \text{D-current}) & Pr(\text{D-next} \mid \text{R-current}) \\ Pr(\text{R-next} \mid \text{D-current}) & Pr(\text{R-next} \mid \text{R-current}) \end{bmatrix}
$$

The columns sum to 1 because they represent all possible outcomes in a sample space. Hence a transition matrix satisfies the following properties.

—————————————————————— **Definition** ——————————————————————
The defining properties for a transition matrix are

1. It is a square matrix since the rows and columns represent the same states. That is, there are the same number of states at each stage so there are the same number of rows and columns.
2. Its entries are greater than or equal to 0.
3. The entries in each column sum to 1.

———

EXAMPLE 2 | **Problem**

Which of the following matrices could be transition matrices for a Markov process?

(a) $\begin{bmatrix} .2 & .3 & .6 \\ .8 & .7 & .4 \end{bmatrix}$ (b) $\begin{bmatrix} .1 & .4 \\ .9 & .7 \end{bmatrix}$ (c) $\begin{bmatrix} .2 & .9 & .5 \\ .6 & .0 & .2 \\ .2 & .1 & .3 \end{bmatrix}$

SOLUTION (a) The matrix is not square, so it is not a transition matrix. Note that if it were a transition matrix, the current stage would have three states corresponding to the three columns, while the next stage would have two states corresponding to the two rows. This contradicts the assumption of a Markov process, which requires the states to be the same from one stage to the next.
(b) The second column does not sum to 1, so this matrix is not a transition matrix.
(c) This matrix satisfies all three properties, so it is a transition matrix.

The next example illustrates how to construct a transition matrix with three states.

EXAMPLE 3 | The manager of an investment fund has a conservative portfolio containing only blue chip stocks that are intended for future growth. Because it is very stable the daily management decision depends primarily on the previous day's performance. Each day the manager makes one of three decisions for each stock in the portfolio: buy 5% new shares, sell 5% of the shares, or make no transition. The manager employs the following strategy:

1. If a stock issue was purchased in one day's trading, then 10% of the time it will be purchased the next day, 70% of the time it will be sold, and 20% of the time it will be left unchanged.
2. If a stock issue was sold in one day's trading, then 60% of the time it will be purchased the next day, 10% of the time it will be sold, and 30% of the time it will be left unchanged.

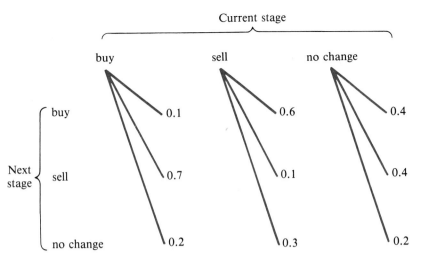

FIGURE 2

3. If a stock issue was left unchanged in one day's trading, then 40% of the time it will be purchased the next day, 40% of the time it will be sold, and 20% of the time it will be left unchanged.

Problem Find the transition matrix for this Markov process.

SOLUTION We use modified tree diagrams in Figure 2 to determine the columns of the transition matrix T. We refer to the three states as the three outcomes: (1) buy, (2) sell, and (3) no change. Hence the transition matrix T is

$$
\begin{array}{c}
\text{Current Stage} \\
\text{state} \longrightarrow
\begin{array}{ccc}
(1) & (2) & (3) \\
\text{buy} & \text{sell} & \text{no} \\
 & & \text{change}
\end{array}
\end{array}
$$

$$
\begin{array}{c}
\text{Next} \\
\text{Stage}
\end{array}
\left\{
\begin{array}{l}
(1)\ \text{buy} \\
(2)\ \text{sell} \\
(3)\ \text{no change}
\end{array}
\right.
\begin{bmatrix}
.1 & .6 & .4 \\
.7 & .1 & .4 \\
.2 & .3 & .2
\end{bmatrix} = T
$$

In general the transition matrix for a process with three outcomes looks like this:

$$
\begin{array}{c}
\text{Current Stage} \\
\begin{array}{ccc}
\text{state 1} & \text{state 2} & \text{state 3}
\end{array}
\end{array}
$$

$$
\begin{array}{c}
\text{state 1} \\
\text{state 2} \\
\text{state 3}
\end{array}
\begin{bmatrix}
p_{11} & p_{12} & p_{13} \\
p_{21} & p_{22} & p_{23} \\
p_{31} & p_{32} & p_{33}
\end{bmatrix}
$$

where p_{ij} is the conditional probability

$$p_{ij} = Pr(\text{process will be in state } i \mid \text{process is in state } j)$$

Distribution Matrices

The transition matrix enables the process to go from one stage of the process to the next. Now we will see how the process gets started. In Example 1 the process starts by looking at one particular election and measuring the percentage of people who voted Democratic and Republican, respectively. It is called a **distribution.** In the 1948 presidential election 80% of Elmira's Christians voted Democratic, while 20% voted Republican. This is called the **initial distribution** and is represented by the column matrix, $V = \begin{bmatrix} .8 \\ .2 \end{bmatrix}$. Let W be the distribution of voters in the next election. Then W is the matrix product TV, where T is the transition matrix, $T = \begin{bmatrix} .9 & .3 \\ .1 & .7 \end{bmatrix}$. This means that to find the distribution of voters in the next election, compute TV. Hence

$$W = TV = \begin{bmatrix} .9 & .3 \\ .1 & .7 \end{bmatrix} \begin{bmatrix} .8 \\ .2 \end{bmatrix} = \begin{bmatrix} .78 \\ .22 \end{bmatrix}$$

To see why $W = TV$, we compute the percentage of (1) the Democratic vote, which is the (1, 1) entry of W, and (2) the Republican vote, which is the (2, 1) entry of W.

1. There are two types of people who will vote Democratic in the next election: (a) those who in the previous election voted Democratic and (b) those who in the previous election voted Republican.
   (a) Of the 80% who voted Democratic in the first election, 90% will vote Democratic in the next election, so $.8 \times .9 = .72$ or 72% will vote Democratic.
   (b) Of the 20% who voted Republican in the first election, 30% will vote Democratic in the next election, so $.2 \times .3 = .06$ or 6% will vote Democratic.
   Therefore in all, $.72 + .06 = .78$ or 78% will vote Democratic in the next election. This is the (1, 1) entry of TV.

2. How many will vote Republican in the next election?
   (a) Of the 80% who voted Democratic in the first election, 10% will vote Republican in the next election.
   (b) Of the 20% who voted Republican in the first election, 70% will vote Republican in the next election.
   Thus $(.8 \times .1) + (.2 \times .7) = .08 + .14$ will vote Republican. Therefore in all, $.08 + .14 = .22$ or 22% will vote Republican in the next election.
   This shows that the computations used to compute W are the the the same as those in the matrix product TV. Thus $TV = W$. This says that to find the next distribution from the previous one, take the product of the transition matrix and the distribution matrix.
   Let us apply this technique to the process in Example 3.

EXAMPLE 4   **Problem**

Given the transition matrix T in Example 3 and the initial distribution

$$V = \begin{bmatrix} .2 \\ .3 \\ .5 \end{bmatrix},$$

find the distribution of the next stage, W.

SOLUTION Compute TV.

$$\begin{bmatrix} .1 & .6 & .4 \\ .7 & .1 & .4 \\ .2 & .3 & .2 \end{bmatrix} \begin{bmatrix} .2 \\ .3 \\ .5 \end{bmatrix} = \begin{bmatrix} .40 \\ .37 \\ .23 \end{bmatrix}$$

Thus if the manager on one day bought shares for 20% of the fund's stocks, sold shares for 30%, and left 50% unchanged, then on the next day 40% would be bought, 37% would be sold, and 23% would be left unchanged.

Further Stages

To go from one distribution V to the next stage's distribution W, we compute $W = TV$. Applying the same reasoning, the following stage's distribution, say, U, is obtained by computing $U = TW$. Substituting for W yields $U = TW = T(TV) = T^2V$, so the distribution after two stages is T^2V. Similarly, the third stage's distribution is T^3V, and in general, the distribution of n stages after the initial stage is T^nV.

EXAMPLE 5

Problem

Suppose a Markov process has transition matrix $T = \begin{bmatrix} .4 & .2 \\ .6 & .8 \end{bmatrix}$ and initial distribution $V = \begin{bmatrix} .3 \\ .7 \end{bmatrix}$. Find the distributions for the next three stages.

SOLUTION We need to compute TV, T^2V, and T^3V. First

$$TV = \begin{bmatrix} .4 & .2 \\ .6 & .8 \end{bmatrix} \begin{bmatrix} .3 \\ .7 \end{bmatrix} = \begin{bmatrix} .26 \\ .74 \end{bmatrix}$$

Since $T^2V = T(TV)$, the matrix T^2V is obtained by multiplying T by TV.

$$T^2V = T(TV) = \begin{bmatrix} .4 & .2 \\ .6 & .8 \end{bmatrix} \begin{bmatrix} .26 \\ .74 \end{bmatrix} = \begin{bmatrix} .252 \\ .748 \end{bmatrix}$$

Similarly

$$T^3V = T(TTV) = \begin{bmatrix} .4 & .2 \\ .6 & .8 \end{bmatrix} \begin{bmatrix} .252 \\ .748 \end{bmatrix} = \begin{bmatrix} .2504 \\ .7496 \end{bmatrix}$$

There are two ways to compute successive distributions from the present stage. We can use powers of the transition matrix.

$$TV, T^2V, T^3V, \ldots, T^nV, \ldots$$

The other way is to use the sequence

$$TV, T(TV), T(T(TV)), \ldots$$

The sequences are the same; only the method of computation differs. The latter method is preferred in this section, but in the next section we will use the sequence $T^n V$ to determine the long-term trends of Markov processes.

EXERCISE SET 9–1

• In Problems 1 and 2 determine the transition matrix of the Markov process.

1. Researchers determine that of the people who voted Democratic in a given election, 85% voted Democratic in the next election and 15% voted Republican (only those who voted in every election were considered). Of those who voted Republican in the current election, 63% voted Republican in the next election and 37% voted Democratic.

2. The manager of the investment fund has determined that the following strategy will be employed for the 100 blue chip stock portfolio: (1) If a stock issue was purchased in one day's trading, then 20% of the time it will be purchased the next day, 50% of the time it will be sold, and 30% of the time it will be left unchanged. (2) If a stock issue was sold in one day's trading, then 60% of the time it will be purchased the next day, 15% of the time it will be sold, and 25% of the time it will be left unchanged. (3) If a stock issue was left unchanged in one day's trading, then 35% of the time it will be purchased the next day, 45% of the time it will be sold, and 20% of the time it will be left unchanged.

In Problems 3 to 6 determine whether the matrix could be a transition matrix of a Markov process.

3. $\begin{bmatrix} .9 & .2 \\ .1 & .8 \end{bmatrix}$

4. $\begin{bmatrix} .1 & 0 \\ .9 & 1 \end{bmatrix}$

5. $\begin{bmatrix} .9 & .4 & .1 \\ 0 & .3 & .6 \\ .1 & .3 & .3 \end{bmatrix}$

6. $\begin{bmatrix} .3 & .7 & .1 \\ .5 & .1 & .1 \\ .2 & .2 & 1.8 \end{bmatrix}$

In Problems 7 and 8 determine a, b, and c so that the matrix could be a transition matrix.

7. $\begin{bmatrix} .3 & .4 & .5 \\ .2 & .5 & .3 \\ a & b & c \end{bmatrix}$

8. $\begin{bmatrix} .8 & .5 & c \\ .2 & .2 & .4 \\ a & b & 0 \end{bmatrix}$

In Problems 9 and 10 the transition matrix of a Markov process is given. Determine the probability that the process will be in state 2 in the next stage, given that its current stage is in (a) state 1, (b) state 2.

9. $\begin{bmatrix} .2 & .6 \\ .8 & .4 \end{bmatrix}$

10. $\begin{bmatrix} .1 & .3 & .6 \\ .4 & .2 & .1 \\ .5 & .5 & .3 \end{bmatrix}$

In Problems 11 to 14 determine whether the column matrix could be a probability distribution for a Markov process.

11. $\begin{bmatrix} .3 \\ .7 \end{bmatrix}$

12. $\begin{bmatrix} .4 \\ .2 \end{bmatrix}$

13. $\begin{bmatrix} .2 \\ .7 \\ .2 \end{bmatrix}$

14. $\begin{bmatrix} .4 \\ .6 \\ 0 \end{bmatrix}$

In Problems 15 to 20 let T represent the transition matrix of a Markov process. For the given probability distribution V find the distribution of the next three stages of the process.

$$T = \begin{bmatrix} .1 & .5 \\ .9 & .5 \end{bmatrix}$$

15. $V = \begin{bmatrix} .5 \\ .5 \end{bmatrix}$

16. $V = \begin{bmatrix} .6 \\ .4 \end{bmatrix}$

17. $V = \begin{bmatrix} .2 \\ .8 \end{bmatrix}$

18. $V = \begin{bmatrix} .3 \\ .7 \end{bmatrix}$

19. $V = \begin{bmatrix} .1 \\ .9 \end{bmatrix}$

20. $V = \begin{bmatrix} .4 \\ .6 \end{bmatrix}$

•• Problems 21 to 26 refer to the following transition matrix T of the Markov process determined by a weather bureau. The first state represents the likelihood of rain in the forecast and the second state represents the likelihood of no rain in the forecast. Each stage represents a daily forecast. That is, the present stage refers to today's weather forecast and the next stage refers to tomorrow's forecast.

$$T = \begin{array}{cc} & \begin{array}{cc} \text{rain} & \text{no rain} \end{array} \\ \begin{array}{c} \text{rain} \\ \text{no rain} \end{array} & \left[\begin{array}{cc} .3 & .2 \\ .7 & .8 \end{array} \right] \end{array}$$

In Problems 21 to 24 find the conditional probability of the event.

21. It will rain tomorrow given that it is raining today.

22. It will rain tomorrow given that it is not raining today.

23. It will not rain tomorrow given that it is raining today.

24. It will not rain tomorrow given that it is not raining today.

25. Find the probability that it will rain tomorrow if the probability of rain today is .5.

26. Find the probability it will not rain tomorrow if the probability of rain today is .3.

In Problems 27 to 34 a Markov process is described. Find its transition matrix.

27. A manufacturer classifies outstanding debts as high priority (H) and low priority (L). In any given month 20% of the debts that are high priority will become low priority (because all or most of the debt will have been paid) and 40% of the low priority debts will remain low priority.

28. A city determines that in a given year 10% of the city population moves to the suburbs and 5% of the suburban populace moves to the city.

29. An audio equipment retailer advertises each week in either television, radio, or the newspaper. The retailer's strategy for the next week is influenced primarily by the previous week's choice. There are three possibilities: (1) If in the current week a television ad was used, then for the next week's

ad 10% of the time television will be used, 30% of the time the radio will be used, and 60% of the time the newspaper will be used; (2) if in the current week a radio ad was used, then for the next week's ad 45% of the time television will be used, 5% of the time the radio will be used, and 50% of the time the newspaper will be used; (3) if in the current week a newspaper ad was used, for the next week's ad 55% of the time television will be used, 35% of the time the radio will be used, and 10% of the time the newspaper will be used.

30. An audio equipment retailer advertises each week in either television, radio, or the newspaper. The retailer is influenced primarily by the previous week's choice and so, if in the current week (1) a television ad was used, for the next week's ad 0% of the time the television will be used, 30% of the time the radio will be used, and 70% of the time the newspaper will be used; (2) a radio ad was used, for the next week's ad 35% of the time television will be used, 15% of the time the radio will be used, and 50% of the time the newspaper will be used; (3) a newspaper ad was used, for the next week's ad 100% of the time television will be used, 0% of the time the radio will be used, and 0% of the time the newspaper will be used.

31. Three brands of beer, A, B, and C, share a particular market. An advertising agency determines that because of a new ad campaign, Brand A beer will gain 20% of Brand B's customers, and 30% of Brand C's customers each month over the next few months while retaining 90% of its own customers. During the same period it is expected that Brand B will retain 70% of its customers and Brand C will retain 60%. Thus each will lose the usual 10% to each other. Brand A will lose only 5% to each of the others.

32. In a manufacturing process the state of the assembly line is monitored hourly as either being ahead of schedule, on schedule, or behind schedule. If the line is on schedule, the probability that it will be ahead of schedule the next hour is .2, on schedule is .7, and behind sched-

ule is .1. If the line is ahead of schedule, the probability that it will be ahead of schedule the next hour is .3, on schedule is .6, and behind schedule is .1. If the line is behind schedule, the probability that it will be ahead of schedule the next hour is .1, on schedule is .5, and behind schedule is .4.

33. Each week the manager of a supply house checks the inventory of stock. The amount ordered depends primarily on how much stock was ordered the previous week. The order is either large, medium, or small. If in the current week the order was large, then in the next week 10% of the time the order will be large, 45% of the time it will be medium, and otherwise it will be small. If in the current week the order was medium, then in the next week 10% of the time the order will be large, and 60% of the time the order will be medium. If in the current week the order was small, then in the next week 65% of the time the order will be large, and 30% of the time the order will be medium.

34. A population study of a state has determined that in any given year the percentage of the population that remains urban, rural, or suburban depends primarily on the previous year's figures. Of the urban population, 85% remains urban, 10% moves to rural areas, and 5% moves to suburbs. Of the rural population, 95% remains rural, 4% moves to urban areas, and 1% moves to suburbs. Of the suburban population, 85% remains suburban, 10% moves to rural areas, and 5% moves to urban areas.

In Problems 35 and 36 let T represent the transition matrix of a Markov process. For the given probability distribution V find the distribution of the next stage of the process.

$$T = \begin{bmatrix} .3 & .1 & .7 \\ .6 & .4 & .2 \\ .1 & .5 & .1 \end{bmatrix}$$

35. $V = \begin{bmatrix} .1 \\ .1 \\ .8 \end{bmatrix}$ 36. $V = \begin{bmatrix} .3 \\ .3 \\ .4 \end{bmatrix}$

In Problems 37 to 40 let T represent the transition matrix of the Markov process in Problems 31 to 34, respectively. For the given probability distribution V find the distribution of the next stage of the process.

37. $V = \begin{bmatrix} .2 \\ .5 \\ .3 \end{bmatrix}$ 38. $V = \begin{bmatrix} .3 \\ .2 \\ .5 \end{bmatrix}$

39. $V = \begin{bmatrix} .3 \\ .3 \\ .4 \end{bmatrix}$ 40. $V = \begin{bmatrix} .2 \\ .1 \\ .7 \end{bmatrix}$

••• 41. Show that $T^2V = TTV$ for

$$T = \begin{bmatrix} .3 & .1 & .7 \\ .6 & .4 & .2 \\ .1 & .5 & .1 \end{bmatrix} \qquad V = \begin{bmatrix} .1 \\ .1 \\ .8 \end{bmatrix}$$

42. In an article* on urban ecology within a large city researchers used Markov processes to analyze community change in Chicago. They used four states based on economic status (ES) and family status (FS); for each community each status was rated as high or low. Thus the four states were HES/HFS (high ES and high FS), HES/LFS (high ES and low FS), LES/HFS, and LES/LFS. They constructed the following transition matrix, where the first stage corresponded to 1950 and the next stage to 1960.

		First Stage		
		HES/HFS	HES/LFS	LES/HFS	LES/LFS
Next Stage	HES/HFS	.81	.06	.26	.00
	HES/LFS	.07	.70	.00	.09
	LES/HFS	.12	.00	.68	.35
	LES/LFS	.00	.24	.06	.56

*Albert Hunter, "Community Change: A Stochastic Analysis of Chicago's Local Communities, 1930–1960," *American Journal of Sociology,* 79:4, pp. 923–947.

(a) Find the probability that a community went from state HES/LFS to LES/LFS.

(b) If the initial distribution was

$$\begin{bmatrix} 0.2 \\ 0.3 \\ 0.4 \\ 0.1 \end{bmatrix},$$

find the next distribution.

43. It has been predicted that in the next decade there will be a significant increase in the number of physicians in the United States. The delivery of primary health care is dependent not only on the number of physicians available but also on how those physicians are dispersed within a given geographic region. In a paper modelling the transitions of physicians, researchers use Markov processes to predict the percentages of physicians that will move from one type of practice to another.* They consider six states and two types of geographical areas, urban and rural. Each area has three types of practices: private practice (P), hospital practice (H), and clinics (C), including health maintenance organizations and government-funded clinics. Data from Washington state in 1982 were used to generate the following transition matrix.

Current Stage

		urban P	urban H	urban C	rural P	rural H	rural C
Next Stage urban	P	0	.5	.4	.1	.05	0
	H	.3	0	.2	.1	.05	0
	C	.4	.3	0	.2	.1	.2
rural	P	.2	.05	.05	0	.4	.5
	H	0	.05	.3	.4	0	.3
	C	.1	.1	.05	.2	.4	0

(a) Find the probability that a physician went from an urban hospital practice to a rural private practice.

(b) State why the main diagonal of the matrix has all zeros.

(c) If the initial distribution is

$$\begin{bmatrix} .4 \\ .4 \\ .2 \\ 0 \\ 0 \\ 0 \end{bmatrix},$$

find the next distribution.

44. Consumers switch brands of a product for many reasons. Sometimes a new brand appears on the market or perhaps a different brand is on sale. One of the most important marketing tools used by marketing directors is the analysis of how and why consumers switch brands. In an article† studying purchasing behavior of various types of consumers, Blattberg and Sen classify consumers into four categories: (1) last-purchase-loyal, LPL, which stands for a person who is likely to purchase the same brand for a while but then switches to a new brand, which is purchased for a while, and then switches to still another brand; (2) low-brand-loyal, LBL, which stands for a person who remains loyal to a brand, then switches to another brand but usually returns to the favorite brand; (3) high-brand-loyal, HBL, which stands for a person who is very loyal to the favorite brand, hardly ever purchasing another brand; (4) deal-not-loyal, DNL, which stands for a person who purchases a brand that offers the best deal, such as a brand on sale. The data used are from the *Chicago Tribune's* Panel of Women in the greater Chicago Area. The purchasing patterns of 200 consumers of aluminum foil were studied over a four-year period. The data relating how the consumers changed from one state to another are given by the following transition matrix.

---

*Vandan Trivedi, et al., "A Semi-Markov Model for Primary Health Care Manpower Supply Predictions," *Management Science*, 33, 1987, pp. 149–160.

†Robert Blattberg and Sbrata K. Sen, "A Bayesian Technique to Discriminate Between Stochastic Models of Brand Choice," *Management Science*, 21, 1975, pp. 682–696.

Current Stage

		LPL	LBL	HBL	DNL
Next Stage	LPL	.61	.06	0	.07
	LBL	.17	.74	.15	.03
	HBL	.02	.17	.85	0
	DNL	.20	.03	0	.90

(a) Find the probability that a consumer went from the HBL category to the DNL category.

(b) State why the main diagonal of the matrix has the four highest values.

9–2 REGULAR MARKOV PROCESSES

In this section we study the long-term trend of Markov processes. The basic question is, what happens to the distribution in the nth stage of the process as n gets larger and larger? Certain transition matrices have a predictable trend but others do not. In this section we study one type that exhibits a stable long-term trend.

Long-Term Trends

In the Elmira voting study the percentage of people who vote Democratic will become stable in the long run, that is, after many elections. The transition matrix and initial probability distribution for the Markov process are

$$T = \begin{bmatrix} .9 & .3 \\ .1 & .7 \end{bmatrix} \quad \text{and} \quad V = \begin{bmatrix} .8 \\ .2 \end{bmatrix}$$

What will the distribution of voters be after four additional elections? From the last section this is done by calculating T^nV for $n = 1$ to 4. (We round off the matrix entries to two decimal places.)

$$TV = \begin{bmatrix} .9 & .3 \\ .1 & .7 \end{bmatrix} \begin{bmatrix} .8 \\ .2 \end{bmatrix} = \begin{bmatrix} .78 \\ .22 \end{bmatrix}$$

$$T^2V = \begin{bmatrix} .84 & .48 \\ .16 & .52 \end{bmatrix} \begin{bmatrix} .8 \\ .2 \end{bmatrix} = \begin{bmatrix} .768 \\ .232 \end{bmatrix}$$

$$T^3V = \begin{bmatrix} .80 & .59 \\ .20 & .41 \end{bmatrix} \begin{bmatrix} .8 \\ .2 \end{bmatrix} = \begin{bmatrix} .758 \\ .242 \end{bmatrix}$$

$$T^4V = \begin{bmatrix} .78 & .65 \\ .22 & .35 \end{bmatrix} \begin{bmatrix} .8 \\ .2 \end{bmatrix} = \begin{bmatrix} .754 \\ .246 \end{bmatrix}$$

The pattern seems to indicate that the percentage that will vote Democratic approaches 75%. As we take higher powers of T, the distribution approaches $\begin{bmatrix} .75 \\ .25 \end{bmatrix}$. This is what is meant by the **long-term trend.** Example 1 explores some further ideas about the long-term trend of Markov processes and determines whether the initial probability distribution affects the long-term trend.

EXAMPLE 1 | **Problem**

Given the transition matrix T in the Elmira voting study, let $V = \begin{bmatrix} .4 \\ .6 \end{bmatrix}$. How is the long-term trend affected?

SOLUTION We can use the same powers of T to get

$$TV = \begin{bmatrix} .9 & .3 \\ .1 & .7 \end{bmatrix} \begin{bmatrix} .4 \\ .6 \end{bmatrix} = \begin{bmatrix} .54 \\ .46 \end{bmatrix}$$

$$T^2V = \begin{bmatrix} .84 & .48 \\ .16 & .52 \end{bmatrix} \begin{bmatrix} .4 \\ .6 \end{bmatrix} = \begin{bmatrix} .624 \\ .376 \end{bmatrix}$$

$$T^3V = \begin{bmatrix} .80 & .59 \\ .20 & .41 \end{bmatrix} \begin{bmatrix} .4 \\ .6 \end{bmatrix} = \begin{bmatrix} .674 \\ .326 \end{bmatrix}$$

$$T^4V = \begin{bmatrix} .78 & .65 \\ .22 & .35 \end{bmatrix} \begin{bmatrix} .4 \\ .6 \end{bmatrix} = \begin{bmatrix} .702 \\ .298 \end{bmatrix}$$

The first term is increasing and the second is decreasing, but the long-term trend is not clear yet because the distribution has not stabilized. We must use higher powers of T. After some tedious labor we see that the powers of T all become equal. For instance

$$T^{12} = \begin{bmatrix} .75 & .75 \\ .25 & .25 \end{bmatrix}$$

All powers greater than 12 are the same. Hence for n greater than 12

$$T^nV = \begin{bmatrix} .75 & .75 \\ .25 & .25 \end{bmatrix} \begin{bmatrix} .4 \\ .6 \end{bmatrix} = \begin{bmatrix} .75 \\ .25 \end{bmatrix}$$

Thus the long-term trend with this new initial distribution is the same as with the first one—75% vote Democratic.

Example 1 computed the long-term trend for the initial distribution. The same long-term trend will hold for any initial distribution $\begin{bmatrix} a \\ b \end{bmatrix}$ because $a + b = 1$ implies that for all n greater than or equal to 12:

$$\begin{bmatrix} .75 & .75 \\ .25 & .25 \end{bmatrix} \begin{bmatrix} a \\ b \end{bmatrix} = \begin{bmatrix} .75(a + b) \\ .25(a + b) \end{bmatrix} = \begin{bmatrix} .75 \\ .25 \end{bmatrix}$$

Regular Matrices

Example 1 indicates that the long-term trend for this particular transition matrix T does not depend on the choice of the initial probability distribution. However, only certain types of matrices have this property. The entries of every transition matrix are either positive or zero. No entry can be negative because it represents a probability. The special property that we focus on now requires a matrix to have all positive entries, that is, no zero entries. But we can relax that requirement a little, as we see in the next definition.

Definition

A transition matrix is **regular** if some power of the matrix contains all positive entries. A Markov process is *regular* if its transition matrix is regular.

EXAMPLE 2

Problem

Which of the following matrices are regular?

$$\text{(a) } A = \begin{bmatrix} .3 & .4 \\ .7 & .6 \end{bmatrix} \quad \text{(b) } B = \begin{bmatrix} .9 & 1 \\ .1 & 0 \end{bmatrix} \quad \text{(c) } C = \begin{bmatrix} .8 & 0 \\ .2 & 1 \end{bmatrix}$$

SOLUTION (a) A has all positive entries so it is regular.

(b) B has a zero entry so it may not be regular, but we must determine whether its powers have zeros. We compute $B^2 = \begin{bmatrix} .91 & .9 \\ .09 & .1 \end{bmatrix}$. So B is regular since one of its powers has all positive entries.

(c) This looks very similar to B. Compute $C^2 = \begin{bmatrix} .64 & 0 \\ .36 & 1 \end{bmatrix}$. The answer is still not clear. Compute higher powers of C. The same second column appears in all C^n. Hence every power of C will have a zero in the (1, 2) entry. Thus C is not regular since no power of it will have all positive entries.

The key property of regular Markov processes is that long-range predictions are possible. For instance, in Example 1 the transition matrix is regular, so the long-term trend of the process can be determined. The higher powers of T approached the matrix

$$\begin{bmatrix} .75 & .75 \\ .25 & .25 \end{bmatrix}$$

This matrix is called the *stable matrix* for T. No matter what the initial distribution is, after a sufficient number of stages in the process the distribution approaches the limiting distribution $\begin{bmatrix} .75 \\ .25 \end{bmatrix}$. This is called the *stable distribution*. These are important ideas so we record them here.

Definition

If T is a regular matrix, then the powers of T will approach a limiting matrix, called the **stable matrix** of T. This means that T^n gets arbitrarily close to the stable matrix as n gets large. Once the process has reached the stage of the stable matrix, any initial distribution will eventually lead to a distribution that remains the same in all succeeding stages. This limiting distribution is called the **stable distribution.**

Computing Stable Matrices

Finding the stable matrix by computing T^nV for large enough n is often time-consuming. A faster method exists. Suppose V is the initial distribution and after n stages the stable distribution W is reached. Thus $T^nV = W$. The next stage's distribution is $TT^nV = TW$. But the definition of the stable distribution implies that further stages' distributions are equal to W. Thus $TW = W$. This means that in order to compute the stable distribution, we solve the equation $TX = X$. For example, start with the matrices $T = \begin{bmatrix} .9 & .3 \\ .1 & .7 \end{bmatrix}$ and $X = \begin{bmatrix} x \\ y \end{bmatrix}$ and solve $TX = X$.

$$TX = \begin{bmatrix} .9 & .3 \\ .1 & .7 \end{bmatrix} \begin{bmatrix} x \\ y \end{bmatrix} = \begin{bmatrix} .9x + .3y \\ .1x + .7y \end{bmatrix} = X = \begin{bmatrix} x \\ y \end{bmatrix}$$

Thus we must solve the system of equations

$$.9x + .3y = x$$
$$.1x + .7y = y$$

Subtract x from each side of the first equation and subtract y from each side of the second equation. This yields

$$-.1x + .3y = 0$$
$$.1x - .3y = 0$$

The equations are multiples of each other so there is an infinite number of solutions. Solving for x in terms of y yields $x = 3y$. Since x and y represent probabilities they must add up to 1 because they are the only outcomes. Hence

$$x + y = 1$$

Combining the last equation with $x = 3y$ yields $3y + y = 1$, so $4y = 1$. Thus $y = \frac{1}{4}$ and $x = \frac{3}{4}$. Hence the stable distribution is

$$W = \begin{bmatrix} \frac{3}{4} \\ \frac{1}{4} \end{bmatrix}$$

Here is the procedure for finding the stable distribution of a regular matrix.

Procedure for Finding the Stable Distribution of a Regular Matrix

Let T be a regular $n \times n$ matrix and let X be a variable column matrix with n rows. The system of equations $TX = X$ has n equations in n unknowns. The sum of the entries of X must equal 1 because X is a probability distribution. Since X has n rows the "sum of the entries of $X = 1$" is an equation with n variables. The stable distribution W is the solution of the system of $n + 1$ equations in n unknowns.

$$TX = X$$
$$\text{sum of the entries of } X = 1$$

The next two examples illustrate this procedure.

EXAMPLE 3 A retail store has 55% of the market versus 45% for the lone competitor. An advertisement firm predicts that an ad campaign will draw 20% of the competitor's customers to the store, while the store will lose only 15% of their own customers to the competitor.

Problem

What will be the long-term trend of the ad campaign?

SOLUTION Refer to the store as A and the competitor as C. The results of the ad campaign can be represented as a Markov process whose transition matrix T is

$$T = \begin{matrix} & A & C \\ A & \\ C & \end{matrix} \begin{bmatrix} .85 & .2 \\ .15 & .8 \end{bmatrix}$$

Notice that T is a regular matrix. Finding the long-term trend is the same as finding the stable distribution. Solve $TX = X$ and $x + y = 1$. First write $TX = X$ as a system of equations.

$$TX = \begin{bmatrix} .85 & .2 \\ .15 & .8 \end{bmatrix} \begin{bmatrix} x \\ y \end{bmatrix} = \begin{bmatrix} x \\ y \end{bmatrix}$$

$$\begin{array}{ll} .85x + .2y = x & \text{so} \\ .15x + .8y = y & \end{array} \qquad \begin{array}{l} -.15x + .2y = 0 \\ .15x - .2y = 0 \end{array}$$

Adding these equations eliminates one of them since they are multiples of each other. Combining the remaining equation with $x + y = 1$ yields

$$\begin{array}{l} .15x - .2y = 0 \\ x + y = 1 \end{array}$$

Using elimination, we get $.35x = .2$. Hence $x = .2/.35 = 20/35 = 4/7$. Since $x + y = 1$ we get $y = 3/7$. The long-term trend is $\begin{bmatrix} 4/7 \\ 3/7 \end{bmatrix}$. Since $4/7 \approx 57\%$ the ad campaign would increase the store's advantage by 2 percentage points, from 55% to 57%.

To illustrate further the procedure, consider the Markov process used by the investment manager whose strategy was described by the transition matrix T.

		Current Stage		
		buy	sell	no change
Next Stage	buy	.1	.6	.4
	sell	.7	.1	.4
	no change	.2	.3	.2

$= T$

EXAMPLE 4 | **Problem**

Find the long-term trend of the investment strategy.

SOLUTION We introduce the matrix X, which has three entries since T has three columns, and then compute $TX = X$.

$$TX = \begin{bmatrix} .1 & .6 & .4 \\ .7 & .1 & .4 \\ .2 & .3 & .2 \end{bmatrix} \begin{bmatrix} x \\ y \\ z \end{bmatrix} = \begin{bmatrix} .1x + .6y + .4z \\ .7x + .1y + .4z \\ .2x + .3y + .2z \end{bmatrix} = \begin{bmatrix} x \\ y \\ z \end{bmatrix}$$

Coupling this with "the sum of the entries of $X = 1$" gives us the system of equations

$$.1x + .6y + .4z = x$$
$$.7x + .1y + .4z = y$$
$$.2x + .3y + .2z = z$$
$$x + y + z = 1$$

Rewriting the equations after gathering like terms and expressing the system in matrix form as in Chapter 2 yields

$$\begin{aligned} -.9x + .6y + .4z &= 0 \\ .7x - .9y + .4z &= 0 \\ .2x + .3y - .8z &= 0 \\ x + y + z &= 1 \end{aligned} \qquad \begin{bmatrix} 1 & 1 & 1 & 1 \\ -.9 & .6 & .4 & 0 \\ .7 & -.9 & .4 & 0 \\ .2 & .3 & -.8 & 0 \end{bmatrix}$$

Using elementary row operations to reduce the matrix yields

$$\begin{bmatrix} 1 & 0 & 0 & 60/163 \\ 0 & 1 & 0 & 64/163 \\ 0 & 0 & 1 & 39/163 \\ 0 & 0 & 0 & 0 \end{bmatrix}$$

If the fractions are converted to decimals and rounded to the nearest percent, we have $x = 37\%$, $y = 39\%$, and $z = 24\%$. This means that over the long haul the manager will buy 37% of the time, sell 39% of the time, and leave the stocks unchanged 24% of the time.

In Chapters 1 and 2 we said that the skill of solving systems of equations is an important tool that is used often in many areas of mathematics. Here we see that it is an indispensable tool in working with Markov processes.

EXERCISE SET 9–2

• In Problems 1 to 4 compute T^nV for $n = 1, 2,$ and 3.

1. $T = \begin{bmatrix} .9 & .3 \\ .1 & .7 \end{bmatrix}$, $V = \begin{bmatrix} .6 \\ .4 \end{bmatrix}$

2. $T = \begin{bmatrix} .9 & .3 \\ .1 & .7 \end{bmatrix}$, $V = \begin{bmatrix} .1 \\ .9 \end{bmatrix}$

3. $T = \begin{bmatrix} .5 & .2 \\ .5 & .8 \end{bmatrix}$, $V = \begin{bmatrix} .5 \\ .5 \end{bmatrix}$

4. $T = \begin{bmatrix} .2 & .7 \\ .8 & .3 \end{bmatrix}, \quad V = \begin{bmatrix} .3 \\ .7 \end{bmatrix}$

In Problems 5 to 8 find the long-term trend of the Markov process whose transition matrix is T and initial distribution is V by computing $T^n V$ for $n = 1$ to 6.

5. $T = \begin{bmatrix} .9 & .3 \\ .1 & .7 \end{bmatrix}, \quad V = \begin{bmatrix} .6 \\ .4 \end{bmatrix}$

6. $T = \begin{bmatrix} .9 & .3 \\ .1 & .7 \end{bmatrix}, \quad V = \begin{bmatrix} .1 \\ .9 \end{bmatrix}$

7. $T = \begin{bmatrix} .5 & .2 \\ .5 & .8 \end{bmatrix}, \quad V = \begin{bmatrix} .5 \\ .5 \end{bmatrix}$

8. $T = \begin{bmatrix} .2 & .7 \\ .8 & .3 \end{bmatrix}, \quad V = \begin{bmatrix} .3 \\ .7 \end{bmatrix}$

In Problems 9 to 12 determine whether the matrix is regular.

9. $T = \begin{bmatrix} 0 & .2 & .1 \\ 0 & .8 & .1 \\ 1 & 0 & .8 \end{bmatrix}$

10. $T = \begin{bmatrix} 1 & .2 & .7 \\ 0 & .8 & .2 \\ 0 & 0 & .1 \end{bmatrix}$

11. $T = \begin{bmatrix} .2 & .1 & 0 \\ .8 & .1 & 0 \\ 0 & .8 & 1 \end{bmatrix}$

12. $T = \begin{bmatrix} .1 & .7 & 0 \\ .8 & .2 & 1 \\ .1 & .1 & 0 \end{bmatrix}$

In Problems 13 to 22 find the stable matrix for the regular transition matrix T.

13. $T = \begin{bmatrix} .4 & .5 \\ .6 & .5 \end{bmatrix}$

14. $T = \begin{bmatrix} .7 & .2 \\ .3 & .8 \end{bmatrix}$

15. $T = \begin{bmatrix} .9 & .3 \\ .1 & .7 \end{bmatrix}$

16. $T = \begin{bmatrix} .8 & .3 \\ .2 & .7 \end{bmatrix}$

17. $T = \begin{bmatrix} .5 & .2 \\ .5 & .8 \end{bmatrix}$

18. $T = \begin{bmatrix} .2 & .7 \\ .8 & .3 \end{bmatrix}$

19. $T = \begin{bmatrix} 0 & .2 & .1 \\ 0 & .7 & .1 \\ 1 & .1 & .8 \end{bmatrix}$

20. $T = \begin{bmatrix} .4 & .2 & .7 \\ .3 & .8 & .2 \\ .3 & 0 & .1 \end{bmatrix}$

21. $T = \begin{bmatrix} .2 & .1 & 1 \\ .8 & .1 & 0 \\ 0 & .8 & 0 \end{bmatrix}$

22. $T = \begin{bmatrix} .1 & .7 & .3 \\ .8 & .2 & .2 \\ .1 & .1 & .5 \end{bmatrix}$

In Problems 23 to 26 find the long-term trend of the Markov process.

23. A retail store with one main competitor has 65% of the market as compared to 35% for the competitor. The store is considering running a weekly advertisement campaign claiming that each week 25% of the competitor's customers will switch allegiance to the store. It is estimated that the ads will drive 10% of the store's customers to the competitor. What will be the long-term trend if the ad campaign is run?

24. Researchers found that of the people who voted Democratic in a given election, 85% voted Democratic in the next election and 15% voted Republican. Of those who voted Republican in the current election, 63% voted Republican in the next election and 37% voted Democratic.

25. A weather forecaster determines that if it is raining on one day, then there is a 30% chance it will rain the next day, and if it is not raining on one day, there is a 75% chance it will not rain the next day.

26. A weather forecaster determines that if it is raining on one day, then for the following day there is a 30% chance it will rain, a 5% chance it will snow, and a 65% chance of no precipitation. If it is snowing on one day, then for the following day there is a 10% chance it will rain, a 15% chance it will snow, and a 75% chance of no precipitation. If there is no precipitation on one day, then for the following day there is a 10% chance it will rain, a 5% chance it will snow, and an 85% chance of no precipitation.

•• In Problems 27 to 36 a Markov process is described. Find its long-term trend.

27. The manager of an investment fund has determined that the following strategy will be employed for the 100 blue chip stock portfolio: (1) If a stock issue was purchased in one day's trading, then 20% of the time it will be purchased the next day, 50% of the time it will be sold, and 30% of the time it will be left unchanged. (2) If a stock issue was sold in one day's trading, then 60% of the time it will be purchased the next day, 15% of the time it will be sold, and 25% of the time it will be left unchanged. (3) If a stock issue was left unchanged in one day's trading, then 35% of the time it will be purchased the next day, 45% of the time it will be sold, and 20% of the time it will be left unchanged.

28. Each week a manager reports on the work of her staff. Her ratings are poor, average, and excellent. If an individual was rated poor the previous week, the probability that in the next week the individual will be rated poor is .2, average is .7, and excellent is .1. If an individual was rated average the previous week, the probability that in the next week the individual will be rated poor is .3, average is .4, and excellent is .3. If an individual was rated excellent the previous week, the probability that in the next week the individual will be rated poor is .1, average is .7, and excellent is .2. Determine the long-term trend of the percentage of people rated poor, average, and excellent.

29. An audio equipment retailer advertises each week in either television, radio, or the newspaper. The retailer is influenced primarily by the previous week's choice and so, if in the current week (1) a television ad was used, for the next week's ad 10% of the time television will be used, 30% of the time the radio will be used, and 60% of the time the newspaper will be used; (2) a radio ad was used, for the next week's ad 45% of the time television will be used, 5% of the time the radio will be used, and 50% of the time the newspaper will be used; (3) a newspaper ad was used, for the next week's ad 55% of the time television will be used, 35% of the time the radio will be used, and 10% of the time the newspaper will be used.

30. An audio equipment retailer advertises each week in either television, radio, or the newspaper. The retailer is influenced primarily by the previous week's choice and so, if in the current week (1) a television ad was used, for the next week's ad 0% of the time television will be used, 30% of the time the radio will be used, and 70% of the time the newspaper will be used; (2) a radio ad was used, for the next week's ad 35% of the time television will be used, 15% of the time the radio will be used, and 50% of the time the newspaper will be used; (3) a newspaper ad was used, for the next week's ad 100% of the time television will be used, 0% of the time the radio will be used, and 0% of the time the newspaper will be used.

31. A manufacturer classifies outstanding debts as high priority (H) and low priority (L). In any given month 20% of the debts that are high priority will become low priority (because all or most of the debt will have been paid) and 40% of the low priority debts will remain low priority.

32. A city determines that in a given year 10% of the city population moves to the suburbs and 5% of the suburban populace moves to the city.

33. Three brands of beer, A, B, and C, share a particular market. An advertising agency determines that because of a new ad campaign, Brand A beer will gain 20% of Brand B's customers and 30% of Brand C's customers each month over the next few months while retaining 90% of its own customers. During the same period it is expected that Brand B will retain 70% of its customers and Brand C will retain 60%. Thus each will lose the usual 10% to each other. Brand A will lose only 5% to each of the others.

34. In a manufacturing process the state of the assembly line is monitored hourly as either being ahead of schedule, on schedule, or behind

schedule. If the line is on schedule, the probability that it will be ahead of schedule the next hour is .2, on schedule is .7, and behind schedule is .1. If the line is ahead of schedule, the probability that it will be ahead of schedule the next hour is .3, on schedule is .6, and behind schedule is .1. If the line is behind schedule, the probability that it will be ahead of schedule the next hour is .1, on schedule is .5, and behind schedule is .4.

35. Each week the manager of a supply house checks the inventory of stock. The amount ordered depends primarily on how much stock was ordered the previous week, and the order is either large, medium, or small. If in the current week the order was large, then in the next week 10% of the time the order will be large, 45% of the time it will be medium, and 45% of the time it will be small. If in the current week the order was medium, then in the next week 10% of the time the order will be large, 60% of the time the order will be medium, and 30% of the time the order will be small. If in the current week the order was small, then in the next week 65% of the time the order will be large, 30% of the time the order will be medium, and 5% of the time the order will be small.

36. A population study of a small midwestern state has found that in any given year the percentage of the population that remains urban, rural, or suburban depends primarily on the previous year's figures. Of the urban population, 85% remains urban, 10% moves to rural areas, and 5% moves to suburbs. Of the rural population, 95% remains rural, 4% moves to urban areas, and 1% moves to suburbs. Of the suburban population, 85% remains suburban, 10% moves to rural areas, and 5% moves to urban areas.

●●● 37. Show that if a transition matrix has all positive entries, then every power of the matrix has all positive entries.

38. Suppose a transition matrix for a Markov process is $T = \begin{bmatrix} a & b \\ c & d \end{bmatrix}$. Find a relationship between a and c and between b and d. Find the stable matrix for T. When will this matrix be regular?

In Problems 39 to 41 a Markov process is described. Find the long-term trend.

39. Most large department stores encourage customers to purchase products with the store's credit card. An important accounting consideration is to estimate the average number of months it takes for certain types of classifications of credit accounts to be paid up. In an article studying how to compute profit from credit card purchases, Wort and Zumwalt classify credit card accounts in three categories according to the average number of months an account is carried before it is paid up: (1) less than one month, (2) during the second month, and (3) more than 2 months.* In one part of the article, data from the DAK corporation were studied. The findings were†: if an account was less than one month old, there was a 70% chance it would be paid up within the month, a 27% chance it would be paid up within the second month, and a 3% chance it would be paid up after the second month; if an account was between one and two months old, there was a 75% chance it would

*Donald H. Wort and J. Kenton Zumwalt, ''Determining the Impact of Alternate Trade Discount Policies,'' *Decision Sciences*, 16, 1985, pp. 48–56.

†The ''total balance method'' of accounting is used here, and it assumes that all dollars in the account balance are put in the age category of the oldest dollar. This means that a current purchase charge by customer with a two-month past due balance would be classified in category 2 rather than category 1. This accounts for the fact that some dollars can move from category 2 to category 1 or stay in category 2.

be paid up within the first month, a 23% chance it would be paid up within the second month, and a 2% chance it would be paid up after the second month; if an account was more than two months old, there was a 90% chance it would be paid up within the first month, a 9% chance it would be paid up within the second month, and a 1% chance it would be paid up after the second month. Assume that all accounts are eventually paid in full. (The Case Study deals with the situation in which an account is never paid up; it becomes a ''bad debt.'')

40. Most businesses use various means to forecast the market share of their product, where market share is their product's sales as a percentage of the total market sales. In an article studying the likelihood of households continuing to use the same type of heating unit they used in the recent past or switching to another type, Ezzati uses Markov processes to forecast the market share for three types of home heating units: oil burner, gas burner and electric heat.* The data were gathered from the Federal Energy Office. Suppose a household decided to purchase a new heating unit. The data show that: (1) if the household had an oil burner, there is an 85% chance it would purchase an oil burner, a 10% chance it would purchase a gas burner, and a 5% chance it would purchase an electric heater; (2) if the household had a gas burner, there is an 6% chance it would purchase an oil burner, an 88% chance it would purchase a gas burner, and a 6% chance it would purchase an electric heater; (3) if the household had an electric heater, there is an 1% chance it would purchase an oil burner, an 6% chance it would purchase a gas burner, and a 93% chance it would purchase an electric heater.

41. What is a stable culture? How likely is it for a tribe to remain in a certain social and demographic nature? These are important questions raised by anthropologists that apply not only to tribes but to our own culture as well. In an article applying Markov processes to these questions, Hoffman studies data from the Galla tribes of Ethiopia.† In one part of the study, a mature male tribe member is classified in one of three states based on his age when his first son is born. The states are the age groups, youth (ages 13–19), adult (ages 20–29), senior (ages 30 or more). The transition from one stage of the process to the next constitutes the probability that a son enters a state (by having his own first son when he is in a particular age group) given that his father is classified in a particular state. The data show that: (1) given that a male's father is classified in the youth state, the probability is .16 that he will be classified in the youth state, .38 that he will be classified in the adult state, and .46 that he will be classified in the senior state (for example, this means that if a male's father had his first son at age 18, the probability that the son will have his first son at age 30 or more is .46.); (2) given that a male's father is classified in the adult state, the probability is .37 that he will be classified in the youth state, .40 that he will be classified in the adult state, and .23 that he will be classified in the senior state; (3) given that a male's father is classified in the senior state, the probability is .20 that he will be classified in the youth state, .60 that he will be classified in the adult state, and .20 that he will be classified in the senior state.

*Ali Ezzati, ''Forecasting Market Shares of Alternative Home-heating Units by Markov Processes Using Transition Probabilities Estimated from Aggregate Time Series Data,'' *Management Science*, 21, 1974, pp. 462–473.

†Hans Hoffman, ''Markov Chains in Ethiopia,'' in *Explorations in Mathematical Anthropology*, Paul Lay, (ed.), Cambridge, Mass., MIT Press, 1971, pp. 181–190.

9–3 ABSORBING MARKOV PROCESSES

Regular transition matrices are important because they exhibit a long-term trend. In this section we study another type of transition matrix that enables the process to stabilize in the long run.

Absorbing Matrices

Suppose a transition matrix has a column with 1 on the main diagonal and 0's elsewhere. When a higher power is computed, that column will stay the same. To see why, let column i of transition matrix T have a 1 in the (i, i) entry and 0's elsewhere. When computing the (i, i) entry of T^2, the only nonzero contribution will be when the two 1's are multiplied together, producing 1. This means that the (i, i) entry is 1. Therefore the other values in column i must be 0 because T^2 is a transition matrix. Let us look at a specific matrix.

Consider the transition matrix T whose states are labeled 1 to 4. Remember that the columns represent the current stage and the rows represent the next stage of the experiment.

$$
\begin{array}{cc}
 & \text{Current Stage} \\
\begin{array}{ccc}
\text{Next} \\
\text{Stage}
\end{array}
\left\{
\begin{array}{c}
\text{state} \\
1 \\
2 \\
3 \\
4
\end{array}
\right.
&
\begin{array}{cccc}
1 & 2 & 3 & 4 \\
\begin{bmatrix}
1 & .4 & .3 & 0 \\
0 & .3 & .2 & 0 \\
0 & .2 & 0 & 0 \\
0 & .1 & .5 & 1
\end{bmatrix}
\end{array}
\end{array}
$$

It is not regular because the 1's in the 1-1 and 4-4 entries will never change when we take higher powers of T. If a column has an entry that is 1, then the other entries in that column must be 0's. If the 1 is on the main diagonal, as both of the 1's are in T above, then once the process enters that state the process never leaves it. This is because higher powers of T have 1's on the same main diagonals, as described previously.

In the matrix T above, once state 1 or state 4 is entered the process will always stay in that state. For this reason these states are called **absorbing states.** Notice that state 2 is not absorbing because the process can go from state 2 to another state; in fact it can go to states 1, 3, or 4 (with corresponding probabilities .4, .2, and .1). This is called a **nonabsorbing state.** State 3 is nonabsorbing because the process can go from state 3 to states 1, 2, or 4. Thus state n is an absorbing state if the (n, n) entry is 1; otherwise it is nonabsorbing.

There is a second property that is important in this section. In the preceding matrix T the process can go from any nonabsorbing state into some absorbing state. For example, state 2, a nonabsorbing state, can go to the absorbing state, state 1, because the $(1, 2)$ entry is not zero. In fact state 2 can go to state 1 with probability .4 since the $(1, 2)$ entry is .4. State 2 can also go to state 4 with probability .1 because the $(4, 2)$ entry is .1. Similarly, the nonabsorbing state 3 can go to state 1 (with probability .3) or state 4 (with probability .5).

Thus the second property that is important in this section is the ability of each nonabsorbing state to go to some absorbing state. We can now give a formal definition of this type of matrix.

_____ **Definition** _____

An **absorbing matrix** is a transition matrix in which

1. There is at least one absorbing state, and
2. It is possible to go from any nonabsorbing state to an absorbing state after a finite number of stages.

An **absorbing Markov process** is one whose transition matrix is absorbing.

_____

It may take more than one step in the process for a nonabsorbing state to reach an absorbing one. Example 1 will help clarify this definition.

EXAMPLE 1 | **Problem**

Which of the matrices are absorbing?

(a)
$$A = \begin{bmatrix} 0 & .5 & .3 \\ 0 & .2 & .1 \\ 1 & .3 & .6 \end{bmatrix}$$

(b)
$$B = \begin{bmatrix} 1 & 0 & .1 & .2 \\ 0 & 0 & .3 & .3 \\ 0 & 0 & .1 & 0 \\ 0 & 1 & .5 & .5 \end{bmatrix}$$

(c)
$$C = \begin{bmatrix} .8 & 0 & .7 & 0 \\ 0 & 1 & 0 & 0 \\ .2 & 0 & .3 & 0 \\ 0 & 0 & 0 & 1 \end{bmatrix}$$

SOLUTION (a) A does not have an absorbing state so it is not an absorbing matrix. The only 1 in the matrix is in the column corresponding to state 1, but since it is in the third row the process will go from state 1 to state 3. So state 1 is not an absorbing state. The 1 must be on the main diagonal for the state to be absorbing.

(b) B has one absorbing state, state 1, so the first property of the definition is satisfied. To be an absorbing process, it must be possible to go from each nonabsorbing state into state 1. State 3 can go to state 1 directly (with probability .1). State 4 can also go to state 1 directly (with probability .2). To see that state 2 can go to state 1 is a bit more complicated. It cannot go directly because the (1, 2) entry is zero. In fact if the process is in state 2, it must go to state 4 in the next stage because the (4, 2) entry is 1. But then state 4 can go to state 1 in the following stage. Thus state 2 can go first to state 4 and then to state 1. So state 2 can go to state 1 after two stages. This means that each nonabsorbing state of matrix B can go to an absorbing state. Therefore B is an absorbing matrix.

(c) C has two absorbing states, states 2 and 4. For C to be an absorbing matrix, states 1 and 3 must eventually lead to states 2 or 4. But if the process is in state 1 or state 3 it can go to neither state 2 nor state 4. Hence C is not absorbing.

Long-Term Trends—Stable Matrices

Not all transition matrices have long-term trends (see Problem 34). The importance of absorbing matrices is that they have a limiting matrix. As in the last section, the limiting matrix is called the **stable matrix.** In order to compute the stable matrix, we arrange the transition matrix so that the absorbing states come before the nonabsorbing states. The rows and columns might have to be rearranged, but this does not affect the meaning of the transition matrix. For example, consider the matrix T, introduced in the beginning of this section.

$$T = \begin{bmatrix} 1 & .4 & .3 & 0 \\ 0 & .3 & .2 & 0 \\ 0 & .2 & 0 & 0 \\ 0 & .1 & .5 & 1 \end{bmatrix}$$

To put the absorbing states first, keep column 1 in its place but move column 4 to column 2. Equivalently, state 4 becomes state 2. We also keep column 3 where it is but move column 2 to column 4. (Columns 2 and 3 could just as well have been moved to columns 3 and 4, respectively.) The rows change accordingly. The new matrix, which represents the same Markov process but with the names of the states changed, is

$$T = \begin{bmatrix} 1 & 0 & .3 & .4 \\ 0 & 1 & .5 & .1 \\ 0 & 0 & 0 & .2 \\ 0 & 0 & .2 & .3 \end{bmatrix}$$

In this way every absorbing matrix can be arranged so that the absorbing states are to the left of the nonabsorbing states. When in this form the matrix can be divided (sometimes called partitioned) into four submatrices, called I, Q, O, and R in the following manner:

Absorbing   Nonabsorbing
  states   states

$$\begin{bmatrix} I & Q \\ \hline O & R \end{bmatrix}$$

where I is an identity matrix and O is the zero matrix with the same number of rows as R and columns as I. For the previous matrix the partition is

$$\begin{bmatrix} 1 & 0 & .3 & .4 \\ 0 & 1 & .5 & .1 \\ \hline 0 & 0 & 0 & .2 \\ 0 & 0 & .2 & .3 \end{bmatrix}$$

where

$$I = \begin{bmatrix} 1 & 0 \\ 0 & 1 \end{bmatrix} \quad Q = \begin{bmatrix} .3 & .4 \\ .5 & .1 \end{bmatrix} \quad R = \begin{bmatrix} 0 & .2 \\ .2 & .3 \end{bmatrix} \quad 0 = \begin{bmatrix} 0 & 0 \\ 0 & 0 \end{bmatrix}$$

Computing the Stable Matrix

One way to determine the long-term trend of absorbing matrices is to compute their powers to find the stable matrix. But as we saw in the previous section, this computation can be laborious and time-consuming. A more direct method involves finding the inverse of the matrix $I - R$, then multiplying by Q, so the matrix to compute is $Q(I - R)^{-1}$. Then the stable matrix S is defined by its partition in this way.

$$S = \left[\begin{array}{c|c} I & Q(I - R)^{-1} \\ \hline O & O \end{array}\right]$$

In the original matrix I and R may be different sizes. This means that in the matrix S the two identity matrices I and the two zero matrices O may be different sizes. As an illustration of the procedure, let us calculate S from this partitioned transition matrix:

$$T = \left[\begin{array}{cc|c} 1 & 0 & .2 \\ 0 & 1 & .7 \\ \hline 0 & 0 & .1 \end{array}\right]$$

where

$$I = \begin{bmatrix} 1 & 0 \\ 0 & 1 \end{bmatrix} \quad Q = \begin{bmatrix} .2 \\ .7 \end{bmatrix} \quad O = [0 \quad 0] \quad R = [.1]$$

Since R is a 1×1 matrix we form $I - R = [1] - [.1] = [.9]$, where I is the same size as R. Then $(I - R)^{-1} = [.9]^{-1} = [1/.9]$. Therefore

$$Q(I - R)^{-1} = \begin{bmatrix} .2 \\ .7 \end{bmatrix} [1/.9] = \begin{bmatrix} .2/.9 \\ .7/.9 \end{bmatrix} = \begin{bmatrix} 2/9 \\ 7/9 \end{bmatrix}$$

The stable matrix is

$$S = \left[\begin{array}{cc|c} 1 & 0 & 2/9 \\ 0 & 1 & 7/9 \\ \hline 0 & 0 & 0 \end{array}\right]$$

This means that state 3 leads to state 1 two-ninths of the time, while it leads to state 2 seven-ninths of the time. Example 2 illustrates this procedure for a 4×4 matrix.

EXAMPLE 2

Problem

Find the stable matrix of the transition matrix T.

$$T = \left[\begin{array}{cc|cc} 1 & 0 & 0 & .2 \\ 0 & 1 & .2 & 0 \\ \hline 0 & 0 & .5 & 0 \\ 0 & 0 & .3 & .8 \end{array}\right]$$

SOLUTION We have

$$I = \begin{bmatrix} 1 & 0 \\ 0 & 1 \end{bmatrix} \quad Q = \begin{bmatrix} 0 & .2 \\ .2 & 0 \end{bmatrix} \quad R = \begin{bmatrix} .5 & 0 \\ .3 & .8 \end{bmatrix}$$

Compute

$$I - R = \begin{bmatrix} 1 & 0 \\ 0 & 1 \end{bmatrix} - \begin{bmatrix} .5 & 0 \\ .3 & .8 \end{bmatrix} = \begin{bmatrix} .5 & 0 \\ -.3 & .2 \end{bmatrix}$$

Then

$$(I - R)^{-1} = \begin{bmatrix} .2/.1 & -0/.1 \\ .3/.1 & .5/.1 \end{bmatrix} = \begin{bmatrix} 2 & 0 \\ 3 & 5 \end{bmatrix}$$

So

$$Q(I - R)^{-1} = \begin{bmatrix} 0 & .2 \\ .2 & 0 \end{bmatrix} \begin{bmatrix} 2 & 0 \\ 3 & 5 \end{bmatrix}$$

$$= \begin{bmatrix} .6 & 1 \\ .4 & 0 \end{bmatrix}$$

Hence the stable matrix is

$$S = \left[\begin{array}{cc|cc} 1 & 0 & .6 & 1 \\ 0 & 1 & .4 & 0 \\ \hline 0 & 0 & 0 & 0 \\ 0 & 0 & 0 & 0 \end{array} \right]$$

This means that state 4 will eventually always lead to state 1, while state 3 will lead to state 1 60% of the time and to state 2 40% of the time.

The Effect of the Initial Distribution

To compute the long-term trend of regular matrices, we use systems of equations, and to find the long-term trend of absorbing matrices, we use the inverse and multiplication of matrices. Another difference between regular and absorbing matrices is that for regular matrices different initial distributions yield the same stable distribution, but for absorbing matrices different initial distributions yield different distributions in the long run. The next example demonstrates the latter difference.

EXAMPLE 3 | **Problem**

Find the distribution of the long-term trend of the Markov process in Example 2 for the two initial probability distributions

(a)
$$V = \begin{bmatrix} .1 \\ 0 \\ .5 \\ .4 \end{bmatrix} \quad \text{and} \quad (b) \quad W = \begin{bmatrix} .2 \\ .8 \\ 0 \\ 0 \end{bmatrix}$$

SOLUTION In Example 2 the stable matrix of this process was calculated to be

$$S = \begin{bmatrix} 1 & 0 & .6 & 1 \\ 0 & 1 & .4 & 0 \\ 0 & 0 & 0 & 0 \\ 0 & 0 & 0 & 0 \end{bmatrix}$$

Therefore to find the distribution of the long-term trend, we compute SV and SW.

(a)
$$SV = \begin{bmatrix} 1 & 0 & .6 & 1 \\ 0 & 1 & .4 & 0 \\ 0 & 0 & 0 & 0 \\ 0 & 0 & 0 & 0 \end{bmatrix} \begin{bmatrix} .1 \\ 0 \\ .5 \\ .4 \end{bmatrix} = \begin{bmatrix} .8 \\ .2 \\ 0 \\ 0 \end{bmatrix}$$

and

(b)
$$SW = \begin{bmatrix} 1 & 0 & .6 & 1 \\ 0 & 1 & .4 & 0 \\ 0 & 0 & 0 & 0 \\ 0 & 0 & 0 & 0 \end{bmatrix} \begin{bmatrix} .2 \\ .8 \\ 0 \\ 0 \end{bmatrix} = \begin{bmatrix} .2 \\ .8 \\ 0 \\ 0 \end{bmatrix}$$

This means that if the process starts with the initial distribution V, that is, with state 1 comprising 10%, state 2 0%, state 3 50%, and state 4 40%, then in the long run 80% will be in state 1 and 20% will be in state 2. But if the initial distribution is W, these percentages are reversed.

Example 3 shows that the long-term trend of the distribution depends on the initial distribution. Let us summarize the facts that we have presented concerning absorbing Markov processes. The proof of this result is beyond the scope of this text.

Absorbing Markov Processes

1. The powers of the transition matrix approach some particular matrix, called the stable matrix.
2. The long-term trend depends on the initial distribution.
3. In a finite number of steps, the process will enter an absorbing state and stay there.
4. If T is the transition matrix for an absorbing Markov process, then the rows and columns of T can be rearranged so that the absorbing states come first and the matrix will have the form

$$T = \left[\begin{array}{c|c} I & Q \\ \hline O & R \end{array} \right]$$

where I is an identity matrix (with the same number of rows as absorbing states) and O is a zero matrix. Then the stable matrix S is

$$S = \begin{bmatrix} I & Q(I-R)^{-1} \\ \hline O & O \end{bmatrix}$$

where the identity matrix in the expression $Q(I - R)^{-1}$ is the same size as R.

5. The matrix $Q(I - R)^{-1}$ gives the probabilities that the nonabsorbing states will lead to particular absorbing states.

EXERCISE SET 9–3

• In Problems 1 to 4 determine the absorbing and nonabsorbing states of each matrix.

1. $\begin{bmatrix} .2 & .3 & 0 \\ .8 & .2 & 1 \\ 0 & .5 & 0 \end{bmatrix}$ 2. $\begin{bmatrix} 1 & .3 & 0 \\ 0 & .2 & 0 \\ 0 & .5 & 1 \end{bmatrix}$

3. $\begin{bmatrix} .2 & .1 & 0 & 0 \\ .6 & .5 & 1 & 0 \\ .1 & .2 & 0 & 0 \\ .1 & .2 & 0 & 1 \end{bmatrix}$ 4. $\begin{bmatrix} 0 & .1 & 0 & 0 \\ 0 & .5 & 0 & 1 \\ .5 & .2 & 1 & 0 \\ .5 & .2 & 0 & 0 \end{bmatrix}$

In Problems 5 to 12 determine whether the matrix is absorbing.

5. $\begin{bmatrix} 1 & .3 & 0 \\ 0 & .2 & 1 \\ 0 & .5 & 0 \end{bmatrix}$ 6. $\begin{bmatrix} 0 & .3 & 1 \\ 1 & .2 & 0 \\ 0 & .5 & 0 \end{bmatrix}$

7. $\begin{bmatrix} 0 & .3 & 0 \\ 1 & .6 & 1 \\ 0 & .1 & 0 \end{bmatrix}$ 8. $\begin{bmatrix} 1 & .3 & 0 \\ 0 & .7 & 0 \\ 0 & 0 & 1 \end{bmatrix}$

9. $\begin{bmatrix} .1 & .1 & 0 & 0 \\ .7 & .5 & 0 & 0 \\ .2 & .4 & 0 & 0 \\ 0 & 0 & 1 & 1 \end{bmatrix}$ 10. $\begin{bmatrix} 1 & .1 & 0 & 0 \\ 0 & .5 & 0 & 1 \\ 0 & .2 & 1 & 0 \\ 0 & .2 & 0 & 0 \end{bmatrix}$

11. $\begin{bmatrix} 0 & .1 & 1 & 0 \\ 0 & .8 & 0 & 0 \\ 1 & .1 & 0 & 0 \\ 0 & 0 & 0 & 1 \end{bmatrix}$ 12. $\begin{bmatrix} 1 & 0 & 0 & 0 \\ 0 & .5 & 0 & 1 \\ 0 & .2 & 1 & 0 \\ 0 & .2 & 0 & 0 \end{bmatrix}$

In Problems 13 and 14 compute T^4 and determine the stable matrix of the Markov process whose transition matrix is T.

13. $T = \begin{bmatrix} 1 & 0 & 0 \\ 0 & 1 & .9 \\ 0 & 0 & .1 \end{bmatrix}$

14. $T = \begin{bmatrix} .1 & 0 & 0 & 0 \\ .7 & 1 & 0 & 0 \\ .2 & 0 & 1 & 0 \\ 0 & 0 & 0 & 1 \end{bmatrix}$

In Problems 15 to 20 arrange the absorbing matrix so that the absorbing states come before the nonabsorbing ones.

15. $\begin{bmatrix} 1 & .3 & 0 \\ 0 & .2 & 1 \\ 0 & .5 & 0 \end{bmatrix}$ 16. $\begin{bmatrix} 0 & .3 & 0 \\ 1 & .2 & 0 \\ 0 & .5 & 1 \end{bmatrix}$

17. $\begin{bmatrix} .2 & .3 & 0 & 0 \\ .8 & .2 & 0 & 0 \\ 0 & .4 & 1 & 0 \\ 0 & .1 & 0 & 1 \end{bmatrix}$ 18. $\begin{bmatrix} .1 & .1 & 0 & 0 \\ .7 & .5 & 0 & 0 \\ .2 & .4 & 0 & 0 \\ 0 & 0 & 1 & 1 \end{bmatrix}$

19. $\begin{bmatrix} 1 & .1 & 0 & 0 \\ 0 & .5 & 0 & 1 \\ 0 & .2 & 1 & 0 \\ 0 & .2 & 0 & 0 \end{bmatrix}$ 20. $\begin{bmatrix} .2 & 0 & .1 & 0 \\ .6 & 1 & .2 & 0 \\ .1 & 0 & .3 & 0 \\ .1 & 0 & .4 & 1 \end{bmatrix}$

In Problems 21 and 22 compute $(I - R)^{-1}$.

21. $\begin{bmatrix} 1 & 0 & .1 & 0 \\ 0 & 1 & .5 & 1 \\ 0 & 0 & .2 & 0 \\ 0 & 0 & .2 & 0 \end{bmatrix}$ 22. $\begin{bmatrix} 1 & 0 & .1 & .4 \\ 0 & 1 & .5 & .1 \\ 0 & 0 & .4 & .4 \\ 0 & 0 & 0 & .1 \end{bmatrix}$

In Problems 23 to 26 find the stable matrix.

23. $\begin{bmatrix} 1 & 0 & .3 & .1 \\ 0 & 1 & .2 & 0 \\ 0 & 0 & .5 & .6 \\ 0 & 0 & 0 & .3 \end{bmatrix}$ 24. $\begin{bmatrix} 1 & 0 & .2 & .3 \\ 0 & 1 & 0 & .1 \\ 0 & 0 & .8 & .5 \\ 0 & 0 & 0 & .1 \end{bmatrix}$

25. $\begin{bmatrix} 1 & 0 & .1 & 0 \\ 0 & 1 & .5 & 1 \\ 0 & 0 & .2 & 0 \\ 0 & 0 & .2 & 0 \end{bmatrix}$

26. $\begin{bmatrix} 1 & 0 & 0 & .1 & 0 \\ 0 & 1 & 0 & .5 & 0 \\ 0 & 0 & 1 & .2 & .8 \\ 0 & 0 & 0 & .2 & .1 \\ 0 & 0 & 0 & 0 & .1 \end{bmatrix}$

In Problems 27 and 28 show that the probability distribution of the long-term trend depends on the initial distribution by calculating SV and SW, where S is the stable matrix and V and W are two initial distributions.

27. $S = \begin{bmatrix} 1 & 0 & .3 \\ 0 & 1 & .7 \\ 0 & 0 & 0 \end{bmatrix}$, $V = \begin{bmatrix} .2 \\ .3 \\ .5 \end{bmatrix}$,

$W = \begin{bmatrix} .5 \\ .4 \\ .1 \end{bmatrix}$

28. $S = \begin{bmatrix} 1 & 0 & .2 & .1 \\ 0 & 1 & .8 & .9 \\ 0 & 0 & 0 & 0 \\ 0 & 0 & 0 & 0 \end{bmatrix}$, $V = \begin{bmatrix} .5 \\ .2 \\ .2 \\ .1 \end{bmatrix}$,

$W = \begin{bmatrix} .1 \\ .1 \\ .5 \\ .3 \end{bmatrix}$

•• In Problems 29 to 32 a game is given that can be described by a Markov process. Find the transition matrix, find its stable matrix, and then answer the given question.

29. Two players, A and B, engage in a coin tossing game where the coin is not fair and the probability of heads is $\frac{2}{3}$. Player A always bets on heads. If A wins B gives A \$1, and if A loses A gives B \$1. The players have a total of \$3 between them, and they continue to play until one has accumulated all the money. What is the probability that A will win \$3 if A starts with \$2?

30. Two players, A and B, engage in a coin tossing game where the coin is not fair and the probability of heads is $\frac{2}{5}$. Player A always bets on heads. If A wins B gives A \$2, and if A loses A gives B \$1. The players have a total of \$4 between them, and they continue to play until one has accumulated all the money. What is the probability that A will win \$4 if A starts with \$2?

31. Consider a game where A continually bets \$1 on each play. If A wins A receives \$1 in addition to getting back the bet of \$1. A starts with \$2 and plays until either going broke or winning \$3. The probability of winning on each play is $\frac{2}{5}$. What is the probability of A going broke?

32. Consider a game where A continually bets \$1 on each play. If A wins A receives \$2 in addition to getting back the bet of \$1. A starts with \$1 and plays until either going broke or winning \$3. The probability of winning on each play is $\frac{2}{5}$. What is the probability of A going broke?

••• 33. Show that an absorbing transition matrix is not regular.

34. Show that the transition matrix $\begin{bmatrix} 0 & 1 \\ 1 & 0 \end{bmatrix}$ does not have a stable matrix.

35. For the transition matrix

$T = \begin{bmatrix} 1 & 0 & .2 \\ 0 & 1 & .7 \\ 0 & 0 & .1 \end{bmatrix}$

show that the limiting matrix of T^n is equal to the stable matrix computed in the text.

36. In a paper* that studied the movement of coronary patients within a hospital to obtain length-of-stay and patient-day statistics, the researchers used data from St. Raphael Hospital, New Haven, Conn., to define the following Markov

*Edward P. C. Kao, "Modeling the Movement of Coronary Patients Within a Hospital by Semi-Markov Processes," *Operations Research*, July–August 1974, pp. 683–699.

transition matrix:

$$
\begin{array}{c}
\text{CCU} \\
\text{PCCU} \\
\text{ICU} \\
\text{MED} \\
\text{SURG} \\
\text{AMB} \\
\text{ECF} \\
\text{HOME} \\
\text{DIED}
\end{array}
\begin{bmatrix}
0 & .02 & 0 & 0 & 0 & 0 & 0 & 0 & 0 \\
.74 & 0 & .58 & .01 & 0 & 0 & 0 & 0 & 0 \\
.01 & .01 & 0 & .04 & 0 & 0 & 0 & 0 & 0 \\
.12 & .02 & .17 & 0 & 0 & 0 & 0 & 0 & 0 \\
.01 & .01 & .08 & .02 & 0 & 0 & 0 & 0 & 0 \\
.01 & .01 & 0 & .03 & 0 & 0 & 0 & 0 & 0 \\
0 & .06 & 0 & .08 & 0 & 0 & 1 & 0 & 0 \\
.01 & .83 & 0 & .70 & 1 & 1 & 0 & 1 & 0 \\
.10 & .04 & .17 & .12 & 0 & 0 & 0 & 0 & 1
\end{bmatrix}
$$

The abbreviations stand for the following: CCU (coronary care unit), PCCU (postcoronary care unit), ICU (intensive care unit), MED (a medical unit), SURG (surgical unit), AMB (ambulatory unit), ECF (extended care unit), HOME (the patient was discharged), DIED (the patient died).

(a) Find the absorbing states.

(b) Is the matrix an absorbing matrix?

(c) Rearrange the matrix so that the absorbing states are to the left of the nonabsorbing states.

(d) Find the probability that a patient was discharged from a medical unit.

37. Many psychological theories of learning are based on probability theory. One such theory, presented in a book by Wickens on the mathematical study of human behavior, is based on the "all-or-none" type of learning experiment.* A subject is asked to learn a list of 100 paired-associate items, such as nonsense syllables paired with digits from 1 to 5. The first part of the pair is called the stimulus, the second is the response. There are 100 different syllables each paired with a digit. For example, suppose one such pair is "poj-4." The subject is given the syllable "poj" and asked for its corresponding digit. The correct response is "4." Incorrect responses would be 1, 2, 3 or 5. The first time the subject

goes through the list he or she is guessing which response corresponds with each syllable. After the subject chooses a response, the correct pair is shown, thereby giving feedback and a chance to study the pair. Then another, usually different, stimulus is given. The experiment continues until all pairs are learned or a specific amount of time has elapsed. There are two states in this model. An item is either learned, so it is in the L state, or the subject must guess, so it is in the G state. Initially, all items are in the G state. An item can change from the G state to the L state, but once it is in the L state it remains there. The transition probabilities are

Pr(error on presentation $k + 1$ given an error on presentation k)

The next refinement of the model separates state G into two states, a guess that is correct, state GC, and a guess that is an error, state GE. For a given item and a given subject, an example of a transition matrix is presented below:

$$
\begin{array}{cc}
& \text{Current Stage} \\
& \overbrace{\begin{array}{ccc} \text{L} & \text{GC} & \text{GE} \end{array}} \\
\begin{array}{c} \text{Next} \\ \text{Stage} \end{array} \begin{cases} \text{L} \\ \text{GC} \\ \text{GE} \end{cases} &
\begin{bmatrix}
1 & .1 & .2 \\
0 & .6 & .3 \\
0 & .3 & .5
\end{bmatrix}
\end{array}
$$

(a) Show that this transition matrix is an absorbing transition matrix. (b) Find the absorbing and nonabsorbing states. (c) Find the stable matrix.

38. Many public and private enterprises are based on a service system requiring multiple service stations. Examples include banks, grocery stores and government application offices. In an article describing inventory levels for such systems, Serfozo studies the process that the Bell System uses to transmit television programs, such as news, sports and special events, from local sites without permanent transmission equipment.† In

*Thomas D. Wickens, *Models for Behavior: Stochastic Processes in Psychology,* San Francisco, W. H. Freeman and Co., 1982.

†Richard Serfozo, "Allocation of Servers for Stochastic Service Stations with One Overflow Station," *Management Science,* 31, 1985, pp. 1011–1018.

this case the server is the transmission equipment and the service station is the local site, say, a speech or a sporting event. Portable transmission equipment is used to relay the program to the regional pool. As an example, data are taken from the regional pool in Chicago. Requests for service come from northern Illinois, northern Indiana and southern Wisconsin. Transmission of a program from a local site consists of a series of microwave or cable transmission links, usually under 25 miles, that are set up to transmit the program to a nearby entry into the permanent Bell System network. Sometimes the network can handle the demand, and at other times it routes the transmission to the Chicago pool. The Markov process has four states, each corresponding to a geographic area in the network: Chicago, northern Illinois, northern Indiana and southern Wisconsin. We refer to these states as Chi, Ill, Ind and Wis. The probabilities in the transition matrix represent the chance that a program originating in the current state (geographical area) will be transmitted to the next state, that is, transmitted to another geographical area. An example of a typical transition matrix is:

$$
\begin{array}{c}
\quad\quad\quad\quad\text{Current Stage} \\
\quad\quad\quad\begin{array}{cccc} \text{Chi} & \text{Ill} & \text{Ind} & \text{Wis} \end{array} \\
\begin{array}{c} \text{Next} \\ \text{Stage} \end{array}
\begin{array}{c} \text{Chi} \\ \text{Ill} \\ \text{Ind} \\ \text{Wis} \end{array}
\begin{bmatrix}
1 & .6 & .4 & .3 \\
0 & .4 & 0 & 0 \\
0 & 0 & .6 & 0 \\
0 & 0 & 0 & .7
\end{bmatrix}
\end{array}
$$

(a) Show that this transition matrix is an absorbing transition matrix. (b) Find the absorbing and nonabsorbing states. (c) Find the stable matrix.

CASE STUDY　　Allowance for Doubtful Accounts

In retail businesses, especially large firms, a very important quantity called "bad debts" has a significant effect on income. The term "bad debts" refers to the amount of money due that the firm determines will never be collected. The total amount of money owed to the firm is referred to as "accounts receivable." Thus bad debts is that fraction of accounts receivable that will not be collected. These uncollectible debts must be estimated and an allowance made for them in the accounting process. Otherwise estimated income will be much greater than actual income received. When a large portion of a firm's assets is tied up in accounts receivable, an accurate estimation of the bad debt allowance is a crucial accounting technique. Businesses that rely heavily on credit purchases, such as large department stores and automobile dealers, are examples of such firms.

Before the 1960s bad debt allowances were made on the basis of the accounting department's best guess. Then R. M. Cyert* and others started investigating more accurate and scientific methods of estimating the allowances. Their approach treats the process of how some debts eventually become bad debts as a Markov process.

In order to describe their methods, we must define the *allowance for doubtful accounts* that, at any point in time, is the estimated dollar amount of accounts receivable that will prove uncollectible in the future. For example, if a

*R. M. Cyert et al., "Estimation of the Allowance for Doubtful Accounts by Markov Chains," *Management Science* Vol. 8, April 1962, pp. 287–303.

firm has a total amount of $1,000,000 accounts receivable, and it determines that 5% of that amount will never be collected, then the allowance for doubtful accounts is $1,000,000 $\times$ .05 = $50,000. The crucial issue is the accuracy of the estimate. The problem is that the total amount of accounts receivable continually changes with time. In a given time period most debts are paid but some are not. In the next period new debts are added to the accounts receivable, some of which are not paid, and some of the previous period's accounts are paid but some are not. So the allowance for doubtful accounts also changes with time.

Thus the method for determining an accurate amount is a process whose stages are specific time periods. The time periods vary from firm to firm. They might be months, quarters, or even years. For any choice of time periods the dollar amounts of accounts receivable are classified into various age categories. For instance, at a given time there are debts that are due in that time period, some are overdue by one period, some are overdue by two periods, and so on until a time period, say, n, is determined to be that period in which any debt overdue by n periods or more is called a bad debt.

We define for a balance of accounts receivable at a given time

P_0 = % of accounts receivable 0 periods overdue
(these are the current debts)

P_1 = % of accounts receivable one period overdue
P_2 = % of accounts receivable two periods overdue
P_3 = % of accounts receivable three periods overdue
P_n = % of accounts receivable n periods overdue

One category is chosen as the last, and so all dollars in it and those following it are referred to as bad debts. A delinquent account may eventually be repaid, but it is customary in accounting procedure to lump all debts in both this age category and those after it into one bad debt classification.

For example, suppose a firm determines that its time periods will be months and any debt overdue by three or more months is a bad debt. Suppose it has $1,000,000 in accounts receivable, with $700,000 (or 70%) currently due in the present month, $200,000 (or 20%) overdue by one month, $60,000 (or 6%) overdue by two months and $40,000 (or 4%) overdue by three or more months. Then

$$P_0 = .70, \quad P_1 = .20, \quad P_2 = .06, \quad P_3 = .04$$

Now consider the firm's accounts receivables in the next month. Many of these debts will be paid while some will not, and, of course, new debts will be accrued since the previous period. The P_i for the next period will usually be different from the previous ones. Those accounts that are in the current month's P_0 and will not be paid by the next period will move to the next month's P_1. Similarly, the current month's accounts in P_1 that are not paid will move to the next month's P_2. The firm determines that in a given month the percentage of accounts receivable in each category depends primarily on the previous month's percentages. Thus the process is a Markov process. For this firm there are five categories if we let 0* represent the accounts that were paid during the time period, and let 0, 1, 2, and 3 represent the accounts 0 months overdue (current), one month overdue, two months overdue, and three or more months overdue, respectively.

Suppose it is determined that of those accounts receivable during the current month that are in (1) category 0, 10% were paid, 60% stayed in category 0, 30% became one month overdue; (2) category 1, 25% were paid, 35% moved to category 0, 20% stayed in category 1, 20% moved to category 2; (3) category 2, 10% were paid, 10% moved to category 0, 20% moved to category 1, 15% stayed in category 2, 45% moved to category 3.†

Therefore the transition matrix T is

$$
\begin{array}{c}
\text{To} \\
\text{Category}
\end{array}
\begin{array}{cc}
 & \overset{\textstyle\text{From Category}}{\overbrace{\begin{array}{ccccc} 0^* & 0 & 1 & 2 & 3 \end{array}}} \\
\begin{array}{c} 0^* \\ 0 \\ 1 \\ 2 \\ 3 \end{array} &
\left[\begin{array}{ccccc}
1 & .1 & .25 & .1 & 0 \\
0 & .6 & .35 & .1 & 0 \\
0 & .3 & .2 & .2 & 0 \\
0 & 0 & .2 & .15 & 0 \\
0 & 0 & 0 & .45 & 1
\end{array}\right]
\end{array}
$$

Here is the transition matrix that Cyert and his colleagues studied. They prepared the matrix from a random sample of approximately 1000 department stores in order to test their theory.

From Category

Months Overdue

		0*	0	1	2	3	4	5	6
	0*	1	.21	.13	.13	.10	.14	.09	0
	0	0	.67	.19	.08	.01	.02	.01	0
	1	0	.12	.44	.20	.04	0	.02	0
To	2	0	0	.24	.36	.17	.09	.01	0
Category	3	0	0	0	.23	.29	.20	.10	0
	4	0	0	0	0	.39	.41	.12	0
	5	0	0	0	0	0	.14	.47	0
	6	0	0	0	0	0	0	.18	1

Let us turn to a simpler example to see how the theory of Markov processes can help businesses determine an accurate allowance for doubtful accounts. The transition matrix T is the one used by Cyert for illustrative purposes.

---

†The "total balance method" of accounting is used here, and it assumes that all dollars in the account balance are put in the age category of the oldest dollar. This means that a current purchase charge by customer with a two-month past due balance would be classified in category 2 rather than in category 0. This accounts for the fact that some dollars can move from category 1 to category 0 or stay in category 1.

EXAMPLE 1

Suppose the time period is months and accounts two months old are bad debts. Let

$$
T = \begin{matrix} & \begin{matrix} 0^* & 2 & 0 & 1 \end{matrix} \\ \begin{matrix} 0^* \\ 2 \\ 0 \\ 1 \end{matrix} & \begin{bmatrix} 1 & 0 & .3 & .5 \\ 0 & 1 & 0 & .1 \\ 0 & 0 & .5 & .3 \\ 0 & 0 & .2 & .1 \end{bmatrix} \end{matrix}
$$

We have arranged the rows and columns to put the absorbing states first. Let the distribution matrix be

$$
V = \begin{bmatrix} 0 \\ 0 \\ .7 \\ .3 \end{bmatrix}
$$

Problem

What is the allowance for bad debts?

SOLUTION The transition matrix has the partition

$$
T = \left[\begin{array}{cc|cc} 1 & 0 & .3 & .5 \\ 0 & 1 & 0 & .1 \\ \hline 0 & 0 & .5 & .3 \\ 0 & 0 & .2 & .1 \end{array} \right]
$$

where

$$
I = \begin{bmatrix} 1 & 0 \\ 0 & 1 \end{bmatrix}, \quad Q = \begin{bmatrix} .3 & .5 \\ 0 & .1 \end{bmatrix}, \quad R = \begin{bmatrix} .5 & .3 \\ .2 & .1 \end{bmatrix}
$$

We compute

$$
I - R = \begin{bmatrix} 1 & 0 \\ 0 & 1 \end{bmatrix} - \begin{bmatrix} .5 & .3 \\ .2 & .1 \end{bmatrix} = \begin{bmatrix} .5 & -.3 \\ -.2 & .9 \end{bmatrix}
$$

and so

$$
(I - R)^{-1} = \begin{bmatrix} .9/.39 & .3/.39 \\ .2/.39 & .5/.39 \end{bmatrix} = \begin{bmatrix} 2.31 & .77 \\ .51 & 1.28 \end{bmatrix}
$$

$$
Q(I - R)^{-1} = \begin{bmatrix} .3 & .5 \\ 0 & .1 \end{bmatrix} \begin{bmatrix} 2.31 & .77 \\ .51 & 1.28 \end{bmatrix}
$$

$$
= \begin{bmatrix} .95 & .87 \\ .05 & .13 \end{bmatrix}
$$

Hence the stable matrix is

$$S = \left[\begin{array}{cc|cc} 1 & 0 & .95 & .87 \\ 0 & 1 & .05 & .13 \\ \hline 0 & 0 & 0 & 0 \\ 0 & 0 & 0 & 0 \end{array}\right]$$

The long-term trend is SV.

$$SV = \left[\begin{array}{cc|cc} 1 & 0 & .95 & .87 \\ 0 & 1 & .05 & .13 \\ \hline 0 & 0 & 0 & 0 \\ 0 & 0 & 0 & 0 \end{array}\right] \left[\begin{array}{c} 0 \\ 0 \\ .7 \\ .3 \end{array}\right]$$

$$= \left[\begin{array}{c} 0 \\ 0 \\ .926 \\ .074 \end{array}\right]$$

Thus the allowance for bad debts is 7.4%. This means that for every $100 of accounts receivable the firm should reserve $7.40 for bad debts.

This is very close to the amount Cyert calculated in his study of 1000 retail stores. The allowance for doubtful accounts for most of the firms was about 7%.

Case Study Exercises

In Problems 1 to 4 compute the allowance for bad debts using the transition matrix T in Example 1 and the given distribution matrix V.

1. $V = \left[\begin{array}{c} 0 \\ 0 \\ .6 \\ .4 \end{array}\right]$
2. $V = \left[\begin{array}{c} 0 \\ 0 \\ .8 \\ .2 \end{array}\right]$

3. $V = \left[\begin{array}{c} 0 \\ 0 \\ .5 \\ .5 \end{array}\right]$
4. $V = \left[\begin{array}{c} 0 \\ 0 \\ .1 \\ .9 \end{array}\right]$

In Problems 5 to 8 compute the allowance for bad debts using the following transition matrix T and the given distribution matrix V.

$$T = \left[\begin{array}{cc|cc} 1 & 0 & .2 & .3 \\ 0 & 1 & 0 & .1 \\ \hline 0 & 0 & .6 & .4 \\ 0 & 0 & .2 & .2 \end{array}\right]$$

5. $V = \left[\begin{array}{c} 0 \\ 0 \\ .6 \\ .4 \end{array}\right]$
6. $V = \left[\begin{array}{c} 0 \\ 0 \\ .8 \\ .2 \end{array}\right]$

7. $V = \left[\begin{array}{c} 0 \\ 0 \\ .5 \\ .5 \end{array}\right]$
8. $V = \left[\begin{array}{c} 0 \\ 0 \\ .1 \\ .9 \end{array}\right]$

CHAPTER REVIEW

Key Terms	Notation	Formulas

9–1 Basic Properties
Markov Process
Stage
Current Stage
Next Stage
States
Initial Stage
Transition Matrix $\quad T$
Distribution $\quad V$
Initial Distribution

9–2 Regular Markov Processes
Long-Term Trend
Regular Matrices $\qquad\qquad TX = X$
Stable Distribution $\qquad\quad x + y = 1$

9–3 Absorbing Markov Processes
Absorbing States
Nonabsorbing States
Absorbing Matrix
Absorbing Markov Process
Partitioned Matrix

$$\left[\begin{array}{c|c} I & Q \\ \hline O & R \end{array}\right]$$

Stable Matrix $\qquad S$

$$\left[\begin{array}{c|c} I & Q(I - R)^{-1} \\ \hline O & 0 \end{array}\right]$$

REVIEW PROBLEMS

In Problems 1 and 2 determine the transition matrix of the Markov process.

1. Researchers determine that of the people who voted Democratic in a given election, 45% voted Democratic in the next election and 55% voted Republican (only those who voted in every election were considered). Of those who voted Republican in the current election, 60% voted Republican in the next election and 40% voted Democratic.

2. The manager of an investment fund has determined that the following strategy will be em-

ployed for the 100 blue chip stock portfolio: (1) If a stock issue was purchased in one day's trading, then 10% of the time it will be purchased the next day, 60% of the time it will be sold, and 30% of the time it will be left unchanged. (2) If a stock issue was sold in one day's trading, then 50% of the time it will be purchased the next day, 35% of the time it will be sold, and 15% of the time it will be left unchanged. (3) If a stock issue was left unchanged in one day's trading, then 35% of the time it will be purchased the next day, 25% of the time it will be sold, and 40% of the time it will be left unchanged.

3. Determine whether the matrix could be a transition matrix of a Markov process.

$$\begin{bmatrix} .3 & .4 & .1 \\ .4 & .3 & .6 \\ .3 & .1 & .3 \end{bmatrix}$$

4. Determine a, b, and c so that the matrix could be a transition matrix.

$$\begin{bmatrix} .1 & .5 & .1 \\ .2 & .5 & .3 \\ a & b & c \end{bmatrix}$$

5. For the transition matrix T and the probability distribution V find the distribution of the next three stages of the process.

$$T = \begin{bmatrix} .1 & .2 & .4 \\ .1 & .3 & .4 \\ .8 & .5 & .2 \end{bmatrix} \qquad V = \begin{bmatrix} .3 \\ .2 \\ .5 \end{bmatrix}$$

Problems 6 and 7 refer to the following transition matrix T of the Markov process determined by a weather bureau. The first state represents the likelihood of snow in the forecast and the second state represents the likelihood of no snow in the forecast. Each stage represents a daily forecast. That is, the present stage refers to today's weather forecast and the next stage refers to tomorrow's forecast.

$$\begin{array}{cc} & \text{snow} \quad \text{no snow} \end{array}$$
$$\begin{array}{c} \text{snow} \\ \text{no snow} \end{array} T = \begin{bmatrix} .1 & .2 \\ .9 & .8 \end{bmatrix}$$

6. Find the probability that it will snow tomorrow if the probability of snow today is .5.

7. Find the probability that it will not snow tomorrow if the probability of snow today is .3.

8. Determine whether the matrix is regular.

$$T = \begin{bmatrix} 0 & .3 & .2 \\ 0 & .7 & .2 \\ 1 & 0 & .6 \end{bmatrix}$$

9. Compute $T^n V$ for $n = 1, 2,$ and 3.

$$T = \begin{bmatrix} .1 & .7 \\ .9 & .3 \end{bmatrix} \qquad V = \begin{bmatrix} .6 \\ .4 \end{bmatrix}$$

10. Find the long-term trend of the Markov process whose transition matrix is T and initial distribution is V by computing $T^n V$ for $n = 1$ to 6.

$$T = \begin{bmatrix} .9 & .3 \\ .1 & .7 \end{bmatrix} \qquad V = \begin{bmatrix} .6 \\ .4 \end{bmatrix}$$

In Problems 11 and 12 find the stable matrix for the regular transition matrix T.

11. $T = \begin{bmatrix} .8 & .1 \\ .2 & .9 \end{bmatrix}$

12.
$$T = \begin{bmatrix} .3 & .8 & .6 \\ .2 & .2 & .3 \\ .5 & 0 & .1 \end{bmatrix}$$

In Problems 13 and 14 find the long-term trend of the Markov process.

13. A weather forecasting service has found that (1) if it rains on a given day, then the probability that it will rain the next day is .3, and (2) if it does not rain on a given day, then the probability that it will rain the next day is .4.

14. A weather forecaster determines that if it is raining on one day, then for the following day there is a 20% chance it will rain, a 5% chance it will snow, and a 75% chance of no precipitation. If it is snowing on one day, then for the following day there is a 10% chance it will rain, a 10% chance it will snow, and an 80% chance of no precipitation. If there is no precipitation on one day, then for the following day there is a 5% chance it will rain, a 5% chance it will snow, and a 90% chance of no precipitation.

15. Determine the absorbing and nonabsorbing states of the matrix.

$$\begin{bmatrix} .2 & .1 & 0 & 0 \\ .5 & .4 & 1 & 0 \\ .1 & .2 & 0 & 0 \\ .2 & .3 & 0 & 1 \end{bmatrix}$$

16. Determine if the matrix is absorbing.

$$\begin{bmatrix} 1 & .1 & 0 & 0 \\ 0 & .5 & 0 & 0 \\ 0 & .3 & 1 & 0 \\ 0 & .1 & 0 & 1 \end{bmatrix}$$

17. Find the stable matrix.

$$\begin{bmatrix} 1 & 0 & .5 & 0 \\ 0 & 1 & .1 & 1 \\ 0 & 0 & .3 & 0 \\ 0 & 0 & .1 & 0 \end{bmatrix}$$

18. Arrange the absorbing matrix so that the absorbing states come before the nonabsorbing ones.

$$\begin{bmatrix} .1 & .3 & 0 & 0 \\ .8 & .2 & 0 & 0 \\ .1 & .1 & 1 & 0 \\ 0 & .4 & 0 & 1 \end{bmatrix}$$

19. Compute $(I - R)^{-1}$.

$$\begin{bmatrix} 1 & 0 & .2 & 0 \\ 0 & 1 & .4 & 1 \\ 0 & 0 & .2 & 0 \\ 0 & 0 & .2 & 0 \end{bmatrix}$$

20. Show that the probability distribution of the long-term trend depends on the initial distribution by calculating SV and SW, where S is the stable matrix and V and W are two initial distributions.

$$S = \begin{bmatrix} 1 & 0 & .2 \\ 0 & 1 & .8 \\ 0 & 0 & 0 \end{bmatrix} \qquad V = \begin{bmatrix} .2 \\ .4 \\ .4 \end{bmatrix}$$

$$W = \begin{bmatrix} .5 \\ .4 \\ .1 \end{bmatrix}$$

21. The following game can be described by a Markov process. Find the transition matrix, find its stable matrix, and then answer the given question.

Two players, A and B, engage in a coin tossing game where the coin is not fair and the probability of heads is $\frac{3}{7}$. Player A always bets on heads. If A wins B gives A \$1, and if A loses A gives B \$1. The players have a total of \$3 between them, and they continue to play until one has accumulated all the money. What is the probability that A will win \$3 if A starts with \$2?

10 Game Theory

CHAPTER OVERVIEW   Game theory's novel mathematical properties were created in the 1930s by John von Neumann and Oskar Morgenstern to solve problems in economics and management. Their book, *The Theory of Games and Economic Behavior,** showed how social events can be described by models of certain types of games.

Game theory is in an active state of development. During 1983–1984 the program at the Institute for Mathematics and its Applications in Minneapolis was devoted to economics, and the primary tool was game theory. Game theory has been applied to many disciplines, from economics to animal conflict, but it has also been applied to some surprising areas such as evolution.

The thrust of the theory is to derive strategies that maximize an individual's expected gain in a broad spectrum of competitive situations. Game theory's approach to explaining social interaction is significantly different from previous approaches. The physical sciences study the behavior of inanimate objects. Their mathematical techniques fall short when applied to the highly complex social behavior. Game theory assumes that human behavior is sometimes cooperative and sometimes combative. The key to game theory's approach is the blend of probability with the theory of matrices.

*John von Neumann and Oskar Morgenstern, *The Theory of Games and Economic Behavior*, Princeton, New Jersey; Princeton University Press, 1953.

CASE STUDIES PREVIEW

Game theory has been most fully developed for the two-person situation—when two individuals or groups are in conflict. Since key military decisions fit this model, it is not surprising that game theory can readily be applied to military decisions. But many business conflicts are also two-person or two-group competitions.

The first case study following Section 10–1 is taken from a classic paper written by Colonel O. G. Haywood, Jr. in 1954. It shows that game theory can be used to analyze military decisions by centering on a key battle in World War II.

The second case study at the end of the chapter tells the story of the very stormy merger of Norway's three cement corporations in the 1960s. Even though three groups were involved in the negotiation, the conflict was resolved by using a two-person-game-theoretic approach.

10–1 STRICTLY DETERMINED GAMES

In his book *Game Theory and Politics** Steven J. Brams shows how game theory can help politicians make important decisions, not just concerning elections but also in the military, environmental, and judicial arenas. One example deals with local district elections in Illinois. Each district of the Illinois legislature elects three representatives. Each party can nominate one, two, or three candidates in the general election, and the three with the largest number of votes win. The party's decision of how many candidates to run is made under uncertainty because neither party knows what the other will do. The decision is based on many factors including the percentage of vote the party expects to receive and how that percentage will be split among the candidates. For example, suppose in a particular district the party that anticipates getting the majority of votes chooses to run three candidates and the minority party chooses to run two. Then in the election the majority party gets 54% of the vote such that the three majority party candidates split the vote evenly, so that each candidate gets 18%. If the minority party's 46% is also split evenly, the two minority candidates would each get 23% of the vote. This would result in both minority candidates having more votes than each of the majority candidates, so the majority party would win only one of the three seats and the minority party would win two seats.

Brams introduces game theory in the following way. Consider each district election a game with two players (the two parties), and each player has three options, to run one, two, or three candidates. The outcome, called the **payoff,** is the number of seats won by the majority party. In most cases one party had a clear majority so it never ran one candidate and the minority party never ran three. Brams studied various categories of voting percentages. If the minority party had a significant percentage of the vote, then the four possible outcomes could be expressed in a matrix, called a **payoff matrix.**

*Steven J. Brams, *Game Theory and Politics,* New York, The Free Press, 1975.

| | Minority Party Options | |
	run 1	run 2
Majority Party Options { run 2	2	2
run 3	3	1

or simply

$$P = \begin{bmatrix} 2 & 2 \\ 3 & 1 \end{bmatrix}$$

This means that if the majority party runs two candidates, the majority party will win two seats (and so the minority party will win one seat), no matter what strategy the minority party chooses. That is, if the minority party runs one candidate while the majority party runs two candidates, the payoff is the (1, 1) entry, 2, meaning that the majority party wins two seats. Likewise, the (1, 2) entry, 2, means the majority party wins two seats if it runs two candidates and the minority party runs two candidates. But if the majority party runs three candidates, the majority party will either win three seats if the minority party runs one candidate (since the (2, 1) entry is 3), or it will win one seat if the minority party runs two candidates (since the (2, 2) entry is 1).

The (2, 2) entry, 1, is the result previously described when the majority party evenly splits 54% of the vote among three candidates and the minority party evenly splits 46% of the vote between two candidates.

This is a typical example of a mathematical game. The term **game** in game theory can refer to very serious examples of competition and conflict as well as recreations. In such a game there are two or more players and each player is allowed to make a move. As a result of their moves, there is a payoff (positive, negative, or zero) to each player. The mathematical problem of the game is to determine the best strategy for each player in order for each to get the largest possible payoff.

——————————— **The Fundamental Goal of Game Theory** ———————————
For each player, determine the strategy that maximizes the player's gain.

Two-Person Games

Throughout this chapter we consider only **two-person games,** meaning that there are only two players. Brams' voting study is an example of a two-person game. In a two-person game the outcomes can always be expressed in a matrix, called the payoff matrix, where one player's decisions are represented by the rows and the other player's decisions are represented by the columns. The payoff matrix consists of rows corresponding to the options for the row player, which we refer to as R, and columns corresponding to the moves for the column player, referred to as C. The entries of the payoff matrix are the amount won (or lost in the case of a negative entry) by the row player, R. Thus if R chooses row 2 and C chooses column 3 the payoff of the game to R is the (2, 3) entry. For example, suppose a game has payoff matrix

$$R \begin{array}{c} C \\ \begin{bmatrix} 2 & -1 & 3 \\ -1 & 3 & -2 \end{bmatrix} \end{array}$$

The entries in the matrix are the possible payoffs to R. That is, if the number is positive, R wins that many units and C loses that many, and if the number is negative, R loses that many while C wins that many. For example, if R chooses row 2 and C chooses column 3, the payoff is -2, which means that R loses two units (and C wins two units). Let us look at an example.

EXAMPLE 1

Consider the game with the following payoff matrix:

$$R \begin{array}{c} C \\ \begin{bmatrix} 3 & 4 & -7 \\ -6 & 2 & -2 \end{bmatrix} \end{array}$$

Problem

Suppose R chooses row 1. What is the payoff if C chooses (a) column 2, (b) column 3?

SOLUTION R has two choices corresponding to the two rows. C has three choices corresponding to the three columns.
(a) If R chooses row 1 and C chooses column 2, then the payoff is the $(1, 2)$ entry, 4. This means that R wins four units and C loses four units.
(b) The payoff is the $(1, 3)$ entry, -7, meaning that R loses seven units and C wins seven units.

This is an example of a **zero-sum game,** which means that the payoff to one player is the loss of the other. After the game is over the sum of the winnings is zero (where a loss is a negative win). Most poker games, for example, are zero-sum games. In a nonzero-sum game both players may lose simultaneously. For instance, in some union-management negotiations both sides can lose—management from lost profits during a strike and union members from lost wages.

Pure and Mixed Strategies

Some games are played repeatedly. The players adopt **strategies** that allow them to maximize their gains or minimize their losses. The simplest type of strategy consists of choosing the same row (for R, or the same column for C) on every play of the game. This is called a **pure strategy.** When a player chooses various options on successive moves it is called a **mixed strategy.**

Let us look at a simple example of a two-person, zero-sum game in which each player has two options.

EXAMPLE 2

Rose and Collin play a game with cards. Each has an ace and a king. They choose 1 card and place it on the table. The ace beats the king and it pays $2. If both play aces, then Rose wins $1, and if both play kings, then Collin wins $3. Here is the payoff matrix.

$$
\begin{array}{c}
\quad\ \ \overbrace{\phantom{\text{ace king}}}^{C} \\
\quad\ \ \text{ace}\quad \text{king} \\
R \begin{cases} \text{ace} \\ \text{king} \end{cases} \begin{bmatrix} 1 & 2 \\ -2 & -3 \end{bmatrix}
\end{array}
$$

Problem What strategy should each player choose?

SOLUTION If Rose plays the ace, she wins either $1 or $2, but if she plays the king, she loses $2 or $3. Therefore, Rose should always play the ace. The option for Collin that maximizes his gain (or loss) is not so clear. He either (1) loses $1 or wins $2 if he plays the ace or (2) loses $2 or wins $3 if he plays the king. Neither option seems preferable. But look at the game from the following perspective: If Collin determines that Rose will always play the ace, then he should also always play the ace in order to minimize his loss. Hence in this game each player's optimum strategy is a pure strategy: Each should always play the ace.

Let us generalize the reasoning in Example 2. R wants to choose the row that will maximize her payoff. But once she chooses a row she recognizes that C will want to choose the column that contains the smallest number in that row. R wants this number to be as large as possible. Therefore if R is looking for a pure strategy that will maximize her gain she will choose her row as follows:

The pure strategy that maximizes the gain for the row player:

1. Determine the smallest entry in each row and circle it.
2. Determine the largest of the circled entries and choose that row.

The first step for R is to identify, for each row, the column that C will choose if R selects that row, that is, the smallest entry. Once each smallest entry is identified, R chooses the row that has the largest such number. Thus R wants to choose the maximum of the row minima, which is called the **maximin**. In a similar way, C will choose the **minimax**.

How should C determine the best pure strategy? C wants to find the column that will minimize the payoff to R. But once C chooses a column, R will select the row in which the largest entry of the column lies. C wants this number to be as small as possible since that will be the payoff to R. We put squares about the numbers that C identifies to distinguish them from R's selections.

The pure strategy that maximizes the gain for the column player:

1. Determine the largest number in each column and put a square about it.
2. Determine the smallest of these numbers and choose that column.

Let us look at an example that illustrates these pure strategies.

EXAMPLE 3   Consider the payoff matrix

$$R \begin{bmatrix} & C & \\ 2 & 1 \\ 3 & -2 \\ -3 & 0 \end{bmatrix}$$

Problem

Choose the optimum pure strategies for R and C.

SOLUTION First R circles the smallest numbers in each row and C puts a square about the largest numbers in each column.
Step 1.

R's strategy C's strategy

$$\begin{bmatrix} 2 & ① \\ 3 & ⊖2 \\ ⊖3 & 0 \end{bmatrix} \qquad \begin{bmatrix} 2 & \boxed{1} \\ \boxed{3} & -2 \\ -3 & 0 \end{bmatrix}$$

Step 2. R chooses the largest circled number, the (1, 2) entry, 1. Therefore, R's optimum pure strategy is to choose row 1. C chooses the smaller of the numbers in squares, the (1, 2) entry, 1. Therefore, C's optimum pure strategy is to choose column 2.

Saddle Points and Strictly Determined Games

Now view the pure strategies for both R and C in the same matrix. The circled numbers in Example 3 are those that R identified, the smallest in each row, while those with a square about them are the ones that C identified, the largest in each column.

$$R \begin{bmatrix} & C & \\ 2 & \boxed{①} \\ \boxed{3} & ⊖2 \\ ⊖3 & 0 \end{bmatrix}$$

The optimum choice is the one with both a circle and a square. Notice that the (1, 2) entry, 1, is simultaneously the smallest number in its row and the largest number in its column. Such an entry is called a **saddle point** for the game. Not all games have a saddle point. But if a game does have a saddle point, then the optimum pure strategy for R is to choose the row containing the saddle point. Likewise, C should choose the column containing the saddle point. Hence the payoff will always be the saddle point. For example, in the preceding game R will always gain one unit because it is the saddle point.

A game with a saddle point is called a **strictly determined game** because the saddle point, v, will always be the payoff if the players choose the optimum strategy. The number, v, is called the **value** of the game.

The next example demonstrates how to determine the value of a strictly determined game. In general you cannot tell whether a payoff matrix yields a strictly determined game until you find the players' optimum strategies.

EXAMPLE 4

Problem

Determine the optimum strategies for the payoff matrix and show that it is a strictly determined game.

$$R \begin{bmatrix} & C & \\ 1 & -4 & 0 \\ -1 & -2 & 1 \end{bmatrix}$$

SOLUTION We circle the smallest numbers in each row and put a square about the largest numbers in each column.

$$R \begin{bmatrix} & C & \\ \boxed{1} & \boxed{-4} & 0 \\ -1 & \boxed{-2} & \boxed{1} \end{bmatrix}$$

Note that the (2, 2) entry, -2, is the saddle point of the game. Hence $v = -2$ is the value of the game and the game is strictly determined.

The next example shows how game theory is used in decision making.

EXAMPLE 5

A politician is planning a reelection campaign. There are three types of advertising: radio, newspaper, and television. The campaign committee has raised $100,000 to be apportioned among the three media. The decision depends on the opponents' strategy. If the opponent decides to attack the incumbent's past record, then 70% of the money should go to television and the remainder divided equally between radio and newspaper advertising. But if the opponent's strategy is to center on future issues, then only 50% should be spent on television, 30% on newspaper ads, and 20% on the radio.

Problem

Express the possible expenditures in a payoff matrix. Is the game strictly determined?

SOLUTION We let the opponent's strategies be represented by the columns and the type of media by the rows. (The numbers are in $1000.)

	Opponent's strategy	
Media Type	**Past**	**Future**
Radio	15	20
Newspaper	15	30
Television	70	50

The optimum pure strategies are row 3 and column 2, so it is a strictly determined game with value 50.

From a game theory point of view Example 5 indicates that the incumbent should spend 50% of the money on television and the opponent should center on future issues. As we will see later, however, in the real world people often choose options that are not indicated by game theory analysis, often to their detriment.

EXERCISE SET 10–1

● For the payoff matrices in Problems 1 to 4 suppose R chooses row 1 and C chooses column 2. Determine which player wins the game and find the payoff of the game.

1. $\begin{bmatrix} 1 & 2 \\ 3 & -4 \end{bmatrix}$

2. $\begin{bmatrix} 4 & -2 \\ 0 & 4 \\ -1 & 0 \end{bmatrix}$

3. $\begin{bmatrix} 3 & -1 & 2 \\ -2 & 0 & -4 \end{bmatrix}$

4. $\begin{bmatrix} 0 & -1 & 4 \\ 3 & -2 & -3 \\ -4 & 2 & -1 \end{bmatrix}$

For the payoff matrices in Problems 5 to 8 suppose R chooses row 2 and C chooses column 3. Determine which player wins the game and find the payoff of the game.

5. $\begin{bmatrix} 1 & -3 & 6 \\ 0 & -4 & 7 \end{bmatrix}$

6. $\begin{bmatrix} 4 & -2 & -2 \\ 0 & 4 & 1 \\ -1 & 0 & 0 \end{bmatrix}$

7. $\begin{bmatrix} -2 & 1 & -2 \\ 3 & -4 & 4 \end{bmatrix}$

8. $\begin{bmatrix} 2 & -1 & -4 \\ -3 & 5 & 3 \\ 3 & -2 & 1 \end{bmatrix}$

The payoff matrices in Problems 9 to 16 are the matrices of a strictly determined game. Determine the optimum strategies for R and C and the value of the game.

9. $\begin{bmatrix} -5 & 0 \\ 3 & 2 \end{bmatrix}$

10. $\begin{bmatrix} 4 & 2 \\ -1 & -3 \end{bmatrix}$

11. $\begin{bmatrix} 1 & 2 \\ -3 & -2 \\ 0 & 1 \end{bmatrix}$

12. $\begin{bmatrix} 1 & 2 \\ -4 & -1 \\ 4 & 3 \end{bmatrix}$

13. $\begin{bmatrix} 1 & -1 & 0 \\ 2 & 3 & 5 \end{bmatrix}$ 14. $\begin{bmatrix} -3 & -2 & 4 \\ 3 & -1 & 1 \end{bmatrix}$

15. $\begin{bmatrix} 1 & 0 & -3 \\ 6 & -3 & -1 \\ 3 & 1 & 2 \end{bmatrix}$

16. $\begin{bmatrix} -3 & 1 & -2 \\ 4 & -2 & -1 \\ 3 & 2 & 1 \end{bmatrix}$

In Problems 17 to 26 determine if the game with the given payoff matrix is a strictly determined game. If so, find its value.

17. $\begin{bmatrix} 1 & -2 \\ -3 & 5 \end{bmatrix}$ 18. $\begin{bmatrix} 1 & -2 \\ 3 & -5 \end{bmatrix}$

19. $\begin{bmatrix} 1 & -2 \\ -3 & -2 \\ 0 & 3 \end{bmatrix}$ 20. $\begin{bmatrix} 0 & 2 \\ -4 & -1 \\ 5 & 3 \end{bmatrix}$

21. $\begin{bmatrix} 1 & 2 \\ 5 & 1 \\ 6 & 3 \end{bmatrix}$ 22. $\begin{bmatrix} -3 & -2 & 4 \\ 0 & 1 & -1 \end{bmatrix}$

23. $\begin{bmatrix} 4 & 2 & 5 \\ 0 & 1 & 1 \end{bmatrix}$ 24. $\begin{bmatrix} 2 & 0 & -4 \\ 7 & -3 & -2 \\ 2 & 1 & 2 \end{bmatrix}$

25. $\begin{bmatrix} -5 & 0 & -1 \\ 4 & -3 & -1 \\ 4 & 2 & 1 \end{bmatrix}$ 26. $\begin{bmatrix} 0 & 5 & 4 \\ 9 & 2 & 4 \\ 9 & 7 & 5 \end{bmatrix}$

27. A politician is planning a reelection campaign. There are three types of advertising: radio, newspaper, and television. The campaign committee has raised $100,000 to be apportioned among the three media. The decision depends on the opponent's strategy. The opponent has three types of strategies to choose from: (1) to attack the incumbent's past record, (2) to center on future issues, or (3) to attack the incumbent's intention to seek higher office during the present term of office. If the opponent decides to attack the incumbent's past record, then 60% of the money should go to television, 25% to radio, and 15% to newspaper ads. If the opponent's strategy is to center on future issues, then 20% should be spent on television, 50% on news-

paper ads, and 30% on the radio. If the opponent's strategy is to attack the incumbent's desire for higher office, then 80% should be spent on television, 15% on newspaper ads, and 5% on the radio. Express the percentages in a payoff matrix and determine if the game is strictly determined.

28. In Problem 27 double the percentage of money spent on radio and divide the remaining amount evenly between television and newspaper ads. Express the percentages in a payoff matrix and determine if the game is strictly determined.

•• In Problems 29 to 32 determine if the game with the given payoff matrix is a strictly determined game. If so, find its value.

29. $\begin{bmatrix} 0 & 2 & 1 & 3 \\ -4 & -1 & 2 & -1 \\ -5 & 3 & 0 & -2 \end{bmatrix}$

30. $\begin{bmatrix} 1 & 3 & 2 & 4 \\ -3 & 0 & 3 & 0 \\ -4 & 4 & 1 & -1 \\ 2 & 1 & -3 & -1 \end{bmatrix}$

31. $\begin{bmatrix} 1 & -2 & -5 & 0 & -1 \\ 4 & 0 & -3 & 5 & -1 \\ 4 & 2 & 2 & 2 & 1 \end{bmatrix}$

32. $\begin{bmatrix} -2 & 3 & 5 & -1 & -3 \\ 3 & 2 & -3 & -5 & -1 \\ 5 & -2 & 3 & 1 & -1 \\ 2 & 2 & 3 & 1 & 0 \end{bmatrix}$

In Problems 33 to 39 a game is described. Find a payoff matrix for the game and decide if it is a strictly determined game. If it is strictly determined, find the value of the game.

33. Each of two players, A and B, has two cards, a 2 and a 5. Each player selects a card and they simultaneously show their cards. If the cards match, then A pays B $1. If A's card is greater than B's, then A wins $5 from B. If B's card is greater than A's, then B wins $3.

34. Player A has two cards, a 3 and a 5. Player B has three cards, a 2, a 4, and a 7. Each player selects a card and they simultaneously show the

cards. If the sum is even, then A pays B \$4. If the sum is odd, then B pays A \$3.

35. Player A has two cards, a 3 and a 6. Player B has three cards, a 2, a 4, and a 7. Each player selects a card and they simultaneously show the cards. If the sum is even, then A pays B \$4. If the sum is odd, then B pays A \$5.

36. Each of two players has a stone, a scissors, and a paper. Each player selects one, and they simultaneously show their choice. The rules are: a stone beats a scissors, a scissors beats a paper, and a paper beats a stone. The winner receives \$1 from the loser, and the payoff is \$0 in case of a tie.

37. Each of two players has a coin. They match their coins by turning up either heads or tails. If the same side appears on each coin, then A pays B \$1. If A turns up heads, and B turns up tails, then A pays B \$2. If A turns up tails and B turns up heads, then B pays A \$3.

38. An automobile agency places an order for new cars and vans quarterly. The number of cars ordered depends on projected sales during the specified period. Also, a key factor is the type of cars ordered. There are four types: large, midsized, and compact cars, and recreational vans. The state of the economy plays a significant role in the decision. If the economy is strong during the period, not only will more total cars and vans be purchased, but more large cars and vans will be sold. On the other hand, if the economy is slow, fewer cars will be sold and most of them will be compacts. Here is the agency's prediction of the number of each type of car and van to be sold depending on the economy. (1) If the economy is strong, then 45 large cars, 20 midsized cars, 25 compact cars, and 15 vans will be sold. (2) If the economy is slow, then 20 large cars, 15 midsized cars, 45 compact cars, and 5 vans will be sold. Given that the profit on each large car is \$3000, on each midsized car is \$2000, on each compact

is \$1000, and on each van is \$4000, express the profit as a payoff matrix. Determine if the game is strictly determined.

39. The owner of a small video store stocks two types of blank video tapes: Beta and VHS. In the past the owner has stocked them in equal amounts. The industries that make each type of tape have announced a forthcoming breakthrough that will have a dramatic effect on tape sales. The owner must decide among three strategies: order (1) mostly Beta tapes, (2) mostly VHS tapes, or (3) the same amount of each. The profit from tape sales depends on which industry has the more significant breakthrough. The owner determines that profit on the sales will be the following: Suppose the Beta industry has the superior breakthrough: If mostly Beta tapes are ordered, the profit will be \$8000; if mostly VHS tapes are ordered, the profit will be \$2000; and if an equal amount of each type of tape is ordered, the profit will be \$5000. Suppose the VHS industry has the superior breakthrough: If mostly Beta tapes are ordered, the profit will be \$3000; if mostly VHS tapes are ordered, the profit will be \$9000; and if an equal amount of each type of tape is ordered, the profit will be \$5500. Express the profit in a payoff matrix and determine if the game is strictly determined.

●●● 40. In a study designed to define optimal strategies for various types of voting situations, Gerald Glasser used game theory to model corporate voting.* Voting for corporate directors often pits one faction against another. Each side has a certain number of votes to divide among several candidates. Suppose a firm has three directorships open and the two opposing factions, A and B, have 16 votes and 8 votes, respectively, to divide among three candidates. A strategy is an ordered triple. For example, faction A could choose the strategy (8, 7, 1), meaning that their first candidate would get 8 votes, the second 7

*Adapted from Gerald Glasser, "Game Theory and Cumulative Voting for Corporate Directors," *Management Science,* January 1959, pp. 151–156.

votes and the third 1 vote. Let the entries in the payoff matrix be the number of directors won by faction *A*. Find the payoff matrix if *A* chooses between the two strategies (8, 7, 1) and (6, 5, 5), while *B* chooses between (6, 1, 1) and (4, 3, 1). Is the game strictly determined?

41. Military science is a field that utilizes game theory in a wide variety of settings. In a paper studying the peacetime use of aircraft carriers, Grotte and Brooks measure naval presence of large fleets versus small fleets in several areas.* They study two forces, force R and force C, which deploy fleets in one or two areas of an ocean. Force R has three small fleets at its disposal, and force C has two large fleets. The fleets are allocated simultaneously into the two areas, with force R deploying 0, 1, 2 or 3 fleets to an area (with the sum of R's fleets being 3) and force C deploying 0, 1 or 2 fleets to an area (with the sum of C's fleets being 2). The outcome in each area is decided by which force has the greater presence. In a given area, force R is awarded 1 point if its presence is greater than C's, −1 point if its presence is less than C's, and 0 points are awarded if its presence is equal to C's. The payoff of the game for a given choice of options for R and for C is the sum of the outcomes in the two areas. The options are expressed as ordered pairs: the first number represents the number of fleets deployed to the first area, and the second number is the number of fleets deployed to the second area. Thus force R has four options: (3, 0), (2, 1), (1, 2) and (0, 3). Force C has three options: (2, 0), (1, 1) and (0, 2). There are 12 entries in the payoff matrix. The rules of the game for a given area are as follows: if R has 3 fleets, its presence is greater if C has 0 or 1 fleet (so R gets 1 point in that area), and it is a draw if C has 2 fleets (so the outcome is 0 points in that area); if R

has 2 fleets, its presence is greater if C has 0 fleets, it is a draw if C has 1 fleet, and C's presence is greater if C has 2 fleets; if R has 1 fleet, its presence is greater if C has 0 fleets and C's presence is greater if C has 1 or 2 fleets; if R has 0 fleets, it is a draw if C has 0 fleets, and C's presence is greater if C has 1 or 2 fleets. Find the payoff matrix and determine if the game is strictly determined.

42. Adam Smith is often referred to as the founder of modern economics. His major book, *The Wealth of Nations,* deals with the problem of how social order and human progress can be possible in a society where individuals follow their own self-interest.† Tullock shows how the common game theory problem, the "prisoner's dilemma," can be used to explain some of Smith's theory. In the game, two prisoners have committed a crime together and are separated for interrogation. The police claim that they have enough evidence to convict them. Each has the choice of confessing, thereby squealing on the other, or not confessing. If only one confesses, he or she will receive a suspended sentence while the other will receive two years in jail. If both confess or neither confesses they both will receive one year in jail. Tullock compares this game to an ordinary competitive market with two conglomerates deciding whether to confess to having signed secret agreements together. The payoff would be loss of reputation in the business community, corresponding to the number of years in jail for the prisoners. The corresponding payoff matrix is:

	confess	not confess
confess	1	0
not confess	2	1

Is the game strictly determined?

*J. H. Grotte and P. S. Brooks, "Measuring Naval Presence Using Blotto Games," *International Journal of Game Theory,* 12, 1983, pp. 225–236.

†Gordon Tullock, "Adam Smith and the Prisoner's Dilemma," *Quarterly Journal of Economics,* 100, 1985, pp. 1073–1081.

In August 1944, two months after the Allies first landed at Normandy on D-day, they had amassed a force of one million soldiers. They broke out of their beachhead at Cherbourg on the Cotentin peninsula in Normandy, the northwestern part of France. (See Figure 1.) The German Ninth Army was instructed by Hitler to seal off the peninsula ''at all costs,'' thus halting the Allied advance. Hitler felt this could be done easily because he assumed the invasion of Normandy was a trick designed to mask the bulk of the Allies' invasion of Europe, which would come later in northern France near Calais.

The German line at the mouth of the peninsula held strong until August when the Allies broke through along a narrow gap by the sea at the town of Avranches. This threatened the west flank of the German position. The German commander, General von Kluge, was forced to choose between two options: to attack the advancing Allied front in the west or to retreat and take a fortified defensive position in the east by the Seine River.

General Omar Bradley, commander of the Allies, ordered the U.S. Third Army, under General George Patton, to sweep west and south from the gap at Avranches. The First Army threatened the German Ninth to the east. Bradley's problem was how to deploy his reserve consisting of four divisions. He had three options: (1) he could send them to reinforce the First Army at the gap in case the Germans attacked, (2) he could send them east to try to encircle the Germans, or (3) he could hold them back for 24 hours to see whether the First Army was attacked by the Germans. If so, he could use them to reinforce the gap or, if the gap held, he could then send them east.

In his paper ''Military Decision and Game Theory,'' O. G. Haywood* used game theory to analyze the options of the two generals by assigning appropriate numbers to the outcomes. Haywood reasoned that for each of the two German options there were three Allied alternatives—totaling six possibilities altogether. They are listed with their most probable outcomes in Figure 2. The outcomes in the table are expressed in terms of the Allies' point of view. For example, the (1, 1) entry, ''win one battle,'' means that the Allies would win the battle if each general chose the corresponding strategy. The term ''draw'' means that neither side gains an immediate advantage.

Thus the situation is a two-person, zero-sum game with the Allied commander representing the row player and the German commander the column player. There were two possible battles: (1) the Germans attack the Allies at the gap and (2) the Allies harass the Germans in the east. The payoff matrix entries are the number of battles won by the Allies. Thus the (1, 1) entry, corresponding to the Allies reinforcing their position and the Germans attacking, is 1 because the Allies would win the battle. The (2, 1) entry, corresponding to the Allies moving their reserves east while the Germans attack, is -2 because

*O. G. Haywood, ''Military Decision and Game Theory,'' *Journal of Operations Research Society of America,* Vol. 2, 1954, pp. 365–385.

FIGURE 1

		German Decision	
		attack	retreat
Allied Decision	reinforce position	win one battle	draw
	proceed east	lose two battles	win one battle
	hold in reserve	win two battles	draw

FIGURE 2

the Allies would lose both battles. That is, the Germans would harass the Allies' reserves and defeat the weakened Allied force at the gap. The (3, 1) is 2 because the Allies would hold the gap since reinforcements would be available, and the reserve would be free to move east to harass the German rear. In the second column the German retreat would be successful if the Allies reinforced their position with the reserve or if they held the reserve in place. These possibilities would result in a draw, so there is a 0 in the (1, 2) and (3, 2) entries. If the Allies sent the reserve east, they would be able to harass the German retreat. Hence the (2, 2) entry is 1. Here is the payoff matrix.

		German Decision	
		attack	retreat
Allied Decision	reinforce position	1	0
	proceed east	-2	1
	hold in reserve	2	0

or simply

$$\begin{bmatrix} 1 & 0 \\ -2 & 1 \\ 2 & 0 \end{bmatrix}$$

Which strategy do you think each side chose? The Allies chose option 3—to hold back their reserves 24 hours. Von Kluge decided to retreat—his second option. This would have resulted in a relative stalemate corresponding to the entry 0 in the (3, 2) entry of the payoff matrix. The Germans then could have held their position near the gap of the peninsula to harass the Allied advance from their beachhead. At the last moment, however, Hitler, the "Austrian corporal," 300 miles from the battlefield, countermanded von Kluge's decision and ordered him to attack the gap. The Allied forces withstood the attack the first day. Then Bradley deployed his four divisions held in reserve. By the next day the Germans were almost entirely surrounded. Von Kluge was able to escape with only a remnant of his army. A day later he committed suicide.

Case Study Exercises

1. Find the optimum pure strategy for the Allies.

2. Find the optimum pure strategy for the Germans.

3. Is the game defined by the payoff matrix in the Case Study strictly determined?

4. If the (2, 2) entry were changed to -1, would the game be strictly determined?

10–2 MIXED STRATEGIES

In the remainder of this chapter we consider games that are not strictly determined. The players of this type of game usually vary their choice of a row (or column). The mathematics increases in complexity as probability theory plays a key role.

The Even-and-Odd Game

Let us look at the age-old ''even-and-odd'' child's game. One person places some marbles in his hand and the other player guesses whether there is an even or odd number. If she guesses correctly she wins a marble. Otherwise she gives the other player a marble. If R is the player who guesses and C is the player who puts marbles in his hand, the payoff matrix is

$$
\begin{array}{c}
\qquad\qquad C \\
\qquad\quad \overbrace{\text{even}\quad \text{odd}}
\end{array}
$$

$$
R \left\{ \begin{array}{l} \text{even} \\ \text{odd} \end{array} \right. \begin{bmatrix} 1 & -1 \\ -1 & 1 \end{bmatrix}
$$

The game is not strictly determined, so the players cannot have an optimum pure strategy. They must vary their choices if they hope to win. The question is, what is the best way to vary the choices?

In a delightful account of this game Edgar Allen Poe recounted the story of an 8-year-old with psychic powers.* He could read minds and therefore knew precisely how many marbles his opponent had in hand. He won all the marbles in the school. It seemed impossible to win even once against such an invincible foe. Later in the story Poe gave a hint as to how to thwart the ''brain-child'' as he described one opponent's strategy:

> As poet and mathematician, he would reason well; as mere mathematician, he should not reason at all.

The key is not to reason at all! Leave the choice to chance. Flip a coin, roll a die, or use some other chance device to determine how many marbles to hold so the brain-child cannot predict your strategy.

Mixed Strategies

When strategies in game theory are determined by chance, that is, by probability methods, they are called **mixed strategies.** Consider the game with payoff matrix

*From Edgar Allen Poe, *Poe's Tales,* New York, 1845.

$$P = \begin{bmatrix} 1 & -2 \\ -3 & 5 \end{bmatrix}$$

The game is not strictly determined so the players must vary their choices. Suppose R decides to choose row 1 30% of the time and row 2 70% of the time. And suppose C decides to favor column 1 over column 2 by choosing it 80% of the time versus 20% of the time for column 2. These mixed strategies are expressed in matrix form as follows: R's strategy is a row matrix, $A = [.3 \quad .7]$ and C's strategy is a column matrix $B = \begin{bmatrix} .8 \\ .2 \end{bmatrix}$. In this way the game can be considered a probability experiment. The sample space consists of the four outcomes $\{1, -2, -3, 5\}$, which are the entries in the payoff matrix P. The probability assigned to each element in the sample space is the frequency with which the players choose that entry. For instance, the $(1, 1)$ entry, 1, occurs when R chooses row 1 and C chooses column 1. These choices are independent, so we multiply their probabilities to find the probability that both happen at the same time. Thus

$$\begin{aligned} Pr(1) &= Pr(R \text{ chooses row 1 and } C \text{ chooses column 1}) \\ &= Pr(\text{row 1}) \times Pr(\text{column 1}) = .3 \times .8 = .24 \end{aligned}$$

Hence $Pr(1) = .24$. This means that the $(1, 1)$ entry will occur with probability .24. Another way of seeing this is to note that R will choose row 1 30% of the time and, of that 30%, C will choose column 1 80% of the time. So row 1 and column 1 will occur together $30\% \times 80\% = 24\%$ of the time. We similarly compute the other three probabilities where we abbreviate "R chooses row n" by "row n" and "C chooses column n" by "column n."

$$\begin{aligned} Pr(-2) &= Pr(\text{row 1 and column 2}) = .3 \times .2 = .06 \\ Pr(-3) &= Pr(\text{row 2 and column 1}) = .7 \times .8 = .56 \\ Pr(5) &= Pr(\text{row 2 and column 2}) = .7 \times .2 = .14 \end{aligned}$$

We express these probabilities in a probability distribution.

x	p
1	.24
-2	.06
-3	.56
5	.14

Example 1 illustrates how strategies for R and C produce the probabilities of the entries of a payoff matrix.

EXAMPLE 1 Suppose the strategy for R is $A = [0.3 \quad 0.5 \quad 0.2]$ and the strategy for C is $B = \begin{bmatrix} .6 \\ .4 \end{bmatrix}$.

Problem

Determine the size of the payoff matrix and the probability of each entry.

SOLUTION There are three choices for R and two for C, so the payoff matrix must be 3×2. Hence there are six entries. The following probability distribution gives the probabilities of each entry of the payoff matrix:

x	p
(1, 1) entry	.18
(1, 2) entry	.12
(2, 1) entry	.30
(2, 2) entry	.20
(3, 1) entry	.12
(3, 2) entry	.08

Expected Value

Let us return to the game whose payoff matrix P and probability distribution are

$$P = \begin{bmatrix} 1 & -2 \\ -3 & 5 \end{bmatrix}$$

x	p
1	.24
-2	.06
-3	.56
5	.14

What is the long-term trend of this game? That is, if the game is played many times, which player will win the most? The expected value, E, of a probability distribution answers this question. To compute E, sum the products xp. For the above game the expected value is

$$E = (1)(.24) + (-2)(.06) + (-3)(.56) + 5(.14) \tag{1}$$
$$= -.86$$

This means that R can expect to lose .86 units per play. That is, if the game were played 100 times, R would most likely lose about 86 units.

Another way to compute the expected value of a game is to use matrix multiplication, which is why the strategies are expressed as matrices. Consider the three previous matrices and multiply them as follows:

$$APB = [.3 \quad .7] \begin{bmatrix} 1 & -2 \\ -3 & 5 \end{bmatrix} \begin{bmatrix} .8 \\ .2 \end{bmatrix}$$

$$= [.3 \quad .7] \begin{bmatrix} (1)(.8) + (-2)(.2) \\ (-3)(.8) + (5)(.2) \end{bmatrix}$$

$$= [(.3)(1)(.8) + (.3)(-2)(.2) + (.7)(-3)(.8) + (.7)(5)(.2)] \tag{2}$$

$$= [.24 - .12 - 1.68 + .7] = [-.86]$$

This is the same expected value that was computed in equation 1. Look at the numbers in equations 1 and 2. The numbers are the same, only their order is different. As seen in Chapter 2, matrices are often used to establish a specific order to a set of numbers. By using matrices to define the strategies, we have the numbers in a convenient form for computing the expected value. This computation is a particular instance of an important result that we single out.

_____ **The Expected Value of a Game with Mixed Strategies** _____

If a payoff matrix P is an $n \times m$ matrix and the mixed strategy for R is the $1 \times n$ matrix A and the mixed strategy for C is the $m \times 1$ matrix B, then the expected value E of the game is equal to

$$E = APB$$

Example 2 illustrates this result.

EXAMPLE 2   Suppose a game has payoff matrix P and strategies A and B.

$$P = \begin{bmatrix} 1 & 0 & -1 \\ 0 & -2 & 3 \\ -3 & 4 & 0 \end{bmatrix} \quad A = [.1 \quad .5 \quad .4] \quad B = \begin{bmatrix} .7 \\ .1 \\ .2 \end{bmatrix}$$

We compute the expected value as follows:

$$APB = [.1 \quad .5 \quad .4] \begin{bmatrix} 1 & 0 & -1 \\ 0 & -2 & 3 \\ -3 & 4 & 0 \end{bmatrix} \begin{bmatrix} .7 \\ .1 \\ .2 \end{bmatrix}$$

$$= [.1 \quad .5 \quad .4] \begin{bmatrix} .7 \\ .1 \\ .2 \end{bmatrix} = [-.43]$$

Hence the expected value is $-.43$. Therefore, in the long run R can expect to lose .43 units per play with these strategies.

Suppose R learns that C will play the mixed strategy in Example 2. Is there a better strategy for R to use? We could try some different numbers and compute the product of the matrices, but that would be a hit-and-miss approach. Instead, express R's strategy as a row matrix in general form.

$A = [a \quad b \quad c]$

Here a, b, and c are probabilities with $a + b + c = 1$. We compute the expected value and choose the values that maximize E.

$$E = APB = [a \quad b \quad c] \begin{bmatrix} 1 & 0 & -1 \\ 0 & -2 & 3 \\ -3 & 4 & 0 \end{bmatrix} \begin{bmatrix} .7 \\ .1 \\ .2 \end{bmatrix} \tag{3}$$

$$= [.5a + .4b - 1.7c]$$

Since the third term has a negative sign, and since a, b, and c are nonnegative, R must make c as small as possible. R should also choose a and b to be almost equal because their coefficients are close. For instance, if R chooses the strategy $[.5 \quad .4 \quad .1]$, then

$$E = .5(.5) + .4(.4) - 1.7(.1) = .25 + .16 - .17 = .24$$

With this new strategy R changes the losing game of Example 2 into a winning one. So far so good, but is this R's optimum strategy? Example 3 investigates this question further.

EXAMPLE 3

Problem

Given the game with payoff matrix P and strategies A and B above, find values for a, b, and c that maximize E.

SOLUTION From equation 3, $E = .5a + .4b - 1.7c$. Since the coefficient of c is negative, c should be made as small as possible, so let $c = 0$. Since the coefficient of a, .5, is larger than that of b, .4, E will be maximized by letting a be as large as possible. So let $a = 1$ and $b = 0$. Therefore, E is maximized with the strategy $[1 \quad 0 \quad 0]$. Then $E = .5(1) + .4(0) - 1.7(0) = .5$.

The value of E in Example 3 is more than double the previous expected value of .24, but it is a pure strategy and C would easily uncover it. Then C would certainly change strategies accordingly. Thus to maximize R's gain in a nonstrictly determined game, we must keep in mind that C will most likely choose the mixed strategy that minimizes R's chances. This will be the starting point of the next section.

Matrices with Positive Entries

Thus far most of the payoff matrices that were studied have had both positive and negative entries. However, many important results in game theory are derived for matrices with only positive entries. Therefore, in the next two sections where some of these results are presented, we will stipulate that every entry in the payoff matrix must be positive. This seems at first to be a significant restriction because it rules out many possible matrices; in fact it rules out almost all that we have studied so far. But we are primarily interested in computing the expected values of these matrices, and with this view the stipulation is not strict. For if a matrix has a zero or negative entry, then we can add a large enough number to each entry to make all the entries positive. The payoff matrix is changed—but how is the expected value changed? If the number n is added to the entries of a payoff matrix, the expected value of the new matrix is equal to the original expected value plus n. Let us look at an example.

EXAMPLE 4

Suppose the payoff matrix is

$$P = \begin{bmatrix} 1 & -3 \\ -4 & 5 \end{bmatrix}$$

Problem

Construct a matrix Q from P by adding the smallest positive integer to all the entries of P so that Q has all positive entries. Then compare the expected values of P and Q.

SOLUTION Since the smallest entry in P is -4, add 5 to all entries to get Q.

$$Q = \begin{bmatrix} 6 & 2 \\ 1 & 10 \end{bmatrix}$$

Compare the expected values of P and Q using some specific strategies. Let $A = [.6 \quad .4]$ and $B = \begin{bmatrix} .2 \\ .8 \end{bmatrix}$. The expected value for P is

$$APB = [.6 \quad .4] \begin{bmatrix} 1 & -3 \\ -4 & 5 \end{bmatrix} \begin{bmatrix} .2 \\ .8 \end{bmatrix} = [-.04]$$

The expected value for Q is

$$AQB = [.6 \quad .4] \begin{bmatrix} 6 & 2 \\ 1 & 10 \end{bmatrix} \begin{bmatrix} .2 \\ .8 \end{bmatrix} = [4.96]$$

Thus the expected value for Q is 5 greater than the one for P since $4.96 = -.04 + 5$.

In the remainder of this chapter we deal only with payoff matrices with positive entries. Example 4 shows that no generality is lost because if a game has a payoff matrix with negative entries, a new payoff matrix can be constructed with all positive entries and their expected values differ by the constant. Section 10–4 also shows that the optimal strategies remain the same.

The general result is given here.

If P and Q are payoff matrices with $Q = P + c$ for some constant c, then the optimum strategies are the same for P and Q, and for any strategies A and B the expected values for P and Q differ by c; that is

$$AQB = APB + c$$

EXERCISE SET 10–2

- In Problems 1 to 6 determine whether the games are strictly determined.

1. $\begin{bmatrix} 3 & 2 \\ 2 & -1 \end{bmatrix}$

2. $\begin{bmatrix} 0 & 2 \\ 2 & -1 \end{bmatrix}$

3. $\begin{bmatrix} 1 & -1 \\ 2 & 0 \\ 1 & 3 \end{bmatrix}$

4. $\begin{bmatrix} 1 & -1 & 6 \\ -3 & 2 & 1 \end{bmatrix}$

5. $\begin{bmatrix} 1 & 2 & 4 \\ 0 & 1 & 3 \end{bmatrix}$

6. $\begin{bmatrix} -3 & -2 & 1 \\ 2 & -1 & 0 \\ 0 & -3 & 4 \end{bmatrix}$

In Problems 7 to 10 the strategies for R and C are given. Determine the size of the payoff matrix and compute the probability of each entry being played.

7. $[.4 \quad .6]$, $\begin{bmatrix} .5 \\ .5 \end{bmatrix}$

8. $[.5 \quad .5]$, $\begin{bmatrix} .2 \\ .8 \end{bmatrix}$

9. $[.8 \quad .1 \quad .1]$, $\begin{bmatrix} .4 \\ .6 \end{bmatrix}$

10. $[.6 \quad .4]$, $\begin{bmatrix} .4 \\ .1 \\ .5 \end{bmatrix}$

In Problems 11 to 14 the game has payoff matrix

$$P = \begin{bmatrix} 1 & 2 \\ 2 & -4 \end{bmatrix}$$

Calculate the expected value for each of these strategies.

11. $[.5 \quad .5]$, $\begin{bmatrix} .4 \\ .6 \end{bmatrix}$ 12. $[.6 \quad .4]$, $\begin{bmatrix} .7 \\ .3 \end{bmatrix}$

13. $[.1 \quad .9]$, $\begin{bmatrix} .2 \\ .8 \end{bmatrix}$ 14. $[.3 \quad .7]$, $\begin{bmatrix} .5 \\ .5 \end{bmatrix}$

In Problems 15 to 18 the game has payoff matrix

$$P = \begin{bmatrix} 1 & -2 \\ 0 & -1 \\ -2 & 4 \end{bmatrix}$$

Compute the expected value of the game with the given strategies.

15. $[.1 \quad .2 \quad .7]$, $\begin{bmatrix} .5 \\ .5 \end{bmatrix}$

16. $[.1 \quad .7 \quad .2]$, $\begin{bmatrix} .5 \\ .5 \end{bmatrix}$

17. $[.1 \quad .1 \quad .8]$, $\begin{bmatrix} .2 \\ .8 \end{bmatrix}$

18. $[.2 \quad .1 \quad .7]$, $\begin{bmatrix} .2 \\ .8 \end{bmatrix}$

In Problems 19 to 22 add the smallest possible integer to the matrix to get a matrix with all positive entries; that is, choose that number such that the smallest entry is 1.

19. $\begin{bmatrix} 3 & -4 \\ -2 & 6 \end{bmatrix}$ 20. $\begin{bmatrix} 3 & -4 \\ -2 & 6 \\ -4 & 5 \end{bmatrix}$

21. $\begin{bmatrix} -1 & 4 & -5 \\ 2 & -6 & 7 \end{bmatrix}$

22. $\begin{bmatrix} -7 & 4 & -3 \\ 2 & -6 & 5 \\ 2 & -8 & -7 \end{bmatrix}$

•• In Problems 23 to 26 the game has payoff matrix

$$P = \begin{bmatrix} 1 & 0 & -2 \\ -1 & 1 & 3 \end{bmatrix}$$

Let C play the given strategy and let R play the strategy $[a \quad b]$. Determine what values of a and b will maximize the payoff to R.

23. $\begin{bmatrix} .7 \\ .2 \\ .1 \end{bmatrix}$ 24. $\begin{bmatrix} .2 \\ .4 \\ .4 \end{bmatrix}$

25. $\begin{bmatrix} .3 \\ .2 \\ .5 \end{bmatrix}$ 26. $\begin{bmatrix} .4 \\ .3 \\ .3 \end{bmatrix}$

In Problems 27 to 30 the game has payoff matrix

$$P = \begin{bmatrix} -2 & 1 & 2 \\ 1 & -2 & -3 \end{bmatrix}$$

Let C play the given strategy and let R play the strategy $[a \quad b]$. Determine what values of a and b will maximize the payoff to R.

27. $\begin{bmatrix} .1 \\ .8 \\ .1 \end{bmatrix}$ 28. $\begin{bmatrix} .4 \\ .2 \\ .4 \end{bmatrix}$

29. $\begin{bmatrix} .4 \\ .1 \\ .5 \end{bmatrix}$ 30. $\begin{bmatrix} 0 \\ .3 \\ .7 \end{bmatrix}$

In Problems 31 to 34 add the smallest possible integer to the payoff matrix P to obtain the payoff matrix Q with all positive entries and compute the expected values using the matrices A and B, the strategies for R and C, respectively. How are the two expected values related?

31. $P = \begin{bmatrix} 2 & -1 \\ -2 & 1 \end{bmatrix}$, $A = [.4 \quad .6]$,

   $B = \begin{bmatrix} .3 \\ .7 \end{bmatrix}$

32. $P = \begin{bmatrix} 1 & -4 \\ -5 & 2 \end{bmatrix}$, $A = [.6 \quad .4]$,

   $B = \begin{bmatrix} .8 \\ .2 \end{bmatrix}$

33. $P = \begin{bmatrix} -1 & 4 & 9 \\ -2 & 5 & -6 \end{bmatrix}$, $A = [.2 \quad .8]$,

   $B = \begin{bmatrix} .7 \\ .2 \\ .1 \end{bmatrix}$

34.
$$P = \begin{bmatrix} -3 & 3 \\ -2 & 4 \\ 1 & -5 \end{bmatrix}, \quad A = [.6 \quad .1 \quad .3],$$

$$B = \begin{bmatrix} .6 \\ .4 \end{bmatrix}$$

In Problems 35 and 36 the game has payoff matrix

$$P = \begin{bmatrix} 2 & 3 & -4 \\ 1 & 0 & -2 \\ -1 & 1 & 3 \end{bmatrix}$$

Compute the expected value of the game with the given strategies.

35. $[.1 \quad .2 \quad .7], \quad \begin{bmatrix} .5 \\ .4 \\ .1 \end{bmatrix}$

36. $[.1 \quad .1 \quad .8], \quad \begin{bmatrix} .2 \\ .5 \\ .3 \end{bmatrix}$

In Problems 37 and 38 the game has payoff matrix

$$P = \begin{bmatrix} 0 & -2 & 3 & 0 \\ 1 & 1 & 1 & 2 \\ -2 & 2 & 1 & 1 \end{bmatrix}$$

Compute the expected value of the game with the given strategies.

37. $[.2 \quad .5 \quad .3], \quad \begin{bmatrix} .4 \\ .4 \\ .1 \\ .1 \end{bmatrix}$

38. $[.1 \quad .1 \quad .8], \quad \begin{bmatrix} .1 \\ .5 \\ .3 \\ .1 \end{bmatrix}$

39. Action Sportswear store places a new brand of sportswear on the market at the same time that its primary competitor, Bargain Sportswear, introduces its own new line. Each store will use newspaper and television (TV) ads to market the sportswear. The payoff matrix shows the

increase in sales (in millions of dollars) for Action depending on which choice each company makes for its ad campaign.

$$\text{Action} \left\{ \begin{matrix} \text{paper} \\ \text{TV} \end{matrix} \right. \overset{\overset{\text{Bargain}}{\overbrace{\text{paper} \quad \text{TV}}}}{\begin{bmatrix} 3 & -1 \\ -2 & 2 \end{bmatrix}}$$

Suppose Action learns that Bargain will put 20% of its advertising budget into newspaper ads and 80% into TV ads. Determine the optimum strategy for Action and the value of the game.

●●● 40. Most observers of the 1980 strike by the Major League Players Association felt that the very unusual "discontinuous" strike threat by the players' union was strange at best and futile at worst. The players voted on April 1, 1980 to strike the remaining games of the exhibition season (for which they do not get paid in any case), but to return to work for the first six weeks of the regular season. Then, if negotiations were still deadlocked on May 23, they would strike again. Few people at the time saw the logic in the players' strategy. The key is that May 23 represented the start of the owners' peak revenue period. A study by two economists, Lawrence DeBrack and Alvin Roth, applied game theory to show that the players' strategy was an optimum threat to the owners.* The following game theory model was used to describe the players' union strategy. Let the union be the column player and the owners the row player. The union had to choose between two strategies, a continuous strike from the beginning of the season or a discontinuous strike. The owners had to choose between the threat of a lockout (in which case they would not have to pay those not on strike and save that cost) or not to lockout. The entries in the payoff matrix represent revenue of the owners in relative terms—the larger the number is, the more the revenue will be. The payoff matrix is

$$\begin{array}{cc} & \text{Union} \\ & \overbrace{\hspace{4cm}} \\ & \begin{array}{cc} \text{continuous} & \text{discontinuous} \\ \text{strike} & \text{strike} \end{array} \end{array}$$

$$\text{Owners} \begin{cases} \text{lockout} \\ \text{no lockout} \end{cases} \begin{bmatrix} 4 & 3 \\ 2 & 1 \end{bmatrix}$$

Is the game strictly determined? What is the optimum strategy for the players' union?

41. In an article by T. Ingold that studied the problems of reindeer owners in Finland a game theory model was used to help herdsmen decide what to do with reindeer that were expected to calve the following season.* (See matrix that follows.) That is, how many should be slaughtered for food and how many should be left in the wild for possible future profit, either through offspring or slaughter. The entries of the payoff matrix represented revenue to the owner in dollars per deer. The entries took into account the value of the cow, the value of the calf, and an estimate of the future value of each. There were three possibilities for the column player, which could be regarded as Nature: (1) cow and calf found next season, (2) cow not found but still alive, and (3) cow dies a natural death. The two possibilities for the row player, the owner, were (1) leave the cow alive or (2) slaughter the cow. The payoff matrix was

$$\begin{array}{cc} & \begin{array}{ccc} & \text{not found} & \text{natural} \\ \text{found} & \text{but alive} & \text{death} \end{array} \end{array}$$

$$\begin{array}{c} \text{leave alive} \\ \text{slaughter} \end{array} \begin{bmatrix} 300 & 100 & 0 \\ 200 & 200 & 200 \end{bmatrix}$$

Find the expected value of the game with the strategies $[.2 \quad .8]$ and $\begin{bmatrix} .5 \\ .3 \\ .2 \end{bmatrix}$.

42. Fellingham and Newman illustrate how the exercise of an accountant auditing a client can be considered a strategic game between the auditor and the client.† The authors assume that the auditor is the row player and the client is the column player. Each has two strategies. The auditor can "extend auditing procedures" or "not extend auditing procedures." This means the auditor must decide whether ordinary procedures are enough to generate a complete and accurate audit or whether further study and procedures are necessary. The latter produces a significant increase in cost. The client chooses between "more effort" or "less effort" when preparing for the audit. The latter produces a smaller cost to the client. The payoff is the relative profit of the auditor. It is a "relative" profit because there are certain considerations that are computed in the profit for each choice of strategies that cannot be expressed in real dollars, such as the cost of a poor audit causing loss of further work for the client as well as loss of reputation. For one particular auditor and client, the following payoff matrix is given:

$$\begin{array}{cc} & \begin{array}{cc} \text{lower} & \text{higher} \\ \text{effort} & \text{effort} \end{array} \end{array}$$

$$\begin{array}{c} \text{extend} \\ \text{not extend} \end{array} \begin{bmatrix} 47 & 72 \\ 85 & 67 \end{bmatrix}$$

What is the optimum strategy for the auditor?

*T. Ingold, "Statistical Husbandry: Chance, Probability, and Choice in a Reindeer Management Economy," *Science* Vol. 32, 1986; pp. 1029–1038.

†John C. Fellingham and D. Paul Newman, "Strategic Considerations in Auditing," *The Accounting Review*, 60, 1985, pp. 634–648.

10–3 OPTIMAL MIXED STRATEGIES

Section 10–2 showed how closely aligned game theory is with probability. Strategies are expressed in terms of probabilities, and the expected value measures whether the game favors one player. This section is based on the fact that **optimal mixed**

strategies take into account the opponent's **counterstrategy.** We compute R's optimal strategy given that C chooses the column that minimizes the expected value. C's optimal strategy will be considered in Section 10–4.

Linear Equations

For the time being we consider only 2×2 payoff matrices. Let R's strategy be unknown and let us refer to it by the matrix $[x \quad y]$, where x and y are nonnegative numbers such that $x + y = 1$. We look for the values of x and y that maximize the expected value for different strategies of C. Example 1 demonstrates how to do this.

EXAMPLE 1

Let a game have payoff matrix P and let A be R's strategy where

$$P = \begin{bmatrix} 2 & 1 \\ 1 & 3 \end{bmatrix} \qquad A = [x \quad y]$$

Problem

Find the expected value E of the game if C chooses (a) column 1 and (b) column 2.

SOLUTION (a) Since C chooses column 1, then C's strategy is $B = \begin{bmatrix} 1 \\ 0 \end{bmatrix}$. The expected value is

$$E = APB = [x \quad y] \begin{bmatrix} 2 & 1 \\ 1 & 3 \end{bmatrix} \begin{bmatrix} 1 \\ 0 \end{bmatrix} = [x \quad y] \begin{bmatrix} 2 \\ 1 \end{bmatrix} = [2x + y]$$

Since $x + y = 1$ we have $E = 2x + y = 2x + (1 - x) = x + 1$.

(b) C's strategy is $B = \begin{bmatrix} 0 \\ 1 \end{bmatrix}$ and so the expected value is

$$E = APB = [x \quad y] \begin{bmatrix} 2 & 1 \\ 1 & 3 \end{bmatrix} \begin{bmatrix} 0 \\ 1 \end{bmatrix} = [x \quad y] \begin{bmatrix} 1 \\ 3 \end{bmatrix} = [x + 3y]$$

Since $x + y = 1$ we have $E = x + 3y = x + 3(1 - x) = 3 - 2x$. Hence if C chooses column 1, we have $E = x + 1$, and if C chooses column 2, we have $E = 3 - 2x$.

The expressions for the expected value in Example 1 are linear equations in x. Figures 1 and 2 show the graphs of the equations $E = x + 1$ and $E = 3 - 2x$. Each is a linear segment drawn only between the values $x = 0$ and $x = 1$ because x is a probability and thus is a number between 0 and 1. The graphs help us to visualize R's strategy. Remember that a choice for x determines the value of y since $y = 1 - x$, so it gives a complete strategy for R. For instance, if $x = \frac{1}{4}$, then $y = 1 - \frac{1}{4} = \frac{3}{4}$ so $A = [\frac{1}{4} \quad \frac{3}{4}]$. Table 1 refers to the game in Example 1. It lists some choices for x, the corresponding strategies for R, and the expected values of the game depending on C's choice of column.

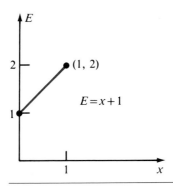

FIGURE 1

TABLE 1

| | | C Chooses | |
| | | Column 1 | Column 2 |
x	$A = [x \quad y]$	$E = x + 1$	$E = 3 - 2x$
$\frac{1}{4}$	$[\frac{1}{4} \quad \frac{3}{4}]$	$\frac{5}{4}$	$\frac{5}{2}$
$\frac{1}{3}$	$[\frac{1}{3} \quad \frac{2}{3}]$	$\frac{4}{3}$	$\frac{7}{3}$
.4	$[.4 \quad .6]$	1.4	2.2
.7	$[.7 \quad .3]$	1.7	1.6

Notice that as x increases the expected value increases when C chooses column 1, but decreases when C chooses column 2. C seeks to minimize E, which implies that C will choose the column that produces the smaller value of E. For instance, if R chooses $x = \frac{1}{4}$, then C will choose column 1 because $\frac{5}{4}$ is less than $\frac{5}{2}$. But if R chooses $x = .7$, then C will choose column 2 because 1.6 is smaller than 1.7.

Here is the problem that R faces. Which strategy will maximize E subject to the fact that C can select the smaller of the two values of E? The graphs will help

FIGURE 2

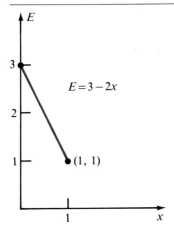

answer this question. In Figure 3 we draw them in the same coordinate system. Here is how to view C's strategy geometrically. If R chooses $x = \frac{1}{4}$, then C draws a vertical line through $x = \frac{1}{4}$. It intersects both lines corresponding to E. C then chooses the value of E corresponding to the lower line. It corresponds to column 1, so $E = \frac{5}{4}$ since the point of intersection is $(\frac{1}{4}, \frac{5}{4})$. However, if R chooses $x = .7$, the lower line intersecting $x = .7$ corresponds to column 2, so $E = 1.6$ since the point of intersection is $(.7, 1.6)$.

Optimal Strategies

The point of intersection is a key to the optimal strategy. It is obtained by setting the two expressions for E equal to each other.

Set

$$E = x + 1 \qquad \text{equal to} \qquad E = 3 - 2x$$
$$x + 1 = 3 - 2x$$
$$3x = 2$$
$$x = \tfrac{2}{3}$$

Therefore, the point of intersection is $(\frac{2}{3}, \frac{5}{3})$, graphed in Figure 4. Notice that if R chooses a value of x between 0 and $\frac{2}{3}$, then C will choose the line $E = x + 1$ (column 1), because it is always the lower line from $x = 0$ to $\frac{2}{3}$. But if R chooses a value of x between $\frac{2}{3}$ and 1, then C will choose the line $E = 3 - 2x$ (column 2) since in that interval it is the lower line. This describes C's counterstrategy.

FIGURE 3

FIGURE 4

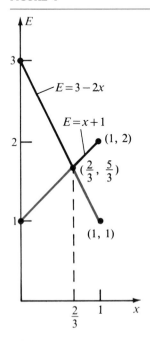

C's Optimal Counterstrategy

Choose

column 1, $E = x + 1$,   if $0 \le x \le \frac{2}{3}$

column 2, $E = 3 - 2x$,   if $\frac{2}{3} \le x \le 1$

In Figure 4 the lower lines are shaded; they indicate C's strategy. Since the second coordinates represent the expected value and since R seeks to maximize E, the largest value of E on the shaded lines occurs at the point of intersection. Therefore, given that C will choose the lower line for each value of x, R maximizes E by choosing the value of x at the point of intersection. This means that R chooses $x = \frac{2}{3}$, and so R's optimal strategy is $[\frac{2}{3} \quad \frac{1}{3}]$. The expected value computed for R's optimal strategy is called the **payoff of the game.** For this game it is $E = \frac{5}{3}$.

Let us review our procedure.

Optimal Mixed Strategy for R

To find the optimal mixed strategy for R that maximizes the expected value, assuming that C's counterstrategy will always result in the smallest possible expected value, we let P be the playoff matrix and $A = [x \quad y]$ be R's strategy. Then

1. Compute $E = AP \begin{bmatrix} 1 \\ 0 \end{bmatrix}$ and $E = AP \begin{bmatrix} 0 \\ 1 \end{bmatrix}$ corresponding to C's choice of either column 1 or 2. This yields two equations in E, x and y.
2. Make the substitution $y = 1 - x$ to get two linear equations in E and x.
3. Graph the lines and shade the lower portions.
4. R's optimal strategy is the value of x that produces the highest shaded point. It will occur either at the point of intersection of the two lines or on the E-axis or on the vertical line $x = 1$.
5. The expected value computed for R's optimal strategy is the payoff of the game.

The next two examples illustrate this procedure.

EXAMPLE 2

Let $P = \begin{bmatrix} 2 & 5 \\ 4 & 3 \end{bmatrix}$ be the payoff matrix of a game.

Problem

Find the optimal mixed strategy for R and the payoff of the game.

SOLUTION   Follow the steps in the procedure.

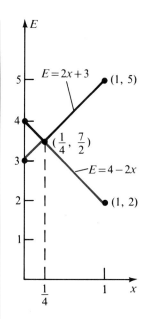

FIGURE 5

1.
$$E = [x \quad y] \begin{bmatrix} 2 & 5 \\ 4 & 3 \end{bmatrix} \begin{bmatrix} 1 \\ 0 \end{bmatrix} = [x \quad y] \begin{bmatrix} 2 \\ 4 \end{bmatrix} = [2x + 4y]$$

$$E = [x \quad y] \begin{bmatrix} 2 & 5 \\ 4 & 3 \end{bmatrix} \begin{bmatrix} 0 \\ 1 \end{bmatrix} = [x \quad y] \begin{bmatrix} 5 \\ 3 \end{bmatrix} = [5x + 3y]$$

2. Substitute $y = 1 - x$ into each expression.

$$E = 2x + 4y = 2x + 4(1 - x) = 4 - 2x$$
$$E = 5x + 3y = 5x + 3(1 - x) = 2x + 3$$

3. The graph is drawn in Figure 5.
4. The highest part of the shaded portion of the graph occurs at the point of intersection.
    Set

$$E = 4 - 2x \qquad\qquad \text{equal to} \qquad\qquad E = 2x + 3$$
$$4 - 2x = 2x + 3$$
$$4x = 1$$
$$x = \tfrac{1}{4}$$

   Therefore, the optimal mixed strategy for R is $[\tfrac{1}{4} \quad \tfrac{3}{4}]$.
5. The point of intersection is $(\tfrac{1}{4}, \tfrac{7}{2})$, and so the payoff of the game is $E = \tfrac{7}{2}$.

   The next example shows that R's optimal mixed strategy does not always occur at the point of intersection.

EXAMPLE 3

Let $P = \begin{bmatrix} 2 & 1 \\ 3 & 4 \end{bmatrix}$ be the payoff matrix of a game.

Problem

Find the optimal mixed strategy for R and the payoff of the game.

SOLUTION We follow the steps in the procedure.

1.
$$E = [x \quad y] \begin{bmatrix} 2 & 1 \\ 3 & 4 \end{bmatrix} \begin{bmatrix} 1 \\ 0 \end{bmatrix} = [x \quad y] \begin{bmatrix} 2 \\ 3 \end{bmatrix} = [2x + 3y]$$

$$E = [x \quad y] \begin{bmatrix} 2 & 1 \\ 3 & 4 \end{bmatrix} \begin{bmatrix} 0 \\ 1 \end{bmatrix} = [x \quad y] \begin{bmatrix} 1 \\ 4 \end{bmatrix} = [x + 4y]$$

2. Substitute $y = 1 - x$ into each expression.

$$E = 2x + 3y = 2x + 3(1 - x) = 3 - x$$
$$E = x + 4y = x + 4(1 - x) = 4 - 3x$$

3. The graph is drawn in Figure 6.
4. The highest part of the shaded portion of the graph occurs at the point $(0, 3)$. Hence the optimal mixed strategy for R is $[0 \quad 1]$.
5. The payoff of the game is $E = 3 - 0 = 3$.

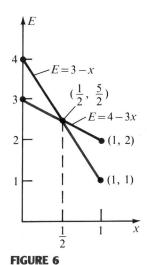

FIGURE 6

The solution to Example 3 has an important implication. The fact that R's optimal strategy occurs at $x = 0$ means that R plays the pure strategy $[0 \quad 1]$, always selecting row 2. Certainly C should select the first column, using the strategy

$\begin{bmatrix} 1 \\ 0 \end{bmatrix}$. Therefore the game has a saddle point, the (2, 1) entry, 3. Thus the game is strictly determined. This says that the procedure for calculating optimal mixed strategies works for strictly determined games as well.

EXERCISE SET 10–3

• In Problems 1 to 10 the payoff matrix is given. Compute the expected value of the game if R's strategy is $[x \quad y]$ and C chooses (a) column 1 and (b) column 2.

1. $\begin{bmatrix} 0 & 1 \\ 2 & 0 \end{bmatrix}$ 2. $\begin{bmatrix} 1 & 2 \\ 0 & 0 \end{bmatrix}$

3. $\begin{bmatrix} 1 & 2 \\ -1 & 0 \end{bmatrix}$ 4. $\begin{bmatrix} 2 & 0 \\ -1 & 1 \end{bmatrix}$

5. $\begin{bmatrix} 2 & 1 \\ -1 & -2 \end{bmatrix}$ 6. $\begin{bmatrix} 3 & 1 \\ -2 & 1 \end{bmatrix}$

7. $\begin{bmatrix} 2 & -1 \\ 1 & -2 \end{bmatrix}$ 8. $\begin{bmatrix} 3 & -1 \\ -2 & 1 \end{bmatrix}$

9. $\begin{bmatrix} 5 & 2 \\ 3 & 4 \end{bmatrix}$ 10. $\begin{bmatrix} 3 & 6 \\ 7 & 4 \end{bmatrix}$

In Problems 11 to 16 two expressions for E are given. The first corresponds to C choosing column 1 and the second corresponds to C choosing column 2. Graph each experssion and find the point of intersection.

11. $E = x + 2y$ and $E = x + 3y$

12. $E = 3x + y$ and $E = 2x + 4y$

13. $E = 2x + 5y$ and $E = 2x + y$

14. $E = x + 2y$ and $E = 2x + 3y$

15. $E = 6x + 3y$ and $E = 8x + 2y$

16. $E = 4x + 2y$ and $E = 2x + 5y$

In Problems 17 to 32 the payoff matrix is given. Compute the optimal mixed strategy for R that maximizes the expected value, given that C will select the best counterstrategy. Then compute the payoff.

17. $\begin{bmatrix} 1 & 2 \\ 3 & 1 \end{bmatrix}$ 18. $\begin{bmatrix} 5 & 2 \\ 3 & 4 \end{bmatrix}$

19. $\begin{bmatrix} 5 & 2 \\ 3 & 1 \end{bmatrix}$ 20. $\begin{bmatrix} 10 & 2 \\ 3 & 11 \end{bmatrix}$

21. $\begin{bmatrix} 11 & 22 \\ 23 & 11 \end{bmatrix}$ 22. $\begin{bmatrix} 2 & 3 \\ 3 & 1 \end{bmatrix}$

23. $\begin{bmatrix} 2 & 6 \\ 7 & 1 \end{bmatrix}$ 24. $\begin{bmatrix} 1 & 3 \\ 7 & 6 \end{bmatrix}$

25. $\begin{bmatrix} 11 & 3 \\ 2 & 16 \end{bmatrix}$ 26. $\begin{bmatrix} 21 & 23 \\ 2 & 6 \end{bmatrix}$

•• 27. $\begin{bmatrix} 3 & \frac{1}{2} \\ 2 & 1 \end{bmatrix}$ 28. $\begin{bmatrix} 2 & \frac{3}{2} \\ 4 & 1 \end{bmatrix}$

29. $\begin{bmatrix} \frac{2}{3} & 3 \\ 1 & 4 \end{bmatrix}$ 30. $\begin{bmatrix} 5 & 3 \\ 1 & \frac{1}{2} \end{bmatrix}$

31. $\begin{bmatrix} \frac{1}{3} & 5 \\ 2 & \frac{2}{3} \end{bmatrix}$ 32. $\begin{bmatrix} 5 & \frac{2}{5} \\ \frac{1}{4} & 4 \end{bmatrix}$

In Problems 33 to 38 find the optimum strategy for the game with the given payoff matrix whose entries are not all positive, given that C will select the best counterstrategy. First add a large enough positive integer to each entry.

33. $\begin{bmatrix} 3 & -2 \\ 2 & 1 \end{bmatrix}$ 34. $\begin{bmatrix} 2 & 5 \\ 4 & -3 \end{bmatrix}$

35. $\begin{bmatrix} 1 & 3 \\ 5 & -4 \end{bmatrix}$ 36. $\begin{bmatrix} -3 & 5 \\ 2 & -3 \end{bmatrix}$

37. $\begin{bmatrix} -7 & 2 \\ 0 & -1 \end{bmatrix}$ 38. $\begin{bmatrix} 0 & 3 \\ 1 & -1 \end{bmatrix}$

In Problems 39 to 42 a game is described. Find a payoff matrix for the game and find the optimum strategy for the row player, given that the column player will select the best counterstrategy.

39. Each of two players, A and B, has two cards: a 2 and a 4. Each player selects a card and they

simultaneously show their cards. If the cards are both 2's, then A pays B \$1. If they are both 4's, then B pays A \$1. If A's card is greater than B's, then A wins \$5 from B. If B's card is greater than A's, then B wins \$3.

40. Player A has two cards: a 3 and a 5. Player B has two cards: a 2 and a 6. Each player selects a card and they simultaneously show the cards. If the sum is even, then A pays B \$4. If the sum is odd, then B pays A \$3.

41. Each of two players has a coin. They match their coins by turning up either heads or tails. If the same side appears on each coin, then A pays B \$1. If A turns up heads and B turns up tails, then A pays B \$2. If A turns up tails and B turns up heads, then B pays A \$3.

42. Two players simultaneously show one or two fingers. If the number of fingers shown by each player matches, then A wins the amount of dollars equal to the sum of the fingers shown. If the number of fingers shown by each player does not match, then B wins the amount of dollars equal to the number of fingers shown.

43. Contingent valuation (CV) has become a commonly used technique for assigning a value to assets such as natural resources that are not traded in markets. CV's are used in many areas such as tax assessments and insurance rates. A common way of conducting a CV is to interview people and ask their opinion of the value of an asset. Boyle et al. use a game theory approach to model a CV of the scenic beauty of the Lower Wisconsin River.* An interviewer telephones individuals at random and offers an initial bid of the value of the scenic beauty, say \$50. The respondent then gives a counter bid, either higher, lower or the same. The authors study whether the starting bid of the interviewer affects the final bid of the respondent. They let the inter-

viewer be the row player and the respondent the column player. The interviewer has two strategies: offer either a high initial bid or a low initial bid. The respondents are classified in terms of two strategies, whether they would offer a bid higher than or lower than (including equal to) the interviewer's bid. The goal of the interviewer is to find the percentage of times a high initial bid should be given in order to maximize the payoff, which is the final bid by the respondent. In one part of the study the following payoff matrix is given.

Respondent's Bid

		higher	lower
Interviewer's	high	50	30
Initial Bid	low	40	45

What is the optimum strategy for the interviewer?

44. Game theory is often used to model diverse markets where both buyers and sellers have potential market power. Karp and McCalla apply these techniques to the world corn market.† The seller, represented by the row player, is the United States and the buyer is Japan. In one model, the United States must choose between including an export tax in the price or not. Japan must choose between ordering a large or small amount of corn from the United States. The payoff is the price of corn in dollars per metric ton. The payoff matrix is:

Buyer

		large	small
Seller	export tax	124	113
	no export tax	116	118

Find the optimum strategy for the row player.

*Kevin J. Boyle, et al., ''Starting Point Bias in Contingent Valuation Bidding Games,'' *Land Economics*, 61, 1985, pp. 188–194.

†Larry S. Karp and Alex F. McCalla, ''Dynamic Games and International Trade: An Application to the World Corn Market,'' *American Journal of Agricultural Economics*, 65, 1983, pp. 641–650.

45. The main objective in cash management is referred to as "collect fast—pay slow." This means speed up collections of accounts receivable and slow down accounts payable. Many firms use lock boxes serviced by local depository banks to reduce the amount of funds tied up in the collection process. Firms also use various disbursement banks to handle accounts payable. Each bank has a different cost for its services. Orgler and Tauman use game theory to model the interaction of a seller and a buyer, the former wanting to collect funds quickly and the latter wanting to delay payment as long as possible.*

The seller, the row player, has two options corresponding to two depository banks, A and B. The buyer has two options corresponding to two disbursement banks, a and b. The payoff is the relative cost to the buyer to enact the transaction. In one example the payoff matrix is

$$\text{Seller} \left\{ \begin{matrix} A \\ B \end{matrix} \right. \overset{\overset{\displaystyle \text{Buyer}}{\overbrace{}}}{\begin{bmatrix} \overset{a}{100} & \overset{b}{205} \\ 77 & 270 \end{bmatrix}}$$

Find the optimum strategy for both players.

*Yair Orgler and Yair Tauman, "A Game Theoretic Approach to Collections and Disbursements," *Management Science,* 32, 1986, pp. 1029–1038.

10–4   GAME THEORY AND LINEAR PROGRAMMING

This section reveals the close link between game theory and linear programming. Because the connection might seem surprising at first, the chapter's epilogue will recount a historical meeting between George Dantzig and John von Neumann that forged the common bond between these two areas.

In this section the players plan their optimal strategies simultaneously. All of the game-theoretic problems are reduced to linear programming problems that can be solved by either the geometric method (Chapter 3) or the simplex method (Chapter 4).

R's Optimal Mixed Strategy

Consider the game with this payoff matrix.

$$P = \begin{bmatrix} 3 & 2 \\ 1 & 3 \end{bmatrix}$$

We will show how to write it as a linear programming problem. Let *R*'s optimal strategy be $A = [x \quad y]$ and let *C*'s optimal strategy be $B = \begin{bmatrix} u \\ v \end{bmatrix}$. The expected value of the game is

$$\begin{aligned} E = APB &= [x \quad y] \begin{bmatrix} 3 & 2 \\ 1 & 3 \end{bmatrix} \begin{bmatrix} u \\ v \end{bmatrix} \\ &= [3x + y \quad 2x + 3y] \begin{bmatrix} u \\ v \end{bmatrix} \\ &= (3x + y)u + (2x + 3y)v \end{aligned}$$

Player C must choose either column 1 or column 2. If the choice is column 1, then $u = 1$ and $v = 0$, so

$$E = 3x + y \tag{1}$$

If the choice is column 2, then $u = 0$ and $v = 1$, so

$$E = 2x + 3y \tag{2}$$

The goal for R is to choose the row that maximizes these two values of E. Let e be the smaller of the two values. Then $E \geq e$, so R's goal will be to maximize e since C will always choose the column producing the smaller value of E. Moreover, $e > 0$ since the entries of P are all positive.

The inequality $E \geq e$ leads to the objective function and constraints of a linear programming problem. Equations 1 and 2 become inequalities.

$$3x + y \geq e$$
$$2x + 3y \geq e$$

Divide each inequality by e.

$$3\left(\frac{x}{e}\right) + \left(\frac{y}{e}\right) \geq 1$$

$$2\left(\frac{x}{e}\right) + 3\left(\frac{y}{e}\right) \geq 1$$

Substitute $X = x/e$ and $Y = y/e$.

$$3X + Y \geq 1 \tag{3}$$
$$2X + 3Y \geq 1 \tag{4}$$

Since R's strategy is $A = [x \quad y]$ it follows that $x \geq 0$ and $y \geq 0$. Then $x/e \geq 0$ and $y/e \geq 0$, hence

$$X \geq 0 \tag{5}$$
$$Y \geq 0 \tag{6}$$

Equations 3 to 6 represent the constraints of a linear programming problem.

Since R's strategy is $[x \quad y]$ it follows that $x + y = 1$. Divide both sides by e, then substitute $X = x/e$ and $Y = y/e$ to obtain the objective function

$$X + Y = 1/e \tag{7}$$

Here is a crucial observation that will be explained in the exercises.

Maximizing e is the same thing as minimizing $1/e$

R's goal of maximizing e translates into minimizing $1/e$. From equation 7 the objective function is

$$\text{Minimize } X + Y \tag{8}$$

Statements 8 and 3 to 6 define a linear programming problem in two variables, X and Y.

Minimize $X + Y$ subject to
$$3X + Y \geq 1$$
$$2X + 3Y \geq 1$$
$$X, Y \geq 0$$

The problem can be solved by the geometric method of Chapter 3 or the dual method of Chapter 4. Figure 1 shows the solution $X = \frac{2}{7}$ and $Y = \frac{1}{7}$ by the geometric method.

However, R's optimal strategy involves x and y. It follows from equation 7 that $1/e = \frac{2}{7} + \frac{1}{7} = \frac{3}{7}$, so the expected value is $e = \frac{7}{3}$. Since $X = x/e$ it follows that $x = eX = (\frac{7}{3})(\frac{2}{7}) = \frac{2}{3}$. Similarly, $y = eY = \frac{1}{3}$. Therefore, the optimal strategy for R is $A = [\frac{2}{3} \quad \frac{1}{3}]$.

The strategy for R is called the *optimal mixed strategy* because it also takes into account C's optimal strategy. The general procedure is described in the display.

Procedure to Determine the Optimal Mixed Strategy for R

Let the payoff matrix for a game with positive entries be

$$P = \begin{bmatrix} q & r \\ s & t \end{bmatrix}$$

1. Solve the linear programming problem.

   Minimize $X + Y$ subject to
   $$qX + sY \geq 1$$
   $$rX + tY \geq 1$$
   $$X, Y \geq 0$$

2. The expected value is $e = 1/(X + Y)$.
3. The optimal mixed strategy for R is $[x \quad y]$, where $x = eX$ and $y = eY$.

Note that the coefficient matrix in step 1 is the transpose of P, not P. Example 1 illustrates the entire procedure.

EXAMPLE 1

Problem

Compute R's optimal mixed strategy for the game with payoff matrix P, where

$$P = \begin{bmatrix} 4 & 3 \\ 2 & 4 \end{bmatrix}$$

SOLUTION First solve the linear programming problem.

Minimize $X + Y$ subject to
$$4X + 2Y \geq 1$$
$$3X + 4Y \geq 1$$
$$X, Y \geq 0$$

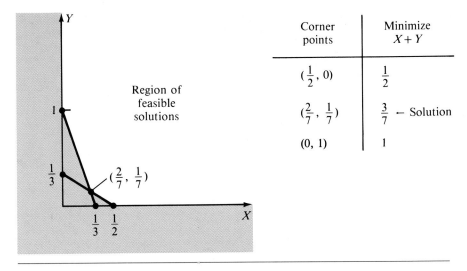

FIGURE 1

The solution is $X = \frac{2}{10}$ and $Y = \frac{1}{10}$. Then $X + Y = \frac{3}{10}$, so $e = \frac{10}{3}$. R's optimal mixed strategy is $[x \quad y]$, where $x = eX = (\frac{10}{3})(\frac{2}{10}) = \frac{2}{3}$ and $y = eY = (\frac{10}{3})(\frac{1}{10}) = \frac{1}{3}$.

C's Optimal Mixed Strategy

Consider the same payoff matrix P that began the section, this time from C's viewpoint.

$$P = \begin{bmatrix} 3 & 2 \\ 1 & 3 \end{bmatrix}$$

The expected value is

$$E = (3x + y)u + (2x + 3y)v$$

Player R must choose either row 1 or row 2. If the choice is row 1, then $x = 1$ and $y = 0$, so

$$E = 3u + 2v \tag{9}$$

If the choice is row 2, then $x = 0$ and $y = 1$, so

$$E = u + 3v \tag{10}$$

The goal for C is to choose the column that minimizes these two values of E. This time let e be the larger of the two values. Then $E \leq e$, so C's goal is to minimize e. Equations 9 and 10 become

$$3u + 2v \leq e$$
$$u + 3v \leq e$$

Divide by e and substitute $U = u/e$ and $V = v/e$.

$$3U + 2V \le 1 \tag{11}$$

$$U + 3V \le 1 \tag{12}$$

Since $u \ge 0$, $v \ge 0$, and $e > 0$, it follows that

$$U \ge 0 \tag{13}$$

$$V \ge 0 \tag{14}$$

Divide both sides of the formula $u + v = 1$ by e, then substitute $U = u/e$ and $V = v/e$.

$$U + V = \frac{1}{e} \tag{15}$$

Since minimizing e is the same thing as maximizing $1/e$, C's goal of minimizing e translates into maximizing $1/e$, so the objective function becomes

$$\text{Maximize } U + V \tag{16}$$

Statements 16 and 11 to 14 define a linear programming problem whose solution is $U = \frac{1}{7}$ and $V = \frac{2}{7}$. It follows from equation 15 that $1/e = \frac{3}{7}$, so $e = \frac{7}{3}$. Then $u = eU = (\frac{7}{3})(\frac{1}{7}) = \frac{1}{3}$ and $v = eV = \frac{2}{3}$. The optimal strategy for C is $\begin{bmatrix} \frac{1}{3} \\ \frac{2}{3} \end{bmatrix}$.

Procedure to Determine the Optimal Mixed Strategy for C

Let the payoff matrix for a game with positive entries be

$$P = \begin{bmatrix} q & r \\ s & t \end{bmatrix}$$

1. Solve the linear programming problem.

   Maximize $U + V$ subject to
   $$qU + rV \le 1$$
   $$sU + tV \le 1$$
   $$U, V \ge 0$$

2. The expected value is $e = 1/(U + V)$.

3. The optimal mixed strategy for C is $\begin{bmatrix} u \\ v \end{bmatrix}$ where $u = eU$ and $v = eV$.

The coefficient matrix in the linear programming problem is the payoff matrix P.

EXAMPLE 2

Problem

Compute C's optimal mixed strategy for the game with payoff matrix P, where

$$P = \begin{bmatrix} 4 & 3 \\ 2 & 4 \end{bmatrix}$$

SOLUTION First solve the linear programming problem.

Maximize $U + V$ subject to
$$4U + 3V \le 1$$
$$2U + 4V \le 1$$
$$U, V \ge 0$$

The solution is $U = \frac{1}{10}$, $V = \frac{2}{10}$. Then $U + V = \frac{3}{10}$, so $e = \frac{10}{3}$. C's optimal mixed strategy is $\begin{bmatrix} u \\ v \end{bmatrix}$, where $u = eU = (\frac{10}{3})(\frac{1}{10}) = \frac{1}{3}$ and $v = eV = (\frac{10}{3})(\frac{2}{10}) = \frac{2}{3}$.

Combined Strategies

Two players' optimal mixed strategies occur when they carry out their strategies simultaneously. The expected value of the game using R's strategy is the same as the expected value of the game using C's strategy. This common number is called the *value* of the game. The value of the game in Examples 1 and 2 is $\frac{10}{3}$. Example 3 considers the two players' optimal mixed strategies simultaneously.

EXAMPLE 3

Let the payoff matrix for a game be $P = \begin{bmatrix} 2 & 5 \\ 4 & 3 \end{bmatrix}$.

Problem

Find the optimal mixed strategies for R and C, and find the value of the game.

SOLUTION The optimal mixed strategy for R is obtained by solving the linear programming problem.

Minimize $X + Y$ subject to
$$2X + 4Y \ge 1$$
$$5X + 3Y \ge 1$$
$$X, Y \ge 0$$

The solution is $X = \frac{1}{14}$ and $Y = \frac{3}{14}$. The expected value is $e = 1/(X + Y) = \frac{7}{2}$. Thus the value of the game is $\frac{7}{2}$. Then $x = eX = \frac{1}{4}$ and $y = eY = \frac{3}{4}$, so R's optimal mixed strategy is $[\frac{1}{4} \quad \frac{3}{4}]$.

For C solve the linear programming problem.

Maximize $U + V$ subject to
$$2U + 5V \le 1$$
$$4U + 3V \le 1$$
$$U, V \ge 0$$

The solution is $U = \frac{1}{7}$ and $V = \frac{1}{7}$. Then $u = eU = (\frac{7}{2})(\frac{1}{7}) = \frac{1}{2}$ and $v = eV = \frac{1}{2}$. The optimal mixed strategy for C is $\begin{bmatrix} \frac{1}{2} \\ \frac{1}{2} \end{bmatrix}$.

All of the games in this section have consisted of payoff matrices with positive entries. This assumption was necessary to guarantee that the expected value of the game was positive. What can be done when some of the entries of the payoff matrix P are negative?

As we saw in Section 10–2, the procedure is to add a sufficiently large number n to the entries of P to produce a payoff matrix Q with positive entries. The strategies and value of the game with payoff matrix Q are solved by the methods of this section. The optimal mixed strategies for a game with payoff matrix P are the same as the optimal mixed strategies for a game with payoff matrix Q. However, the value of the game with payoff matrix P is equal to the value of the game with payoff matrix Q minus n. This procedure is verified in the exercises.

EXERCISE SET 10–4

• In Problems 1 to 4 compute APB.

1. $A = [x \quad y]$, $P = \begin{bmatrix} 2 & 3 \\ -1 & 0 \end{bmatrix}$, $B = \begin{bmatrix} 1 \\ 0 \end{bmatrix}$

2. $A = [1 \quad 0]$, $P = \begin{bmatrix} 2 & 3 \\ -1 & 0 \end{bmatrix}$, $B = \begin{bmatrix} u \\ v \end{bmatrix}$

3. $A = [x \quad y]$, $P = \begin{bmatrix} 2 & 0 \\ -1 & 4 \end{bmatrix}$, $B = \begin{bmatrix} u \\ v \end{bmatrix}$

4. $A = [x \quad y]$, $P = \begin{bmatrix} -1 & 2 \\ 3 & -4 \end{bmatrix}$,

$B = \begin{bmatrix} u \\ v \end{bmatrix}$

In Problems 5 and 6 substitute $X = x/e$, $Y = y/e$, and $1/e = X + Y$ into the stated linear programming problem to form a linear programming problem in the variables X and Y.

5. Maximize e subject to
   $4x + y \le e$
   $5x + 2y \le e$
   $x, y \ge 0$

6. Maximize e subject to
   $2x + 5y \le e$
   $6x + y \le e$
   $x, y \ge 0$

In Problems 7 to 10 find the optimal mixed strategy for R for the game with the stated payoff matrix.

7. $\begin{bmatrix} 1 & 0 \\ 2 & 4 \end{bmatrix}$ 8. $\begin{bmatrix} 4 & 5 \\ 5 & 2 \end{bmatrix}$

9. $\begin{bmatrix} 2 & 3 \\ 3 & 0 \end{bmatrix}$ 10. $\begin{bmatrix} 4 & 2 \\ 1 & 5 \end{bmatrix}$

In Problems 11 to 14 find the optimal mixed strategy for C for the game with the payoff matrices in Problems 7 to 10.

In Problems 15 to 18 find the optimal mixed strategies for R and C simultaneously for the game with the given payoff matrix.

15. $\begin{bmatrix} 3 & 1 \\ 2 & 4 \end{bmatrix}$ 16. $\begin{bmatrix} 1 & 2 \\ 3 & 1 \end{bmatrix}$

17. $\begin{bmatrix} 5 & 2 \\ 3 & 1 \end{bmatrix}$ 18. $\begin{bmatrix} 5 & 2 \\ 3 & 4 \end{bmatrix}$

•• In Problems 19 to 24 find the optimal mixed strategies for R and C simultaneously for the game with the given payoff matrix, some of whose entries are negative.

19. $\begin{bmatrix} -2 & 1 \\ 0 & -1 \end{bmatrix}$ 20. $\begin{bmatrix} -4 & -1 \\ -2 & -3 \end{bmatrix}$

21. $\begin{bmatrix} 1 & 2 \\ -1 & 0 \end{bmatrix}$
22. $\begin{bmatrix} 2 & 0 \\ -1 & 1 \end{bmatrix}$

23. $\begin{bmatrix} 2 & 1 \\ -1 & -2 \end{bmatrix}$
24. $\begin{bmatrix} 3 & 1 \\ -2 & 1 \end{bmatrix}$

In Problems 25 to 28 verify the stated solution to the given example in the text.

25. In Example 1 the solution is $X = \frac{2}{10}$ and $Y = \frac{1}{10}$.

26. In Example 2 the solution is $U = \frac{1}{10}$ and $V = \frac{2}{10}$.

27. In Example 3 the solution for player R is $X = \frac{1}{14}$ and $Y = \frac{3}{14}$.

28. In Example 3 the solution for player C is $U = \frac{1}{7}$ and $V = \frac{1}{7}$.

Problems 29 to 34 examine the assertion, "maximizing e is the same thing as minimizing $1/e$." Let S be the given set. Find the (a) maximum value of e and (b) the minimum value of $1/e$, for all numbers e in S.

29. $S = \{3, 4, 5, 6, 7\}$

30. $S = \{10, 100, 100\}$

31. $S = \{\frac{1}{2}, \frac{1}{3}, \frac{1}{4}\}$

32. $S = \{1.1, 2.2, 3.3\}$

33. $S = \{\frac{1}{2}, \frac{1}{3}, \frac{1}{4}, \ldots\}$

34. $S = \{\frac{2}{1}, \frac{3}{2}, \frac{4}{3}, \ldots\}$

In Problems 35 to 38 let S be the given set. Find the (a) minimum value of e and (b) the maximum value of $1/e$, for all numbers e in S.

35. $S = \{3, 4, 5, 6, 7\}$

36. $S = \{1, 2, \ldots, 100\}$

37. $S = \{.3, .33, .333, \ldots\}$

38. $S = \{\frac{1}{2}, \frac{2}{3}, \frac{3}{4}, \ldots\}$

●●● 39. In an article describing competitive encounters among firms, Teng and Thompson use a game theoretic approach to model how two firms use level of advertising to attempt to increase market share.* The players are two firms, and their strategies involve whether or not to increase advertising. In one part of the article they model two firms that manufacture pocket calculators.

The payoff is the percent of market gained by firm 1, the row player. A typical payoff matrix is

$$\text{Firm 1} \begin{cases} \text{increase advertising} \\ \text{keep the same level} \end{cases} \overset{\displaystyle \overbrace{\text{increase} \quad \text{same}}^{\text{Firm 2}}}{\begin{bmatrix} 40 & 60 \\ 55 & 45 \end{bmatrix}}$$

Find the optimal mixed strategies for each player.

40. An agricultural cooperative represents an attempt by farmers, each of whom has a different set of resources and goals, to compete together in the food market. Staatz uses game theory to illustrate how cooperatives can assess fees equitably, depending upon the effort and support given to the cooperative by each farmer.† In one example, a cooperative consisting of four farmers must assess a fee to each member. The farmers form two coalitions, A and B, to increase their bargaining power. The members in each coalition will pay the same fee. The coalitions try to influence cost allocations through threats and counterthreats. The ability of a coalition to obtain concessions from other members depends upon the costs the coalition could impose on the other members if they were to exit the cooperative or cut back on the level of cooperation and effort extended. Each threat has an implied counterthreat, the cost the threatening coalition imposes on itself by withdrawing partially or totally from the cooperative. Each coalition has two strategies: threaten to withdraw or do not exercise the threat. The payoff in the game is the relative gain to coalition A, which is the relative cost to coalition A. A typical payoff matrix is

$$A \begin{cases} \text{exercise threat} \\ \text{do not threaten} \end{cases} \overset{\displaystyle \overbrace{\text{threat} \quad \text{no threat}}^{B}}{\begin{bmatrix} 5 & 15 \\ 20 & 10 \end{bmatrix}}$$

Find the optimal mixed strategies for each player.

*Jinn-tsair Teng and Gerald L. Thompson, "Oligopoly Models for Optimal Advertising When Production Costs Obey a Learning Curve," *Management Science*, 29, 1983, pp. 1087–1100.

†John M. Staatz, "The Cooperative as a Coalition: A Game-theoretic Approach," *American Journal of Agricultural Economics*, 65, 1983, pp. 1084–1089.

Game theory has been applied in many areas involving human competition. One of the most useful applications has been in the resolution of conflicts in various types of corporate mergers. When two companies explore the possibility of a merger it is often the division of the assets—who gets what—that is the major stumbling block.

A classic example is the celebrated case of the merger of Norway's three independent cement companies in the 1960s. The companies, which we will call Alpha, Beta, and Delta, were asked to consider consolidating into one large national company. The central question was how to divide the ownership of the new company. What payoff, that is, what share of the new company, should be assigned to each of the participants?

Another significant consideration was that each company had the right to remain independent or form a coalition with one other company, leaving the third company independent. For example, it was possible that Alpha and Beta would form one company and Delta would remain independent. The latter development would have been devastating for Delta because it was by far the smallest company. However, that was very unlikely to occur, because the owners of Alpha and Beta could not even come close to agreement on the division of the new company. Thus Delta, being much more flexible in its negotiations, had significant bargaining power by threatening to form a coalition with one company against the other.

Even though it was clear to all of the owners that each would gain more from merging all three companies, the negotiations dragged on and almost collapsed because they could not agree upon an ownership split that was acceptable to all. Alpha argued that the present value of future earnings should be the main criteria, in which case it should receive 52% of the ownership. Beta claimed that production capacity was the best measure to use to split ownership, in which case it would get 47% (as opposed to 38% under Alpha's plan). Delta wanted the average production capacity over the previous decade to be the primary measurement, in which case it would get 13% (versus 8% and 9% under the other plans).

The breakthrough occurred when the government's consultant used a complicated formula, called the *Shapley value,* to split the ownership. It used all three measurements. But its real significance was that it took into account each company's bargaining power—the ability to threaten to enter into a coalition with one other company, and thereby potentially affecting the third.

The consultant explained the situation by using the following game-theoretic approach. Consider the merger negotiation from Alpha's point of view. The owners of Alpha had three options: (1) to remain independent (I), (2) to merge with Beta (M/B), or (3) to merge with Delta (M/D). Let the owners of Alpha be the row player. The column "player" is then Beta and Delta, in the sense that the column player has two choices—to remain independent (I) or to merge (M). Hence there are six options. They can be expressed in a 3 × 2 payoff matrix, where the entries represent the percentage of business that Alpha would receive. For example, the (1, 1) entry, 52, represents the consultant's calculation that Alpha would get 52% of Norway's cement business if all three remained independent. If Alpha remained independent but Beta and Delta merged, then Alpha would receive 40% of the business, so the (1, 2) entry is 40.

Beta and Delta's Decision

		I	M
	I	52	40
Alpha's Decision	M/B	44	48
	M/D	46	48

Note that the (2, 2) and (3, 2) entries represent the same eventuality, that all three merge, and Alpha would then get 48% ownership in the company. Given this description of the merger, what should Alpha decide? First notice that Alpha would not even consider the second option, because both entries in the third row are greater than or equal to the entries in the second row. The game-theoretic terminology is that row 3 "dominates" row 2, so if Alpha sought a merger it would be with Delta and not with Beta. Thus we can ignore row 2 and consider the payoff matrix P.

Beta and Delta

$$\text{Alpha} \begin{cases} I \\ M/D \end{cases} \quad P = \begin{bmatrix} 52 & 40 \\ 46 & 48 \end{bmatrix} \begin{matrix} I \\ M \end{matrix}$$

We can now solve the problem by using our procedure to find Alpha's optimum mixed strategy, $[x \quad y]$. If one number is significantly greater than the other, then Alpha should choose the option corresponding to the larger number. However, when we translate the problem into a linear programming problem it becomes a minimization problem.

The problem can be expressed as the linear programming problem.

Minimize $X + Y$ subject to
$$52X + 46Y \geq 1$$
$$40X + 48Y \geq 1$$
$$X, Y \geq 0$$

The inequalities are graphed in Figure 1.

We then test the corner points.

Corner Points	$X + Y$
(1/40, 0)	1/40
(1/328, 3/164)	7/328
(0, 1/46)	1/46

The smallest number is 7/328, so the solution is (1/328, 3/164). Remember that

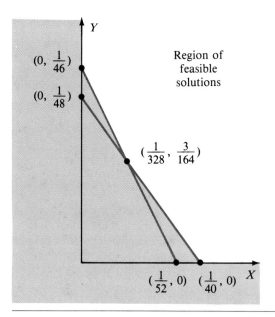

FIGURE 1

we want to maximize the expected value e. The corresponding linear programming problem finds the minimum of $X + Y = 1/e$. Therefore

$$e = \frac{1}{X + Y} = \frac{328}{7}$$

and so

$$x = eX = \left(\frac{328}{7}\right)\left(\frac{1}{328}\right) = \frac{1}{7}$$

$$y = eY = \left(\frac{328}{7}\right)\left(\frac{3}{164}\right) = \frac{6}{7}$$

Alpha's decision was clear-cut. The second option was the most favorable, so they decided to merge. The result was similar from Beta and Delta's perspective and the companies formed one national firm. From the Shapley value the owners of Alpha received 48% of the new company. Similar reasoning gave 40% to Beta and 12% to Delta.

EPILOGUE Dantzig Meets Von Neumann

In a delightful article entitled "Reminiscences About the Origins of Linear Programming," George Dantzig gave us an account of how John von Neumann and he pooled their resources to tackle two intriguing problems.* Dantzig had traveled

*Arthur Schlissel (ed.), *Essays in the History of Mathematics, Memoirs of the American Mathematical Society,* 298, 1984, pp. 2–16.

across the country in the late 1940s eliciting help in solving problems that later became known as linear programming problems. In 1946 he was the Mathematics Adviser to the U.S. Air Force Comptroller, having just completed his Ph.D. He was trying to mechanize the planning process for the Air Force. This led him to formulate a model based on the work of the well-known economist, Wassily Leontief. The solution to the problem eluded him because of the extraordinary number of calculations involved. For example, the simple model of assigning 70 people to 70 jobs involves 70! possible solutions—and 70! is larger than 10^{100}. By mid-1947 the problem had been stated with sufficient clarity that he could seek help from many diverse quarters. But it was his conversation with von Neumann that was particularly fruitful.

Imagine a young mathematician, barely out of graduate school, calling on one of the greatest living scientists. In his own words, here is Dantzig's recollection of the dramatic meeting:

> I decided to consult with the ''great'' Johnny von Neumann to see what he could suggest in the way of solution techniques. He was considered by many as the leading mathematician in the world. On October 3, 1947, I visited him for the first time at the Institute for Advanced Study at Princeton. I remember trying to describe to von Neumann, as I would an ordinary mortal, the Air force problem. I began with the formulation of the linear programming model in terms of activities and items, etc. Von Neumann did something which I believe was uncharacteristic of him. ''Get to the point,'' he said impatiently. Having at times a somewhat low kindling-point, I said to myself ''O.K., if he wants a quicky, then that's what he'll get.'' In under one minute I slapped the geometric and the algebraic version of the problem on the blackboard. Von Neumann stood up and said ''Oh that!.'' Then for the next hour and a half he proceeded to give me a lecture on the mathematical theory of linear programs.

> At one point, seeing me sitting there with my eyes popping and my mouth open (after all I had searched the literature and found nothing), von Neumann said: ''I don't want you to think that I am pulling all this out of my sleeve on the spur of the moment like a magician. I have just recently completed a book with Oscar Morgenstern on the theory of games. What I am doing is conjecturing that the two problems are equivalent. The theory that I am outlining for your problem is an analogue to the one we have developed for games.''

CHAPTER REVIEW

Key Terms

10–1 Strictly Determined Games

Payoff	Mixed Strategy
Payoff Matrix	Maximin
Game	Minimax
Two-Person Game	Saddle Point
Zero-Sum Game	Strictly Determined Game
Strategy	Value
Pure Strategy	

REVIEW PROBLEMS

1. Determine the optimum strategies for R and C and the value of the strictly determined game whose payoff matrix is given.

$$\begin{bmatrix} -3 & 1 & 0 \\ -1 & 6 & -3 \\ 2 & 3 & 1 \end{bmatrix}$$

In Problems 2 and 3 determine if the game with the given payoff matrix is a strictly determined game. If so, find its value.

2. $\begin{bmatrix} 0 & 2 & -1 \\ -4 & -1 & -2 \\ 5 & 3 & 0 \end{bmatrix}$

3. $\begin{bmatrix} 1 & -2 & -2 & 4 \\ -3 & 0 & 0 & -1 \\ -4 & -4 & 1 & -1 \\ 3 & 4 & 2 & 3 \end{bmatrix}$

In Problems 4 and 5 a game is described. Find a payoff matrix for the game and decide if it is a strictly determined game. If it is strictly determined, find the value of the game.

4. Each of two players, A and B, has 2 cards, a 2 and a 6. Each player selects a card and they simultaneously show their cards. If the cards match, then A pays B \$2. If A's card is greater than B's, then A wins \$4 from B. If B's card is greater than A's, then B wins \$5 from A.

5. Player A has 2 cards, a 2 and a 5. Player B has 3 cards, a 2, a 5, and a 7. Each player selects a card and they simultaneously show the cards. If the sum is even, then A pays B \$3. If the sum is odd, then B pays A \$8.

6. The strategies for R and C are given. Determine the size of the payoff matrix and compute the probability of each entry being played.

$$R = [.7 \quad .1 \quad .2], \qquad C = \begin{bmatrix} .4 \\ .6 \end{bmatrix}$$

In Problems 7 and 8 calculate the expected value for the game with payoff matrix P and the given strategies.

7. $P = \begin{bmatrix} 1 & 2 \\ 2 & -4 \end{bmatrix}$

$[.5 \quad .5], \qquad \begin{bmatrix} .4 \\ .6 \end{bmatrix}$

8. $P = \begin{bmatrix} -1 & 3 \\ 0 & -1 \\ -1 & 2 \end{bmatrix}$

$[.1 \quad .7 \quad .2], \qquad \begin{bmatrix} .5 \\ .5 \end{bmatrix}$

9. A game has payoff matrix

$$P = \begin{bmatrix} 1 & -2 & 3 \\ 0 & -1 & 2 \end{bmatrix}$$

Let C play the given strategy and let R play the strategy $[a \quad b]$. Determine what values of a and b will maximize the payoff to R.

$$\begin{bmatrix} .7 \\ .2 \\ .1 \end{bmatrix}$$

10. Add the smallest possible integer to the matrix to get a matrix with all positive entries; that is, choose the number such that the smallest entry is 1.

$$\begin{bmatrix} -5 & 2 & -1 \\ 1 & -6 & 4 \\ 2 & -9 & -1 \end{bmatrix}$$

11. Add the smallest possible integer to the payoff

matrix P to obtain the payoff matrix Q with all positive entries and compute the expected value using the matrices A and B, the strategies for R and C, respectively. How are the two expected values related?

$$P = \begin{bmatrix} -1 & 3 \\ -2 & 2 \\ 3 & -4 \end{bmatrix}, \qquad A = [.5 \quad .2 \quad .3],$$

$$B = \begin{bmatrix} .6 \\ .4 \end{bmatrix}$$

12. The payoff matrix of a game is given. Compute the expected value of the game if C chooses (a) column 1 and (b) column 2.

$$\begin{bmatrix} 3 & 1 \\ 2 & 4 \end{bmatrix}$$

In Problems 13 and 14 two expressions for E are given. The first corresponds to C choosing column 1 and the second corresponds to C choosing column 2. Graph each expression and find the point of intersection.

13. $E = x + 3y$ and $E = x + y$

14. $E = 4x + 2y$ and $E = 3x + 5y$

15. For the given payoff matrix compute the optimal mixed strategy for R that maximizes the expected value, given that C will select the best counterstrategy. Then compute the payoff.

$$\begin{bmatrix} 3 & 1 \\ 1 & 2 \end{bmatrix}$$

16. Find the optimum strategy for both R and C for

the game with the given payoff matrix.

$$\begin{bmatrix} 4 & 2 \\ 1 & \frac{1}{2} \end{bmatrix}$$

17. Find a payoff matrix for the game and find the optimum strategy for the row player and the column player. Each of two players has a coin. They match their coins by turning up either heads or tails. If the same side appears on each coin, then A pays B \$3. If A turns up heads and B turns up tails, then A pays B \$2. If A turns up tails and B turns up heads, then B pays A \$1.

18. Substitute $X = x/e$, $Y = y/e$, and $1/e = X + Y$ into the stated linear programming problem to form a linear programming problem in the variables X and Y.

Maximize e subject to
$$3x + 4y \le e$$
$$2x + 7y \le e$$
$$x, y \ge 0$$

19. Find the optimal mixed strategy for R for the game with the stated payoff matrix.

$$\begin{bmatrix} 1 & 4 \\ 2 & 3 \end{bmatrix}$$

20. Find the optimal mixed strategies for R and C for the game with the following payoff matrix by translating the problem into a linear programming problem and solving it.

$$\begin{bmatrix} 4 & 5 \\ 1 & 2 \end{bmatrix}$$

Appendix:
Tables

TABLE 1   Compound Interest Table

n	2%	4%	6%	8%	10%	12%	14%
1	1.02000	1.04000	1.06000	1.08000	1.10000	1.12000	1.14000
2	1.04040	1.08160	1.12360	1.16640	1.21000	1.25440	1.29960
3	1.06121	1.12486	1.19102	1.25971	1.33100	1.40493	1.48154
4	1.08243	1.16986	1.26248	1.36049	1.46410	1.57352	1.68896
5	1.10408	1.21665	1.33823	1.46933	1.61051	1.76234	1.92541
6	1.12616	1.26532	1.41852	1.58687	1.77156	1.97382	2.19497
7	1.14869	1.31593	1.50363	1.71382	1.94872	2.21068	2.50227
8	1.17166	1.36857	1.59385	1.85093	2.14359	2.47596	2.85258
9	1.19509	1.42331	1.68948	1.99900	2.35795	2.77308	3.25194
10	1.21900	1.48024	1.79085	2.15892	2.59374	3.10584	3.70722
11	1.24338	1.53945	1.89830	2.33164	2.85312	3.47854	4.22622
12	1.26824	1.60103	2.01220	2.51817	3.13843	3.89597	4.81790
13	1.29361	1.66507	2.13293	2.71962	3.45227	4.36349	5.49240
14	1.31948	1.73167	2.26090	2.93719	3.79750	4.88710	6.26133
15	1.34587	1.80094	2.39656	3.17217	4.17725	5.47355	7.13792
16	1.37279	1.87298	2.54035	3.42594	4.59497	6.13038	8.13723
17	1.40024	1.94790	2.69277	3.70002	5.05447	6.86602	9.27644
18	1.42825	2.02581	2.85434	3.99602	5.55992	7.68995	10.57510
19	1.45681	2.10684	3.02560	4.31570	6.11591	8.61274	12.05570
20	1.48595	2.19112	3.20713	4.66095	6.72750	9.64627	13.74340
21	1.51567	2.27876	3.39956	5.03383	7.40025	10.80380	15.66750
22	1.54598	2.36991	3.60354	5.43654	8.14027	12.10030	17.86100
23	1.57690	2.46471	3.81975	5.87146	8.95430	13.55230	20.36150
24	1.60844	2.56330	4.04893	6.34117	9.84973	15.17860	23.21210
25	1.64061	2.66583	4.29187	6.84847	10.83470	17.00000	26.46180
26	1.67342	2.77246	4.54938	7.39634	11.91820	19.04000	30.16640
27	1.70689	2.88336	4.82234	7.98805	13.11000	21.32480	34.38970
28	1.74103	2.99869	5.11168	8.62709	14.42100	23.88380	39.20430
29	1.77585	3.11864	5.41839	9.31726	15.86310	26.74980	44.69290
30	1.81136	3.24339	5.74349	10.06260	17.44940	29.95980	50.94990
31	1.84759	3.37312	6.08810	10.86770	19.19430	33.55500	58.08290
32	1.88454	3.50804	6.45338	11.73710	21.11380	37.58160	66.21450
33	1.92223	3.64837	6.84059	12.67600	23.22510	42.09130	75.48450
34	1.96068	3.79430	7.25102	13.69010	25.54770	47.14230	86.05230
35	1.99989	3.94607	7.68608	14.78530	28.10240	52.79940	98.09960
36	2.03989	4.10391	8.14725	15.96810	30.91270	59.13530	111.83300
37	2.08069	4.26807	8.63608	17.24560	34.00390	66.23150	127.49000
38	2.12230	4.43879	9.15425	18.62520	37.40430	74.17930	145.33900
39	2.16475	4.61634	9.70350	20.11530	41.14480	83.08080	165.68600
40	2.20804	4.80100	10.28570	21.72450	45.25920	93.05040	188.88200
41	2.25220	4.99304	10.90290	23.46240	49.78520	104.21600	215.32600
42	2.29725	5.19276	11.55700	25.33940	54.76370	116.72200	245.47100
43	2.34319	5.40047	12.25040	27.36660	60.24000	130.72900	279.83700
44	2.39006	5.61648	12.98550	29.55590	66.26410	146.41700	319.01400
45	2.43786	5.84114	13.76460	31.92040	72.89050	163.98700	363.67600
46	2.48662	6.07479	14.59050	34.47400	80.17950	183.66500	414.59100
47	2.53635	6.31778	15.46590	37.23190	88.19750	205.70500	472.63300
48	2.58708	6.57049	16.39390	40.21050	97.01720	230.38900	538.80200
49	2.63882	6.83331	17.37750	43.42730	106.71900	258.03600	614.23400
50	2.69160	7.10664	18.42010	46.90150	117.39100	289.00000	700.22700

TABLE 2   Present Value I Table
(Compounded Annually)

| | | | | r | | | |
n	2%	4%	6%	8%	10%	12%	14%
1	0.980392	0.961539	0.943396	0.925926	0.909091	0.892857	0.877193
2	0.961169	0.924556	0.889996	0.857339	0.826446	0.797194	0.769468
3	0.942322	0.888997	0.839619	0.793832	0.751315	0.711781	0.674972
4	0.923845	0.854805	0.792094	0.735030	0.683014	0.635519	0.592081
5	0.905731	0.821928	0.747258	0.680583	0.620921	0.567427	0.519369
6	0.887971	0.790315	0.704961	0.630170	0.564474	0.506632	0.455587
7	0.870560	0.759918	0.665057	0.583491	0.513158	0.452350	0.399638
8	0.853490	0.730691	0.627413	0.540269	0.466507	0.403884	0.350560
9	0.836755	0.702588	0.591899	0.500249	0.424098	0.360611	0.307508
10	0.820348	0.675565	0.558395	0.463194	0.385543	0.321974	0.269744
11	0.804263	0.649582	0.526788	0.428883	0.350494	0.287477	0.236618
12	0.788493	0.624598	0.496969	0.397114	0.318631	0.256676	0.207560
13	0.773032	0.600575	0.468839	0.367698	0.289664	0.229175	0.182070
14	0.757875	0.577476	0.442301	0.340461	0.263331	0.204620	0.159710
15	0.743014	0.555266	0.417265	0.315242	0.239392	0.182697	0.140097
16	0.728445	0.533909	0.393646	0.291891	0.217629	0.163122	0.122892
17	0.714162	0.513374	0.371365	0.270269	0.197845	0.145645	0.107800
18	0.700159	0.493629	0.350344	0.250249	0.179859	0.130040	0.094561
19	0.686430	0.474644	0.330513	0.231712	0.163508	0.116107	0.082949
20	0.672971	0.456388	0.311805	0.214548	0.148644	0.103667	0.072762
21	0.659775	0.438835	0.294156	0.198656	0.135131	0.092560	0.063826
22	0.646838	0.421957	0.277505	0.183941	0.122846	0.082643	0.055988
23	0.634155	0.405728	0.261797	0.170315	0.111678	0.073788	0.049112
24	0.621721	0.390123	0.246979	0.157700	0.101526	0.065882	0.043081
25	0.609530	0.375118	0.232999	0.146018	0.092296	0.058824	0.037790
26	0.597579	0.360690	0.219810	0.135202	0.083905	0.052521	0.033149
27	0.585861	0.346818	0.207368	0.125187	0.076278	0.046894	0.029078
28	0.574374	0.333479	0.195630	0.115914	0.069343	0.041869	0.025507
29	0.563112	0.320653	0.184557	0.107328	0.063039	0.037383	0.022375
30	0.552070	0.308320	0.174110	0.099378	0.057309	0.033378	0.019627
31	0.541245	0.296461	0.164255	0.092016	0.052099	0.029802	0.017217
32	0.530633	0.285059	0.154957	0.085200	0.047362	0.026609	0.015102
33	0.520228	0.274095	0.146186	0.078889	0.043057	0.023758	0.013248
34	0.510027	0.263553	0.137912	0.073045	0.039143	0.021212	0.011621
35	0.500027	0.253417	0.130105	0.067635	0.035584	0.018940	0.010194
36	0.490222	0.243670	0.122741	0.062625	0.032349	0.016910	0.008942
37	0.480610	0.234298	0.115793	0.057986	0.029408	0.015099	0.007844
38	0.471186	0.225287	0.109239	0.053691	0.026735	0.013481	0.006880
39	0.461947	0.216622	0.103056	0.049714	0.024304	0.012036	0.006036
40	0.452890	0.208290	0.097222	0.046031	0.022095	0.010747	0.005294
41	0.444009	0.200279	0.091719	0.042621	0.020086	0.009595	0.004644
42	0.435303	0.192576	0.086527	0.039464	0.018260	0.008567	0.004074
43	0.426768	0.185169	0.081630	0.036541	0.016600	0.007649	0.003574
44	0.418400	0.178047	0.077009	0.033834	0.015091	0.006830	0.003135
45	0.410196	0.171199	0.072650	0.031328	0.013719	0.006098	0.002750
46	0.402153	0.164615	0.068538	0.029007	0.012472	0.005445	0.002412
47	0.394267	0.158284	0.064658	0.026859	0.011338	0.004861	0.002116
48	0.386537	0.152196	0.060998	0.024869	0.010307	0.004340	0.001856
49	0.378958	0.146342	0.057546	0.023027	0.009370	0.003875	0.001628
50	0.371527	0.140714	0.054288	0.021321	0.008519	0.003460	0.001428

TABLE 3   Present Value II Table
(Compounded Daily)

				r			
n	2%	4%	6%	8%	10%	12%	14%
1	0.980188	0.960870	0.941776	0.923131	0.904856	0.886942	0.869396
2	0.960768	0.923271	0.886942	0.852171	0.818764	0.786666	0.755849
3	0.941733	0.887143	0.835301	0.786666	0.740863	0.697727	0.657133
4	0.923076	0.852429	0.786666	0.726196	0.670374	0.618844	0.571308
5	0.904788	0.819073	0.740863	0.670374	0.606592	0.548878	0.496693
6	0.886862	0.787022	0.697727	0.618844	0.548878	0.486823	0.431823
7	0.869291	0.756226	0.657103	0.571274	0.496656	0.431784	0.375425
8	0.852069	0.726634	0.618844	0.527361	0.449402	0.382967	0.326393
9	0.835187	0.698201	0.582812	0.486823	0.406644	0.339670	0.283765
10	0.818641	0.670880	0.548878	0.449402	0.367954	0.301267	0.246704
11	0.802422	0.644628	0.516921	0.414857	0.332945	0.267207	0.214484
12	0.786524	0.619404	0.486823	0.382967	0.301267	0.236997	0.186471
13	0.770941	0.595166	0.458478	0.353529	0.272604	0.210203	0.162117
14	0.755667	0.571877	0.431784	0.326354	0.246667	0.186437	0.140944
15	0.740696	0.549500	0.406644	0.301267	0.223198	0.165359	0.122536
16	0.726021	0.527997	0.382967	0.278109	0.201962	0.146664	0.106533
17	0.711637	0.507337	0.360670	0.256732	0.182747	0.130083	0.092619
18	0.697538	0.487485	0.339670	0.236997	0.165359	0.115376	0.080523
19	0.683718	0.468409	0.319893	0.218779	0.149626	0.102331	0.070006
20	0.670172	0.450080	0.301267	0.201962	0.135390	0.090762	0.060863
21	0.656895	0.432468	0.283726	0.186437	0.122509	0.080501	0.052914
22	0.643880	0.415546	0.267207	0.172106	0.110853	0.071399	0.046003
23	0.631124	0.399285	0.251649	0.158877	0.100306	0.063327	0.039995
24	0.618620	0.383661	0.236997	0.146664	0.090762	0.056168	0.034772
25	0.606364	0.368648	0.223198	0.135390	0.082127	0.049817	0.030230
26	0.594350	0.354223	0.210203	0.124983	0.074313	0.044185	0.026282
27	0.582575	0.340362	0.197964	0.115376	0.067242	0.039190	0.022849
28	0.571033	0.327044	0.186437	0.106507	0.060845	0.034759	0.019865
29	0.559719	0.314246	0.175582	0.098320	0.055056	0.030829	0.017271
30	0.548630	0.301950	0.165359	0.090762	0.049817	0.027344	0.015015
31	0.537761	0.290134	0.155731	0.083785	0.045078	0.024252	0.013054
32	0.527106	0.278781	0.146664	0.077345	0.040789	0.021510	0.011349
33	0.516663	0.267873	0.138125	0.071399	0.036908	0.019078	0.009867
34	0.506427	0.257391	0.130083	0.065911	0.033396	0.016921	0.008578
35	0.496394	0.247319	0.122509	0.060845	0.030219	0.015008	0.007458
36	0.486559	0.237641	0.115376	0.056168	0.027344	0.013312	0.006484
37	0.476919	0.228342	0.108658	0.051850	0.024742	0.011807	0.005637
38	0.467471	0.219407	0.102331	0.047864	0.022388	0.010472	0.004901
39	0.458209	0.210822	0.096373	0.044185	0.020258	0.009288	0.004261
40	0.449131	0.202572	0.090762	0.040789	0.018330	0.008238	0.003704
41	0.440233	0.194645	0.085478	0.037653	0.016586	0.007306	0.003221
42	0.431511	0.187029	0.080501	0.034759	0.015008	0.006480	0.002800
43	0.422962	0.179710	0.075814	0.032087	0.013580	0.005748	0.002434
44	0.414582	0.172678	0.071399	0.029621	0.012288	0.005098	0.002116
45	0.406368	0.165921	0.067242	0.027344	0.011119	0.004522	0.001840
46	0.398317	0.159429	0.063327	0.025242	0.010061	0.004010	0.001600
47	0.390425	0.153190	0.059640	0.023301	0.009104	0.003557	0.001391
48	0.382690	0.147196	0.056168	0.021510	0.008238	0.003155	0.001209
49	0.375108	0.141436	0.052897	0.019857	0.007454	0.002798	0.001051
50	0.367677	0.135902	0.049817	0.018330	0.006745	0.002482	0.000914

TABLE 4   Annuity Table

n	i 2	4	6	8	10	12
1	1.00000	1.00000	1.00000	1.00000	1.00000	1.00000
2	2.02000	2.04000	2.06000	2.08000	2.10000	2.12000
3	3.06040	3.12160	3.18360	3.24640	3.31000	3.37440
4	4.12161	4.24646	4.37462	4.50611	4.64100	4.77933
5	5.20404	5.41632	5.63709	5.86660	6.10510	6.35285
6	6.30812	6.63297	6.97532	7.33593	7.71561	8.11519
7	7.43428	7.89829	8.39384	8.92280	9.48717	10.08901
8	8.58297	9.21423	9.89747	10.63663	11.43589	12.29969
9	9.75463	10.58279	11.49131	12.48756	13.57948	14.77566
10	10.94972	12.00610	13.18079	14.48657	15.93743	17.54874
11	12.16871	13.48635	14.97164	16.64549	18.53117	20.65458
12	13.41209	15.02580	16.86994	18.97713	21.38429	24.13313
13	14.68033	16.62683	18.88213	21.49530	24.52272	28.02911
14	15.97394	18.29191	21.01506	24.21493	27.97499	32.39260
15	17.29341	20.02358	23.27596	27.15212	31.77249	37.27972
16	18.63928	21.82452	25.67252	30.32429	35.94974	42.75328
17	20.01207	23.69750	28.21287	33.75024	40.54471	48.88368
18	21.41231	25.64540	30.90564	37.45026	45.59918	55.74972
19	22.84055	27.67122	33.75997	41.44628	51.15910	63.43968
20	24.29737	29.77807	36.78557	45.76199	57.27502	72.05245
21	25.78331	31.96919	39.99270	50.42295	64.00252	81.69874
22	27.29898	34.24795	43.39226	55.45678	71.40277	92.50259
23	28.84496	36.61787	46.99579	60.89333	79.54305	104.60290
24	30.42186	39.08258	50.81554	66.76480	88.49736	118.15525
25	32.03029	41.64589	54.86447	73.10599	98.34709	133.33388
26	33.67090	44.31172	59.15633	79.95447	109.18181	150.33395
27	35.34431	47.08419	63.70571	87.35083	121.09999	169.37402
28	37.05120	49.96755	68.52805	95.33890	134.20999	190.69890
29	38.79222	52.96625	73.63973	103.96601	148.63099	214.58277
30	40.56807	56.08490	79.05811	113.28330	164.49410	241.33271
31	42.37943	59.32830	84.80159	123.34597	181.94351	271.29263
32	44.22702	62.70143	90.88968	134.21365	201.13787	304.84775
33	46.11155	66.20948	97.34305	145.95075	222.25166	342.42948
34	48.03379	69.85786	104.18363	158.62682	245.47683	384.52102
35	49.99446	73.65217	111.43464	172.31697	271.02452	431.66354
36	51.99435	77.59825	119.12072	187.10233	299.12698	484.46317
37	54.03423	81.70218	127.26795	203.07053	330.03968	543.59876
38	56.11492	85.97026	135.90402	220.31618	364.04366	609.83061
39	58.23722	90.40907	145.05826	238.94148	401.44803	684.01028
40	60.40196	95.02543	154.76174	259.05681	442.59284	767.09152
41	62.61000	99.82644	165.04744	280.78137	487.85214	860.14251
42	64.86219	104.81950	175.95028	304.24389	537.63736	964.35961
43	67.15944	110.01227	187.50728	329.58341	592.40111	1081.08277
44	69.50263	115.41276	199.75771	356.95010	652.64124	1211.81271
45	71.89268	121.02927	212.74316	386.50612	718.90538	1358.23024
46	74.33053	126.87043	226.50774	418.42663	791.79593	1522.21788
47	76.81714	132.94524	241.09819	452.90078	871.97554	1705.88403
48	79.35348	139.26305	256.56407	490.13286	960.17312	1911.59012
49	81.94055	145.83357	272.95789	530.34351	1057.19045	2141.98094
50	84.57936	152.66690	290.33535	573.77101	1163.90952	2400.01867

TABLE 5   Amortization Table

			i			
n	2	4	6	8	10	12
1	0.98039	0.96154	0.94340	0.92593	0.90909	0.89286
2	1.94156	1.88609	1.83339	1.78326	1.73554	1.69005
3	2.88388	2.77509	2.67301	2.57710	2.48685	2.40183
4	3.80773	3.62990	3.46511	3.31213	3.16987	3.03735
5	4.71346	4.45182	4.21236	3.99271	3.79079	3.60478
6	5.60143	5.24214	4.91733	4.62288	4.35526	4.11141
7	6.47199	6.00206	5.58238	5.20637	4.86842	4.56376
8	7.32548	6.73275	6.20980	5.74664	5.33493	4.96764
9	8.16224	7.43533	6.80169	6.24689	5.75902	5.32825
10	8.98259	8.11090	7.36009	6.71008	6.14457	5.65022
11	9.78685	8.76048	7.88688	7.13896	6.49506	5.93770
12	10.57534	9.38508	8.38385	7.53608	6.81369	6.19437
13	11.34838	9.98565	8.85269	7.90377	7.10336	6.42355
14	12.10625	10.56313	9.29499	8.24423	7.36669	6.62817
15	12.84927	11.11839	9.71225	8.55948	7.60608	6.81086
16	13.57771	11.65230	10.10590	8.85137	7.82371	6.97399
17	14.29187	12.16567	10.47726	9.12164	8.02155	7.11963
18	14.99203	12.65930	10.82761	9.37188	8.20141	7.24967
19	15.67846	13.13394	11.15812	9.60360	8.36492	7.36578
20	16.35144	13.59033	11.46993	9.81814	8.51356	7.46944
21	17.01121	14.02916	11.76408	10.01680	8.64869	7.56200
22	17.65805	14.45112	12.04159	10.20074	8.77154	7.64465
23	18.29221	14.85685	12.30339	10.37106	8.88322	7.71843
24	18.91393	15.24697	12.55036	10.52875	8.98474	7.78432
25	19.52346	15.62209	12.78336	10.67477	9.07704	7.84314
26	20.12104	15.98278	13.00317	10.80997	9.16094	7.89566
27	20.70690	16.32959	13.21054	10.93516	9.23722	7.94255
28	21.28128	16.66307	13.40617	11.05107	9.30656	7.98442
29	21.84439	16.98372	13.59073	11.15840	9.36960	8.02181
30	22.39646	17.29204	13.76484	11.25778	9.42691	8.05518
31	22.93771	17.58850	13.92909	11.34979	9.47901	8.08499
32	23.46834	17.87356	14.08405	11.43499	9.52637	8.11159
33	23.98857	18.14765	14.23024	11.51388	9.56943	8.13535
34	24.49860	18.41121	14.36815	11.58693	9.60857	8.15656
35	24.99863	18.66462	14.49826	11.65456	9.64416	8.17550
36	25.48885	18.90829	14.62100	11.71719	9.67651	8.19241
37	25.96946	19.14259	14.73679	11.77517	9.70591	8.20751
38	26.44065	19.36787	14.84603	11.82886	9.73265	8.22099
39	26.90260	19.58450	14.94909	11.87858	9.75695	8.23303
40	27.35549	19.79279	15.04631	11.92461	9.77905	8.24378
41	27.79950	19.99306	15.13803	11.96723	9.79913	8.25337
42	28.23480	20.18564	15.22455	12.00669	9.81740	8.26194
43	28.66157	20.37081	15.30618	12.04323	9.83400	8.26959
44	29.07997	20.54885	15.38319	12.07707	9.84909	8.27642
45	29.49017	20.72005	15.45584	12.10840	9.86281	8.28252
46	29.89232	20.88467	15.52438	12.13740	9.87528	8.28796
47	30.28659	21.04295	15.58904	12.16426	9.88662	8.29282
48	30.67313	21.19514	15.65004	12.18913	9.89692	8.29716
49	31.05209	21.34149	15.70758	12.21216	9.90629	8.30104
50	31.42362	21.48220	15.76187	12.23348	9.91481	8.30450

TABLE 6 **Areas Under the Standard Normal Curve**

z	A(z)	z	A(z)	z	A(z)
−3.00	.0013	−2.48	.0066	−1.96	.0250
−2.99	.0014	−2.47	.0068	−1.95	.0256
−2.98	.0014	−2.46	.0069	−1.94	.0262
−2.97	.0015	−2.45	.0071	−1.93	.0268
−2.96	.0015	−2.44	.0073	−1.92	.0274
−2.95	.0016	−2.43	.0075	−1.91	.0281
−2.94	.0016	−2.42	.0078	−1.90	.0287
−2.93	.0017	−2.41	.0080	−1.89	.0294
−2.92	.0017	−2.40	.0082	−1.88	.0300
−2.91	.0018	−2.39	.0084	−1.87	.0307
−2.90	.0019	−2.38	.0087	−1.86	.0314
−2.89	.0019	−2.37	.0089	−1.85	.0322
−2.88	.0020	−2.36	.0091	−1.84	.0329
−2.87	.0020	−2.35	.0094	−1.83	.0336
−2.86	.0021	−2.34	.0096	−1.82	.0344
−2.85	.0022	−2.33	.0099	−1.81	.0351
−2.84	.0023	−2.32	.0102	−1.80	.0359
−2.83	.0023	−2.31	.0104	−1.79	.0367
−2.82	.0024	−2.30	.0107	−1.78	.0375
−2.81	.0025	−2.29	.0110	−1.77	.0384
−2.80	.0026	−2.28	.0113	−1.76	.0392
−2.79	.0026	−2.27	.0116	−1.75	.0401
−2.78	.0027	−2.26	.0119	−1.74	.0409
−2.77	.0028	−2.25	.0122	−1.73	.0418
−2.76	.0029	−2.24	.0125	−1.72	.0427
−2.75	.0030	−2.23	.0129	−1.71	.0436
−2.74	.0031	−2.22	.0132	−1.70	.0446
−2.73	.0032	−2.21	.0135	−1.69	.0455
−2.72	.0033	−2.20	.0139	−1.68	.0465
−2.71	.0034	−2.19	.0143	−1.67	.0475
−2.70	.0035	−2.18	.0146	−1.66	.0485
−2.69	.0036	−2.17	.0150	−1.65	.0495
−2.68	.0037	−2.16	.0154	−1.64	.0505
−2.67	.0038	−2.15	.0158	−1.63	.0515
−2.66	.0039	−2.14	.0162	−1.62	.0526
−2.65	.0040	−2.13	.0166	−1.61	.0537
−2.64	.0041	−2.12	.0170	−1.60	.0548
−2.63	.0043	−2.11	.0174	−1.59	.0559
−2.62	.0044	−2.10	.0179	−1.58	.0570
−2.61	.0045	−2.09	.0183	−1.57	.0582
−2.60	.0047	−2.08	.0188	−1.56	.0594
−2.59	.0048	−2.07	.0192	−1.55	.0606
−2.58	.0049	−2.06	.0197	−1.54	.0618
−2.57	.0051	−2.05	.0202	−1.53	.0630
−2.56	.0052	−2.04	.0207	−1.52	.0643
−2.55	.0054	−2.03	.0212	−1.51	.0655
−2.54	.0055	−2.02	.0217	−1.50	.0668
−2.53	.0057	−2.01	.0222	−1.49	.0681
−2.52	.0059	−2.00	.0227	−1.48	.0694
−2.51	.0060	−1.99	.0233	−1.47	.0708
−2.50	.0062	−1.98	.0238	−1.46	.0721
−2.49	.0064	−1.97	.0244	−1.45	.0735

TABLE 6 Areas Under the Standard Normal Curve (*continued*)

z	A(z)	z	A(z)	z	A(z)
−1.44	.0749	−0.92	.1788	−0.40	.3446
−1.43	.0764	−0.91	.1814	−0.39	.3483
−1.42	.0778	−0.90	.1841	−0.38	.3520
−1.41	.0793	−0.89	.1867	−0.37	.3557
−1.40	.0808	−0.88	.1894	−0.36	.3594
−1.39	.0823	−0.87	.1921	−0.35	.3632
−1.38	.0838	−0.86	.1949	−0.34	.3669
−1.37	.0853	−0.85	.1977	−0.33	.3707
−1.36	.0869	−0.84	.2004	−0.32	.3745
−1.35	.0885	−0.83	.2033	−0.31	.3783
−1.34	.0901	−0.82	.2061	−0.30	.3821
−1.33	.0918	−0.81	.2090	−0.29	.3859
−1.32	.0934	−0.80	.2119	−0.28	.3897
−1.31	.0951	−0.79	.2148	−0.27	.3936
−1.30	.0968	−0.78	.2177	−0.26	.3974
−1.29	.0985	−0.77	.2206	−0.25	.4013
−1.28	.1003	−0.76	.2236	−0.24	.4052
−1.27	.1020	−0.75	.2266	−0.23	.4090
−1.26	.1038	−0.74	.2296	−0.22	.4129
−1.25	.1056	−0.73	.2327	−0.21	.4168
−1.24	.1075	−0.72	.2358	−0.20	.4207
−1.23	.1093	−0.71	.2388	−0.19	.4246
−1.22	.1112	−0.70	.2420	−0.18	.4286
−1.21	.1131	−0.69	.2451	−0.17	.4325
−1.20	.1151	−0.68	.2482	−0.16	.4364
−1.19	.1170	−0.67	.2514	−0.15	.4404
−1.18	.1190	−0.66	.2546	−0.14	.4443
−1.17	.1210	−0.65	.2578	−0.13	.4483
−1.16	.1230	−0.64	.2611	−0.12	.4522
−1.15	.1251	−0.63	.2643	−0.11	.4562
−1.14	.1271	−0.62	.2676	−0.10	.4602
−1.13	.1292	−0.61	.2709	−0.09	.4641
−1.12	.1314	−0.60	.2742	−0.08	.4681
−1.11	.1335	−0.59	.2776	−0.07	.4721
−1.10	.1357	−0.58	.2810	−0.06	.4761
−1.09	.1379	−0.57	.2843	−0.05	.4801
−1.08	.1401	−0.56	.2877	−0.04	.4840
−1.07	.1423	−0.55	.2912	−0.03	.4880
−1.06	.1446	−0.54	.2946	−0.02	.4920
−1.05	.1469	−0.53	.2981	−0.01	.4960
−1.04	.1492	−0.52	.3015	0.00	.5000
−1.03	.1515	−0.51	.3050	0.01	.5040
−1.02	.1539	−0.50	.3085	0.02	.5080
−1.01	.1562	−0.49	.3121	0.03	.5120
−1.00	.1587	−0.48	.3156	0.04	.5160
−0.99	.1611	−0.47	.3192	0.05	.5199
−0.98	.1635	−0.46	.3228	0.06	.5239
−0.97	.1660	−0.45	.3264	0.07	.5279
−0.96	.1685	−0.44	.3300	0.08	.5319
−0.95	.1711	−0.43	.3336	0.09	.5359
−0.94	.1736	−0.42	.3372	0.10	.5398
−0.93	.1762	−0.41	.3409	0.11	.5438

TABLE 6 Areas Under the Standard Normal Curve (*continued*)

z	A(z)	z	A(z)	z	A(z)
0.12	.5478	0.64	.7389	1.16	.8770
0.13	.5517	0.65	.7422	1.17	.8790
0.14	.5557	0.66	.7454	1.18	.8810
0.15	.5596	0.67	.7484	1.19	.8830
0.16	.5636	0.68	.7518	1.20	.8849
0.17	.5675	0.69	.7549	1.21	.8869
0.18	.5714	0.70	.7580	1.22	.8888
0.19	.5754	0.71	.7612	1.23	.8907
0.20	.5793	0.72	.7642	1.24	.8925
0.21	.5832	0.73	.7673	1.25	.8944
0.22	.5871	0.74	.7704	1.26	.8962
0.23	.5910	0.75	.7734	1.27	.8980
0.24	.5948	0.76	.7764	1.28	.8997
0.25	.5987	0.77	.7794	1.29	.9015
0.26	.6026	0.78	.7823	1.30	.9032
0.27	.6064	0.79	.7852	1.31	.9049
0.28	.6103	0.80	.7881	1.32	.9066
0.29	.6141	0.81	.7910	1.33	.9082
0.30	.6179	0.82	.7939	1.34	.9099
0.31	.6217	0.83	.7967	1.35	.9115
0.32	.6255	0.84	.7996	1.36	.9131
0.33	.6293	0.85	.8023	1.37	.9147
0.34	.6331	0.86	.8051	1.38	.9162
0.35	.6368	0.87	.8079	1.39	.9177
0.36	.6406	0.88	.8106	1.40	.9192
0.37	.6443	0.89	.8133	1.41	.9207
0.38	.6480	0.90	.8159	1.42	.9222
0.39	.6517	0.91	.8186	1.43	.9236
0.40	.6554	0.92	.8212	1.44	.9251
0.41	.6591	0.93	.8238	1.45	.9265
0.42	.6628	0.94	.8264	1.46	.9279
0.43	.6664	0.95	.8289	1.47	.9292
0.44	.6700	0.96	.8315	1.48	.9306
0.45	.6736	0.97	.8340	1.49	.9319
0.46	.6772	0.98	.8365	1.50	.9332
0.47	.6808	0.99	.8389	1.51	.9345
0.48	.6844	1.00	.8413	1.52	.9357
0.49	.6879	1.01	.8438	1.53	.9370
0.50	.6915	1.02	.8461	1.54	.9382
0.51	.6950	1.03	.8485	1.55	.9394
0.52	.6985	1.04	.8508	1.56	.9406
0.53	.7019	1.05	.8531	1.57	.9418
0.54	.7054	1.06	.8554	1.58	.9430
0.55	.7088	1.07	.8577	1.59	.9441
0.56	.7123	1.08	.8599	1.60	.9452
0.57	.7157	1.09	.8621	1.61	.9463
0.58	.7190	1.10	.8643	1.62	.9474
0.59	.7224	1.11	.8665	1.63	.9485
0.60	.7258	1.12	.8686	1.64	.9495
0.61	.7291	1.13	.8708	1.65	.9505
0.62	.7324	1.14	.8729	1.66	.9515
0.63	.7357	1.15	.8749	1.67	.9525

TABLE 6   Areas Under the Standard Normal Curve (*continued*)

z	A(z)	z	A(z)	z	A(z)
1.68	.9535	2.13	.9834	2.58	.9951
1.69	.9545	2.14	.9838	2.59	.9952
1.70	.9554	2.15	.9842	2.60	.9953
1.71	.9564	2.16	.9846	2.61	.9955
1.72	.9573	2.17	.9850	2.62	.9956
1.73	.9582	2.18	.9854	2.63	.9957
1.74	.9591	2.19	.9857	2.64	.9959
1.75	.9599	2.20	.9861	2.65	.9960
1.76	.9608	2.21	.9865	2.66	.9961
1.77	.9616	2.22	.9868	2.67	.9962
1.78	.9625	2.23	.9871	2.68	.9963
1.79	.9633	2.24	.9875	2.69	.9964
1.80	.9641	2.25	.9878	2.70	.9965
1.81	.9649	2.26	.9881	2.71	.9966
1.82	.9656	2.27	.9884	2.72	.9967
1.83	.9664	2.28	.9887	2.73	.9968
1.84	.9671	2.29	.9890	2.74	.9969
1.85	.9678	2.30	.9893	2.75	.9970
1.86	.9686	2.31	.9896	2.76	.9971
1.87	.9693	2.32	.9898	2.77	.9972
1.88	.9700	2.33	.9901	2.78	.9973
1.89	.9706	2.34	.9904	2.79	.9974
1.90	.9713	2.35	.9906	2.80	.9974
1.91	.9719	2.36	.9909	2.81	.9975
1.92	.9726	2.37	.9911	2.82	.9976
1.93	.9732	2.38	.9913	2.83	.9977
1.94	.9738	2.39	.9916	2.84	.9977
1.95	.9744	2.40	.9918	2.85	.9978
1.96	.9750	2.41	.9920	2.86	.9979
1.97	.9756	2.42	.9922	2.87	.9980
1.98	.9762	2.43	.9925	2.88	.9980
1.99	.9767	2.44	.9927	2.89	.9981
2.00	.9773	2.45	.9929	2.90	.9981
2.01	.9778	2.46	.9931	2.91	.9982
2.02	.9783	2.47	.9932	2.92	.9983
2.03	.9788	2.48	.9934	2.93	.9983
2.04	.9793	2.49	.9936	2.94	.9984
2.05	.9798	2.50	.9938	2.95	.9984
2.06	.9803	2.51	.9940	2.96	.9985
2.07	.9808	2.52	.9941	2.97	.9985
2.08	.9812	2.53	.9943	2.98	.9986
2.09	.9817	2.54	.9945	2.99	.9986
2.10	.9821	2.55	.9946	3.00	.9987
2.11	.9826	2.56	.9948		
2.12	.9830	2.57	.9949		

Answers to Odd-Numbered Exercises

CHAPTER 1

Exercise Set 1–1

1.

x	-2	-1	0	1	2	3
y	-7	-5	-3	-1	1	3
point	$(-2, -7)$	$(-1, -5)$	$(0, -3)$	$(1, -1)$	$(2, 1)$	$(3, 3)$

3.

x	-2	-1	0	1	2	3
y	6	5	4	3	2	1
point	$(-2, 6)$	$(-1, 5)$	$(0, 4)$	$(1, 3)$	$(2, 2)$	$(3, 1)$

5.

7.

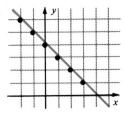

9. $2x + y = 6$ $y = 6 - 2x$

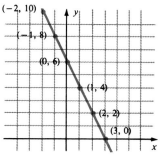

x	-2	-1	0	1	2	3
y	10	8	6	4	2	0

11. $x - y = 3$

x	y
0	-3
3	0

13. $3x - y = -1$

x	0	$-\dfrac{1}{3}$
y	1	0

15.

17.

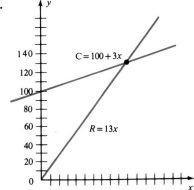

$C = 100 + 3x$

$R = 13x$

19. (a) $58,100
   (b) 10%

21.

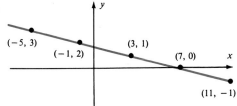

$(-5, 3)$
$(-1, 2)$
$(3, 1)$
$(7, 0)$
$(11, -1)$

23.

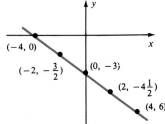

$(-4, 0)$
$(-2, -\frac{3}{2})$
$(0, -3)$
$(2, -4\frac{1}{2})$
$(4, 6)$

25.

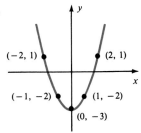

$(-2, 1)$
$(2, 1)$
$(-1, -2)$
$(1, -2)$
$(0, -3)$

27.

29.

31.

33.

35.

37.

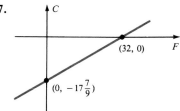

39. 37° Celsius
41. $50,000

43. (a)

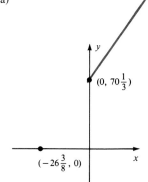

  (b) 169 cm
  (c) 43 cm
45. This is the break-even point.
47. (a) 72.4%
  (b) 1978

Exercise Set 1–2

1.

3.

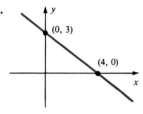

5. $m = 2$
7. $m = -\frac{3}{2}$
9. $m = 2$

11.

13.

15. not parallel
17. (a) undefined   (b) $m = 0$
19. $y = -7x + 15$

21.

23.

25.

27.

29.

31.

33.

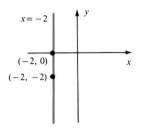

35. (a) positive   (b) zero   (c) negative
37. (a) $y = -\frac{3}{2}x + 6$ (b) $m = -\frac{3}{2}$ (c) y-intercept $= 6$
39. (a) $y = -\frac{5}{2}x + 5$ (b) $y = -\frac{5}{2}x - 5$
41. $y = -4x + 1$

43. (a)

(b) $15,000
(c) The y-intercept represents the worth of the work station when it was purchased.
45. no, for example, $xy = 1$
47. (a) $y = .5x + 7.5$

(b)

(c) 43.5

Exercise Set 1–3

1. (4, 1)

3.

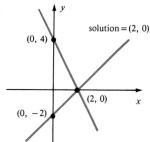

solution = (2, 0)

(0, 4)

(2, 0)

(0, −2)

5. (1, 1)
7. (1, −2)
9. ($\frac{7}{5}$, $\frac{3}{5}$)
11. (1, −3)
13. (7/2, 3/2)
15. inconsistent
17. dependent
19. 20,000 boxes of mints
21. ($\frac{2}{3}$, −$\frac{5}{3}$)
23. (4, 2)
25. (20, 100)
27. (2, 0)
29. (100, 20)
31. (0, 0)
33. no solution
35. (10, 4)
37. two 747's and four 707's
39. $1500 in school bonds and $3500 in public utility bonds
41. (a) 12.2 million cars sold
     (b) 8.9 million cars sold
     (c) 3.3 million cars
43. (a) 9%   (b) 9%

Exercise Set 1–4

1. (1, 1, 1)
3. none
5. (1, 1, 1)
7. (0, 0, 0)
9. (2, 1, 1)
11. (3, 1, 0)
13. (2, 0, 3)
15. (0, 1, 2)
17. 5 small disks, 1 medium disk, and 4 large disks
19. (1, 2, −1)
21. (4, 0, −1)
23. (2, 0, 3)
25. no unique solution

27. no unique solution
29. one solo artist and two groups
31. 4 oz fish, 2 oz corn, and 1 oz rice
33. $x = (ed - bf)/(ad - bc), y = (af - ce)/(ad - bc)$
35. (a) (0, 0, 0)
    (b) no unique solution
    (c) no unique solution
37. false
39. true
41. (70, 70, 40)

Exercise Set 1–5

1. $y = 0.3x + 2$
3. $y = -0.6x + 4$
5. $y = 2x + 2.75$
7. $y = -3.5x + 22.8$
9. $y = 0.316x + 0.163$
11. $y = 2.036x + 6.214$
13. $y = -0.46x - 0.57$
15. $y = \frac{12}{5}x + \frac{92}{3}$
17. $71\frac{1}{2}$ in.

19. (a)

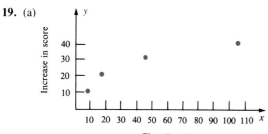

(b) no
(c) $4b + \quad 179\,m = \quad 100$
       $179b + 13899\,m = 6090$
(d) $y = \dfrac{1292}{4711}x + \dfrac{59{,}958}{4711}$
(e) $\dfrac{318{,}358}{4711} \approx 67.6$
(f) $\dfrac{411{,}142}{1292} \approx 318.2$

21. The points are linear: $y = -0.00356x + 58.62$

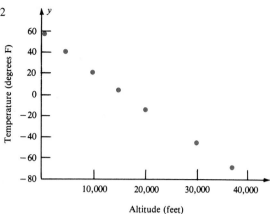

23. (a) least squares (b) $y = 1.25t + 20.5$
(c) 45.5 (d) 10

25. (a) $y = .8857t - 54.24$
(b) about 34%

Case Study Exercises

1. $5b + 376m = 1148.4$
$376b + 28380m = 86341.6$

3. $y = -0.173x + 242.65$

5. $3:45.35$

7. 2262

9. (a) $3:42$ (b) 2006

11. (a) $-.32$ (b) $\approx\frac{1}{3}$ sec (c) 32 sec

Review Exercises

1.

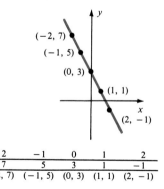

x	−2	−1	0	1	2
y	7	5	3	1	−1
point	(−2, 7)	(−1, 5)	(0, 3)	(1, 1)	(2, −1)

5.

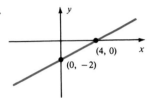

(4, 0) x
(0, −2)

7. $y = 3x - 3$
9. $C = 0.22x + 25$
11. $(-1\frac{2}{3}, 1\frac{1}{3})$
13. $(1, -1), (3, 1)$, other solutions possible
15. $(-3, -1, -2)$
17. $y = 0.3x + 4.6$

CHAPTER 2

Exercise Set 2–1

1. The size of $A = 2 \times 3$; the size of $C = 3 \times 1$; the size of $E = 2 \times 2$
3. (a) $a_{22} = 4$ (b) $b_{31} = -1$ (c) $d_{13} = 3$ (d) $e_{11} = 1$
5. (a) a_{12} (b) a_{23} (c) a_{22}
7. $x = 5$, $y = 6$, $z = 7$

9. (a) $\begin{bmatrix} -2 & 5 \\ 12 & 6 \\ 7 & -3 \end{bmatrix}$ (b) $\begin{bmatrix} 9 & 1 \\ 9 & 16 \\ 2 & -4 \end{bmatrix}$ (c) same as (b)

11. (a) $\begin{bmatrix} 3 & -3 \\ 6 & -6 \\ 9 & -9 \end{bmatrix}$ (b) $\begin{bmatrix} 6 & -12 \\ -20 & -16 \\ -8 & 0 \end{bmatrix}$ (c) $\begin{bmatrix} -11 & 4 \\ 3 & -10 \\ 5 & 1 \end{bmatrix}$

13. (a) $\begin{bmatrix} -7 & 16 \\ 34 & 20 \\ 18 & -6 \end{bmatrix}$ (b) $\begin{bmatrix} 11 & -20 \\ -26 & -28 \\ -6 & -6 \end{bmatrix}$

17. (a) 2×3 (b) 3×3

19. $\begin{bmatrix} -1 & 6 & -7 \\ 2 & -4 & -3 \end{bmatrix}$

21.

	March	
	cotton	synthetics
Fall River	3	1.5
Passaic	0.5	3
L.A.	5	2

23.

	March and April	
	cotton	synthetics
Fall River	8	3.5
Passaic	4	5
L.A.	6.5	4.5

25. $x = 6$, $y = 4$
27. $x = -1$, $y = -3$, $z = -8$
29. $x = -3$, $y = -21$, $z = -7$

31. $x = 2, \quad y = 3, \quad z = 1$

33. (a) $\begin{bmatrix} -7 & -3 \\ -5 & -20 \\ 4 & -2 \end{bmatrix}$ (b) $\begin{bmatrix} -7 & -3 \\ -5 & -20 \\ 4 & -2 \end{bmatrix}$

35. $R - (S + T) = (R - S) - T$

37. $x = 3, \quad y = 1$

39. Impossible, because the system of equations is inconsistent

41. (a) $K + U = \begin{bmatrix} 165 & 18 \\ 345 & 23 \\ 730 & 28 \end{bmatrix}$ (b) $K - U = \begin{bmatrix} 135 & 6 \\ 305 & 7 \\ 670 & 8 \end{bmatrix}$ (c) $\frac{1}{2}K + 2U = \begin{bmatrix} 105 & 18 \\ 202.5 & 23.5 \\ 410 & 29 \end{bmatrix}$

43. $\begin{bmatrix} 0 & 706 & 2823 \\ 706 & 0 & 3497 \\ 2823 & 3497 & 0 \end{bmatrix}$

45. (a) 34×8 (the "individual" column is not included)
(b) each entry is either 0 or $\frac{1}{99}$
(c) the diagonal entries are $\frac{20}{99}, \frac{17}{99}, \frac{8}{99}, \frac{7}{99}, \frac{16}{99}, \frac{11}{99}, \frac{11}{99}$, and $\frac{9}{99}$; all other entries are 0
(d) the diagonal entries are
$\frac{4}{99}, \frac{2}{99}, \frac{2}{99}, \frac{3}{99}, \frac{2}{99}, \frac{3}{99}, \frac{4}{99}, \frac{5}{99}, \frac{5}{99}, \frac{3}{99}, \frac{3}{99}, \frac{2}{99}, \frac{4}{99}, \frac{2}{99}, \frac{3}{99}, \frac{2}{99}, \frac{2}{99}, \frac{3}{99}, \frac{2}{99}, \frac{4}{99}, \frac{2}{99}, \frac{2}{99}, \frac{3}{99}, \frac{3}{99}, \frac{3}{99}, \frac{2}{99}, \frac{3}{99}, \frac{2}{99}, \frac{3}{99},$
$\frac{3}{99}, \frac{3}{99}, \frac{4}{99};$
all other entries are 0

47. $X = (-1)M$

Exercise Set 2–2

1. $\begin{bmatrix} -2 & -5 & -3 \\ 0 & 60 & 30 \\ 0 & 6 & 1 \end{bmatrix}$

3. $\begin{bmatrix} -1 & -5 & -3 & 3 \\ 20 & 50 & 20 & 30 \\ 6 & 6 & 5 & 4 \end{bmatrix}$

5. $\begin{bmatrix} 1 \\ -10 \\ -3 \end{bmatrix}$

7. $j_{12} = 24$

9. $j_{22} = 10$

11. (a) 4×3 (b) undefined

13. $A = \begin{bmatrix} -1 & 4 & -3 \\ 2 & -2 & 7 \end{bmatrix}$ $X = \begin{bmatrix} x \\ y \\ z \end{bmatrix}$ $C = \begin{bmatrix} 11 \\ -4 \end{bmatrix}$

15. $\begin{aligned} x \qquad\quad + 6z &= 0 \\ -2x + 3y + 4z &= 0 \\ x + 2y \qquad &= 0 \end{aligned}$

17. $\begin{aligned} 2x + 6y + 7z &= 0 \\ 3x + 5y - 2z &= 0 \end{aligned}$

19. No
$AB = \begin{bmatrix} 8 & 10 \\ 26 & 22 \end{bmatrix}$ $BA = \begin{bmatrix} 16 & 28 \\ 11 & 14 \end{bmatrix}$

21. No

$$AB = \begin{bmatrix} 30 & 6 \\ 2 & -8 \end{bmatrix} \qquad BA = \begin{bmatrix} 30 & -4 \\ -3 & -8 \end{bmatrix}$$

23. $\begin{bmatrix} 2.0 & 4.5 \\ 2.5 & 4.0 \\ 3.5 & 1.5 \end{bmatrix} \begin{bmatrix} 1.0 \\ 1.5 \end{bmatrix} = \begin{bmatrix} 8.75 \\ 8.50 \\ 5.75 \end{bmatrix}$

25. (a) $\begin{bmatrix} -28 & 41 \\ 32 & -52 \end{bmatrix}$ (b) $\begin{bmatrix} -28 & 41 \\ 32 & -52 \end{bmatrix}$ (c) $(AB)C = A(BC)$ This verifies the associative law.

    (d) same as (a) and (b)

27. (a) $\begin{bmatrix} -10 & 10 \\ 8 & 1 \end{bmatrix}$ (b) $\begin{bmatrix} -10 & 10 \\ 8 & 1 \end{bmatrix}$ (c) $A(B + C) = AB + AC$ This verifies the distributive law.

29. (a) $\begin{bmatrix} 4 & 5 & 6 \\ 0 & 0 & 0 \\ 0 & 0 & 0 \end{bmatrix}$ (b) $\begin{bmatrix} 0 & 0 & 0 \\ 7 & 8 & 9 \\ 0 & 0 & 0 \end{bmatrix}$

    (c) This procedure moves the jth row of A to the ith row. Zeros everywhere else.

31. AB is a 3×4 matrix whose bottom row consists entirely of 0's.

33. How much protein, carbohydrate, and fat is consumed by the three age groups from milk.

35. F1:$3,632,400; F2:$2,701,400; F3:$2,028,600

37. $690

39. A:$13.00

    B:$10.75

41. (a) $AS = SA = \begin{bmatrix} as & bs \\ cs & ds \end{bmatrix}$

    (b) Let $B = \begin{bmatrix} u & v \\ w & x \end{bmatrix}$. If $C = \begin{bmatrix} 0 & 1 \\ 1 & 0 \end{bmatrix}$ then $BC = CB$ implies that $u = x$ and $v = w$. If $D = \begin{bmatrix} 0 & 1 \\ 0 & 0 \end{bmatrix}$ then

    $BD = DB$ implies that $w = 0$. Therefore $B = \begin{bmatrix} x & 0 \\ 0 & x \end{bmatrix}$.

43. $A = \begin{bmatrix} 0 & 1 \\ 0 & 0 \end{bmatrix}$ and $B = \begin{bmatrix} 1 & 0 \\ 0 & 0 \end{bmatrix}$; other answers are possible

45. (a) They add up to one because each entry represents one of the three possible options

    (b) E is the most preferable option

    (c) C is the least preferable option

47. The domestic market D has the higher score, 57 to 35

49. Common stock financing has the higher score, 18 to 10

Exercise Set 2–3

  1. $(2, 1, 1)$

  3. $(3, 1, 0)$

  5. $(-2, 3, 0)$

  7. $(3, -1, 6)$

  9. $x = 6,$ $z = 4,$ $y = $ any number

11. $x = -3z - 2,$ if $z = 0$ $(-2, -2, 0)$

    $y = -z - 2,$ if $z = 1$ $(-5, -3, 1)$

    $z = $ any number,   other solutions possible

13. $x = 4,$ if $z = 0$ $(4, 2, 0)$

    $y = z + 2,$ if $z = 1$ $(4, 3, 1)$

    $z = $ any number,   other solutions possible

15. $x = -y - 2v + 1$ basic variables: x, u
$y = $ any number   free variables: y, v
$u = -3v$
$v = $ any number

17. $x = -v$ basic variables: x, y, u
$y = -2v + 5$ free variable: v
$u = -3v + 4$
$v = $ any number

19. $x = 4z - 5$ basic variables: x, y
$y = 6z - 9$
$z = $ any number

21. $x = 2z + 26$ basic variables: x, y
$y = -15$
$z = $ any number

23. two servings of chicken; one serving of rice

25. $(1, 2, -3)$

27. $x = -5 + z,$ if $z = 0$ $(-5, 4, 0)$
$y = 4 + 2z,$ if $z = 1$ $(-4, 6, 1)$
$z = $ any number,   other solutions possible

29. $(1, -3, -2)$

31. $x = -2z + 5,$ if $z = 0$ $(5, 4, 0)$
$y = 4,$ if $z = 1$ $(3, 4, 1)$
$z = $ any number,   other solutions possible

33. no solution

35. Two servings of chicken and one serving of rice

37. Different combinations are possible

hard disk drives:	6	4	2	0
1.2MB floppy:	0	1	2	3
360K floppy:	7	7	7	7
Total no. of drives:	13	12	11	10

39. $\begin{bmatrix} 1 & 0 & d/(ad - bc) & -b/(ad - bc) \\ 0 & 1 & -c/(ad - bc) & a/(ad - bc) \end{bmatrix}$

41. $x = 480{,}000/49,$ $y = 390{,}000/49,$ $z = 1{,}328{,}000/49,$ $w = 1{,}171{,}000/49$

Exercise Set 2–4

1. $A^{-1} = Z$ **3.** $A^{-1} = Y$

5. $\begin{bmatrix} -2 & 1 \\ 7 & -3 \end{bmatrix}$

7. not invertible

9. $\frac{1}{4} \begin{bmatrix} 8 & 2 \\ 2 & 1 \end{bmatrix}$

11. $\begin{bmatrix} -1 & 1 & 0 \\ 4 & -1 & -1 \\ -2 & 0 & 1 \end{bmatrix}$

13. $\begin{bmatrix} -3 & 10 & -3 \\ 2 & -4 & 1 \\ -1 & 1 & 0 \end{bmatrix}$

15. $\begin{bmatrix} 1 & 2 & -5 \\ 0 & 1 & -4 \\ 0 & 0 & 1 \end{bmatrix}$

17. $\begin{bmatrix} 0 & \frac{1}{2} & 0 \\ -\frac{1}{3} & 0 & 0 \\ 0 & 0 & \frac{1}{4} \end{bmatrix}$

19. $(AB)^{-1} = -\frac{1}{2}\begin{bmatrix} 0 & 1 \\ 2 & 1 \end{bmatrix} = A^{-1}B^{-1}$

21. $\begin{bmatrix} 4 & 1 \\ -1 & 0 \end{bmatrix}$

23. not invertible

25. $\begin{bmatrix} 1 & 0 & 0 \\ 0 & \frac{1}{3} & 0 \\ 0 & 0 & -1 \end{bmatrix}$ 　　**27.** $\begin{bmatrix} 1 & 2 & -7 & 7 \\ 0 & 1 & -2 & 4 \\ 0 & 0 & 1 & 0 \\ 0 & 0 & 0 & 1 \end{bmatrix}$

29. $A^{-1} = \begin{bmatrix} -2 & 1 \\ 7 & -3 \end{bmatrix}$　　$(A^{-1})^{-1} = \begin{bmatrix} 3 & 1 \\ 7 & 2 \end{bmatrix} = A$

31. $A^{-1} = \begin{bmatrix} 2 & \frac{1}{2} \\ \frac{1}{2} & \frac{1}{4} \end{bmatrix}$　　$(A^{-1})^{-1} = \begin{bmatrix} 1 & -2 \\ -2 & 8 \end{bmatrix} = A$

33. $(A^{-1})^{-1} = A$

35. $(AB)^{-1} = \begin{bmatrix} 12 & -5 \\ 26 & -11 \end{bmatrix}$　　$AB = -\frac{1}{2}\begin{bmatrix} -11 & 5 \\ -26 & 12 \end{bmatrix}$

37. (a) no　　(b) $A^{-1} = \frac{1}{2}B$

39. $B = IB = (CA)B = C(AB) = CI = C$

41. (a) A^{-1} means taking off a sock, B^{-1} means taking off a shoe

(b) AB means putting on a sock and then a shoe, while $(AB)^{-1}$ means taking off a shoe and a sock

(c) in order to take off a shoe and a sock, the shoe must be taken off first and the sock second

Exercise Set 2–5

1. $(-30, 9, 3)$

3. $(-5, 1, 4)$

5. $(1, 1)$

7. $(1, -2)$

9. $(\frac{7}{5}, \frac{3}{5})$

11. $(5, -3, -4)$

13. $(2, 5, -4)$

15. (a) $(1, 4)$　　(b) $(3, -5)$　　(c) $(0, 3)$

17. (a) $(-2, 1)$　　(b) $(1, 4)$　　(c) $(-3, 12)$

19. (a) $(5, 4, 0)$　　(b) $(6, 1, -8)$　　(c) $(1, 0, -1)$

21. $(2, 3, -2)$

23. $(-3, 5, 4)$

25. $(5, 8, -1)$

27. no solution

29. $(0, 16, -4, 2)$

31. 20 wool suits; 100 blend suits

33. one red pill; three blue pills

35. 105 first-class passengers; 25 executive passengers; 60 economy class passengers

37. $x = 480,000/49, \quad y = 390,000/49, \quad z = 1,328,000/49, \quad w = 1,171,000/49$

39. (a) $x = \begin{bmatrix} 56,327 \\ 84,898 \\ 71,837 \end{bmatrix}$ (b) $\hat{x} = \begin{bmatrix} 54,594.5 \\ 84,594.5 \\ 71,621.5 \end{bmatrix}$ (c) $\begin{bmatrix} 96.9\% \\ 99.6\% \\ 99.7\% \end{bmatrix}$

Exercise Set 2–6

1. $\begin{bmatrix} 65 \\ 40 \end{bmatrix}$

3. $\begin{bmatrix} 85/2 \\ 140/3 \end{bmatrix}$

5. $\begin{bmatrix} 200 \\ 200 \end{bmatrix}$

7. $\begin{bmatrix} 56 \\ 60 \end{bmatrix}$

9. $\begin{bmatrix} 140 \\ 160 \\ 120 \end{bmatrix}$

11. $\begin{bmatrix} 280,000 \\ 260,000 \end{bmatrix}$

13. $AB = \begin{bmatrix} 1 & 0 & 0 \\ 0 & 1 & 0 \\ 0 & 0 & 1 \end{bmatrix} = I, \quad$ so $A^{-1} = B$ and $B^{-1} = A$

15. $A^{-1} = \dfrac{1}{k} B^{-1}$

Case Study Exercises

1. $\begin{bmatrix} 189,274 \\ 909,976 \\ 1,282,165 \\ 1,624,600 \\ 487,042 \\ 793,055 \end{bmatrix}$

3. $\begin{bmatrix} 399,279 \\ 353,596 \\ 1,121,874 \\ 595,287 \\ 313,933 \\ 1,590,451 \end{bmatrix}$

5. $\begin{bmatrix} 3,827,000,000 \\ 4,929,000,000 \\ 7,577,000,000 \end{bmatrix}$

Review Exercises

1. (a) 5×3 (b) $m_{42} = 5$ (c) $7 = m_{52}$

3. $x = -1, y = 3$

5. $\begin{bmatrix} 29 & 25 & 22 & 23 \\ 0 & 0 & 0 & 0 \\ 33 & 57 & 25 & 53 \end{bmatrix}$

7. $\begin{bmatrix} -6 & 3 \\ 18 & -9 \end{bmatrix}$

9. $\begin{bmatrix} 37 & 54 \\ 81 & 118 \end{bmatrix}$

11.

$$A = \begin{bmatrix} 1 & 1 & -5 & 0 \\ 0 & 3 & 1 & 7 \end{bmatrix} \qquad X = \begin{bmatrix} w \\ x \\ y \\ z \end{bmatrix} \qquad C = \begin{bmatrix} -1 \\ 8 \end{bmatrix}$$

13. $x = -2 - z,$ if $z = 0$ $(-2, 4, 0)$
    $y = 4 + z,$ if $z = 1$ $(-3, 5, 1)$
    $z = $ any number,   other solutions possible

15. infinitely many solutions
    $x = 2y + 26$
    $y = $ any number
    $z = -15$
    basic variables: x, z

17. $\frac{1}{3} \begin{bmatrix} 1 & 1 & 1 \\ -2 & 1 & 1 \\ 2 & 2 & -1 \end{bmatrix}$

19. (a) BA^{-1} (b) $CB^{-1}A^{-1}$

21. $A^{-1} = \frac{1}{3}B$

23. $(2, -5, 3)$

25. $\begin{bmatrix} 1300 \\ 1360 \end{bmatrix}$

CHAPTER 3

Exercise Set 3–1

1. $1 + 2 \not> 4$ no; $2 + 2 \not> 4$ no, $2 + 2$ is not *greater than* 4
3. $-(-1) + 7(1) \geq 8$ yes; $-(0) + 7(0) \not\geq 8$ no
5. $(0, 0)$ is below the line; $(1, 7)$ is above the line
7. $(1, -1)$ is below the line; $(-1, -1)$ is below the line
9.

11.

13.

15.

17.

19.

21.

23.

25.

27.

29.

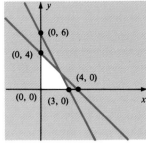

31. $200x + 130y \leq 2600$

33. $40x + 50y \geq 400$

35. $20x + 40y \leq 2400$
     $\quad x + \quad y \leq \quad 100$
     $\quad\quad x \geq 0, \, y \geq 0$

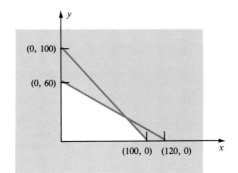

37. $45x + 9y \geq 45$
     $\quad 9x + 6y \geq 15$
     $\quad\quad x \geq 0, \, y \geq 0$

39.

41. no solution

43.

45.

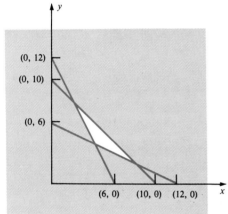

47. $x + \quad y \leq \quad 100$
$30x + 10y \leq 2500$
$x \geq 0,\ y \geq 0$

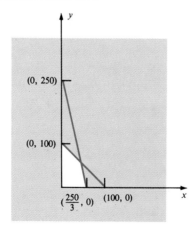

Exercise Set 3–2

1. Maximize: $P = 40x + 30y$
   Subject to: $150x + 100y \leq 12{,}000$
$$x + y \leq 100$$
$$x \geq 0,\ y \geq 0$$

3. Minimize: $F = 4x + 2y$
   Subject to: $45x + 9y \geq 45$
$$9x + 6y \geq 15$$
$$x \geq 0,\ y \geq 0$$

5. Minimize: $C = 132x + 80y$
   Subject to: $3x + 6y \leq 36$
$$3x + 2y \geq 24$$
$$x \geq 0,\ y \geq 0$$

7. Maximize: $P = 500x + 800y$
   Subject to: $100x + 200y \leq 2000$
$$3x + 4y \leq 48$$
$$x \geq 0,\ y \geq 0$$

9. Maximize: $P = .5x + 0.7y$
   Subject to: $.3x + .4y \leq 120$
$$.2x + .1y \leq 60$$
$$x \geq 0,\ y \geq 0$$

11. Maximize: $G = .15x + .12y$
   Subject to: $x + y \leq 15{,}000{,}000$
$$y \geq 2x$$
$$x \geq 2{,}000{,}000$$
$$y \geq 2{,}000{,}000$$

13. Maximize: $R = 1000x + 1200y$
   Subject to: $8000x + 10{,}000y \leq 9{,}000{,}000$
$$x + y \leq 1000$$
$$x \geq 200,\ y \geq 200$$

15. Minimize: $C = 20{,}000x + 15{,}000y$
Subject to: $150x + 100y \geq 1000$
$\qquad\qquad\;\; 50x + 50y \geq 450$
$\qquad\qquad\qquad x \geq 0,\; y \geq 0$

17. Minimize: $C = 9000x + 15{,}000y$
Subject to: $\quad x + \quad y \leq \quad 12$
$\qquad\qquad\; 2x + \quad 3y \leq \quad 30$
$\qquad\qquad 100x + 300y \geq 1800$
$\qquad\qquad\qquad x \geq 0,\; y \geq 0$

19. Minimize: $C = .07x + .03y$
Subject to: $9.2x + 4.6y \geq 165{,}000{,}000$
$\qquad\qquad\; 7.6x + 3.1y \leq 230{,}000{,}000$
$\qquad\qquad\qquad x \geq 0,\; y \geq 0$

21. Maximize: $P = .34x + .22y$
Subject to: $x + \quad y \leq 2000$
$\qquad\qquad\; x - 3y \geq \quad 0$
$\qquad\qquad\; x \geq 0,\; y \geq 0$

Exercise Set 3–3

1.

3.

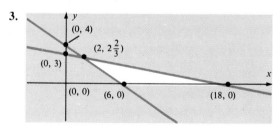

5. no set of feasible solutions; no corner points

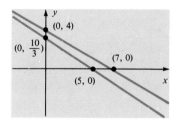

7. $(10, 0)$
9. $(2, 2\frac{2}{3})$
11. no solution
13. $(40, 60)$
15. $(\frac{5}{7}, \frac{10}{7})$
17. $(6, 3)$

19. (8, 6)
21. (0, 300)
23. (5, 10) in millions
25. (500, 500)
27. (2, 7)
29. (0, 6)
31. (3, 1)
33. (0, 8)
35. (4, 4)
37. (2, 8)
39. (40, 40)
41. (600, 200)
43. (0, 90), plant 90 acres of potatoes
45. (420, 80)

Review Exercises

1.

3.

5.

7.

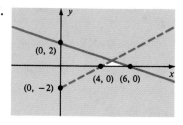

9. $280x + 330y \le 4800$

11. Minimize: $C = 12,000x + 15,000y$
  Subject to: $150x + 200y \ge 2500$
                      $x + \quad y \le 15$
                      $x \ge 0,\, y \ge 0$

13. Minimize: $C = 40x + 60y$
  Subject to: $3x + \quad 2y \ge 800$
                  $60x + 80y \le 24,000$
                      $x \ge 0,\, y \ge 0$

15.

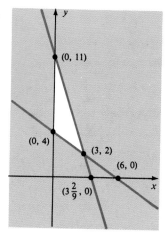

$z = 4x + 5y$

$(0, 11)$	55
$(0, 4)$	20
$(3, 2)$	22

$(0, 11)$ maximizes z

17. $(0,\, \frac{25}{2})$

19. $(\frac{800}{3},\, 0)$

CHAPTER 4

Exercise Set 4–1

1. yes   **3.** yes

5. $x + 2y + s_1 \quad\quad = 4$
$\quad x - y \quad\quad + s_2 = 5$

7. $-3x - 4y + m = 0$

9.

x	y	s_1	s_2	m	
1	2	1	0	0	10
3	1	0	1	0	8
−3	−4	0	0	1	0

11.

x	y	s_1	s_2	s_3	m	
3	8	1	0	0	0	1
4	−5	0	1	0	0	4
2	7	0	0	1	0	6
−1	−5	0	0	0	1	0

13.

x	y	z	s_1	s_2	s_3	m	
1	0	1	1	0	0	0	15
1	2	0	0	1	0	0	8
0	1	5	0	0	1	0	10
−3	−1	−5	0	0	0	1	0

15. basic: y, s_1, m,   nonbasic: x, s_2

17. basic: x, s_2, m,   nonbasic: y, s_1

19. basic: y, z, s_2, m,   nonbasic: x, s_1, s_3

21. $x = 0$, $y = 3$, $s_1 = 1$, $s_2 = 0$, $m = 6$

23. $x = 3$, $y = 0$, $s_1 = 0$, $s_2 = 4$, $m = 8$

25. $x = s_1 = s_3 = 0$, $m = 6$, $y = 4$, $z = 1$, $s_2 = 3$

27. Maximize: $m = x + 3y$
Subject to: $2x + 3y \le 3$
$\quad\quad\quad\quad 4x + y \le 6$
$\quad\quad\quad\quad x \ge 0, y \ge 0$

29. Maximize: $m = 5x + 4y + 7z + 40$
Subject to: $2y + z \le 3$
$\quad\quad\quad\quad 3y + 2z \le 0$
$\quad\quad x + 4y \quad\quad \le 1$
$\quad\quad\quad\quad x \ge 0, y \ge 0, z \ge 0$

31. $s_1 = 7 \quad s_2 = 13$

33. $s_1 = 4 \quad s_2 = 5$

35.

x	y	z	w	s_1	s_2	s_3	s_4	m	
1	0	1	2	1	0	0	0	0	13
1	2	0	3	0	1	0	0	0	8
0	1	3	1	0	0	1	0	0	10
1	0	3	2	0	0	0	1	0	12
−3	−1	−5	−6	0	0	0	0	1	0

37. basic: x, z, s_3, m nonbasic: y, s_1, s_2
basic feasible solution $x = 8$, $y = 0$, $z = 2$, $s_1 = 0$, $s_2 = 0$, $s_3 = 5$, $m = 9$

39. There is a unique slack variable for each equation

41.
$$\left[\begin{array}{ccccccc|c} .5 & .25 & .5 & 1 & 0 & 0 & 40 \\ .5 & .25 & .4 & 0 & 1 & 0 & 25 \\ 0 & .5 & .1 & 0 & 0 & 1 & 30 \\ -2 & -1 & -2.5 & 0 & 0 & 0 & 0 \end{array}\right]$$

43. Maximize: $P = .54x + .44y + .37z$
  Subject to: $x + y + z \le 12$
      $-\frac{1}{2}x + y + z \le 0$
      $x \ge 0, y \ge 0, z \ge 0$

$$\left[\begin{array}{ccccc|c} 1 & 1 & 1 & 1 & 0 & 12 \\ -\frac{1}{2} & 1 & 1 & 0 & 1 & 0 \\ -.54 & -.44 & -.37 & 0 & 0 & 0 \end{array}\right]$$

45. Minimize: $C = 5x + 7y + z$
  Subject to: $.0015x + .0015y + .00015z \le .012$
      $3x + 4y + 5z \le 30$
      $x \ge 0, y \ge 0, z \ge 0$

$$\left[\begin{array}{ccccc|c} .0015 & .0015 & .00015 & 1 & 0 & .012 \\ 3 & 4 & 5 & 0 & 1 & 30 \\ -5 & -7 & -1 & 0 & 0 & 0 \end{array}\right]$$

Exercise Set 4–2

1. no   **3.** yes   **5.** no

7. $y = s_2 = 0$, $x = 15$, $s_1 = 10$, $m = 20$; no

9. $s_1 = s_2 = s_3 = 0$, $x = 28$ $y = 34$, $z = 10$, $m = 22$; yes

11. the 2nd column; y is the entering variable

13. the 5th column; s_2 is the entering variable

15. the 3rd column; z is the entering variable

17.

x	y	s_1	s_2	m	
0	0	0	1	0	1
1	1	1	0	0	3
3	0	2	0	1	17

19.

x	y	z	s_1	s_2	s_3	m	
$\frac{1}{2}$	$\frac{1}{2}$	0	0	1	0	0	$\frac{3}{2}$
$-\frac{1}{2}$	$-\frac{1}{2}$	0	1	0	1	0	$\frac{13}{2}$
0	0	1	3	0	0	0	9
1	3	0	-1	0	0	1	6

21.

x	y	z	s_1	s_2	s_3	m	
$-\frac{1}{3}$	0	0	$-\frac{1}{3}$	$-\frac{14}{3}$	$-\frac{1}{3}$	0	7
$\frac{1}{3}$	0	1	$\frac{1}{3}$	$\frac{2}{3}$	$\frac{1}{3}$	0	3
$\frac{1}{3}$	1	0	$\frac{1}{3}$	$\frac{5}{3}$	$\frac{4}{3}$	0	5
3	0	0	3	1	6	1	72

23.

x	y	z	s_1	s_2	s_3	m	
0	0	1	0	-4	0	0	10
1	0	0	1	14	1	0	-21
0	1	0	0	-3	1	0	12
0	0	0	0	-41	3	1	135

25.

x	y	z	s_1	s_2	s_3	m	
0	4	-3	0	0	4	0	18
1	2	5	1	0	-1	0	5
0	1	-1	0	1	1	0	2
0	5	-14	0	0	8	1	55

27.

x	y	z	s_1	s_2	s_3	m	
0	0	1	0	-4	0	0	10
1	0	3	1	2	1	0	9
0	1	-1	0	1	1	0	2
0	0	-9	0	-5	3	1	45

29.

x	y	s_1	s_2	m	
1	3	1	0	0	12
2	3	0	1	0	18
-5	-7	0	0	1	0

x	y	s_1	s_2	m	
$\frac{1}{3}$	1	$\frac{1}{3}$	0	0	4
1	0	-1	1	0	6
$-\frac{8}{3}$	0	$\frac{7}{3}$	0	1	28

31.

x	y	s_1	s_2	m	
3	5	1	0	0	15
1	5	0	1	0	10
-4	-7	0	0	1	0

x	y	s_1	s_2	m	
2	0	1	-1	0	5
$\frac{1}{5}$	1	0	$\frac{1}{5}$	0	2
$-\frac{13}{5}$	0	0	$\frac{7}{5}$	1	14

33.

x	y	z	s_1	s_2	s_3	m	
1	3	0	1	0	0	0	300
1	0	1	0	1	0	0	150
1	2	1	0	0	1	0	200
-1	-3	-4	0	0	0	1	0

x	y	z	s_1	s_2	s_3	m	
1	3	0	1	0	0	0	300
1	0	1	0	1	0	0	150
0	2	0	0	-1	1	0	50
3	-3	0	0	4	0	1	600

35. Maximize: $m = 2x + 5y + 30z$

Subject to:
$$3x + 5y + 10z \leq 1500$$
$$x + y + z \leq 200$$
$$-x - y + z \leq 0$$
$$x \geq 0, y \geq 0, z \geq 0$$

x	y	z	s_1	s_2	s_3	m	
1	1	1	1	0	0	0	200
3	5	10	0	1	0	0	1500
-1	-1	1	0	0	1	0	0
-2	-5	-30	0	0	0	1	0

x	y	z	s_1	s_2	s_3	m	
1	1	1	1	0	0	0	200
0	2	7	-3	1	0	0	900
0	0	2	1	0	1	0	200
0	-3	-28	2	0	0	1	400

37. Minimize: $c = 44x + 40y + 30z$

Subject to:
$$3x + 2y + z \geq 24$$
$$3x + 6y + 4z \leq 50$$
$$2x + y + z \leq 20$$
$$x \geq 0, y \geq 0, z \geq 0$$

x	y	z	s_1	s_2	s_3	m	
3	2	1	1	0	0	0	24
3	6	4	0	1	0	0	50
2	1	1	0	0	1	0	20
-44	-40	-30	0	0	0	1	0

39.

1	1	1	1	1	0	0	0	25,000
1	1	0	0	0	1	0	0	10,000
1	1	1	0	0	0	1	0	15,000
1	0	0	1	0	0	0	1	20,000
-.08	-.10	-.12	-.11	0	0	0	0	0

41.

1	1	1	1	0	0	60
0	1	1	0	1	0	40
1	1	0	0	0	1	50
-5	-6	-3	0	0	0	0

43. Total capacity for the four machines is 300 units. Machines one and two have a combined capacity of 100 units. Machines two, three, and four have a combined capacity of 230 units. If the profit is $2 per item for machine one, $3 per item for machine two, and $4 per item for machine three, how many items should be made on each machine to maximize profit?

Exercise Set 4–3

1. Row 1; the pivot is 2
3. Row 2; the pivot is 2
5. Row 3; the pivot is 6
7. The pivot is the $(2, 2)$ entry, 1.

$$\begin{array}{ccccc|c} x & y & s_1 & s_2 & m & \\ \left[\begin{array}{ccccc|c} -4 & 0 & 1 & -2 & 0 & 10 \\ \hline 1 & 1 & 0 & 1 & 0 & 15 \\ \hline 1 & 0 & 0 & 1 & 1 & 35 \end{array}\right] \end{array}$$

9. The pivot is the $(1, 1)$ entry, 2.

$$\begin{array}{ccccc|c} x & y & s_1 & s_2 & m & \\ \left[\begin{array}{ccccc|c} 2 & 1 & 0 & 1 & 0 & 10 \\ 0 & -\frac{1}{2} & 1 & -\frac{1}{2} & 0 & 3 \\ \hline 0 & 4 & 0 & 0 & 1 & 60 \end{array}\right] \end{array}$$

11. The pivot is the $(3, 5)$ entry, 1.

$$\begin{array}{ccccccc|c} x & y & z & s_1 & s_2 & s_3 & m & \\ \left[\begin{array}{ccccccc|c} 1 & 1 & -2 & -6 & 0 & 0 & 0 & 2 \\ 0 & 1 & 0 & 1 & 0 & 1 & 0 & 2 \\ 0 & 1 & 1 & 3 & 1 & 0 & 0 & 9 \\ \hline 0 & 4 & 3 & 8 & 0 & 0 & 1 & 30 \end{array}\right] \end{array}$$

13. $x = 9,\quad y = 0,\quad m = 45$
15. $x = \frac{5}{2},\quad y = \frac{3}{2},\quad m = \frac{41}{2}$
17. $x = 0,\quad y = 0,\quad z = 30,\quad m = 90$
19. $x = 15,\quad y = 0,\quad z = 0,\quad m = 45$
21. $x = 5,\quad y = 3,\quad m = 46$
23. $x = 12,\quad y = 12,\quad m = 204$
25. $x = 4.8,\quad y = 4.8,\quad m = 67.2$
27. $x = 0,\quad y = 25,\quad z = 150,\quad m = 675$
29. $x = 0,\quad y = 0,\quad z = 30,\quad w = 20,\quad m = 190$
31. $x = 30,\quad y = 0,\quad z = 0,\quad w = 0,\quad m = 240$
33. $x = 105$ first class, $y = 25$ executive class, $z = 60$ economy class, $m = \$25{,}200$
35. $x = 71$ deluxe, $y = 0$ regular, $z = 44$ economy, $m = \$13{,}900$
37. $x = 120$ small, $y = 80$ medium, $z = 0$ large, $m = \$15{,}600$
39. $x = 0$ type 1, $y = 0$ type 2, $z = 72$ type 3, $m = 288$ zones
41. $x = \$20{,}000$ first payment, $y = \$20{,}000$ second payment, $z = \$10{,}000$ third payment

Exercise Set 4–4

1. (iii) The sense of the inequality is reversed.
3. (i) The objective function is minimized. (iii) The sense of the second inequality is reversed.
5. (i) The objective function is minimized. (iii) The sense of the second and third inequalities is reversed.
7. Maximize: $-m = -x - y$
$$[-1\quad -1\quad 0\quad 0\quad 1\quad 0]$$
9. Maximize: $-m = -x - 3y$
$$[-1\quad -3\quad 0\quad 0\quad 0\quad 1\quad 0]$$
11. Maximize: $-m = -x + y - 2z$
$$[-1\quad 1\quad -2\quad 0\quad 0\quad 0\quad 1\quad 2]$$
13. Maximize: $-m = -x - y$
Subject to: $x + 2y \le 4$
$-x - 3y \le -1$
$x \ge 0,\quad y \ge 0$

15. Maximize: $-m = -2x + y$
Subject to: $-2x - y \leq -2$
$-x + y \leq -4$
$x \geq 0, \quad y \geq 0$

17. Maximize: $-m = -3x - y - z$
Subject to: $x \qquad - z \leq -2$
$-x - y \qquad \leq -4$
$x \geq 0, \quad y \geq 0, \quad z \geq 0$

19. $x = 12, \qquad y = 15, \qquad m = 60$
21. $x = 0, \qquad y = 50, \qquad m = 150$
23. $x = 11, \qquad y = 0, \qquad m = 44$
25. $x = 0, \qquad y = 3, \qquad m = -12$
27. $x = 0, \qquad y = 4, \qquad z = 0, \qquad m = 0$
29. $x = 0, \qquad y = 0, \qquad z = 4, \qquad m = 4$
31. $x = 0, \qquad y = 0, \qquad z = 0, \qquad m = 0$
33. $x = 0, \qquad y = 0, \qquad z = 0, \qquad u = 0, \qquad m = 0$
35. $x = 0$ high quality, $y = 200$ regular quality
37. $x = 10$ chocolate nut bars, $y = 0$ chocolate peanut bars, $z = 30$ chocolate cluster bars
39. $x = 0$ tank 1, $y = 3$ tank 2, $z = 0$ tank 3

Exercise Set 4–5

1. Minimize: $m = RX$
Subject to: $AX \geq C$ and $X \geq 0$, where

$$A = \begin{bmatrix} 3 & 2 \\ 4 & -5 \end{bmatrix} \qquad C = \begin{bmatrix} 4 \\ 9 \end{bmatrix} \qquad R = \begin{bmatrix} 4 & 2 \end{bmatrix} \qquad X = \begin{bmatrix} x \\ y \end{bmatrix}$$

3. Minimize: $m = RX$
Subject to: $AX \geq C$ and $X \geq 0$, where

$$A = \begin{bmatrix} 3 & 5 \\ -1 & 3 \\ 2 & -6 \end{bmatrix} \qquad C = \begin{bmatrix} 8 \\ 4 \\ 7 \end{bmatrix} \qquad R = \begin{bmatrix} 2 & 4 \end{bmatrix} \qquad X = \begin{bmatrix} x \\ y \end{bmatrix}$$

5. $\begin{bmatrix} 5 & 2 \\ -1 & -3 \end{bmatrix}$

7. $\begin{bmatrix} -4 & 2 \end{bmatrix}$

9. $\begin{bmatrix} 1 & 6 & -2 \\ -4 & 15 & 1 \\ 11 & 0 & -2 \end{bmatrix}$

11. Maximize: $m = 4u + 9v$
Subject to: $3u + 4v \leq 4$
$2u - 5v \leq 2$
$u \geq 0, \quad v \geq 0$

13. Maximize: $m = 8u + 4v + 7w$
Subject to: $3u - v + 2w \leq 2$
$5u + 3v - 6w \leq 4$
$u \geq 0, \quad v \geq 0, \quad w \geq 0$

15. $x = 5, \qquad y = 0, \qquad m = 5$
17. $x = 11, \qquad y = 0, \qquad m = 44$
19. $x = \frac{5}{7}$ serving chicken, $y = \frac{10}{7}$ serving rice
21. $x = 8, \qquad y = 0, \qquad z = 0$

23. $x = 2,$ $y = 0,$ $z = 3,$ $w = 6$
25. $x = 2,$ $y = 0,$ $z = 6$
27. $x = 50$ chicken, $y = 40$ rice, $z = 0$ spinach
29. $x = 0$ hopper 1, $y = 100$ hopper 2, $z = 140$ hopper 3

Case Study Exercise

1. From East Chicago to Niles   None
From East Chicago to DesPlains   50,000 barrels
From Hammond to Niles   120,000 barrels
From Hammond to DesPlains   50,000 barrels
at a cost of \$1590 per mile per 1000 barrels

Review Exercises

1. no

3. $4x + 3y + s_1 \qquad\qquad = 30$
$\quad 2x + 7y \qquad + s_2 = 65$

5.

x	y	z	s_1	s_2	s_3	m	
1	2	5	1	0	0	0	30
2	3	1	0	1	0	0	40
4	3	0	0	0	1	0	20
-1	-4	-2	0	0	0	1	0

7. not maximal

9.

x	y	z	s_1	s_2	s_3	m	
0	0	1	-2	0	-2	0	0
1	0	0	1	2	1	0	10
0	1	0	-1	-2	0	0	2
0	0	0	3	9	4	1	85

11. Maximize: $p = 45x + 9y + 7z$
Subject to: $500x + 150y + 100z \leq 1000$
$\qquad\qquad\qquad 4x + \quad 2y + \quad 2z \leq \quad 5$
$\qquad\qquad\qquad 9x + \quad 6y + \quad 5z \leq \quad 15$

x	y	z	s_1	s_2	s_3	m	
10	3	2	1	0	0	0	20
4	2	2	0	1	0	0	5
9	6	5	0	0	0	1	15
-45	-9	-7	0	0	0	1	0

x	y	z	s_1	s_2	s_3	m	
1	.3	.2	.1	0	0	0	2
0	.8	1.2	$-.4$	1	0	0	-3
0	3.3	3.2	$-.9$	0	1	0	-3
0	4.5	2	4.5	0	0	1	90

13.

x	y	z	s_1	s_2	s_3	m	
0	0	1	.6	$-.2$	.8	0	7.2
1	0	0	.4	.2	$-.2$	0	4.8
0	1	0	-1.6	.2	.8	0	5.8
0	0	3	1.6	.8	-2.8	1	61.2

15. $x = 15, s_2 = 35, y = z = s_3 = 0, m = 60$

17. Maximize: $-w = -2x - 5y$
Subject to: $x + 4y \le 4$
$-x - 2y \le -7$
$x \ge 0, y \ge 0$
19. $x = \frac{1}{3}, \quad y = \frac{5}{3}, \quad m = 14$

CHAPTER 5

Exercise Set 5–1
1. (a) $385.00   (b) $32.08   (c) $1.06
3. (a) $3669.85   (b) $494.02   (c) $44.86
5. (a) $869.00   (b) $19.31
7. (a) $1385.00   (b) $1032.08   (c) $1001.06
9. (a) $5657.85   (b) $2482.02   (c) $2032.86
11. (a) $2275.20   (b) $1637.93   (c) $1581.93
13. (a) $2500.00   (b) $4615.38   (c) $4986.15
15. discount = $600.00; proceeds = $1400.00; effective rate = 14%
17. discount = $3.29; proceeds = $172.71; effective rate = 11.4%
19. $1079.28
21. (a) $512.50   (b) $584.25
23. (a) $10,512.50   (b) $10,584.25
25. $28,121.48
27. 6.5%
29. 6.25%
31. $8\frac{1}{3}$ years (or 100 months)
33. 6.06 years
35. 30.8%
37. 11.1%
39. (a) $10,000   (b) $5000 at 50%, $7300 at 27%,   (c) 46%

Exercise Set 5–2
1. $95.51
3. $1189.10
5. $4808.39
7. $3172.65
9. $3148.91
11. $2208.04
13. $1064.40
15. $709.32
17. $1225.35
19. $1083.11
21. 5.654%
23. $7472.58
25. $573.09
27. $7408.36
29. $2790.87
31. (a) $862.22   (b) $870.45   (c) $874.51
35. $7903.15
37. lump sum of $10,000, by $5.00
41. 6.48%

43. $5,080,265.37
45. $1,363,636.36 for 1985
$1,239,669.42 for 1986
$1,126,972.20 for 1987

Case Study Exercises
  1. $(18.9/6.2) - 1 = 2.05$, so 205%
  3. 12.8/9.1 is 41%, 18.4/12.8 is 44%, 24.2/18.4 is 32%, 31.3/24.2 is 29%; the compound growth was 36.2% since $(9.1)(1 + .362)^4 = 31.3$

Exercise Set 5–3
  1. $60.55
  3. $43,178.50
  5. (a) 276,108,118   (b) 336,574,255
  7. 7.4%
  9. $140,712.62
 11. $5.61
 13. $3741.66
 15. (a) 28,430,350   (b) 4,559,573   (c) 3,623,218   (d) 6,851,173   (e) 11,742,236   (f) 16,901,405
 17. 29,388.91 cruzeiros
 19. 11.9 years
 21. $938,972.38
 23. 281.27 pesos
 25. $266,627.43
 27. 4.32%

Exercise Set 5–4
  1. $9897.47
  3. $12,705.70
  5. $4291.48
  7. $2697.35
  9. $4507.74
 11. $4877.32
 13. $2724.32
 15. $1773.96
 17. $952.38
 19. $223.68
 21. $223.06
 23. $362,164.06
 25. $406,140.63
 27. $29,390.74
 29. $262.26
 31. $685.65

33. $3761.39
35. $13,432.70
37. $2.63
39. (a) 1070 trillion barrels   (b) 1144.9 trillion barrels   (c) 1967.15 trillion barrels   (d) 13,816.45 trillion barrels
41. No, Atlanta's market will be 90,362, and Detroit's market will be 91,802

Exercise Set 5–5
  1. $2930.19
  3. $385.44
  5. $1891.39
  7. $5310.85
  9. $949.59
11. $964.21
13. $58.75
15. $124.81
17. $539.84
19. $880.52

21.

Payment	Interest	Principal	Balance
0			4000.00
1	240.00	709.59	3290.41
2	197.42	752.17	2538.24
3	152.29	797.30	1740.95
4	104.46	845.13	895.82
5	53.75	895.84	−0.02

23.

Payment	Interest	Principal	Balance
0			60,000.00
1	450.00	89.84	59,910.16
2	449.33	90.51	59,819.65
3	448.65	91.19	59,728.46
4	447.96	91.88	59,636.58

25. $210.67
27. (a) $5035.18   (b)

Payment	Interest	Principal	Balance
0			600,000.00
1	4500.00	535.18	599,464.82
2	4495.99	539.19	598,925.63
3	4491.94	543.24	598,382.39
4	4487.87	547.31	597,835.08

29.

Payment	Interest	Principal	Balance
0			2500.00
1	12.50	98.30	2401.70
2	12.01	98.79	2302.91
3	11.51	99.29	2203.62
4	11.02	99.78	2103.84
5	10.52	100.28	2003.56
6	10.02	100.78	1902.78
7	9.51	101.29	1801.49
8	9.01	101.79	1699.70
9	8.50	102.30	1597.40
10	7.99	102.81	1494.59
11	7.47	103.33	1391.26
12	6.96	103.84	1287.41
13	6.44	104.36	1183.05
14	5.92	104.88	1078.17
15	5.39	105.41	972.76
16	4.86	105.94	866.82
17	4.33	106.47	760.36
18	3.80	107.00	653.36
19	3.27	107.53	545.82
20	2.73	108.07	437.75
21	2.19	108.61	329.14
22	1.65	109.15	219.99
23	1.10	109.70	110.29
24	0.55	110.25	0.04

31. $6,144,567.11

33. $1755.81

35. $30,578.36

37. $586.51

39. $57,570.67

Review Exercises

1. $262.50

3. $270.40

5. (a) $7938.32   (b) $7866.49

7. $186.44

9. 92.3%

11. $3353.02

13. $109,729.02

15. $10,046.26

17. (a) $617.17   (b)

Payment	Interest	Principal	Balance
0			60,000.00
1	600.00	17.17	59,982.83
2	599.83	17.34	59,965.49
3	599.65	17.52	59,947.97
4	599.48	17.69	59,930.28

CHAPTER 6

Exercise Set 6–1

1. not well-defined
3. well-defined
5. {1, 2, 3, 4, 6, 12}
7. {−3, −2, −1, 0}
9. A: no, B: yes
11. A: yes, B: yes
13. $X' = \{4, 5, 6\}$
15. $X' = \{x \mid x \in N; x > 3\}$
17. $X \cup Y = \{1, 2, 3, 4\}$
19. $X \cup Y = Q$
21. $X \cap Y = \{3, 4\}$
23. $X \cap Y = \{1, 2\}$
25. A: yes B: no
27. A: yes B: yes
29. $X' = \{2, 4, 6, \ldots\}$
31. $X \cup Y = \{-2, -1, 0, 1, 2, 3, \ldots\}$
33. $\{\ldots, -3, -2, -1, 1, 2, 3, \ldots\}$
35. $X \cap Y = \{2, 3, 4, \ldots\}$
37. $X \cap Y = \{1, 2, 3\}$
39. $F \cap X = \{12, 24, 36, 48, \ldots\}$
41. {Denver, Phoenix, Columbus}
43. For each of the 10 major cities, either the cost of housing is more than $100 or the cost of miscellaneous goods and services is more than $100
45. (a) $F = $ {Brazil, Canada, China}   (b) $T = $ {Canada, U.S., U.S.S.R.}
47. $\varnothing$ contains no elements, whereas $\{\varnothing\}$ contains one element
49. (a) $A = $ {Boston, Los Angeles, Miami, New York, Washington}
    (b) $B = $ {Chicago, Los Angeles, New York, San Francisco, Washington}
    (c) every city having one of the five highest CBD rates would also have to have one of the five highest rates outside CBDs
    (d) no

Exercise Set 6–2

1. 5
3. 7
5. 3
7. 20
9. 4
11. 65
13. 35,152,000
15. 5,000
17. (a) 2   (b) 6   (c) 8
19. $n(A) = 6$ and $n(B) = 4$
21. 3
23. 20

25.

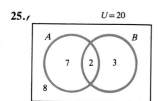

27. 14
29. yes; if and only if $A = B$
31. $2^3 = 8$
33. yes; $n(\varnothing) = 0$ and $n(\{\varnothing\}) = 1$
35. 1429
37. 719
39. 30%
41. (a) yellow or white
    (b) magenta or white
    (c) cyan or white
    (d) white
    (e) black
43. $2^6 - 1 = 63$ regions; see the reference for the Venn diagram

Exercise Set 6–3

1. 16 outcomes

	Juice	Eggs	Drinks

orange
  scrambled — coffee / tea
  poached — coffee / tea
  soft–boiled — coffee / tea
  sunny — coffee / tea

apple
  scrambled — coffee / tea
  poached — coffee / tea
  soft–boiled — coffee / tea
  sunny — coffee / tea

3. four selections

dime — dime / penny
penny — dime / penny

5. 24 choices

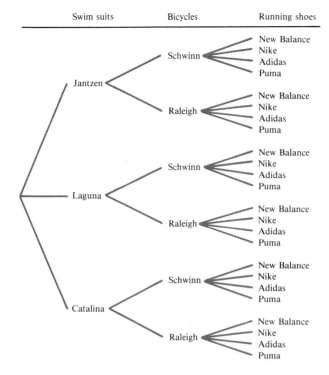

Swim suits	Bicycles	Running shoes

7. 95,040
9. 420
11. 15,600
13. 676,000
15. 1,000
17. 27,000

19.

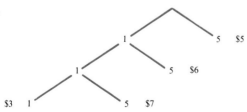

21. $216 = 6^3$
23. $60,466,176 = 6^{10}$
25. $9 = 3^2$
27. $59,049 = 3^{10}$
29. 36^6
31. 8,000,000
33. 24
35. 12 ways: JILL, JLIL, JLLI, IJLL, ILJL, ILLJ, LJIL, LJLI, LIJL, LILJ, LLJI, LLIJ

37. 24

39. 67,600

41. 1,326

43. 1976 (26 with three alike, and 3·650 with two alike)

45. 3,628,800 = 10·9·8·7·6·5·4·3·2·1

47.

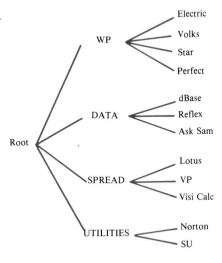

Exercise Set 6–4

1. (a) 20   (b) 10

3. (a) 30   (b) 45

5. (a) 1   (b) 1

7. (a) 1   (b) 120

9. 90

11. 13,800 = 25·24·23

13. 12

15. 720

17. 48

19. 15,504

21. (a) 21   (b) 42

23. 24

25. 56

27. 64

29. (a) 792   (b) 120

31. 1950

33. 845,000

35. 120

37. 900

39. 219 (9 with 3 digits the same, 90 with exactly two digits the same, and 120 with three distinct digits)

41. 72

43. (a) because 5 + 9 = 14
(b) because $r + (n - r) = n$

45. (a) The number of subsets having 0, 1, 2, 3, 4, 5 elements is $\binom{5}{0}$, $\binom{5}{1}$, $\binom{5}{2}$, $\binom{5}{3}$, $\binom{5}{4}$, $\binom{5}{5}$, respectively.

Therefore the number of possible subsets is $\binom{5}{0} + \binom{5}{1} + \binom{5}{2} + \binom{5}{3} + \binom{5}{4} + \binom{5}{5} = 2^5 = 32$.

(b) The reasoning is similar to (a).

47. 1

49. The scam was discovered only because of the large amount bet on the winning number. It would not have been discovered if less money had been bet.

51. $8! = 40{,}320$

53. There are $6 = C(4, 2)$ ways of choosing two numbers, without regard to order, from the set $\{1, 2, 3, 4\}$.

Case Study Exercises

1. 1,352,000
3. 16,208,000
5. 12,000,000
7. no

Review Exercises

1. (a) not well-defined   (b) well-defined
3. (a) $A' = \{f, g, h, i, j, \ldots, z\}$ (b) $V' \cap A' = \{f, g, h, j, k, l, m, n, p, q, r, s, t, v, w, x, y, z\}$
(c) $(V \cup A)' = V' \cap A'$
5. 10
7. X' is the set of irrational numbers
9. (a) 3   (b) 17   (c) 13   (d) 22
11. 324
13. (a) 20,736   (b) 11,880
15. 40,320
17. 635,376
19. (a) 6,840   (b) 680
21. 15,600

CHAPTER 7

Exercise Set 7–1

1. $S = \{\text{odd, even}\}$
3. $S = \{p, r, o, b, a, i, l, t, y\}$
5. $S = \{\text{red, black}\}$
7. $S = \{\text{Democratic, Republican, Independent}\}$
9. let g = gained weight; l = lost weight; s = stayed the same
$S = \{gg, gl, gs, lg, ll, ls, sg, sl, ss\}$
11. let i = increases, d = decreases; u = unchanged
$S = \{iiii, iiid, iiiu, iidi, iidd, iidu, \ldots\}$
There are 81 possible combinations in the sample space.
13. let c = corn; p = potatoes; a = alfalfa
$S = \{cpa, cap, pca, pac, apc, acp\}$
15. $E \cup F = \{1, 2, 3, 4\}; E \cap F = \{2, 3\}; E' = \{4, 5, 6\}$
17. $E \cup F = \{p, r, o, b, a, i, l, t, y\}; E \cap F = \varnothing; E' = \{p, r, b, l, t, y\}$
19. $E \cup F = \{\text{red card}\}; E \cap F = \{\text{heart}\}; E' = \{\text{club, spade, diamond}\}$

21. $E \cup F = \{2, 3, 4, 5, 6, 7\}$; $E \cap F = \{3, 4\}$; $E' = \{5, 6, 7, 8, 9, 10, 11, 12\}$

23. $E \cup F = \{0 \text{ heads}, 1 \text{ head}, 2 \text{ heads}\}$; $E \cap F = \{1 \text{ head}\}$; $E' = \{2 \text{ heads}\}$

25. yes

27. no

29. $S = \{AB, AC, AR, AS, AT, BC, BR, BS, BT, CR, CS, CT, RS, RT, ST\}$;
$E = \{AR, AB, AC, AS, AT, BC, BR, BS, BT, CR, CS, CT\}$; $\quad$ $F = \{CR, CS, CT, RS, RT, ST\}$;
$G = \{AC, AS, BC, BS, CR, CS, CT, RS, ST\}$

31. (a) $E \cup F = S = \{yyy, yyn, yny, ynn, nyy, nyn, nny, nnn\}$
(b) $E \cap F = \{yyn, yny, nyy\}$
(c) $E' = \{ynn, nyn, nny, nnn\}$

33. (a) $E \cap F$ $\quad$ (b) $E \cap F'$ $\quad$ (c) E' $\quad$ (d) $E \cup F$

35. $\{0, 1, 2, 3\}$, $\quad$ (a) $\{0, 2, 3\}$, $\quad$ (b) $\{2\}$, $\quad$ (c) $\{0, 1\}$

37. yes, $E \cap E' = \varnothing$

39. (a) $\varnothing$, $\{1\}$, $\{2\}$, $\{3\}$, $\{1, 2\}$, $\{2, 3\}$, $\{1, 3\}$, S
(b) $\varnothing$ $\{1\}$, $\{2\}$, . . . , $\{n\}$, $\{1, 2\}$, . . . , $\{1, n\}$, $\{2, 3\}$, . . . , $\{2, n\}$, . . . , S

41. The patient had (a) influenza or took aspirin, (b) took aspirin or had an (a-d) diet, (c) had influenza and an a-d diet, (d) took aspirin and had an a-d diet, (e) did not have influenza, (f) had neither influenza nor took aspirin, (g) had influenza, took aspirin, had an a-d diet

Exercise Set 7–2

1. $\frac{1}{6}$

3. $\frac{1}{2}$

5. $\frac{2}{3}$

7. $\frac{1}{2}$

9. $\frac{3}{4}$

11. .43

13. .65

15. $\frac{3}{4}$

17. $\frac{3}{52}$

19. $\frac{25}{26}$

21. yes

23. no, the sum of the probabilities $= 1.1$; must $= 1$

25. (a) $\frac{1}{2}$ $\quad$ (b) $\frac{1}{2}$

27. $\frac{9}{19}$

29. $\frac{3}{10}$

31. (a) $\frac{2}{13}$ $\quad$ (b) $\frac{1}{13}$ $\quad$ (c) $\frac{1}{2}$

33. $Pr(E) = .83$; $Pr(F) = .17$

35. $Pr(E) = .45$; $Pr(F) = .2$, $Pr(G) = .35$

37. $Pr(E') = .55$; $Pr(F') = .8$

39. .7

41. (a) .2 $\quad$ (b) .32 $\quad$ (c) .52

43.

x	a	b	c
p	$\frac{6}{11}$	$\frac{3}{11}$	$\frac{2}{11}$

45. (a) $\dfrac{8442}{12{,}763}$ $\quad$ (b) $\dfrac{4321}{12{,}763}$ $\quad$ (c) $\dfrac{5232}{12{,}763}$ $\quad$ (d) $\dfrac{7531}{12{,}763}$

47. .03, .29

Exercise Set 7–3

1. $\frac{10}{36} = \frac{5}{18}$

3. $\frac{2}{3}$

5. $\frac{8}{13}$

7. $\frac{4}{13}$

9. $\frac{2}{3}$

11. 1

13. .75

15. .33

17. .11

19. $\frac{4}{11}$

21. $\frac{2}{7}$

23. odds in favor of E = 2 to 7; odds against F = 4 to 3

25. odds in favor of E = 13 to 37; odds against F = 31 to 19

27. 1

29. (a) .32   (b) .3

31. (a) $\frac{9}{19}$ (b) $\frac{9}{19}$ (c) $\frac{18}{19}$ (d) $\frac{10}{19}$

33. (a) 7 to 13   (b) 7 to 13

35. .63

37. .15

39. .3

43. .68 and .44

45. .88

Exercise Set 7–4

1. $\frac{6}{64} \simeq .09$

3. $\frac{22}{64} \simeq .34$

5. $\frac{15}{16} \simeq .9375$

7. $\frac{7}{99} \simeq .07$

9. $\frac{42}{99} \simeq .424$

11. $\frac{10}{11} \simeq .91$

13. $\frac{35}{38} \simeq .92$

15. $1/204 \simeq .0049$

17. $226/2118760 \simeq .00011$

19. $12/51 \simeq .24$

21. $22/425 \simeq .052$

23. $62/125 \simeq .496$

25. $2093/5000 \simeq .4186$

27. (a) $15/128 \simeq .117$ (b) $11/64 \simeq .1719$ (c) $121/128 \simeq .9453$

29. (a) $14/323 \simeq .0433$ (b) $94/323 \simeq .291$

31. (a) .0002401   (b) .00394

33. .6032

35. $110/143 \simeq .77$

37. .856

39. $\dfrac{60,460}{1,048,566}$

41. $\dfrac{3}{91}$

43. $\dfrac{19}{989}$

Exercise Set 7–5

1. $\frac{1}{3}$
3. $\frac{2}{13}$
5. $\frac{7}{25}$
7. $\frac{7}{20}$
9. .9366 . . .
11. $\frac{2}{3}$
13. $\frac{64}{67} \simeq .96$
15. $\frac{19}{83} \simeq .23$
17. not independent
19. independent
21. .005
23. $Pr(E \mid F') = \frac{1}{3};$ $Pr(E' \mid F') = \frac{2}{3}$
25. $79/180 \simeq .44$
27. (a) $\frac{1}{7}$ (b) $\frac{12}{35}$ (c) $\frac{3}{7}$
29. (a) 8%   (b) 36%
31. $\frac{18}{43} \simeq .42$
33. $\frac{12}{57} \simeq .21$
35. $Pr(F \mid E) = Pr(F \cap E)/Pr(E) = Pr(F)Pr(E)/Pr(E) = Pr(F)$
37. (b) Use $Pr(E \cap F) + Pr(E' \cap F) = Pr(F)$
    (a) and (c), Use Example 1 to disprove
39. Use the addition principle, the rule for complements and two factorizations
41. (a) $\frac{3}{13}$ (b) $\frac{4}{26} = \frac{2}{13}$ (c) $\frac{17}{26}$
43. $.000000736, Pr(E) \ne Pr(E \mid C),$ not independent

Exercise Set 7–6

1. $\frac{9}{13} \simeq .69$
3. $\frac{2}{23} \simeq .087$
5. $\frac{4}{27} \simeq .15$
7. $Pr(U \cap E) = .18;$ $Pr(H \cap E) = .56$
9. $Pr(U \mid E) = \frac{9}{37} \simeq .24;$ $Pr(H \mid E) = \frac{28}{37} \simeq .76$
11. $Pr(U \cap E) = .18;$ $Pr(H \cap E) = .18;$ $Pr(W \cap E) = .35$
13. $\frac{18}{71} \simeq .25$
15. $\frac{35}{71} \simeq .49$
17. $\frac{12}{29} \simeq .41$
19. $\frac{5}{13} \simeq .38$
21. $\frac{2}{13} \simeq .15$
23. $\frac{24}{59} \simeq .41$
25. $\frac{20}{73} \simeq .27$
27. $25/107 \simeq .23$
29. $\frac{3}{11} \simeq .27$
31. $\frac{2}{5}$
33. .276
35. $60/166 \simeq .36$
39. $\frac{55}{92}$
41. $\dfrac{51}{157}$

Exercise Set 7–7

1. $\frac{15}{64} \simeq .23$
3. $\frac{11}{32} \simeq .32$

5. $15/1024 \simeq .0146$

7. .0046

9. $16/1024 \simeq .0156$

11. 25

13. 2

15. $\frac{2}{13}$, no

17. $\frac{4}{9}$, no

19. Site 2

21. (a) .29　　(b) .8

23. (a) .158　　(b) .178

25. (a) .088　　(b) .104

27. 3

29. $-.499$

31. $-.588$

33. \$19,250; with averages of 5000, 15,000, 25,000, and 40,000

35. 1.0

37. .3

39. E (American) $= -\frac{1}{19}$

x	$-.5$	-1	0	1
p	$\dfrac{1}{1369}$	$\dfrac{684}{1369}$	$\dfrac{18}{1369}$	$\dfrac{18}{37}$

$$E \text{ (European)} = -\frac{18.5}{1369}$$

41. Strategy A

Case Study Exercises

1. .7　　**3.** .5　　**5.** 218.04

Review Exercises

1. {m, a, t, h, e, i, c, s}

3. (a) $E \cup F = F = \{\text{red, blue}\}$　　(b) $E \cap F = E = \{\text{blue}\}$　　(c) $E' = \{\text{red}\}$

5. (a) $\frac{1}{13}$　　(b) $\frac{1}{26}$　　(c) $\frac{3}{26}$

7. (a) .1　　(b) .9

9. $Pr(E \cap F) = .14$;　　$Pr(E' \cup F') = .86$

11. $\frac{6}{13} \simeq .46$

13. $79/4096 \simeq .0019$

15. $\frac{13}{25} = .52$

17. no

19. $\frac{3}{7}$

21. 75

CHAPTER 8

Exercise Set 8–1

1. .2856

3. .4283

5. $\frac{3}{8}$

7. $\frac{1}{2}$

9.

11.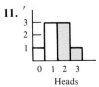

13.
x	f
0	1
1	2
2	1

15.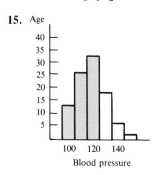

17. There are more lower (100–120) blood pressures among the nonusers group. The pill may elevate blood pressure.

19.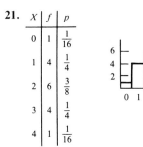

21.
X	f	p
0	1	$\frac{1}{16}$
1	4	$\frac{1}{4}$
2	6	$\frac{3}{8}$
3	4	$\frac{1}{4}$
4	1	$\frac{1}{16}$

$Pr(X \geq 1) = \frac{15}{16}$

23.

X	f	p
12	1	$\frac{1}{36}$
11	2	$\frac{1}{18}$
10	3	$\frac{1}{12}$
9	4	$\frac{1}{9}$
8	5	$\frac{5}{36}$
7	6	$\frac{1}{6}$

X	f	p
6	5	$\frac{5}{36}$
5	4	$\frac{1}{9}$
4	3	$\frac{1}{12}$
3	2	$\frac{1}{18}$
2	1	$\frac{1}{36}$

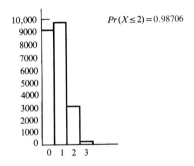

$Pr(6 \le X \le 8) = \frac{4}{9}$

25.

X	f	p
3	286	.01294
2	3042	.13765
1	9633	.43588
0	9139	.41353

$Pr(X \le 2) = 0.98706$

27. $\frac{19}{24} \simeq .76$

29. $\frac{7}{24} \simeq .29$

31.

Score
Quiz 1

Score
Quiz 2

33. 0.7

35. 0.7

37. Attendance increased each year until 1980–1981, and has decreased since then. Ticket prices increased throughout the time period.

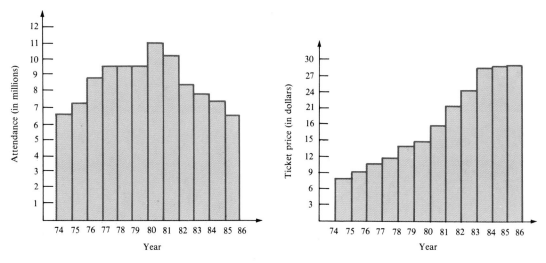

39. Each histogram is close to a normal curve.

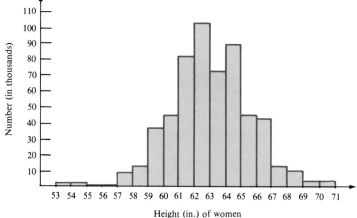

41. Fear of 13
43. No
45. No. The sum of the probabilities is only .95.

Exercise Set 8–2

1. mean $= 161/10 = 16.1$;   median $= 8$
3. mean $= .32/6 \simeq .053$;   median $= 0.055$
5. mean $= \frac{23}{14} \simeq 1.64$;   median $= 1.5$
7. mean $= \frac{47}{7} \simeq 6.7$;   median $= 9$
9. $E(X) = \frac{31}{8}$
11. $E(X) = 3.3$
13. mean $= \frac{37}{22} \simeq 1.68$
15. expected value $= \frac{13}{8} = 1.625$
17. mean $= -.5$;   median $= 0$
19. mean $= 6.695$;   median $= 7.3$
21. mean $= 0$;   median $= 0$
23. $E(X) = -.1$
25. $E(X) = -2$
27. $E(X) = 1.5$
29. mean $= 230.3$ pounds;   median $= 225$ pounds
31. mean $= \frac{43}{13} \simeq 3.3$
33. (a) Each entry is 0   (b) Each entry is 2   (c) no
35. (a) mean $= 4.3/11 \simeq 0.39$ (b) reward
37. (a) 12.09   (b) 8   (c) people argued for median, bank argued for mean
   (d) the payment via the mean is $(20)(\frac{7}{12})(\$9500) = \$166,250$.
      the payment via the median is $(20)(\frac{7}{8})(\$9500) = \$110,833.33$.
      the difference is $55,416.67
39. The expected number is $-.2$ so the expected death month is very close to the birthday.
41. mean is 62.57 inches, median is 62 inches
43. 5.908 (or about 6 games)

Exercise Set 8–3

1. $V(X) = 5.05$ $\sigma = 2.25$
3. $V(X) = 11.89$ $\sigma = 3.45$
5. $V(X) = 107.4$ $\sigma = 10.36$
7. $V(X) = 10,229/900 = 11.37$ $\sigma = 3.37$
9. $V(X) = 12,848/1764 \simeq 7.28$ $\sigma = 2.7$
11. $V(X) = 151/64 \simeq 2.36$ $\sigma = 1.54$
13. $V(X) = .61$ $\sigma = .78$
15. $13,800 $\sigma = 3763$
17. $Pr(40 \leq X \leq 60) = .75$
19. $V(X) = .108$ $\sigma = .329$
21. $V(X) = .836$ $\sigma = .914$
23. $V(X) = 21,417,600/10,201 \simeq 2883.796$ $\sigma = 53.701$
25. $V(X) = 3.29$ $\sigma = 1.814$
27. $V(X) = 41$ $\sigma = 6.403$
29. $\sigma = 1.26$
31. $\sigma = .992$
33. $\sigma = \sqrt{3}/2 \simeq .87$
35. (a) $\mu = 1.87$ (b) $\sigma = 1.26$
39. 3.4
41. $\sqrt{n}/2$

43.

	EF	TS	C	AG
mean	204.0	200.3	198.6	200.0
σ	11.6	8.5	6.4	10.4

This method of comparison is not conclusive; a more exhaustive analysis must be performed.
45. (a) 3.87%   (b) 65.84%   (c) the Standard Line
47. The authors are convinced

Exercise Set 8–4
1. about .475
3. about .34
5. about .16
7. about .025
9. .0668
11. .6826
13. 44.35%
15. 11.51%
17. 24.17%
19. 59.87%
21. about .135
23. about .815
25. about .84
27. .6480
29. .6541
31. 38.21%
33. .3812
35. 0.99% (less than 1%)
37. (a) 9.18%   (b) 84.13%
39. (a) .8413   (b) .7088

Exercise Set 8–5
1. $\mu = 72$ $\sigma = 6$
3. $\mu = 40.5$ $\sigma = 4.5$
5. .1841
7. .4041
9. .1210
11. .6757
13. .9370
15. .1635
17. .1841
19. .7286
21. .9945
23. .9328
25. .7088
27. .6886
29. .8485
31. .2643
33. .9576
35. .9279
37. .5516
39. virtually zero

41. .1788

43. (a) .9972   (b) 166

45. (a) .9987   (b) .0013

Case Study Answers

1.

	H	M	D
$\bar{x}$	.44	.73	.80
σ	.33	.59	.35

Not strong enough

3.

	H	M	D
$\bar{x}$	7.04	11.52	4
σ	2.43	3.44	1.73

Not strong enough

5.

	H	M	D
$\bar{x}$	3.46	7.68	8.5
σ	1.74	2.70	2.47

Madison

7.

	H	M	D
$\bar{x}$	7.67	5.96	6.83
σ	2.13	2.16	2.07

Not strong enough

9.

	H	M	D
$\bar{x}$	.58	1.11	.77
σ	.43	.83	.45

Hamilton

Review Exercises

1.

3.

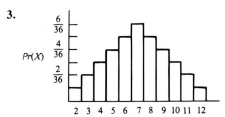

5.

X	probability
3	0.0129
2	0.1376
1	0.4359
0	0.4135

7. .9871

9. mean = 2.4;   median = 2

11. $-\frac{1}{4}$

13. (a) $15,000   (b) $17,916.67

15. average cost = \$148.90; $\sigma = 37.31$
17. (a) .6924   (b) .8531
19. .0694
21. $.1294 = C(18, 14)(2/3)^{14} (1/3)^4$
23. $\sigma = 2$
25. 1200
27. .9505

CHAPTER 9

Exercise Set 9–1

1. $T = \begin{bmatrix} .85 & .37 \\ .15 & .63 \end{bmatrix}$

3. This matrix can be a transition matrix.
5. This matrix can be a transition matrix.
7. $a = .5,$ $b = .1,$ $c = .2$
9. (a) .8   (b) .4
11. This could be a probability distribution.
13. This matrix could not be a probability distribution since the sum of the column entries is not 1.
15.   Stage 1   Stage 2   Stage 3
$\begin{bmatrix} .3 \\ .7 \end{bmatrix}$ $\begin{bmatrix} .38 \\ .62 \end{bmatrix}$ $\begin{bmatrix} .348 \\ .652 \end{bmatrix}$

17.   Stage 1   Stage 2   Stage 3
$\begin{bmatrix} .42 \\ .58 \end{bmatrix}$ $\begin{bmatrix} .332 \\ .668 \end{bmatrix}$ $\begin{bmatrix} .3672 \\ .6328 \end{bmatrix}$

19.   Stage 1   Stage 2   Stage 3
$\begin{bmatrix} .46 \\ .54 \end{bmatrix}$ $\begin{bmatrix} .316 \\ .684 \end{bmatrix}$ $\begin{bmatrix} .3736 \\ .6264 \end{bmatrix}$

21. .3
23. .7
25. .25
27. $T = \begin{bmatrix} .8 & .6 \\ .2 & .4 \end{bmatrix}$

29. $T = \begin{bmatrix} .1 & .45 & .55 \\ .3 & .05 & .35 \\ .6 & .50 & .10 \end{bmatrix}$

31. $T = \begin{bmatrix} .90 & .2 & .3 \\ .05 & .7 & .1 \\ .05 & .1 & .6 \end{bmatrix}$

33. $T = \begin{bmatrix} .10 & .10 & .65 \\ .45 & .60 & .30 \\ .45 & .30 & .05 \end{bmatrix}$

35. $\begin{bmatrix} .60 \\ .26 \\ .14 \end{bmatrix}$

37. $\begin{bmatrix} .37 \\ .39 \\ .24 \end{bmatrix}$

39. $\begin{bmatrix} .320 \\ .435 \\ .245 \end{bmatrix}$

41. $T^2 = \begin{bmatrix} .22 & .42 & .30 \\ .44 & .32 & .52 \\ .34 & .26 & .18 \end{bmatrix}$ $\qquad T^2V = \begin{bmatrix} .304 \\ .492 \\ .204 \end{bmatrix}$ $\qquad TV = \begin{bmatrix} .60 \\ .26 \\ .14 \end{bmatrix}$ $\qquad T(TV) = \begin{bmatrix} .304 \\ .492 \\ .204 \end{bmatrix}$

43. (a) .05   (b) Only those who were transferred were studied   (c) $\begin{bmatrix} .28 \\ .16 \\ .28 \\ .11 \\ .08 \\ .09 \end{bmatrix}$

Exercise Set 9–2

1. $T'V = \begin{bmatrix} .66 \\ .34 \end{bmatrix}$ $\qquad T^2V = \begin{bmatrix} .696 \\ .304 \end{bmatrix}$ $\qquad T^3V = \begin{bmatrix} .7176 \\ .2824 \end{bmatrix}$

3. $T'V = \begin{bmatrix} .35 \\ .65 \end{bmatrix}$ $\qquad T^2V = \begin{bmatrix} .305 \\ .695 \end{bmatrix}$ $\qquad T^3V = \begin{bmatrix} .2915 \\ .7085 \end{bmatrix}$

5. $T^6 V \simeq \begin{bmatrix} .743 \\ .257 \end{bmatrix}$

7. $T^6 V \simeq \begin{bmatrix} .286 \\ .714 \end{bmatrix}$

9. T is regular since $T^2 = \begin{bmatrix} .1 & .16 & .10 \\ .1 & .64 & .16 \\ .8 & .20 & .74 \end{bmatrix}$ Thus one of the powers of T contains all positive entries.

11. not regular, the $(1, 3)$ entry will be 0 in all T^n

13. $\begin{bmatrix} \frac{5}{11} \\ \frac{6}{11} \end{bmatrix}$

15. $\begin{bmatrix} \frac{3}{4} \\ \frac{1}{4} \end{bmatrix}$

17. $\begin{bmatrix} \frac{2}{7} \\ \frac{5}{7} \end{bmatrix}$

19. $\begin{bmatrix} \frac{1}{9} \\ \frac{2}{9} \\ \frac{2}{3} \end{bmatrix}$

21. $\begin{bmatrix} 45/117 \\ 40/117 \\ 32/117 \end{bmatrix}$

23. The retail store will have $\frac{5}{7}$ or approximately 0.71 of the business, while the competitor will have $\frac{2}{7}$ or approximately .29 of the business. Ad campaign will increase business by approximately 6%.

25. Stable matrix $= \begin{bmatrix} \frac{25}{95} \\ \frac{70}{95} \end{bmatrix}$. So in the long run it will rain about 26% (or $\frac{25}{95}$) of the time and not rain 74% of the time on the next day regardless of the previous day's weather.

27. In the long run a stock will be purchased 38% of the time, sold 36% of the time, and left unchanged 26% of the time.

29. In the long run the retailer will use TV 36% of the time, radio 26% of the time, and newspapers 38% of the time.

31. In the long run 75% of the debts becomes high priority, while 25% of the debts becomes low priority.

33. In the long run Brand A has a 71% share of the market, while Brand B has a 16% share and Brand C has a 13% share of the market.

35. In the long run the order will be large 24% of the time, medium 49% of the time, and small 27% of the time.

39. $\begin{bmatrix} .727 \\ .246 \\ .027 \end{bmatrix}$ **41.** $\begin{bmatrix} .26 \\ .45 \\ .28 \end{bmatrix}$

Exercise Set 9–3

1. All three states are nonabsorbing

3. State 4 is absorbing; the others are nonabsorbing

5. absorbing

7. nonabsorbing since it does not have an absorbing state

9. absorbing

11. nonabsorbing. State 1 can only go to state 3, which can only go to state 1. So state 1 cannot go to state 4, which is the absorbing state.

13.
$$T^4 = \begin{bmatrix} 1 & 0 & 0 \\ 0 & 1 & .9999 \\ 0 & 0 & .0001 \end{bmatrix} \quad \begin{matrix} \text{Stable} \\ \text{matrix} \end{matrix} = \begin{bmatrix} 1 & 0 & 0 \\ 0 & 1 & 1 \\ 0 & 0 & 0 \end{bmatrix}$$

15. $\begin{bmatrix} 1 & .3 & 0 \\ 0 & .2 & 1 \\ 0 & .5 & 0 \end{bmatrix}$

17.
$$\begin{array}{c} \\ 3 \\ 4 \\ 1 \\ 2 \end{array} \begin{array}{cccc} 3 & 4 & 1 & 2 \end{array} \\ \begin{bmatrix} 1 & 0 & 0 & .4 \\ 0 & 1 & 0 & .1 \\ 0 & 0 & .2 & .3 \\ 0 & 0 & .8 & .2 \end{bmatrix}$$

19.
$$\begin{array}{c} \\ 1 \\ 3 \\ 2 \\ 4 \end{array} \begin{array}{cccc} 1 & 3 & 2 & 4 \end{array} \\ \begin{bmatrix} 1 & 0 & .1 & 0 \\ 0 & 1 & .2 & 0 \\ 0 & 0 & .5 & 1 \\ 0 & 0 & .2 & 0 \end{bmatrix}$$

21. $\begin{bmatrix} \frac{5}{4} & 0 \\ \frac{1}{4} & 1 \end{bmatrix}$ or $\begin{bmatrix} 1.25 & 0 \\ .25 & 1 \end{bmatrix}$

23. $\begin{bmatrix} 1 & 0 & \frac{6}{10} & \frac{23}{35} \\ 0 & 1 & \frac{4}{10} & \frac{12}{35} \\ 0 & 0 & 0 & 0 \\ 0 & 0 & 0 & 0 \end{bmatrix}$

25. $\begin{bmatrix} 1 & 0 & .125 & 0 \\ 0 & 1 & .875 & 1 \\ 0 & 0 & 0 & 0 \\ 0 & 0 & 0 & 0 \end{bmatrix}$

27.
$$SV = \begin{bmatrix} .35 \\ .65 \\ 0 \end{bmatrix} \qquad SW = \begin{bmatrix} .53 \\ .47 \\ 0 \end{bmatrix}$$

29.

	\$0	\$3	\$1	\$2
\$0	1	0	$\frac{1}{3}$	0
$T = $ \$3	0	1	0	$\frac{2}{3}$
\$1	0	0	0	$\frac{1}{3}$
\$2	0	0	$\frac{2}{3}$	0

	\$0	\$3	\$1	\$2
\$0	1	0	$\frac{3}{7}$	$\frac{1}{7}$
$S = $ \$3	0	1	$\frac{4}{7}$	$\frac{6}{7}$
\$1	0	0	0	0
\$2	0	0	0	0

If Player A starts with \$2, he will end up with \$3 with a probability of $\frac{6}{7}$ or .857.

31.

	\$0	\$3	\$1	\$2
\$0	1	0	$\frac{3}{5}$	0
$T = $ \$3	0	1	0	$\frac{2}{5}$
\$1	0	0	0	$\frac{3}{5}$
\$2	0	0	$\frac{2}{5}$	0

	\$0	\$3	\$1	\$2
\$0	1	0	$\frac{15}{19}$	$\frac{9}{19}$
$S:$ \$3	0	1	$\frac{4}{19}$	$\frac{10}{19}$
\$1	0	0	0	0
\$2	0	0	0	0

If Player A starts with \$2, he will end up with \$3 with a probability of $\frac{10}{19}$ or .526.

35.
$$T^n = \begin{bmatrix} 1 & 0 & .222\ldots \\ 0 & 1 & .777\ldots \\ 0 & 0 & 0 \end{bmatrix}$$

37. (b) State L is absorbing.

(c) $\begin{bmatrix} 1 & 1 & 1 \\ 0 & 0 & 0 \\ 0 & 0 & 0 \end{bmatrix}$

39. (a) State L is absorbing, and each of the other states can reach L in one stage.
(b) State L is absorbing states GC and GE are nonabsorbing.

(c) $\begin{bmatrix} 1 & 1 & 1 \\ 0 & 0 & 0 \\ 0 & 0 & 0 \end{bmatrix}$

Case Study Exercise

1. $\begin{bmatrix} 0 \\ 0 \\ .918 \\ .082 \end{bmatrix}$

3. $\begin{bmatrix} 0 \\ 0 \\ .91 \\ .09 \end{bmatrix}$

5. $SV = \begin{bmatrix} 0 \\ 0 \\ \frac{53}{60} \\ \frac{7}{60} \end{bmatrix}$ Allowance for bad debts $= \frac{7}{60}$ or .1167 or 11.67%.

7. $SV = \begin{bmatrix} 0 \\ 0 \\ \frac{7}{8} \\ \frac{1}{8} \end{bmatrix}$ Allowance for bad debts $= \frac{1}{8}$ or .125 or 12.5%.

Review Problems

1. $\begin{bmatrix} .45 & .40 \\ .55 & .60 \end{bmatrix}$

3. not a transition matrix

5.

1st stage	2nd stage	3rd stage
$\begin{bmatrix} .27 \\ .29 \\ .44 \end{bmatrix}$	$\begin{bmatrix} .261 \\ .290 \\ .449 \end{bmatrix}$	$\begin{bmatrix} .2637 \\ .2927 \\ .4436 \end{bmatrix}$

7. .83

9. $TV = \begin{bmatrix} .34 \\ .66 \end{bmatrix}$ $T^2 = \begin{bmatrix} .496 \\ .504 \end{bmatrix}$ $T^3V = \begin{bmatrix} .4024 \\ .5976 \end{bmatrix}$

11. $\begin{bmatrix} \frac{1}{3} \\ \frac{2}{3} \end{bmatrix}$

13. $\begin{bmatrix} \frac{4}{11} \\ \frac{7}{11} \end{bmatrix}$ So in the long run it will rain $\frac{4}{11}$ or 36.36% of the time and not rain $\frac{7}{11}$ or 63.64% of the time.

15. State 4 is an absorbing state;   states 1, 2, and 3 are nonabsorbing.

17. $\begin{bmatrix} 1 & 0 & \frac{5}{7} & 0 \\ 0 & 1 & \frac{2}{7} & 1 \\ 0 & 0 & 0 & 0 \\ 0 & 0 & 0 & 0 \end{bmatrix}$

19. $\begin{bmatrix} \frac{5}{4} & 0 \\ \frac{1}{4} & 1 \end{bmatrix}$

21.

$$T = \begin{matrix} & \$0 & \$3 & \$1 & \$2 \\ \$0 & 1 & 0 & \frac{4}{7} & 0 \\ \$3 & 0 & 1 & 0 & \frac{3}{7} \\ \$1 & 0 & 0 & 0 & \frac{5}{7} \\ \$2 & 0 & 0 & \frac{3}{7} & 0 \end{matrix} \qquad S = \begin{matrix} & \$0 & \$3 & \$1 & \$2 \\ \$0 & 1 & 0 & \frac{28}{37} & \frac{16}{37} \\ \$3 & 0 & 1 & \frac{9}{37} & \frac{21}{37} \\ \$1 & 0 & 0 & 0 & 0 \\ \$2 & 0 & 0 & 0 & 0 \end{matrix}$$

If Player A starts with $2, he will end up with $3 with a probability of $\frac{21}{37}$ or .5675.

CHAPTER 10

Exercise Set 10–1

1. *R*   2
3. *C*   1
5. *R*   7
7. 4   4
9. *R* chooses row 2;   *C* chooses column 2; $v = 2$
11. *R* chooses row 1;   *C* chooses column 1; $v = 1$
13. *R* chooses row 2;   *C* chooses column 1; $v = 2$
15. *R* chooses row 3;   *C* chooses column 1; $v = 1$
17. no
19. no
21. yes, $v = 3$
23. yes, $v = 2$
25. yes, $v = 1$

27.

	Past	Future	Higher
TV	60%	20%	80%
Radio	25%	30%	5%
Newspaper	15%	50%	15%

Game is not strictly determined.

29. yes, $\quad v = 0$

31. yes, $\quad v = 1$

33.
$$\begin{array}{c} \\ 2 \\ 5 \end{array} \begin{array}{cc} 2 & 5 \\ \begin{bmatrix} -1 & -3 \\ 5 & -1 \end{bmatrix} \end{array} \quad \text{yes,} \quad v = -1$$

35.
$$\begin{array}{c} \\ 3 \\ 6 \end{array} \begin{array}{ccc} 2 & 4 & 7 \\ \begin{bmatrix} 5 & 5 & -4 \\ -4 & -4 & 5 \end{bmatrix} \end{array} \quad \text{no}$$

37.
$$\begin{array}{c} \\ H \\ T \end{array} \begin{array}{cc} H & T \\ \begin{bmatrix} -1 & -2 \\ 3 & -1 \end{bmatrix} \end{array} \quad \text{yes,} \quad v = -1$$

39.
$$\begin{bmatrix} 8 & 2 & 5 \\ 3 & 9 & 5.5 \end{bmatrix} \quad \text{no}$$

41.
$$\begin{array}{c} \\ (3, 0) \\ (2, 1) \\ (1, 2) \\ (0, 3) \end{array} \begin{array}{ccc} (2, 0) & (1, 1) & (0, 2) \\ \begin{bmatrix} 0 & 0 & 0 \\ 0 & -1 & 0 \\ 0 & -1 & 0 \\ 0 & 0 & 0 \end{bmatrix} \end{array} \quad \text{yes,} \quad v = 0$$

Case Study Exercises

1. hold in reserve

3. no

Exercise Set 10–2

1. yes

3. no

5. yes

7. 2×2 $\begin{bmatrix} 0.2 & 0.2 \\ 0.3 & 0.3 \end{bmatrix}$

9. 3×2 $\begin{bmatrix} 0.32 & 0.48 \\ 0.04 & 0.06 \\ 0.04 & 0.06 \end{bmatrix}$

11. 0

13. -2.34

15. .55

17. 2.02

19. 5

21. 7

23. $a = 1, \quad b = 0$

25. $a = 0, \quad b = 1$

27. $a = 1, \quad b = 0$

29. $a = 1, \quad b = 0$

31. 3.02/payoff with matrix Q is 3 more than the payoff with matrix P (3 is integer that was added to P).

33. 6.4/payoff with matrix Q is 7 more than the payoff with matrix P.

35. .38

37. .51

39. [0, 1], 1.2

41. 196

Exercise Set 10–3

1. $E = 2 - 2x$ $E = x$

3. $E = 2x - 1$ $E = 2x$

5. $E = 3x - 1$ $E = 3x - 2$

7. $E = x + 1$ $E = x - 2$

9. $E = 2x + 3$ $E = 4 - 2x$

11.

13.

15.

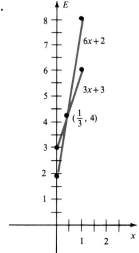

17. $R = [\frac{2}{3} \ \frac{1}{3}]$ $\qquad E = \frac{5}{3}$
19. $R = [1 \ 0]$ $\qquad E = 2$
21. $R = [\frac{12}{23} \ \frac{11}{23}]$ $\qquad E = 385/23$
23. $R = [.6 \ .4]$ $\qquad E = 4$
25. $R = [\frac{7}{11} \ \frac{4}{11}]$ $\qquad E = 85/11$
27. $R = [0 \ 1]$ $\qquad E = 1$
29. $R = [0 \ 1]$ $\qquad E = 1$
31. $R = [\frac{2}{9} \ \frac{7}{9}]$ $\qquad E = 44/27$
33. $R = [0 \ 1]$ $\qquad E = 1$
35. $R = [\frac{9}{11} \ \frac{2}{11}]$ $\qquad E = 72/11$
37. $R = [.1 \ .9]$ $\qquad E = 7.3$

39. $\begin{bmatrix} -1 & -3 \\ 5 & 1 \end{bmatrix}$ $\qquad R = [0 \ 1]$

41. $\begin{bmatrix} -1 & -2 \\ 3 & -1 \end{bmatrix}$ $\qquad R = [0 \ 1]$

43. $R = [.2 \ .8]$

45. $R = [1 \ 0], C = \begin{bmatrix} 0 \\ 1 \end{bmatrix}$

Exercise Set 10–4

1. $E = 2x - y$
3. $E = u(2x - y) + 4vy$
5. Minimize $X + Y$: $\quad 4X + \ Y \le 1$
$\qquad\qquad\qquad\qquad\quad 5X + 2Y \le 1$
$\qquad\qquad\qquad\qquad\quad X \ge 0, \ \ Y \ge 0$

7. $x = 0,$ $\qquad y = 1$
9. $x = \frac{3}{4},$ $\qquad y = \frac{1}{4}$
11. $u = 1,$ $\qquad v = 0$
13. $u = \frac{3}{4},$ $\qquad v = \frac{1}{4}$
15. $x = \frac{1}{2},$ $\qquad y = \frac{1}{2},$ $\qquad u = \frac{3}{4},$ $\qquad v = \frac{1}{4}$

17. $x = 1$,　　$y = 0$,　　$u = 0$,　　$v = 1$
19. $x = \frac{1}{4}$,　　$y = \frac{3}{4}$,　　$u = \frac{1}{2}$,　　$v = \frac{1}{2}$
21. $x = 1$,　　$y = 0$,　　$u = 1$,　　$v = 0$
23. $x = 1$,　　$y = 0$,　　$u = 0$,　　$v = 1$
29. (a) 7　　(b) $\frac{1}{7}$
31. (a) $\frac{1}{2}$　　(b) 2
33. (a) $\frac{1}{2}$　　(b) 2
35. (a) 3　　(b) $\frac{1}{3}$
37. (a) .3　　(b) $\frac{10}{3}$
39. $[\frac{1}{3}, \frac{2}{3}]$　$[\frac{1}{2} \, \frac{1}{2}]$

Review Exercises

1. R chooses row 3; C chooses column 3;　　$v = 1$
3. yes,　　$v = 2$
5. no
7. $E = 0$
9. $a = 1$,　　$b = 0$
11. $Q = P + 5$; E (using P) $= .28$;　　E (using Q) $= 5.28$;　　E (using P) $+ 5 = E$ (using Q)
13.

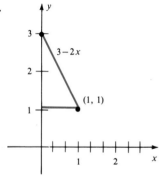

15. $x = \frac{1}{3}$,　　$y = \frac{2}{3}$,　　$u = \frac{1}{3}$,　　$v = \frac{2}{3}$,　　payoff $= \frac{5}{3}$
17. $\begin{bmatrix} -3 & -2 \\ 1 & -3 \end{bmatrix}$　　$R = [\frac{4}{5} \; \frac{1}{5}]$　　$C = \begin{bmatrix} \frac{1}{5} \\ \frac{4}{5} \end{bmatrix}$
19. $[0, 1]$

Index